Australian Guidebook for Structural Engineers

T0174112

A guide to structural engineering
on a multidiscipline project

Australian Guidebook for Structural Engineers

A guide to structural engineering on a multidiscipline project

Lonnie Pack

CRC Press
Taylor & Francis Group
Boca Raton London New York

CRC Press is an imprint of the
Taylor & Francis Group, an **informa** business

A CHAPMAN & HALL BOOK

CRC Press
Taylor & Francis Group
6000 Broken Sound Parkway NW, Suite 300
Boca Raton, FL 33487-2742

First issued in paperback 2019

ISBN-13: 978-1-138-03185-2 (hbk)
ISBN-13: 978-0-367-88500-7 (pbk)

Library of Congress Cataloging-in-Publication Data

Names: Pack, Lonnie, author.
Title: Australian guidebook for structural engineers : a guide to structural
engineering on a multidiscipline project / Lonnie Pack.
Description: Boca Raton : CRC Press, [2017] | Includes bibliographical
references and index.
Identifiers: LCCN 2016049432| ISBN 9781138031852 (hardback : alk. paper) |
ISBN 9781315197326 (ebook)
Subjects: LCSH: Structural engineering--Australia. | Multidisciplinary design
optimization--Australia.
Classification: LCC TA633 .P28 2017 | DDC 624.102/1894--dc23
LC record available at https://lccn.loc.gov/2016049432

Visit the Taylor & Francis Web site at
http://www.taylorandfrancis.com

and the CRC Press Web site at
http://www.crcpress.com

SAI GLOBAL

This text includes detailed references to numerous standards produced by Standards Australia Limited and the International Organization for Standardization. Permission to reproduce the work has graciously been provided by SAI Global Limited. This text provides a guide to the use of numerous standards and should not be used in place of the standards. Full copies of the referenced standards are available online at https://www.saiglobal.com/.

DISCLAIMER

Contents

Preface

This text aims to bridge the gap between Australian Standards, textbooks and industry knowledge. After years of work in the industry, many engineers will have created a library of calculations, specifications and standards that can be used to help complete projects. It is hoped that this compendium of resources helps to create a useful set of references that can aid engineers in the expedient delivery of design projects. Many examples and information in this document are drawn from the mining and oil and gas industries.

A multidiscipline engineering project requires close and coordinated work between engineers and designers of the same and other disciplines. A clear scope and list of deliverables is imperative, along with a structured formwork of how to achieve the desired outcome.

This book details each step of a project to be followed by a design engineer. The sequence of information is provided roughly in the order in which it is required. The creation of project documentation is outlined, including a scope, schedule and deliverables list. Calculation methods and details are shown for actions (wind, seismic, dead and live). Details are then provided for steel, concrete and geotechnical calculations. Design items are explained for typical items of equipment found in the mining and oil and gas industries. Design aids are provided, including guides and examples for popular engineering programs. Finally, vendor product catalogue references are provided for commonly used industry items. This ensures the suitability and availability of products.

An appropriate example is for an engineer tasked with designing a concrete bund (pit) to capture any accidental spillage from a tank. The volume of a pit is sized using legislative and Australian Standards requirements for storage volumes ('Storage and Handling of Workplace Dangerous Goods', AS 1940: 'The Storage and Handling of Flammable and Combustible Liquids' and AS 2067: 'Substations and High Voltage Installations Exceeding 1 kV a.c.'). Liquid pressures (AS 3735: 'Concrete Structures for Retaining Liquids') and geotechnical pressures (AS 4678: 'Earth Retaining Structures') are calculated, along with many supplementary requirements. Additional loads are then calculated and combined in accordance with Australian Standards (AS/NZS 1170: 'Structural Design Actions'), and the strength of the wall is calculated in accordance with the concrete code (AS 3600: 'Concrete Structures'). Typical details and vendor products are then commonly used to complete the design.

The number of books and standards required to thoroughly complete a simple design can be daunting for many engineers. Necessary details and references required for designs, such as technical calculations, legislative requirements and vendor products, make designs like this difficult without extensive research. This often leads to designs which do not consider all necessary requirements. Details within this book are provided to an appropriate level for design engineers to understand the key aspects from each reference. Design items are explained with details presented from local and international standards, supplements and commentaries, common practices and available products. Australian Standards should be referred to in full prior to the completion of each design.

Author

Lonnie Pack holds a bachelor degree in civil engineering with honours from the University of Queensland. He is a chartered professional engineer through Engineers Australia and a registered professional engineer of Queensland. His industry experience includes oil and gas, mining, infrastructure and bridge design, with a focus on coal seam gas plant design. Lonnie has designed and led teams of structural engineers and designers in the greenfield creation of upwards of 20 gas compression facilities, as well as numerous brownfield projects. His projects have been delivered via traditional (stick-built) methods, as well as modular (pre-assembled) designs. Lonnie's experience also includes designing more than 80 machine foundations, including screw, reciprocating and centrifugal compressors.

Chapter 1

Setting up the project

This chapter details the typical requirements for a multidiscipline project in Australia. Specific recommendations are based on mining or oil and gas examples. Although all projects are unique, this set of requirements can be consistently applied. The following aspects should all be clearly documented and issued at the beginning of the project. They are essential for the successful completion of a project.

1. Basis of design
2. Scope of works
3. Deliverables list
4. Budget
5. Schedule
6. Communications plan
7. Structural design criteria
8. Specifications

1.1 BASIS OF DESIGN

Ideally, the basis of design has already been created by the client and is used as the basis for the scope of works. A basis of design should outline the required functions of a project and what is important for the client. Details should be provided for

1. What the project includes
2. How the project fits into the client's company
3. Important aspects of the project
4. Any client-driven requirements for the facility
5. How the facility functions
6. The design life of the facility

1.2 SCOPE OF WORKS

The scope of works should list each item from the basis of design and detail exactly what is going to be delivered by your organisation. A list of inclusions and exclusions is important to clearly delineate scope. Assumptions may also be listed if they are important for the project. A clear deliverables list is necessary to outline exactly what is being delivered to the client. Budget and schedule requirements may also be provided in a scope.

1.3 DELIVERABLES LIST

A clearly defined scope, along with experience in the field, should lead to an accurate and well-set-out deliverables list. A structural deliverables list is generally split into two sections: steel and concrete (Tables 1.1 and 1.2). The easiest way to create a deliverables list is to look at a similar project from the past and use it to estimate the number of drawings and calculations required for each of the items listed in the scope of works. The following is a typical example of what would normally be produced.

A typical document numbering system is adopted using the following identifiers: 1-2-3-4_5 (e.g. A1-ST-IN-001_01) (Table 1.3). A system using sheet numbers, as well as drawing numbers, is often beneficial, as it allows the user to add numbers at a later date without losing sequence.

Care should be taken to ensure documentation (such as signed check prints, calculations and drawings) is filed and scanned to ensure traceability of work.

1.4 BUDGET

Most companies have formal procedures for completing budget estimates. A common method is to estimate the number of engineering hours and the number of drafting hours,

Table 1.1 Steel deliverables list

Document number	Description
A1-ST-DC-0001_01	Structural design criteria
A1-ST-IN-0001_01	Steel drawing index
A1-ST-SD-0001_01	Steel standard notes
A1-ST-SD-0001_02	Steel standard drawing 2 – Bolted connections
A1-ST-SD-0001_03	Steel standard drawing 3 – Base plates
A1-ST-SD-0001_04	Steel standard drawing 4 – Stairs and ladders
A1-ST-SD-0001_05	Steel standard drawing 5 – Grating and handrails
A1-ST-PL-0001_01	Steel plot plan
A1-ST-AR-0001_01	Steel area plan 1
A1-ST-AR-0001_02	Steel area plan 2
A1-ST-DE-0001_01	Steel structure 1 – Sheet 1
A1-ST-DE-0001_02	Steel structure 1 – Sheet 2
A1-ST-DE-0002_01	Steel structure 2 – Sheet 1
A1-ST-DE-0003_01	Miscellaneous steel – Sheet 1
A1-ST-MT-0001_01	Steel material take-off

Table 1.2 Concrete deliverables list

Document number	Description
A1-CN-IN-0001_01	Concrete drawing index
A1-CN-SD-0001_01	Concrete standard notes
A1-CN-SD-0001_02	Concrete standard drawing 2 – Piles and pad footings
A1-CN-SD-0001_03	Concrete standard drawing 3 – Slabs
A1-CN-SD-0001_04	Concrete standard drawing 4 – Anchor bolts
A1-CN-SD-0001_05	Concrete standard drawing 5 – Joints and sealing
A1-CN-PL-0001_01	Concrete plot plan
A1-CN-AR-0001_01	Concrete area plan 1
A1-CN-AR-0001_02	Concrete area plan 2
A1-CN-AR-0001_03	Concrete area plan 3
A1-CN-DE-0001_01	Concrete structure 1 – Sheet 1
A1-CN-DE-0001_02	Concrete structure 1 – Sheet 2
A1-CN-DE-0002_01	Concrete structure 2 – Sheet 1
A1-CN-DE-0003_01	Miscellaneous concrete – Sheet 1
A1-CN-MT-0001_01	Concrete material take-off

Table 1.3 Document numbering system

Identifier location	Identifier	Description
1	A1	Project number
2	ST	Steel discipline
2	CN	Concrete discipline
3	DC	Design criteria
3	IN	Index
3	SD	Standard drawing
3	PL	Plot plan
3	AR	Area plan
3	DE	Detailed drawing
3	MT	Material take-off
4	###	Drawing number
5	##	Sheet number

and then multiply each by the appropriate hourly rate. An allowance should also be included for project management and any meetings and client reviews.

For example, a project may include the following typical allowances:

1. 1 h each week for each team member to attend team meetings
2. 2 h each week for the lead engineer to attend project meetings and model reviews
3. 1 h each week for the lead designer to attend model reviews
4. 2 h each fortnight for the lead engineer to attend client reviews
5. 8 engineering hours per drawing for standard items (40 h for complex items)
6. 12 drafting hours per drawing for standard items (24 h for complex items)

Team members also need to be considered on an individual basis if they have varying rates (i.e. graduate engineer, engineer, senior engineer or principal engineer). Depending on the project setup, an allowance may also need to be made for administration and document control.

1.5 SCHEDULE

The schedule is the most important tracking tool for the project. It should be set up with links to milestones, outputs from other disciplines and receipt of vendor data. Durations need to be based on experience and discussion with engineering and drafting team members. The duration should always allow the budgeted number of hours to be used within the timeframe. It is important to ensure that all members have input to ensure engagement in the project and ownership of deadlines.

Links in the schedule are important, as they create accountability. Without correctly associating items, a delay in the project will often reflect badly on the trailing disciplines (structural) rather than leading disciplines (process and piping). However, if a schedule is correctly built and updated, it will ensure that delays are highlighted early and that the culprit is found and fixed.

For example, the completion of drawings showing a pipe rack may be tied, FF + 5 (finish to finish plus 5 days) to receipt of the final pipe stress for all pipe systems resting on the rack. This means that the drawings will be completed 5 days after receiving the final pipe stress. This is because the rack would normally be designed and drafted using preliminary stress, and then checked against the final loads. However, if there is no consultation in the design, then (perhaps for vendor data) the link may be created as FS (finish to start), meaning that the design cannot be started until the vendor data is received.

1.6 COMMUNICATIONS PLAN

Communication is the key to success. Ensure that a plan is in place for large projects to detail the level and frequency of formal communication. Weekly formal meetings are a good idea at most levels of reporting. The suggested schedule shown in Table 1.4 may be adopted depending on the project requirements.

1.7 STRUCTURAL DESIGN CRITERIA

For established clients, structural design criteria are usually based on previous projects. If this is not the case, it may help to complete a form such as Table 1.5, in conjunction with the client, to formally set out the unknown requirements prior to starting the structural design criteria.

The criteria should list all required design decisions, as well as project documentation, standards and codes which are to be followed during design (refer to Section 2.2) and all relevant specifications. All key decisions for the project need to be documented; therefore, any that occur after completion and client approval of the design criteria should be documented on formal 'technical queries' (TQs) to the client.

Table 1.4 Meeting schedule

Meeting title	Invitees	Suggested recurrence
Project team	Engineering manager, lead engineers, project manager, project engineers	Weekly
Model review	Lead engineers, lead designers/draftsmen, engineering manager	Weekly
Structural team	Structural engineers and structural designers/draftsmen	Weekly
Client model review	Client representatives, lead engineers, lead designers/draftsmen, engineering manager	Fortnightly (depending on client)

Table 1.5 Design criteria agreements

Discussion	Agreement	Follow-up required? (Y/N)
General		
Design life	___ years	
Importance level	2/3	
Seismic hazard factor, Z	0.05	
LODMAT	5°C	
Maximum temperature	50°C	
Installation temperature	20°C–30°C	
Design temperature range	+___/–___°C	
Transportation envelope	___ m × ___ m × ___ m	
Rainfall	___ mm in 24-hour event ___ mm in 72-hour event	
Steel		
Structural sections	AS – 300Plus	
Plate grades	250/250L15/300	
Protective coatings	Painted/galvanised (bath size)	
Platform loading	2.5 kPa/5.0 kPa	
Concrete		
Grade	N32/N40	
Reinforcement	D500N (main), D500N/R250N/ D500L (fitments)	
Slab loading	W80/A160/T44	
Exposure classification	B1 for aboveground A2 for belowground	
Climatic zone	Temperate	
Geotechnical data	Refer report _____, Revision ___, Date _____	
Lead engineer _____	Date _____ Client representative _____	Date _____

Other important issues which should be discussed and detailed in the criteria are

1. Site-specific data and parameters (hazard factor, wind speeds, etc.)
2. Geotechnical report and important parameters
3. Adopted materials (steel, concrete, bars, plates, bolt types and fastening methods)
4. Preferred framing systems and construction methods
5. Bunding requirements
6. Foundation preferences
7. Standardisation and simplification of details, members, connections and so forth
8. Access requirements (900 mm wide walkways, 2000 mm height clearances)
9. Load combinations and factors
10. Specific loads, such as vehicle loading (W80, A160, T44 and SM1600 – refer to AS 5100), platform loading (2.5 and 5.0 kPa) or minimum pipe rack loads (2 and 3 kPa), blast loads, cable tray loads (2.0 kPa), impact loads and thermal loads
11. Friction coefficients (refer to Section 2.4)
12. Deflection limitations

1.7.1 Load factors and combinations

Load factors and combinations presented in this section are based on AS/NZS 1170, 'Process Industry Practices – Structural Design Criteria' (PIP STC01015)[46] and common practice. Further explanations of each item are presented in the sections referenced in Tables 1.6 through 1.8.

1.7.2 Construction category

A construction category should be specified for the project in accordance with the Australian Steel Institute's (ASI) *Structural Steelwork Fabrication and Erection Code of Practice*.[1] This is a new requirement (2014), yet it has already been adopted by some large operators. It specifies the level of quality assurance (testing and record keeping) required for a project or structure. Typical structures in areas of low seismic activity should be CC2 (for importance level 2) or CC3 (for importance level 3). Refer to Appendix C of the full code for further details.

Table 1.6 Strength load combinations

Combination	Title	Reference
General combinations		
1.35 G	Maximum self-weight	AS/NZS 1170.0
1.2 G + 1.5 Q	Live load	AS/NZS 1170.0
1.2 G + W_u + 0.6 Q	Ultimate wind (down)	AS/NZS 1170.0
0.9 G + W_u	Ultimate wind (Up)	AS/NZS 1170.0
G + E_u + 0.6 Q	Ultimate seismic	AS/NZS 1170.0
1.2 G + S_u + 0.6 Q	Ultimate snow	AS/NZS 1170.0
Pipe racks		
1.35 (G + F_f + T + A_f)	Self-weight + friction + thermal + operational	Section 6.1
1.2 G + 1.5 F_{slug}	Slug loading	Section 6.1
1.35 or 1.4 H_t	Hydrotest	Section 6.2
1.2 H_t + $W_{OP/SLS}$*	Hydrotest with wind	Section 6.2
Vessels and tanks		
1.35 or 1.4 H_t	Hydrotest	Section 6.2
1.2 H_t + $W_{OP/SLS}$*	Hydrotest with wind	Section 6.2
1.35 G + 1.35 F_f	Maximum self-weight + thermal	Section 6.6.4
1.2 G + 1.5 Q + 1.2 F_f	Self-weight + live load + thermal	Section 6.6.4
1.35 G + 1.35 F_f	Maximum self-weight + thermal	Section 6.6.4

Notes:
1. A_f – Pipe anchor and guide forces.
2. E_u – Ultimate seismic load.
3. F_{slug} – Slug loading force (no strength factor is required if the load is already based on ultimate circumstances).
4. F_f – Thermal load.
5. G – Dead load (including operating weights for piping and equipment).
6. H_t – Hydrotest weight (including vessel weight).
7. Q – Live load.
8. S_u – Ultimate snow load.
9. T – Thermal load.
10. $W_{OP/SLS}$* – Operating wind or serviceability wind (depending on preference).
11. W_u – Ultimate wind load.
12. Combination factors (ψ) for live load (Q) taken as 0.6; refer to AS/NZS 1170.0, Table 4.1, for full range of values.

Table 1.7 Serviceability load combinations

Combination	Title	Reference
1.0 G	Self-weight	AS/NZS 1170.0
1.0 G + 0.6 Q	Self-weight plus live load	AS/NZS 1170.0
1.0 G + E_s	Serviceability seismic	AS/NZS 1170.0
1.0 G + W_s	Serviceability wind	AS/NZS 1170.0
1.0 H_t	Hydrotest	Section 6.2
1.0 H_t + $W_{OP/SLS}$*	Hydrotest with wind	Section 6.2

Notes:
1. E_s – Serviceability seismic load (if applicable).
2. $W_{OP/SLS}$* – Operating wind or serviceability wind (depending on preference).
3. Combination factors (ψ) for live load (Q) taken as 0.6; refer to AS/NZS 1170.0, Table 4.1, for full range of values.

Table 1.8 Stability load combinations

Combination	Title	Reference
$E_{d,stb}$ = 0.9 G	Self-weight	AS/NZS 1170.0
$E_{d,dst}$ = 1.35 G	Self-weight	AS/NZS 1170.0
$E_{d,dst}$ = 1.2 G + 1.5 Q	Self-weight plus live load	AS/NZS 1170.0
$E_{d,dst}$ = 1.2 G + W_u + 0.6 Q	Ultimate wind (down)	AS/NZS 1170.0
$E_{d,dst}$ = G + E_u + 0.6 Q	Ultimate seismic	AS/NZS 1170.0
$E_{d,dst}$ = 1.2 G + S_u + 0.6 Q	Ultimate snow	AS/NZS 1170.0

Notes:
1. Design shall ensure that stabilising force is greater than any combination of destabilising forces ($E_{d,stb} > E_{d,dst}$).
2. Combination factors (ψ) for live load (Q) taken as 0.6; refer to AS/NZS 1170.0, Table 4.1, for full range of values.

1.8 SPECIFICATIONS

A thorough set of construction specifications is necessary to ensure the successful completion of a project. Most clients (and many design firms) have their own set of specifications. The full set should be provided and agreed upon at the beginning of a project to ensure that the design complies with construction requirements. Specifications include detailed descriptions of construction materials and methodologies. They may have to be created if neither party has existing documents.

Typical structural specifications include

1. Buildings
2. Concrete supply and installation
3. Civil works and earthworks
4. Steel fabrication, supply and erection
5. Protective coatings

Chapter 2

Design

The key task of a structural engineering team on a multidiscipline project is to create a set of deliverables which can be used to fabricate, construct and install a structure in accordance with applicable legislation, Australian Standards, the design criteria and relevant specifications.

2.1 LIMIT STATES DESIGN

Information contained in this book is generally relevant to limit state design. The structural design procedure outlined in AS/NZS 1170.0 – Section 2, should be followed in general principle. Ultimate limit state (ULS) is the design process for a structure to manage the probability of collapse (Table 2.1). Serviceability limit state (SLS) is the design process to manage the probability of a structure remaining fit for use without requiring repairs (Table 2.2). Forces and reactions calculated using USL combinations are denoted using an asterisk (e.g. N^*, V^*, M^*).

Experience allows a design engineer to choose when to consider serviceability limit states (stress or deflection) prior to ultimate limit states (strength). Deflection is often the limiting case for portal frame buildings (especially with cranes), pipe supports and other lightweight steel structures. Strength more commonly governs for rigid structures, especially concrete structures. An experienced designer may also choose to ignore either wind or seismic loading when the other clearly governs. Small, heavy structures are typically governed by seismic loading, and large, lightweight structures are typically governed by wind loading.

2.2 STANDARDS AND LEGISLATION

Knowing and being able to find applicable codes of practice is an important skill for an engineer to gain. It should be noted that Australian Standards are not necessarily compulsory. Regulations such as the National Construction Code (NCC) are required in accordance with state legislation. The NCC includes the two chapters of the Building Code of Australia (BCA), as well as the plumbing code. The BCA then references standards such as AS 4100 to meet requirements. Unless there is a legislative requirement to conform to a standard, it

Table 2.1 Design procedure for ultimate limit states

Step	Description	External reference	Internal reference
a	Select an importance level and annual probabilities of exceedance for wind, snow and earthquake	BCA (or AS/NZS 1170.0, Appendix F if unavailable in BCA)	2.3
b	Determine permanent (dead) and imposed (live) actions	AS/NZS 1170.1	
c	Calculate wind loads	AS/NZS 1170.2 (in conjunction with AS/NZS 1170.0, Appendix F)	
d	Calculate seismic loads	AS 1170.4 (in conjunction with AS/NZS 1170.0, Appendix F)	
e	Calculate snow loads	AS/NZS 1170.3 (in conjunction with AS/NZS 1170.0, Appendix F)	N/A
f	Other loads (such as liquid and earth pressure)	AS/NZS 1170.1	Chapters 5 and 6
g	Determine load combinations	AS/NZS 1170.0, Section 4	1.7.1
h	Analyse structure	AS/NZS 1170.0, Section 5	Chapters 3 and 4
i	Design and detail structure (incorporate robustness and seismic requirements)	AS/NZS 1170.0, Section 6 (and AS/NZS 1170.4)	
j	Determine the design resistance	BCA and applicable standards	
k	Confirm that the design resistance exceeds the calculated combination of actions (strength and stability)	AS/NZS 1170.0, Section 7	

Table 2.2 Design procedure for serviceability limit states

Step	Description	External reference	Internal reference
a	Determine the design conditions for structure and components	N/A	2.3
b	Determine serviceability loads and limits for conditions	AS/NZS 1170.0, Appendix C	
c	Determine permanent (dead) and imposed (live) actions	AS/NZS 1170.1	
d	Calculate wind loads	AS/NZS 1170.2 (in conjunction with AS/NZS 1170.0, Appendix F)	
e	Calculate snow loads	AS/NZS 1170.3 (in conjunction with AS/NZS 1170.0, Appendix F)	N/A
f	Other loads (such as liquid and earth pressure)	AS/NZS 1170.1	Chapters 5 and 6
g	Determine load combinations	AS/NZS 1170.0, Section 4	1.7.1
h	Analyse structure	AS/NZS 1170.0, Section 5	Chapters 3 and 4
i	Determine the serviceability response	BCA and applicable standards	
j	Confirm that the calculated response does not exceed the limiting values	AS/NZS 1170.0, Section 7	

is not technically required. Some standards even state that conformance is not mandatory. AS/NZS 1418 is a good example of a standard that allows the design engineer to deviate from the code, if the deviation is a 'generally accepted method' and procedures or well-documented research is employed. As an engineer, you should always attempt to follow the most applicable code of practice available in order to avoid negligence.

The following is a list of legislation referenced on a typical gas project in Queensland, Australia:

- Environmental Protection Act 1994
- Gas Supply Act 2003
- Mineral Resources Act 1989
- Mineral Resources Regulation 2003
- Mining and Quarrying Safety and Health Act 1999
- Mining and Quarrying Safety and Health Regulation 2001
- National Gas (Queensland) Act 2008
- Petroleum and Gas (Production and Safety) Regulation 2004
- Queensland Work Health and Safety Act 2011
- Queensland Work Health and Safety Regulation 2011

The following is a list of Australian Standards referenced by the structural team on a typical gas project in Australia:

- AS 1101.3–2005: 'Graphical Symbols for General Engineering – Part 3: Welding and Non-Destructive Examination'
- AS/NZS 1170.0–2002: 'Structural Design Actions – General Principles'
- AS/NZS 1170.1–2002: 'Structural Design Actions – Permanent, Imposed and Other Actions'
- AS/NZS 1170.2–2011: 'Structural Design Actions – Wind Actions'
- AS/NZS 1170.3–2003: 'Structural Design Actions – Snow and Ice Actions'
- AS 1170.4–2007: 'Structural Design Actions – Earthquake Actions in Australia'
- NZS 1170.5–2004: 'Structural Design Actions – Earthquake Actions – New Zealand'
- AS 1210–2010: 'Pressure Vessels'
- AS 1379–2007: 'The Specification and Manufacture of Concrete'
- AS/NZS 1418 Set (1–18): 'Cranes, Hoists and Winches'
- AS/NZS 1554 Set (1–7): 'Structural Steel Welding'
- AS 1657–2013: 'Fixed Platforms, Walkways, Stairways and Ladders – Design, Construction and Installation'
- AS 1940–2004: 'The Storage and Handling of Flammable and Combustible Liquids'
- AS 2159–2009: 'Piling – Design and installation'
- AS 2327.1–2003: 'Composite Structures – Simply Supported Beams'
- AS 2741–2002: 'Shackles'
- AS 2870–2011: 'Residential Slabs and Footings'
- AS 3600–2009: 'Concrete Structures'
- AS 3610–1995: 'Formwork for Concrete'
- AS/NZS 3678–2011: 'Structural Steel – Hot Rolled Plates, Floorplates and Slabs'
- AS/NZS 3679.1–2010: 'Structural Steel – Hot Rolled Bars and Sections'
- AS 3700–2011: 'Masonry Structures'
- AS 3735–2001: 'Concrete Structures for Retaining Liquids'
- AS 3850–2003: 'Tilt-Up Concrete Construction'
- AS 3990–1993: 'Mechanical Equipment – Steelwork'
- AS 3995–1994: 'Design of Steel Lattice Towers and Masts'
- AS 4100–1998: 'Steel Structures'
- AS/NZS 4600–2005: 'Cold-Formed Steel Structures'

- AS/NZS 4671–2001: 'Steel Reinforcing Materials'
- AS 4678–2002: 'Earth-Retaining Structures'
- AS 4991–2004: 'Lifting Devices'
- AS 5100 Set (1 to 7): 'Bridge Design'

The following foreign and international standards may also be useful due to a gap in required detail in Australian Standards:

- ACI 351.3R–2004: 'Foundations for Dynamic Equipment'
- API 620–2013: 'Design and Construction of Large, Welded, Low-Pressure Storage Tanks'
- API 650–2013: 'Welded Tanks for Oil Storage'
- BS 7385-2–1993: 'Evaluation and Measurement for Vibration in Buildings. Guide to Damage Levels from Groundborne Vibration'
- CP 2012-1–1974: 'Code of Practice for Foundations for Machinery. Foundations for Reciprocating Machines'
- ISO 1940-1–2003: 'Mechanical Vibration - Balance Quality Requirements for Rotors in a Constant (Rigid) State – Part 1: Specification and Verification of Balance Tolerances'

2.3 ACTIONS

A site classification needs to be chosen in order to calculate design actions, such as wind and seismic parameters. AS/NZS 1170.0, Appendix F should be used for sites that are not covered by the BCA or other more specific design codes. These requirements are summarised in Tables 2.3 and 2.4. Sites within the oil and gas industry typically use an importance level of 3, and a design life of either 25 or 50 years (depending on client preference). It is not uncommon to use an importance level of 2 for less important structures on the site (such as storage sheds). Mine sites more commonly adopt an importance level of 2 for the entire site.

2.3.1 Wind

This section presents wind speeds and shows calculations for pressures on pipes and pipe support structures. It is based on AS/NZS 1170.2–2011, including Amendments 1–3. The

Table 2.3 Structure types for importance levels

Importance level	Consequence of failure	Description	Examples
1	Low	Low consequence to human life and other factors	Storage sheds
2	Ordinary	Medium consequence to human life; considerable economic, social or environmental consequences	Small buildings, mining structures, warehouses
3	High	High consequence to human life; very great economic, social or environmental consequences	Process structures, pipe racks, compressors, vessels
4			Postdisaster structures
5	Exceptional	Case-by-case basis	Exceptional structures

Source: Modified from AS/NZS 1170.0, table F1. Copied by L. Pack with permission from Standards Australia under Licence 1607-c010.

Table 2.4 Annual probability of exceedance of the design events for ultimate limit states

Design working life	Importance level	Design events for safety in terms of annual probability of exceedance		
		Wind	Snow	Earthquake
Construction equipment	2	1/100	1/50	Not required
5 years or less (no risk to human life)	1	1/25	1/25	
	2	1/50	1/50	
	3	1/100	1/100	
25 years	1	1/100	1/25	
	2	1/200	1/50	1/250
	3	1/500	1/100	1/500
	4	1/1000	1/250	1/1000
50 years	1	1/100 (noncyclonic)	1/100	1/250
	1	1/200 (cyclonic)		
	2	1/500	1/150	1/500
	3	1/1000	1/200	1/1000
	4	1/2500	1/500	1/2500
100 years or more	1	1/500	1/200	1/250
	2	1/1000	1/250	1/1000
	3	1/2500	1/500	1/2500
	4	(See AS/NZS 1170.0, paragraph F3)		

Source: Modified from AS/NZS 1170.0, table F2. Copied by L. Pack with permission from Standards Australia under Licence 1607-c010.

Notes:
1. Structures in wind regions C and D (i.e. cyclonic regions, as defined in AS/NZS 1170.2) that are erected and remain erected, only during the period of May to October, may be designed for regional wind speeds given in AS/NZS 1170.2, for Region A, or alternatively from a specific analysis of noncyclonic wind events for the site. A structure not designed for cyclonic wind speeds shall not remain erected during the months of November to April inclusive.

drag force coefficients for steelwork and piping are based on Appendix E of the standard. For details on wind pressures for buildings, refer to AS/NZS 1170.2, Section 5.

A wind speed can be selected from AS/NZS 1170.2 after calculating an annual probability of exceedance (refer to Section 2.3) in accordance with AS/NZS 1170.0. Typically, the ultimate values are used in calculations, and results are factored down to calculate serviceability loads. Serviceability values are generally calculated using an annual probability of exceedance equal to 1/25 (i.e. V_{25}).

First, a wind region is selected from Figure 2.1. The wind region is used to calculate a regional wind speed, V_R, for the appropriate exceedance in Table 2.5.

A site wind speed ($V_{\text{sit},\beta}$) is then calculated incorporating specific details for the actual location and wind directions. Generic values are provided below for each of the variables. For less conservative values, refer to AS/NZS 1170.2.

$$V_{\text{sit},\beta} = V_R M_d \left(M_{z,\text{cat}} M_s M_t \right)$$

where:

M_d = Wind directional multiplier = 1 (any direction)
M_s = Shielding multiplier = 1 (conservative)
M_t = Topographic multiplier = 1 (for flat sites)

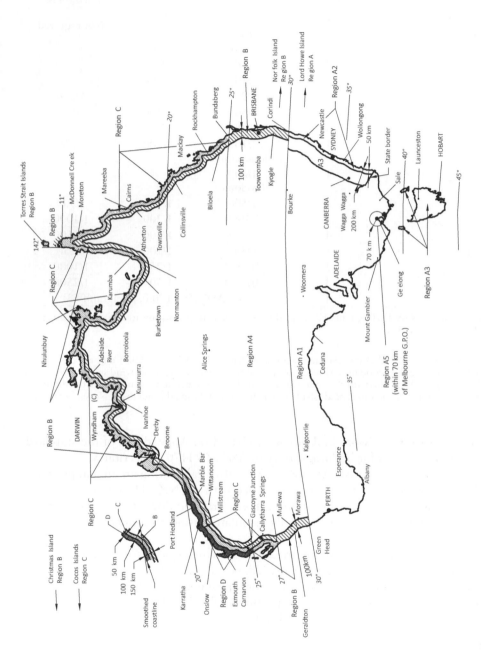

Figure 2.1 Wind regions.

Notes: 1. Regions are marked with the letters A (A1–A5), B, C and D.
2. Coastal region boundaries are smooth lines set in from a smoothed coastline by 50, 100, 150 and 200 km.
3. Islands within 50 km of the area designated in the same region as the adjacent coast.
(From AS/NZS 1170.2, figure 3.1A. Copied by Mr L. Pack with the permission of Standards Australia under Licence 1607-c010.)

Table 2.5 Regional wind speeds

Regional wind speed, V_R (m/s)	Region				
	Noncyclonic			Cyclonic	
	A (1–7)	W	B	C	D
V_1	30	34	26	$23 \times F_C$	$23 \times F_D$
V_5	32	39	28	$33 \times F_C$	$35 \times F_D$
V_{10}	34	41	33	$39 \times F_C$	$43 \times F_D$
V_{20}	37	43	38	$45 \times F_C$	$51 \times F_D$
V_{25}	37	43	39	$47 \times F_C$	$53 \times F_D$
V_{50}	39	45	44	$52 \times F_C$	$60 \times F_D$
V_{100}	41	47	48	$56 \times F_C$	$66 \times F_D$
V_{200}	43	49	52	$61 \times F_C$	$72 \times F_D$
V_{250}	43	49	53	$62 \times F_C$	$74 \times F_D$
V_{500}	45	51	57	$66 \times F_C$	$80 \times F_D$
V_{1000}	46	53	60	$70 \times F_C$	$85 \times F_D$
V_{2000}	48	54	63	$73 \times F_C$	$90 \times F_D$
V_{2500}	48	55	64	$74 \times F_C$	$91 \times F_D$
V_{5000}	50	56	67	$78 \times F_C$	$95 \times F_D$
V_{10000}	51	58	69	$81 \times F_C$	$99 \times F_D$
V_R ($R \geq$ 5 years)	$67 - 41R^{-0.1}$	$104 - 70R^{-0.045}$	$106 - 92R^{-0.1}$	$F_C(122 - 104R^{-0.1})$	$F_D(156 - 142R^{-0.1})$

Source: AS/NZS 1170.2, table 3.1. Copied by L. Pack with permission from Standards Australia under Licence 1607-c010.

Notes:
1. The peak gust has an equivalent moving average time of approximately 0.2 s.
2. Values for V_1 have not been calculated by the formula for V_R.
3. For ULS or SLS, refer to the BCA or AS/NZS 1170.0 for information on values of annual probability of exceedance appropriate for the design of structures.
4. For $R \geq 50$ years, $F_C = 1.05$ and $F_D = 1.1$. For R < 50 years, $F_C = F_D = 1.0$.

Table 2.6 Terrain category

Terrain category	Description
1	Open terrain (water or open plains)
1.5	Open water near shoaling waves
2	Open terrain with well-scattered obstructions from 1.5 to 5 m high
2.5	Developing outer urban areas, scattered trees and houses (<10 per hectare)
3	Numerous closely spaced obstructions (≥10 house-sized obstructions per hectare)
4	Large city centres or industrial complexes (10–30 m tall)

Source: AS/NZS 1170.2, Clause 4.2.1.

$M_{s, cat}$ = Terrain/height multiplier (refer to Tables 2.6 and 2.7)
V_R = Regional wind speed

The design wind speed ($V_{des,\theta}$) is calculated using the maximum site wind speed from within 45° of the direction being considered. However, if generic values have been used, the design wind speed will be equal to the site wind speed. For many projects, conservative values are used so that all structures have the same design speed. This is done by adopting the conservative numbers shown above. If there is a large structure, such as a compressor house, it may be economical to recalculate a specific speed for that calculation.

These calculations should be completed for an ultimate wind speed and a serviceability wind speed.

Table 2.7 Terrain/height multiplier

Height, z (m)	Terrain/height multiplier ($M_{z,cat}$)			
	Terrain category 1	Terrain category 2	Terrain category 3	Terrain category 4
≤3	0.99	0.91	0.83	0.75
5	1.05	0.91	0.83	0.75
10	1.12	1.00	0.83	0.75
15	1.16	1.05	0.89	0.75
20	1.19	1.08	0.94	0.75
30	1.22	1.12	1.00	0.80
40	1.24	1.16	1.04	0.85
50	1.25	1.18	1.07	0.90
75	1.27	1.22	1.12	0.98
100	1.29	1.24	1.16	1.03
150	1.31	1.27	1.21	1.11
200	1.32	1.29	1.24	1.16

Source: Copied by L. Pack with permission from Standards Australia under Licence 1607-c010.

Note: For intermediate values of height, z, and terrain category, use linear interpolation.

2.3.1.1 Wind pressure

The wind pressure is calculated using factors which depend on the shape and dynamic sensitivity of a structure.

$$p = (0.5\rho_{air})[V_{des,\theta}]^2 C_{fig} C_{dyn}$$

where:

ρ_{air} = Density of air = 1.2 kg/m³
C_{fig} = Aerodynamic shape factor
C_{dyn} = Dynamic response factor = 1 (except for wind-sensitive structures, i.e. where frequency < 1 Hz; refer AS1170.2, Section 6)

When calculations are completed using ultimate wind speeds, they can be converted to serviceability speeds by multiplying the resultant pressure by the ratio of the wind velocities squared,

$$p_{SLS} = p_{ULS} \times \frac{V_{des,\theta(SLS)}^2}{V_{des,\theta(ULS)}^2}$$

For example, if the serviceability design wind speed is 34 m/s and the ultimate design wind speed is 42 m/s, the serviceability pressure is equal to [34²/42² = 0.655] times the ultimate pressure. Care should be taken where drag force coefficients are dependent on wind velocities.

2.3.1.2 Wind on piping

Transverse wind on piping is calculated using the aerodynamic shape factor shown below. The calculations are applicable where the length (*l*) of the pipe is at least eight times the diameter (*b*) (Figure 2.2).

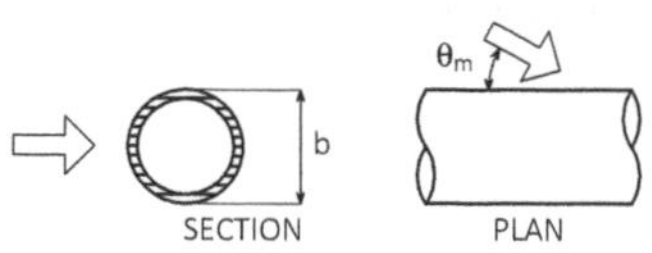

SECTION   PLAN

Figure 2.2 Wind on pipe diagram.

Force per linear metre, $F = p \times b$

$$C_{\text{fig}} = K_{ar}K_iC_d$$

where:

C_d = Drag force coefficient
K_{ar} = Correction factor (refer to Table 2.8)
K_i = Angle of wind = 1.0 (wind normal to pipe)
  = $\sin^2\theta_m$ (wind inclined to pipe)
θ_m = Angle between wind direction and longitudinal pipe axis

The drag force coefficient is dependent on the design wind speed multiplied by the pipe diameter.

$$C_d = 1.2 \qquad \text{for} \quad bV_{\text{des},\theta} < 4 \text{ m}^2/\text{s}$$

$$C_d = \text{MAX}\left\{1.0 + 0.033\left[\log_{10}\left(V_{\text{des},\theta}b_r\right)\right] - 0.025\left[\log_{10}\left(V_{\text{des},\theta}b_r\right)\right]^2, 0.6\right\} \text{ for} \quad bV_{\text{des},\theta} > 10\text{m}^2/\text{s}$$

b_r = Average height of surface roughness (refer to Table 2.9)

Linear interpolation should be used where 4 m²/s $\leq bV_{\text{des},\theta} \leq$ 10 m²/s, adopting a value of $C_d = 0.6$ at $bV_{\text{des},\theta} = 10$ m²/s for the purpose of interpolation only.

2.3.1.3 Wind on exposed steelwork

Wind on exposed steelwork is calculated in a similar method to wind on piping,
Force per linear metre, $F = p \times b$ (refer to Table 2.10).

Note: b is a different variable to that used for calculation of the aspect ratio. It is the larger section dimension and is not dependent on wind direction, as shown in Table 2.10.

Table 2.8 Aspect ratio correction factor, K_{ar}

Aspect ratio, l/b[a]	Correction factor, K_{ar}
≤8	0.7
14	0.8
30	0.9
40 or more	1.0

Source: AS/NZS 1170.2, table E1. Copied by L. Pack with permission from Standards Australia under Licence 1607-c010.

Notes:
1. For intermediate values of *l/b*, use linear interpolation.
[a] For calculation of aspect ratio, b is the section dimension measured perpendicular to the wind direction.

Table 2.9 Surface roughness values

Material	Average height of surface roughness, h_r (m)
Glass or plastic	1.5×10^{-6}
Steel (galvanised)	150×10^{-6}
Steel (light rust)	2.5×10^{-3}
Steel (heavy rust)	15×10^{-3}
Concrete (new, smooth)	60×10^{-6}
Concrete (new, rough)	1×10^{-3}
Metal (painted)	30×10^{-6}
Timber	2×10^{-3}

Source: AS/NZS 1170.2, appendix E.

Table 2.10 Force coefficients for common sections

Section shape	Force coefficient	Wind direction measured clockwise (θ)				
		0	*45*	*90*	*135*	*180*
$(d = 0.5b)$	$C_{F,x}$	2.0	1.8	−2.0	−1.8	−1.9
	$C_{F,y}$	−0.1	0.1	−1.7	−0.8	−0.95
$(d = b)$	$C_{F,x}$	1.8	1.8	−1.0	0.3	−1.4
	$C_{F,y}$	1.8	2.1	−1.9	−2.0	−1.4
$(d = 0.43b)$	$C_{F,x}$	2.05	1.85	0	−1.6	−1.8
	$C_{F,y}$	0	−0.6	−0.6	−0.4	0
$(d = 0.48b)$	$C_{F,x}$	2.05	1.95	±0.5	—	—
	$C_{F,y}$	0	−0.6	−0.9	—	—
$(d = b)$	$C_{F,x}$	1.6	1.5	0	—	—
	$C_{F,y}$	0	−1.5	−1.9	—	—

Source: Modified from AS/NZS 1170.2, table E5. Copied by L. Pack with permission from Standards Australia under Licence 1607-c010.

Note: *b* and *d* are independent from the wind direction and are as shown in the tabulated diagrams.

$$C_{fig} = K_{ar}K_iC_{F,x} \qquad \text{(wind on major axis)}$$

$$C_{fig} = K_{ar}K_iC_{F,y} \qquad \text{(wind on minor axis)}$$

where:

$C_{F,x}$ = Drag force coefficient for x-axis (refer to Table 2.10)
$C_{F,y}$ = Drag force coefficient for y-axis (refer to Table 2.10)
K_{ar} = Correction factor (refer to Table 2.8)
K_i = Angle of wind = 1.0 (wind normal to beam)
$\quad\quad$ = $\sin^2\theta_m$ (wind inclined to beam)

2.3.1.4 Wind on multiple items

For wind on multiple items, such as parallel pipes or beams, the shape factor for the shielded item may be reduced. AS/NZS 1170.2, Clause E2.3 provides detail on shielding factors for frames; it is the most applicable reference for wind on closely spaced items. For items with 100% solidity (i.e. pipes and solid members), the full C_{fig} factor is applicable for the windward item; however, the pressure is reduced by K_{sh} for shielded items.

$$C_{fig\,(shielded)} = K_{sh}C_{fig}$$

where:

K_{sh} = 0.2 for spacing/depth ≤ 4
K_{sh} = 1.0 for spacing/depth ≥ 8

Interpolation may be used for intermediate values.

Example 2.1: Wind loading

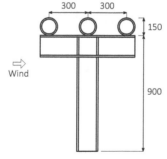

Three pipes are supported on a 900 mm high t-post constructed from 200 × 200 universal column (UC) sections. The pipes are 150 mm in diameter; they are spaced at 300 mm centres and span 3000 mm between supports. Calculate the ultimate force caused by the wind load.

Importance level = 2

Design life = 25 years

Region = A4

Terrain category = 2

Annual probability of exceedance = 1/200

Regional wind speed $\quad\quad\quad\quad V_{200} = 43$ m/s

Site and design wind speed $\quad\quad V_{sit,\beta} = 43 \times 1 \times 0.91 \times 1 \times 1 = 39.13\,m/s = V_{des,\theta}$

WIND ON PIPING

Drag force coefficient $\qquad C_d = 1.01 \qquad$ (linear interpolation)

$$bV_{\text{des},\theta} = 0.15 \times 39.13 = 5.9 \frac{\text{m}^2}{\text{s}}$$

Correction factor $\qquad K_{ar} = 0.84 \quad$ (linear interpolation)

Aspect ratio $\qquad \dfrac{l}{b} = \dfrac{3000}{150} = 20$

Angle of wind $\qquad K_i = 1.0$

Aerodynamic shape factor $\qquad C_{\text{fig}} = K_{ar} K_i C_d = 0.84 \times 1 \times 1.01 = 0.85$

Dynamic response factor $\qquad C_{\text{dyn}} = 1$

Pressure on windward pipe

$$p_1 = (0.5\rho_{\text{air}}) \left[V_{\text{des},\theta} \right]^2 C_{\text{fig}} C_{\text{dyn}} = 0.5 \times 1.2 \frac{\text{kg}}{\text{m}^3} \times (39.13 \text{ m/s})^2 \times 0.85 \times 1 = 0.78 \text{ kPa}$$

Pressure on second pipe $\qquad p_2 = 0.2 p_1 = 0.16 \text{ kPa}$

Pressure on third pipe $\qquad p_3 = 0.2 p_1 = 0.16 \text{ kPa}$

Force on piping $\qquad F_{\text{pipe}} = (0.15 \text{ m} \times 3 \text{ m}) \times (0.78 \text{ kPa} + 0.16 \text{ kPa} + 0.16 \text{ kPa}) = 0.5 \text{ kN}$

WIND ON STEELWORK

Force coefficient $\qquad C_{F,y} = 1.9$

Correction factor $\qquad K_{ar} = 0.7$

Aspect ratio $\qquad \dfrac{l}{b} = \dfrac{900}{200} = 4.5$

Angle of wind $\qquad K_i = 1.0$

Aerodynamic shape factor $\qquad C_{\text{fig}} = K_{ar} K_i C_{F,y} = 0.7 \times 1 \times 1.9 = 1.33$

Dynamic response factor $\qquad C_{\text{dyn}} = 1$

Pressure on steelwork

$$p = (0.5\rho_{\text{air}}) \left[V_{\text{des},\theta} \right]^2 C_{\text{fig}} C_{\text{dyn}} = 0.5 \times 1.2 \frac{\text{kg}}{\text{m}^3} \times (39.13 \text{ m/s})^2 \times 1.33 \times 1 = 1.22 \text{ kPa}$$

Force on steelwork $\qquad F_{\text{steel}} = (0.2 \text{ m} \times 0.9 \text{ m}) \times 1.22 \text{ kPa} = 0.22 \text{ kN}$

TOTAL FORCE

Total ultimate wind load $\qquad F_{\text{total}} = V_y^* = F_{\text{pipe}} + F_{\text{steel}} = 0.72 \text{ kN}$

Ultimate wind moment (at base)

$$M_{\text{total}} = M_x^* = F_{\text{pipe}} \left(900 \text{ mm} + \frac{150 \text{ mm}}{2} \right) + F_{\text{steel}} \left(\frac{900 \text{ mm}}{2} \right) = 0.59 \text{ kNm}$$

2.3.2 Seismic

This section presents the seismic design of structures in accordance with AS 1170.4–2007. The method involves calculating an equivalent static force (a percentage of the vertical load), and applying it horizontally. Seismic loading is calculated using the annual probability of exceedance (refer to Section 2.3).

2.3.2.1 Earthquake design categories

The design is dependent on a classification of earthquake design category (EDC). Typical structures will be classified as EDC1 or EDC2. The required combination of importance level and height for EDC3 rarely occurs in mining or oil and gas sites; therefore, the details are not typically relevant and are not presented in this chapter (refer to AS 1170.4 for details). The EDC can be chosen from Table 2.11 after calculating a site subsoil class, probability factor and hazard factor.

2.3.2.2 Site subsoil class

The site subsoil class is chosen from Table 2.12, based on the geotechnical profile of the site.

2.3.2.3 Probability factor

The probability factor (k_p) is dependent only on the annual probability of exceedance and is chosen from Table 2.13.

Table 2.11 Selection of earthquake design categories

Importance level, type of structure	$(k_p Z)$ for site subsoil class				Structure height, h_n (m)	Earthquake design category
	E_e or D_e	C_e	B_e	A_e		
1	—				—	Not required to be designed for earthquake actions
Domestic structure (housing)	—				Top of roof ≤8.5	Refer to AS 1170.4, Appendix A
					Top of roof >8.5	Design as importance level 2
2	≤0.05	≤0.08	≤0.11	≤0.14	≤12	1
					>12, <50	2
					≥50	3
	>0.05 to ≤0.08	>0.08 to ≤0.12	>0.11 to ≤0.17	>0.14 to ≤0.21	<50	2
					≥50	3
	>0.08	>0.12	>0.17	>0.21	<25	2
					≥25	3
3	≤0.08	≤0.12	≤0.17	≤0.21	<50	2
					≥50	3
	>0.08	>0.12	>0.17	>0.21	<25	2
					≥25	3
4	—				<12	2
					≥12	3

Source: AS 1170.4, table 2.1. Copied by L. Pack with permission from Standards Australia under Licence 1607-c010.

Notes:
1. A higher EDC or procedure may be used in place of that specified.
2. The building height (h_n) is taken as the height of the centre of mass above the base.
3. In addition to the above, a special study is required for importance level 4 structures to demonstrate that they remain serviceable for immediate use following the design event for importance level 2 structures.

Table 2.12 Soil subsoil class definition

Subsoil class	Description
A_e Strong rock	Strong to extremely strong rock, satisfying 1. Unconfined compressive strength greater than 50 MPa or an average shear-wave velocity, over the top 30 m, greater than 15 m/s 2. Not underlain by materials having a compressive strength less than 18 MPa or an average shear-wave velocity less than 600 m/s
B_e Rock	Rock, satisfying 1. A compressive strength between 1 and 50 MPa or an average shear-wave velocity, over the top 30 m, greater than 360 m/s 2. Not underlain by materials having a compressive strength less than 0.8 MPa or an average shear-wave velocity less than 300 m/s A surface layer of no more than 3 m depth of highly weathered or completely weathered rock or soil (a material with a compressive strength less than 1 MPa) may be present.

C_e Shallow soil	A site that is not Class A_e, Class B_e (i.e. not rock site) or Class E_e (i.e. not very soft soil site) and either
	1. The low-amplitude natural site period is less than or equal to 0.6 s.
	2. The depths of soil do not exceed those listed in Table 4.1.
	   Note:
	   The low-amplitude natural site period may be estimated from either
	   a. Four times the shear-wave travel time from the surface to rock
	   b. Nakamura ratios
	   c. Recorded earthquake motions
	or evaluated in accordance with AS 1170.4, Clause 4.1.3 for sites with layered subsoil. Where more than one method is used, the value determined from the most preferred method given in AS 1170.4, Clause 4.1.2 shall be adopted.

D_e Deep or soft soil	A site that is not Class A_e, Class B_e (i.e. not rock site) or Class E_e (i.e. not very soft soil site) and
	1. Underlain by less than 10 m of soil with an undrained shear strength less than 12.5 kPa or soil with standard penetration test (SPT) N values less than 6, and either
	   a. The low-amplitude natural site period is greater than 0.6 s.
	   b. The depths of soil exceed those listed in AS 1170.4, Table 4.1.
	The low-amplitude natural site period is estimated in accordance with AS 1170.4, Clause 4.2.3.

E_e Very soft soil	A site with any one of the following: 1. More than 10 m of very soft soil with undrained shear strength of less than 12.5 kPa 2. More than 10 m of soil with SPT N values of less than 6 3. More than 10 m depth of soil with shear-wave velocities of 150 m/s or less 4. More than 10 m combined depth of soils with properties as described in the three previous points

Source: Compiled from AS 1170.4, Clause 4.2. Copied by L. Pack with permission from Standards Australia under Licence 1607-c010.

Table 2.13 Probability factor (k_p)

Annual probability of exceedance, p	Probability factor, k_p	Annual probability of exceedance, p	Probability factor, k_p
1/2500	1.8	1/250	0.75
1/2000	1.7	1/200	0.7
1/1500	1.5	1/100	0.5
1/1000	1.3	1/50	0.35
1/800	1.25	1/25	0.25
1/500	1.0	1/20	0.20

Source: AS 1170.4, Table 3.1. Copied by L. Pack with permission from Standards Australia under Licence 1607-c010.

2.3.2.4 Hazard factor

A hazard factor (Z) needs to be chosen based on the location of the structure. Values are presented in Table 2.14 and Figure 2.3. Refer to AS 1170.4, Figure 3.2(A) to (F) for more detailed maps of specific regions within Australia. A new map has been created for consideration in the next revision of AS 1170.4. Refer to Geoscience Australia for more details.

2.3.2.5 Design principles

The seismic design of structures requires a clear load path for both vertical and lateral actions. All components of the structure should be considered as part of the structure (with their seismic weight adequately restrained), or separated so that no interaction takes place. Individual foundations such as piles or pads which are founded in weak soils (ultimate bearing capacity <250 kPa) should be connected to prevent differential movement.

Table 2.14 Hazard factor (Z) for specific Australian locations

Location	Z	Location	Z	Location	Z
Adelaide	0.1	Geraldton	0.09	Port Augusta	0.11
Albany	0.08	Gladstone	0.09	Port Lincoln	0.1
Albury/Wodonga	0.09	Gold Coast	0.05	Port Hedland	0.12
Alice Springs	0.08	Gosford	0.09	Port Macquarie	0.06
Ballarat	0.08	Grafton	0.05	Port Pirie	0.1
Bathurst	0.08	Gippsland	0.1	Robe	0.1
Bendigo	0.09	Goulburn	0.09	Rockhampton	0.08
Brisbane	0.05	Hobart	0.03	Shepparton	0.09
Broome	0.12	Karratha	0.12	Sydney	0.08
Bundaberg	0.11	Katoomba	0.09	Tamworth	0.07
Burnie	0.07	Latrobe Valley	0.1	Taree	0.08
Cairns	0.06	Launceston	0.04	Tennant Creek	0.13
Camden	0.09	Lismore	0.05	Toowoomba	0.06
Canberra	0.08	Lorne	0.1	Townsville	0.07
Carnarvon	0.09	Mackay	0.07	Tweed Heads	0.05
Coffs Harbour	0.05	Maitland	0.1	Uluru	0.08
Cooma	0.08	Melbourne	0.08	Wagga Wagga	0.09
Dampier	0.12	Mittagong	0.09	Wangaratta	0.09
Darwin	0.09	Morisset	0.1	Whyalla	0.09
Derby	0.09	Newcastle	0.11	Wollongong	0.09
Dubbo	0.08	Noosa	0.08	Woomera	0.08
Esperance	0.09	Orange	0.08	Wyndham	0.09
Geelong	0.1	Perth	0.09	Wyong	0.1
Meckering region				Islands	
Ballidu	0.15	Meckering	0.2	Christmas Island	0.15
Corrigin	0.14	Northam	0.14	Cocos Islands	0.08
Cunderdin	0.22	Wongan Hills	0.15	Heard Island	0.1
Dowerin	0.2	Wickepin	0.15	Lord Howe Island	0.06
Goomalling	0.16	York	0.14	Macquarie Island	0.6
Kellerberrin	0.14			Norfolk Island	0.08

Source: AS 1170.4, Table 3.2. Copied by L. Pack with permission from Standards Australia under Licence 1607-c010.

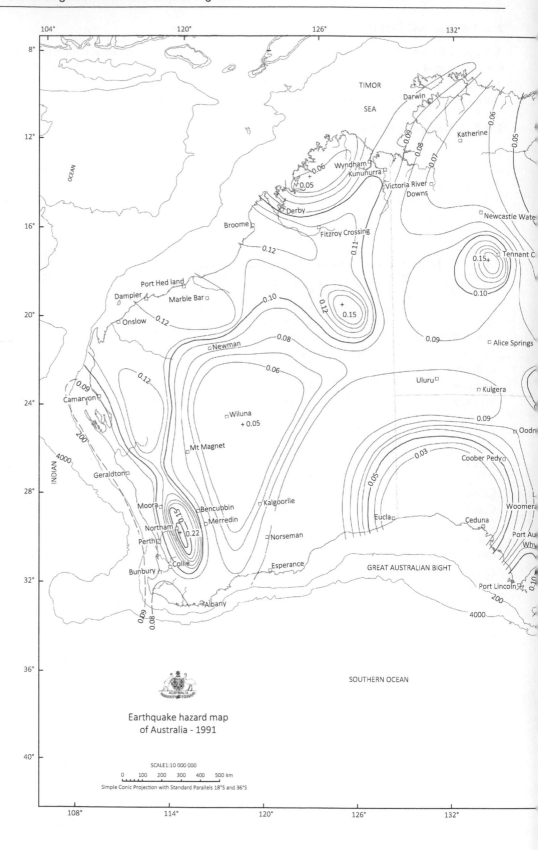

Earthquake hazard map
of Australia - 1991

SCALE 1:10 000 000

0   100   200   300   400   500 km

Simple Conic Projection with Standard Parallels 18°S and 36°S

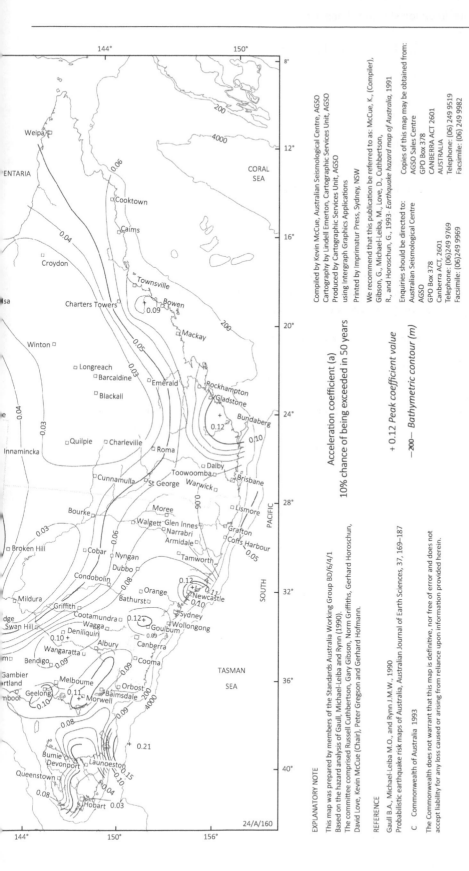

Acceleration coefficient (a)
10% chance of being exceeded in 50 years

+ 0.12 *Peak coefficient value*

–**200**— *Bathymetric contour (m)*

Figure 2.3 Hazard *factor (Z) – Australian Seismological Centre.* (From McCue, K., et al., *Earthquake hazard map of Australia, Australian Seismological Centre, Geoscience Australia, Symonston, 1991.* Copyright Commonwealth of Australia (Geoscience Australia) 2016. This product is released under the Creative Commons Attribution 4.0 International Licence.)

The seismic weight is applied in calculating equivalent static loads:

Seismic weight, $W_i = G_i + \psi_c Q_i$

where:
  G_i = Permanent action at level i
  Q_i = Live action at level i
  ψ_c = 0.6 (for storage applications)
    = 0.3 (for other applications)

2.3.2.6 Earthquake design category 1 (EDC1)

EDC1 requires an equivalent static force equal to 10% of the seismic weight. Each of the major axes of the structure should be checked individually. This is done in programs (such as Space Gass) by creating a load case with self-weight applied in the x-direction, and another with self-weight applied in the z-direction (assuming the y-direction is vertical). The acceleration for these cases should be 0.1 g, or 0.981 m/s². Additional forces may have to be applied for non-structural masses, such as cable trays or piping. For stress-critical pipes, the load should be provided by pipe stress engineers (refer to Section 6.1).

2.3.2.7 Earthquake design category 2 (EDC2)

Each element within the structure may be designed to restrain the seismic load in either one or both orthogonal directions (axes) of the building. If elements are required for restraint in both directions, an additional allowance of 30% of the primary load is required concurrently in the secondary direction. Elements which are designed to restrain seismic load in only one of the major axes shall be designed for only that load. Generally, this is achieved by creating a load combination case with the x-direction load factored and 100% and the z-direction load factored down to 30% for the primary x-direction case. The values are swapped when considering the primary z-direction combination case.

Connections shall be designed for a minimum horizontal load equal to the greater of 5% of the vertical seismic load or 5% of the mass of the supported component.

2.3.2.7.1 Structures not exceeding 15 m

EDC2 structures which do not exceed 15 m in height are required to be designed using an equivalent static analysis:

Equivalent static force, $F_i = K_s \left[k_p Z S_p / \mu \right] W_i$

where:
  K_s = Floor factor (refer to Table 2.15)
  S_p = Structural performance factor (refer to Table 2.16)
  W_i = Weight of structure or component (at level i)
  μ = Structural ductility factor (refer to Table 2.16)

A percentage will be calculated if the variable for weight (W_i) is excluded from the calculation. Values typically range from 5% to 20%; however, higher values can occur. In the absence of specific data from a material standard or from a 'push-over' analysis, the ductility and performance factors are adopted from Table 2.16.

2.3.2.7.2 Structures exceeding 15 m

EDC2 structures which exceed 15 m in height are required to be designed using an equivalent static analysis, as shown in AS 1170.4, Section 6. The design method is dependent on

Table 2.15 Values of K_s for structures not exceeding 15 m

Total number of stories	Subsoil class	K_s factor				
		Storey under consideration				
		5th	4th	3rd	2nd	1st
5	A_e	2.5	1.9	1.4	1.0	0.5
	B_e	3.1	2.5	1.8	1.2	0.6
	C_e	4.4	3.5	2.6	1.7	0.9
	D_e, E_e	6.1	4.9	3.6	2.5	1.2
4	A_e	—	2.7	2.0	1.4	0.6
	B_e	—	3.5	2.6	1.7	0.9
	C_e	—	4.9	3.6	2.5	1.2
	D_e, E_e	—	5.8	4.4	3.0	1.4
3	A_e	—	—	3.1	2.0	1.0
	B_e	—	—	3.9	2.6	1.3
	C_e, D_e, E_e	—	—	5.5	3.6	1.8
2	A_e	—	—	—	3.1	1.6
	B_e	—	—	—	3.9	1.9
	C_e, D_e, E_e	—	—	—	4.9	2.5
1	A_e	—	—	—	—	2.3
	B_e	—	—	—	—	3.0
	C_e, D_e, E_e	—	—	—	—	3.6

Source: AS 1170.4, table 5.4. Copied by L. Pack with permission from Standards Australia under Licence 1607-c010.

the natural frequency of the structure. The standard presents a formula for a shear force at the base; however, a percentage will be calculated if the variable for weight (W_t) is excluded from the calculation.

Horizontal equivalent static shear force at base,

$$V = \left[k_p Z C_h (T_1) S_p / \mu \right] W_t$$

$C_h(T_1)$ = Spectral shape factor for fundamental natural period

All other variables are calculated in the same manner as for structures not exceeding 15 m.

The spectral shape factor is dependent on the period of the structure and can be selected from Table 2.17. The natural period may be calculated using the formula shown or by a rigorous structural analysis. If the analysis method is used, the resultant should not be less than 80% of the value achieved by using the formula.

Natural period, $T_1 = 1.25 k_t h_n^{0.75}$ (for ultimate limit states)

where:
h_n = Height from base to uppermost seismic weight (m)
k_t = 0.11 (for moment-resisting steel frames).
      = 0.075 (for moment-resisting concrete frames).
      = 0.06 (for eccentrically-braced steel frames).
      = 0.05 (for all structures)

Table 2.16 Structural ductility factor (μ) and structural performance factor (S_p)

Structural system	Description	μ	S_p	S_p/μ	μ/S_p
Steel structures					
	Special moment-resisting frames (fully ductile)[a]	4	0.67	0.17	6
	Intermediate moment-resisting frames (moderately ductile)	3	0.67	0.22	4.5
	Ordinary moment-resisting frames (limited ductile)	2	0.77	0.38	2.6
	Moderately ductile concentrically braced frames	3	0.67	0.22	4.5
	Limited ductile concentrically braced frames	2	0.77	0.38	2.6
	Fully ductile eccentrically braced frames[a]	4	0.67	0.17	6
	Other steel structures not defined above	2	0.77	0.38	2.6
Concrete structures					
	Special moment-resisting frames (fully ductile)[a]	4	0.67	0.17	6
	Intermediate moment-resisting frames (moderately ductile)	3	0.67	0.22	4.5
	Ordinary moment-resisting frames	2	0.77	0.38	2.6
	Ductile coupled walls (fully ductile)[a]	4	0.67	0.17	6
	Ductile partially coupled walls[a]	4	0.67	0.17	6
	Ductile shear walls	3	0.67	0.22	4.5
	Limited ductile shear walls	2	0.77	0.38	2.6
	Ordinary moment-resisting frames in combination with limited ductile shear walls	2	0.77	0.38	2.6
	Other concrete structures not listed above	2	0.77	0.38	2.6
Timber structures					
	Shear walls	3	0.67	0.22	4.5
	Braced frames (with ductile connections)	2	0.77	0.38	2.6
	Moment-resisting frames	2	0.77	0.38	2.6
	Other wood or gypsum-based seismic force–resisting systems not listed above	2	0.77	0.38	2.6
Masonry structures					
	Close-spaced reinforced masonry[b]	2	0.77	0.38	2.6
	Wide-spaced reinforced masonry[b]	1.5	0.77	0.5	2
	Unreinforced masonry[b]	1.25	0.77	0.62	1.6
	Other masonry structures not complying with AS 3700	1	0.77	0.77	1.3
Specific structure types					
	Tanks, vessels or pressurised spheres on braced or unbraced legs	2	1	2	0.5
	Cast-in-place concrete silos and chimneys having walls continuous to the foundation	3	1	3	0.33
	Distributed mass cantilever structures, such as stacks, chimneys, silos and skirt-supported vertical vessels	3	1	3	0.33
	Trussed towers (freestanding or guyed), guyed stacks and chimneys	3	1	3	0.33
	Inverted pendulum-type structures	2	1	2	0.5
	Cooling towers	3	1	3	0.33
	Bins and hoppers on braced or unbraced legs	3	1	3	0.33
	Storage racking	3	1	3	0.33
	Signs and billboards	3	1	3	0.33
	Amusement structures and monuments	2	1	2	0.5
	All other self-supporting structures not otherwise covered	3	1	3	0.33

Source: AS 1170.4, table 6.5[A] and [B]. Copied by L. Pack with permission from Standards Australia under Licence 1607-c010.

Note:
[a] The design of structures with μ > 3 is outside the scope of AS 1170.4.
[b] These values are taken from AS 3700.

Table 2.17 Spectral shape factor ($C_h(T)$)

Period (s)	Site subsoil class				
	A_e	B_e	C_e	D_e	E_e
0.0	2.35 (0.8)[a]	2.94 (1.0)[a]	3.68 (1.3)[a]	3.68 (1.1)[a]	3.68 (1.1)[a]
0.1	2.35	2.94	3.68	3.68	3.68
0.2	2.35	2.94	3.68	3.68	3.68
0.3	2.35	2.94	3.68	3.68	3.68
0.4	1.76	2.20	3.12	3.68	3.68
0.5	1.41	1.76	2.50	3.68	3.68
0.6	1.17	1.47	2.08	3.30	3.68
0.7	1.01	1.26	1.79	2.83	3.68
0.8	0.88	1.10	1.56	2.48	3.68
0.9	0.78	0.98	1.39	2.20	3.42
1.0	0.70	0.88	1.25	1.98	3.08
1.2	0.59	0.73	1.04	1.65	2.57
1.5	0.47	0.59	0.83	1.32	2.05
1.7	0.37	0.46	0.65	1.03	1.60
2.0	0.26	0.33	0.47	0.74	1.16
2.5	0.17	0.21	0.30	0.48	0.74
3.0	0.12	0.15	0.21	0.33	0.51
3.5	0.086	0.11	0.15	0.24	0.38
4.0	0.066	0.083	0.12	0.19	0.29
4.5	0.052	0.065	0.093	0.15	0.23
5.0	0.042	0.053	0.075	0.12	0.18

Equations for spectra

$0 < T \le 0.1$	$0.8 + 15.5T$	$1.0 + 19.4T$	$1.3 + 23.8T$	$1.1 + 25.8T$	$1.1 + 25.8T$
$0.1 < T \le 1.5$	$0.704/T$ but ≤ 2.35	$0.88/T$ but ≤ 2.94	$1.25/T$ but ≤ 3.68	$1.98/T$ but ≤ 3.68	$3.08/T$ but ≤ 3.68
$T > 1.5$	$1.056/T^2$	$1.32/T^2$	$1.874/T^2$	$2.97/T^2$	$4.62/T^2$

Source: AS 1170.4, table 6.4. Copied by L. Pack with permission from Standards Australia under Licence 1607-c010.

[a] Values in brackets correspond to values of spectral shape factor for the modal response spectrum and the numerical integration time history methods, and for use in the method of calculation of forces on parts and components (see AS 1170.4, Section 8).

For structures which exceed 15 m in height, a higher portion of the total seismic load is applied at higher floors. The total base shear is not affected by this clause, only the distribution of the load. The distribution is dependent on the period of the structure. A higher percentage of the load is applied at higher floors for structures which are less stiff (lower fundamental natural periods).

Force at the ith floor, $F_i = k_{F,i}V$

where:

$$k_{F,i} = \left(W_i h_i^k\right)/\left(\sum_{j=1}^{n} W_j h_j^k\right)$$
h_i = Height of level i (metres)
k = 1.0 (for $T_1 \le 0.5$)
= 2.0 (for $T_1 \ge 2.5$)(k is linearly interpolated for $0.5 < T_1 < 2.5$)
n = Number of levels

Example 2.2: Seismic loading

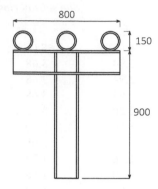

Three 400 kg pipes are supported on a 900 mm high t-post constructed from 200UC46 sections. Calculate the ultimate force caused by the seismic load.

Importance level = 2
Soil type = C_e
Design life = 50 years
Hazard factor, $Z = 0.09$

Annual probability of exceedance = 1/500   (refer to Table 2.4)

Probability factor $k_p = 1.0$ (refer to Table 2.13)

$k_p Z = 0.09$

Earthquake design category = 2 (refer to Table 2.11)

Equivalent static force, $F_i = K_s \left[k_p Z S_p / \mu \right] W_i$

$$= K_s \left[k_p Z S_p / \mu \right] W_i = 3.6 \left[1.0 \times 0.09 \times 0.77 \div 2 \right] W_i = 12.5\% W_i$$

$$K_s = 3.6 \text{ (refer to Table 2.15)}$$

$$S_p = 0.77 \text{ (refer to Table 2.16)}$$

$$\mu = 2 \text{ (refer to Table 2.16)}$$

SEISMIC ON PIPING
Force on piping,

$$F_{pipe} = 12.5\% \ W_{pipe} = 12.5\% \times \left(3 \times 400 \text{ kg} \right) \times 9.81 \text{ m/s}^2 = 1.47 \text{ kN}$$

SEISMIC ON STEELWORK
Force on steel,

$$F_{steel} = 12.5\% W_{steel} = 12.5\% \left(46 \text{ kg/m} \times \left(900 \text{ mm} + 800 \text{ mm} - 200 \text{ mm} \right) \right)$$

$$\times 9.81 \text{ m/s}^2 = 0.08 \text{ kN}$$

TOTAL FORCE

Total ultimate seismic load, $\qquad F_{\text{total}} = V_y^* = V_x^* = F_{\text{pipe}} + F_{\text{steel}} = 1.55\ \text{kN}$

Ultimate seismic moment (at base), $\qquad M_{\text{total}} = M_x^* = M_y^* = F_{\text{pipe}}\ z_{\text{pipe}} + F_{\text{steel}}\ z_{\text{steel}}$

Centroid of pipes, $\qquad z_{\text{pipe}} = 900\ \text{mm} + 150\ \text{mm}/2 = 975\ \text{mm}$

Centroid of steelwork,

$$z_{\text{steel}} = \frac{\left[\dfrac{900\ \text{mm}}{2}\,900\ \text{mm} + (800-200)\ \text{mm}\left(900 - \dfrac{200}{2}\right)\text{mm}\right]}{(900\ \text{mm} + 800\ \text{mm} - 200\ \text{mm})} = 590\ \text{mm}$$

$$M_{\text{total}} = 1.48\ \text{kNm}$$

(The primary seismic load is the same in each direction being considered, with 30% applied concurrently in the transverse direction.)

2.3.3 Dead and live loads

Dead and live loads (actions) are typically in accordance with AS/NZS 1170.1. Dead loads are the known permanent actions which act on a structure, such as the self-weight of the members and permanently attached items. Live loads are imposed actions which represent the expected surcharges acting on a structure, including people and equipment. Live loads adopt a higher factor for strength design due to the increased uncertainty of the surcharge. Specific loads are detailed throughout this text (refer to Table 2.18), and additional loads are detailed below.

2.3.3.1 Dead loads

Various densities are shown in Table 2.19 for the purpose of calculating dead loads for specific materials. Refer to Table 5.14 for soil densities.

2.3.3.2 Live loads

Live loads typically consist of imposed floor actions (refer to Table 2.20) which include uniform pressures and also point loads. Elements should be designed for the worst combination or pattern of loading.

Table 2.18 Other loads

Item	Reference
Piping	Section 6.1
Thermal loads	Section 6.6
Retained earth pressure	Section 5.3
Swelling soil pressure	Section 5.5

Table 2.19 Densities for common materials

Material	Density (kg/m³)
Air	1.2
Aluminium	2,722
Asphalt	2,160
Bitumen	1,020–1,430
Brass	8,512
Concrete	2,400 (typical) 2,400 + 600 for each 1% of reinforcement
Copper	8,797
Iron	7,207
Lead	11,314
Nickel	8,602
Reinforcement (mild steel)	7,850
Stainless steel	8,000 (varies between grades)
Structural (mild) steel	7,850
Timber	400–1,250
Water	1,000 (999.845 at 0°C) (998.213 at 20°C) (988.050 at 50°C)

Note: Values shown are the common ranges for that material. Values are calculated from various product catalogues, limits in Australian Standards and AS/NZS 1170.1, table A1.

Table 2.20 Imposed floor actions

Description	Pressure (kPa)	Point load (kN)
Work rooms (light industrial) without storage	3.0	3.5
Factories, workshops and similar buildings (general industries)	5.0	4.5
Assembly areas (without fixed seating) susceptible to overcrowding	5.0	3.6
Roofs (structural elements)[a]	$(1.8/A + 0.12)$ Not less than 0.25	1.4
Warehousing and storage areas, areas subject to accumulation of goods, areas for equipment and plant		
Mobile stacking, mechanically operated heavy shelving (wheels on rails, e.g. Compactus)	4.0 for each metre of storage height but not less than 10.0	To be calculated
Plant rooms, fan rooms, etc., including weight of machinery	5.0	4.5
Areas around equipment in boiler rooms (weight of equipment to be determined)	5.0	4.5
Vehicle access		
Light vehicle traffic (not exceeding 2,500 kg)	2.5	13
Medium vehicle traffic (not exceeding 10,000 kg)	5.0	31 (area = 0.025 m²)

Source: Excerpts from AS/NZS 1170.1, tables 3.1 and 3.2. Copied by L. Pack with permission from Standards Australia under Licence 1607-c010.

[a] Structural elements supporting more than 200 m² of roof area shall be designed to support 0.25 kPa on the 200 m² of the supported area that gives the worst effect. A = the plan projection of the surface area of roof supported by the member under analysis, in square metres. For structural elements in roofs of houses, the uniform distributed action shall be 0.25 kPa and the concentrated actions shall be 1.1 kN.

2.3.3.3 Buoyancy loads

Forces caused by buoyancy are equal to the volume of fluid displaced, multiplied by the density of the fluid (Archimedes' principle) in which it is submerged. It is suggested that the problem is a stability issue, and therefore the stabilising mass is factored by 0.9 and the buoyant force is factored at 1.2 (for a liquid pressure).

Example 2.3: Buoyancy

An item (1 × 1 × 1 m) with a mass of 200 kg is half buried in a soil with groundwater at the surface. Check the stability of the structure:

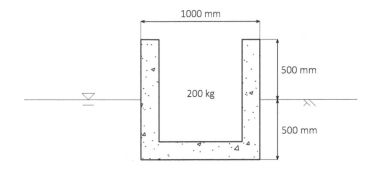

$$F_{buoy} = 1 \text{ m} \times 1 \text{ m} \times 0.5 \text{ m} \times 1000 \frac{\text{kg}}{\text{m}^3} \times 9.81 \frac{\text{m}}{\text{s}^2} = 4.9 \text{ kN}$$

$$F_{sw} = 200 \text{ kg} \times 9.81 \frac{\text{m}}{\text{s}^2} = 1.96 \text{ kN}$$

For stability, $0.9 F_{sw} \geq 1.2\, F_{buoy}$

The structure is not stable, and will be pushed upwards. Mass should be added, or a collar could be used to engage more soil and widen the base.

2.3.3.4 Vehicle loads

Several commonly used vehicle load configurations are shown in Table 2.21 with reference from AS 5100.2 and AS 5100.7. Depending on expected construction and service loading, different values may be chosen. Also, dynamic and ultimate factors are sometimes reduced where loading is controlled. For example, if the vehicle load is inside a building where speed is limited and weights are controlled, the dynamic factor may be reduced to 1.1 and the ultimate factor reduced to 1.5.

Where vehicles can approach the top of a retaining wall, a horizontal load is imparted on the wall based on the mass of the vehicle and the active pressure coefficient. AS 5100.2 allows a reduction in lateral earth pressure from vehicle loads, by adopting an equivalent surcharge of 1 m of fill (instead of the vehicle mass) for a distance equal to the height of the wall. This is typically equivalent to a uniform pressure of 20 kPa.

Table 2.21 Vehicle loading configurations

Title	Factors	Configuration
W80 Wheel load 80 kN AS 5100.2 Cl 6.2.1	Dynamic = 1.4 Ultimate = 1.8	
A160 Axle load 160 kN AS 5100.2 Cl 6.2.2	Dynamic = 1.4 Ultimate = 1.8	
T44[a] Truck load 44 tonnes AS 5100.7 Cl A2.2.2	Dynamic = 1.2[a] Ultimate = 2.0	
S1600 Stationary traffic AS 5100.2 Cl 6.2.4	Dynamic = 1.0 Ultimate = 1.8 (+24 kN/m lane load)	 Spacing, x = 3.75 m (2nd wheel group), 6.25 m (3rd wheel group), 5 m (4th wheel group)
M1600 Moving traffic AS 5100.2 Cl 6.2.3	Dynamic = 1.3 Ultimate = 1.8 (+6 kN/m lane load)	 Spacing, x = 3.75 m (2nd wheel group), 6.25 m (3rd wheel group), 5 m (4th wheel group)

[a] T44 is a superseded configuration; however, it is still used for some purposes where a load of less than SM 1600 is required. The dynamic factor should be increased up to 1.4 for structures with vertical frequencies between 1 and 6 Hz (refer to AS 5100.7, Clause A2.2.10 for details).

Table 2.22 Frictional coefficients

Material 1	Material 2	Friction coefficient, μ
Soil[1]	Concrete	0.5–0.8
Soil[1]	Steel pipelines	0.25–0.4
Concrete[1]	Concrete	0.4–0.8
Concrete[1]	Mild steel	0.4–0.5
Mild steel	Mild steel	0.3–0.4
Mild steel	HDPE	0.2
Mild steel	PTFE	0.1
PTFE	PTFE	0.05

Notes:
1. Highly variable dependent on type and finish, coefficient is indicative only
2. HDPE = High-density polyethylene
3. PTFE = Polytetrafluoroethylene (Teflon)

2.4 FRICTION

Friction between materials should be defined for projects in the structural design criteria (refer to Section 1.7). These values can be variable; however, for the purpose of a project, typical values should be adopted for the design of structural elements. Refer to Table 2.22 for a range of typical values.

2.5 DEFLECTIONS

Deflection limitations are provided in AS/NZS 1170.0, AS 3600 and AS 4100. The limits from AS/NZS 1170.0 are intended to prevent specified forms of damage to individual elements, as listed. AS 3600 and AS 4100 provide limits for design in concrete and steel, respectively.

2.5.1 AS/NZS 1170 deflection requirements

A selection of the deflection requirements of AS/NZS 1170.0 is shown in Table 2.23. The response limits are appropriate for the specific elements and actions listed. The standard includes a much more detailed list.

2.5.2 AS 3600 deflection requirements

The deflection requirements for concrete members are specified in Table 2.24.

2.5.3 AS 4100 deflection requirements

The suggested deflection requirements for steel structures are detailed in Table 2.25 for generic cases and Table 2.26 for industrial portal frame buildings. AS 4100 also notes that the requirements of AS/NZS 1170.0 may be used as a substitute to its requirements.

Table 2.23 Suggested serviceability limit state criteria

Element	Phenomenon controlled	Serviceability parameter	Applied action	Element response
Roof cladding				
Metal roof cladding	Indentation	Residual deformation	$Q = 1$ kN	Span/600, <0.5 mm
	Decoupling	Midspan deflection	$[G, \psi_s Q]$	Span/120
Roof supporting elements				
Roof members (trusses, rafters, etc.)	Sag	Midspan deflection	$[G, \psi_l Q]$	Span/300
Ceiling and ceiling supports				
Ceilings with plaster finish	Cracking	Midspan deflection	$[G, \psi_s Q]$ or $[W_s]$	Span/200
Wall elements				
Columns	Side sway	Deflection at top	$[W_s]$	Height/500
Portal frames (frame racking action)	Roof damage	Deflection at top	$[W_s]$ or $[E_s]$	Spacing/200
Walls – General (face loaded)	Discerned movement	Midheight deflection	W_s	Height/150
		Midheight deflection	$Q = 0.7$ kN	Height/200, <12 mm
	Supported elements rattle	Midheight deflection	W_s	Height/1000
Floors and floor supports				
Beams where line of sight is along invert	Sag	Midspan deflection	$[G, \psi_l Q]$	Span/500
Beams where line of sight is across soffit	Sag	Midspan deflection	$[G, \psi_l Q]$	Span/250
Handrails – Post and rail system	Side sway	Midspan deflection	$Q = 1.5$ kN/m	Height/60 + Span/240

Source: AS/NZS 1170.0, table C1. Copied by L. Pack with permission from Standards Australia under Licence 1607-c010.

Table 2.24 Deflection for concrete members

Type of member	Deflection to be considered	Deflection limitation (Δ/L_{ef}) for spans (Notes 1 and 2)	Deflection limitation (Δ/L_{ef}) for cantilevers (Note 4)
All members	The total deflection	1/250	1/125
Members supporting masonry partitions	The deflection that occurs after the addition or attachment of the partitions	1/500 where provision is made to minimise the effect of movement; otherwise, 1/1000	1/250 where provision is made to minimise the effect of movement; otherwise, 1/500
Members supporting other brittle finishes	The deflection that occurs after the addition or attachment of the finish	Manufacturer's specification, but not more than 1/500	Manufacturer's specification, but not more than 1/250
Members subjected to vehicular or pedestrian traffic	The imposed action (live and dynamic impact) deflection	1/800	1/400
Transfer members	Total deflection	1/500 where provision is made to minimise the effect of deflection of the transfer member on the supported structure; otherwise, 1/1000	1/250

Source: AS 3600, table 2.3.2. Copied by L. Pack with permission from Standards Australia under Licence 1607-c010.:

Notes:
1. In general, deflection limits should be applied to all spanning directions. This includes, but is not limited to, each individual member and the diagonal spans across each design panel. For flat slabs with uniform loadings, only the column strip deflections in each direction need be checked.
2. If the location of masonry partitions or other brittle finishes is known and fixed, these deflection limits need only be applied to the length of the member supporting them. Otherwise, the more general requirements of Note 1 should be followed.
3. Deflection limits given may not safeguard against ponding.
4. For cantilevers, the values of Δ/Lef given in this table apply only if the rotation at the support is included in the calculation of Δ.
5. Consideration should be given by the designer to the cumulative effect of deflections, and this should be taken into account when selecting a deflection limit.
6. When checking the deflections of transfer members and structures, allowance should be made in the design of the supported members and structures for the deflection of the supporting members. This will normally involve allowance for settling supports and may require continuous bottom reinforcement at settling columns.

Table 2.25 Suggested limits on calculated vertical deflections of beams

Type of beam	Deflection to be considered	Deflection limit (Δ) for span (l)[a]	Deflection limit (Δ) for cantilever (l)[b]
Beam supporting masonry partitions	The deflection which occurs after the addition or attachment of partitions	$\dfrac{\Delta}{l} \leq \dfrac{1}{500}$ where provision is made to minimise the effect of movement; otherwise, $\dfrac{\Delta}{l} \leq \dfrac{1}{1000}$	$\dfrac{\Delta}{l} \leq \dfrac{1}{250}$ where provision is made to minimise the effect of movement; otherwise, $\dfrac{\Delta}{l} \leq \dfrac{1}{500}$
All beams	The total deflection	$\dfrac{\Delta}{l} \leq \dfrac{1}{250}$	$\dfrac{\Delta}{l} \leq \dfrac{1}{125}$

Source: AS 4100, table B1. Copied by L. Pack with permission from Standards Australia under Licence 1607-c010.

[a] Suggested deflection limits in this table may not safeguard against ponding.
[b] For cantilevers, the values of Δ/l given in this table apply, provided that the effect of the rotation at the support is included in the calculation of Δ.

Table 2.26 Suggested limits on calculated horizontal deflections for industrial portal frame buildings

Type of construction	Relative deflection limitation	Absolute deflection limitation
Aluminium sheeting and no internal walls or ceilings (no crane)	Spacing/200	Spacing/150
Aluminium sheeting and no internal walls or ceilings (operating gantry crane)	Spacing/250	Spacing/250 (taken at crane height)
External masonry walls supported by steelwork and no internal walls or ceilings (no crane)	Spacing/200	Spacing/250

Source: AS 4100, Clause B2.

Note: Deflections are taken at the eaves height unless noted otherwise.

Chapter 3

Steel design

This section summarises the design of steel structures in accordance with AS 4100, current at the time of writing (AS 4100-1998, incorporating Amendment 1). The status of standards should be checked at the beginning of a project or calculation. Only specific clauses are outlined in this section. The full standard should be followed to ensure that all requirements are met.

3.1 MATERIAL

The material properties presented in this section are based on those presented in AS 4100, Section 2. Project standards or drawings should note that material test certificates should be supplied for all structural materials and that the information should be supplied by the manufacturer or verified by an independent laboratory, accredited by International Laboratory Accreditation Cooperation (ILAC) or Asia Pacific Laboratory Accreditation Cooperation (APLAC).

The general behaviour of mild steelwork is best depicted by graphing stress versus strain (refer to Figure 3.1). The steel behaves elastically while in the linear portion of the graph, meaning that loading and unloading do not permanently distort the material. Once the elastic limit is reached, the material starts to yield and behave plastically; the stress is relatively constant while strain increases. The strain hardening portion of the graph undergoes larger deformations while the stress increases. The steel then starts necking, and the stress decreases until failure is reached. Refer to Section 6.6 for information on material changes due to temperature.

Steel modulus of elasticity, $E_s = 200,000 \text{ MPa}$

Steel shear modulus, $G = \dfrac{E}{2(1+\nu)} \approx 80,000 \text{ MPa}$

Steel coefficient of thermal expansion, $\alpha_T = 11.7 \times 10^{-6} \text{ per } °C \quad \text{(at } 20°C\text{)}$

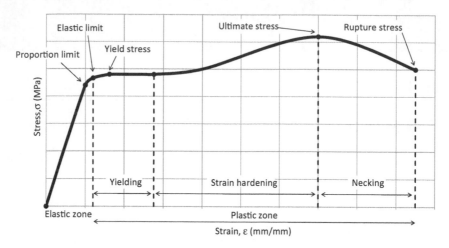

Figure 3.1 Stress vs. strain chart for mild steel.

Steel Poisson's ratio, $\quad v = 0.25$

Steel density, $\quad \rho = 7850 \text{ kg/m}^3$

The slope of the elastic portion of the graph is equal to the Young's modulus of elasticity,

$$E_s = \frac{\sigma}{\varepsilon}$$

ε = Strain
σ = Stress

Therefore, the strain at yield (ε_y) can be calculated as,

$$\varepsilon_y = \frac{f_y}{E_s} = \frac{f_y}{200,000 \text{ MPa}}$$

Slender steel members are designed to stay within the elastic portion of the graph; however, compact and non-compact sections may be designed using a plastic analysis where moment redistribution is ensured (most commonly doubly symmetric universal column sections). High repetitions of any load can lead to fatigue failure, which is dependent on the stress range and the number of cycles.

3.1.1 Cost

An approximate cost should be considered during the design phase. The relative prices of various sections vary between different locations (refer to Table 3.1). A very rough number to use is $5000–$6000/tonne for the supply, fabrication drawings, connections, fabrication, surface protection and installation within Mining or Oil & Gas facilities. Refer to the current edition of the *Australian Construction Handbook*, Rawlinsons [13], for a more detailed estimate. Contact a local supplier when a more accurate estimate is required.

Table 3.1 Approximate costs for supply of various steel sections

Sections	Adelaide ($/t)	Brisbane ($/t)	Melbourne ($/t)	Perth ($/t)	Sydney ($/t)
Supply					
Universal beams	1275	1250	1250	1275	1275
Universal columns	1275	1250	1250	1275	1275
Welded beams	1975	1925	1925	1995	1975
Tapered flanges	1725	1525	1650	1815	1800
Angles	1400	1375	1375	1440	1440
RHS and SHS	1625	1600	1600	1645	1625
CHS	1725	1700	1700	1750	1725
Plates	1300	1275	1275	1360	1300
Bars (flat)	1275	1125	1175	1225	1375
Fabrication, delivery and erection					
Universal beams and columns	3550	4150	3450	4650	3850
Built-up sections	3300	3900	3200	4400	3600
Tapered flanges	3300	3900	3200	4400	3600
Angles	3800	4400	3700	4900	4100
RHS, SHS and CHS	4050	4650	3950	5150	4350

Source: Rawlinsons Quantity Surveyors and Construction Cost Consultants, *Rawlinsons Australian Construction Handbook*, 35th ed., Rawlinsons Publishing, Australia, 2017. Reproduced with permission from Rawlinsons.

Notes:
1. Listed prices are approximate and vary over time. Values are taken from 2017.
2. Listed prices do not include any surface protection (varies from $400/tonne, blast and prime, to as high as $1575/tonne, hot dip galvanising small sections).
3. Estimation is based on a structural steel content of 20–50 tonnes. Price can increase by up to 20% for small orders.
4. Refer to the current edition of the *Australian Construction Handbook*, Rawlinsons, for further details, including welding, bolting, surface treatments and regional price indices.

3.1.2 Steel selection

The most common grades of steel used in Australia are presented in Table 3.2. The main choice to be made is based on yield strength. Cold-formed sections are available in two grades; however, the variation in cost is between 0% and 5%, and therefore either grade can be adopted. Sizes vary slightly. Connections can be fabricated from either plates or flats, and therefore grades can vary between 250, 300 and 350. Grades 250 and 300 are similar prices; however, grade 350 is typically 10%–15% more expensive. The primary constraint with flats is that they are only available in certain widths (refer to Section 7.3); however, plates are supplied in larger sheets and cut to size. Plates are generally used for base plates and end plates, and either may be adopted for side plates, cleats and stiffeners. A common choice is to use grade 250 plates in all locations (this is conservative where the wrong steel is supplied). Open sections are typically produced from grade 300 steel in Australia.

Yield stress is dependent on thickness, and therefore the steel grade is only a nominal yield stress. Table 3.3 shows a selection of yield and tensile strengths of steels in accordance with various Australian Standards; refer to AS 4100 for a more comprehensive list.

Table 3.2 Commonly used steel grades

Item	Grade	Standard
Circular hollow sections (cold formed)	C250 or C350	AS 1163
Rectangular hollow sections (cold formed)	C350 or C450	AS 1163
Plates (hot rolled)	250 or 350	AS/NZS 3678
Open sections (hot-rolled sections and bars)	300	AS/NZS 3679.1
Flats (hot rolled – specific widths only)	300	AS/NZS 3679.1
Open sections (hot rolled – welded sections)	300	AS 3679.2

3.2 FABRICATION AND ERECTION

Standardisation in design has an important influence on reducing cost. The ASI connection suite [6,7] provides standard details and capacities for many typical connections. It is expected that these will become industry standards in the same way that many other countries have adopted standards. Selecting standard sections is also important, as it reduces wastage (i.e. try not to use three different weight 150UCs if the same can be used in all instances). The lightest of each section is generally the most efficient and typically accounts for two-thirds of sales for each beam depth [19].

The fabrication process should be considered during the design phase. The following key decisions need to be outlined to produce an acceptable design:

1. Framing system: braced (pinned) or rigid (fixed)
2. Coating system: painted or galvanised
3. Transportation envelope

Essentially, these decisions are aimed at defining a maximum welded frame size. The design engineer needs to ensure that fabricated sections of the frame can be efficiently coated and transported to site.

3.2.1 Framing system

Braced structures are generally a cheaper option than rigid construction, because foundation loads are smaller and connections are simpler. They also allow more steel to be transported in each truck. Fabrication shops prefer to fabricate welded connections; however, site crews prefer bolted connections.

3.2.2 Coating system

Shop painting is generally the cheapest short-term system, followed by site painting, then galvanising. Galvanising is typically the most durable; however, it adds additional constraints to the design.

For galvanised design, narrow gaps between steelwork should be avoided. Any hollow areas should be provided with vent holes, typically two per corner with minimum diameters of 12 mm. Stiffeners should also be cropped in corners to provide similar gaps. Refer to the ASI guidelines and the galvanising shop for further information.

Table 3.3 Steel strengths for commonly used grades

Steel standard	Form	Steel grade	Thickness of material, t (mm)	Yield stress, f_y (MPa)	Tensile strength, f_u (MPa)
AS/NZS 1163 (Note 3)	Hollow sections	C450	All	450	500
			All	350	430
		C350			
		C250	All	250	320
AS/NZS 3678 (Notes 2 and 3)	Plate and floorplate	350	$t \le 12$	360	450
			$12 < t \le 20$	350	450
			$20 < t \le 80$	340	450
			$80 < t \le 150$	330	450
		300	$t \le 8$	320	430
			$8 < t \le 12$	310	430
			$12 < t \le 20$	300	430
			$20 < t \le 50$	280	430
			$50 < t \le 80$	270	430
			$80 < t \le 150$	260	430
		250	$t \le 8$	280	410
			$8 < t \le 12$	260	410
			$12 < t \le 50$	250	410
			$50 < t \le 80$	240	410
			$80 < t \le 150$	230	410
AS/NZS 3679.1 (Note 3)	Flats and sections	350	$t \le 11$	360	480
			$11 < t \le 40$	340	480
			$t \ge 40$	330	480
		300	$t \le 11$	320	440
			$11 < t \le 17$	300	440
			$t \ge 17$	280	440
	Hexagons, rounds and squares	350	$t \le 50$	340	480
			$50 < t \le 100$	330	480
			$t \ge 100$	320	480
		300	$t \le 50$	300	440
			$50 < t \le 100$	290	440
			$t \ge 100$	280	440

Source: AS 4100 Table 2.1. Copied by Mr L. Pack with the permission of Standards Australia under Licence 1607-c010.

Notes:
1. For design purposes, yield and tensile strengths approximate those of structural Grade HA200. For specific information, contact the supplier.
2. Welded I-sections complying with AS/NZS 3679.2 are manufactured from hot-rolled structural steel plates complying with AS/NZS 3678, so the values listed for steel grades to AS/NZS 3678 shall be used for welded I-sections to AS/NZS 3679.2.
3. AS/NZS 3678, AS/NZS 3679.1 and AS/NZS 1163 all contain, within each grade, a variety of impact grades not individually listed in the table. All impact tested grades within the one grade have the same yield stress and tensile strength as the grade listed.
4. For impact designations and allowable temperatures, refer to Section 6.6.

The selection of a galvanised coating system means that the welded frame size is limited to the size of the galvanising bath. Larger frames can sometimes be galvanised by double-dipping, or separate galvanised frames can be joined; however, the coating needs to be removed prior to welding and retouched with a cold-galv finish. Galvanising bath sizes vary substantially and should always be confirmed with fabricators prior to completing the design. Table 3.4 shows some of the larger available dimensions for galvanising baths in Australia.

Table 3.4 Example galvanising bath dimensions

Location	Length (m)	Width (m)	Depth (m)
Hexham, NSW	14.2	1.8	2.6
Kilburn, SA	14	1.6	2.6
Dandenong Sth, VIC	13.7	1.8	2.95
Nerangba, QLD	13	1.8	3.0
Kewdale, WA	13	1.6	2.6
Tivendale, NT	10.5	1.5	2.6
Launceston, TAS	6.5	1.05	1.4

Note: Dimensions should be checked with shops prior to adopting as a design basis.

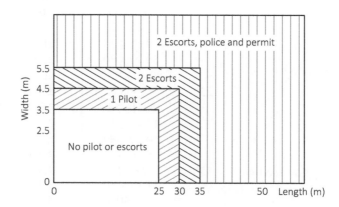

Figure 3.2 Example transport requirements for Queensland. (From Queensland Government Transport and Main Roads, Guideline for Excess Dimension in Queensland, 8th ed., Queensland Government, Australia, 2013. Reproduced with permission from the National Heavy Vehicle Regulator.)

3.2.3 Transportation

The transportation envelope may depend on shipping and/or trucking constraints. Each state has different rules on transportation limits for roads; however, a maximum frame dimension of 15 (long) × 3 (wide) × 2 m (high) is generally used without requiring any special escort [19]. A simplified summary of the maximum vehicle dimensions for Queensland is presented in Figure 3.2; however, additional rules apply for each case, and the legislation should therefore be consulted. Shipping limits are flexible and generally do not govern designs.

3.3 ANALYSIS

Steel structures should be analysed using one of the following methods:

1. First-order elastic analysis, followed by moment amplification (AS 4100, Clause 4.4) (linear)
2. Second-order elastic analysis (AS 4100, Clause 4.4 and Appendix E) (non-linear)
3. Plastic analysis (AS 4100, Clause 4.5) (non-linear with stress strain associations applied)

The first-order elastic analysis can only be used when the calculated amplification factors (δ_b or δ_s) are less than 1.4. Where the second-order effects are greater than 40%, a second-order analysis is required. Second-order analyses include the PΔ effect, which increases bending moment as a frame sways while under axial load.

The quickest and most common choice of methods is the second-order elastic analysis, which can be easily performed in computer packages such as Space Gass, Microstran, STAAD Pro or Strand7.

3.3.1 Section selection

Steel sections should be selected based on the performance of the cross-section. Hollow sections such as circular hollow sections (CHS) and rectangular hollow sections (RHS) are torsionally rigid and therefore perform efficiently in both torsion and also compression. Welded columns (WC) and universal columns (UC) perform well in compression and also bending about both axes. Welded beams (WB), universal beams (UB) and parallel flanged channels (PFC) are ideal in locations with high bending; however, they are poor in torsion and minor axis compression and bending (when not braced). Equal angles (EA) and unequal angles (UA) are often used for bracing and small cantilevers due to their light weight and geometric simplicity.

3.3.2 Notional forces

For multi-storey buildings, a notional force of 0.2% of the vertical load should be allowed for in serviceability and strength limit states, applied in conjunction with dead and live loads only. This is to account for out-of-plumb limits. The notional force is unlikely to affect the design for lateral loads, which are generally governed by wind or seismic events.

AS/NZS 1170.0 increases the notional force to 1.5% for structures taller than 15 m and 1% for all other structures. It also requires connections to be designed for a minimum lateral load of 5% of the sum of $G + \psi_c Q$ for the tributary area.

3.3.3 Bracing

The modelling of bracing is dependent on the slenderness ratio (l/r) (refer to Section 3.7.2), with r selected for the minor axis to result in the maximum ratio. Members may be set to 'tension only' where shown in this sub-section. This usually results in cross-bracing or alternate bracing in each direction due to stability; however, members and deflections are often smaller.

$$\frac{l}{r} \leq 200 \qquad \text{Normal modelling} \quad \text{(tension and compression)}$$

$$200 < \frac{l}{r} \leq 300 \quad \text{Tension-only modelling} \quad \text{(compression causes global buckling at small loads)}$$

$$\frac{l}{r} > 300 \qquad \text{Avoid use} \quad \text{(members are too slender and result in poor serviceability)}$$

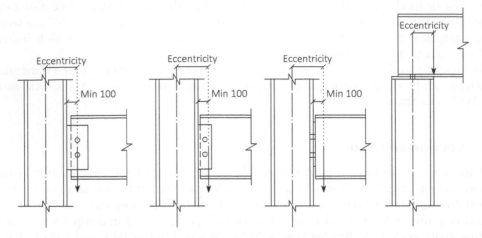

Figure 3.3 Minimum design eccentricities for simple construction.

3.3.4 Connection eccentricity

Columns which support beams using simple connections should be designed to withstand an applied moment equal to the vertical load multiplied by the lever arm from the centre of the column to the load point (refer to AS 4100, Clause 4.3.4). The load point is taken as the edge of the column where supported on an end cap or a minimum of 100 mm from the face of the column when using end plates or web side plates (fin plates). Refer to Figure 3.3 for a diagrammatic representation.

3.4 BENDING

Members subject to bending are required to be assessed for section and member capacities in accordance with AS 4100, Section 5. All moments in the member are required to be smaller than the section moment capacity (M_s), and moments in each segment of the member are required to be lower than the segment's member moment capacity (M_b). Given the complexity of the calculation, values are generally adopted from tables for general design cases (refer to Section 7.3). This section details the calculation about the major (principal) x-axis; however, the same is appropriate about the minor y-axis.

$$M^* \leq \phi M_s \quad \text{and} \quad M^* \leq \phi M_b$$

$$\phi = 0.9 \quad \text{(for members in bending)}$$

3.4.1 Section capacity

The section capacity (M_S) is achieved by multiplying the yield stress by the effective section modulus (Z_e). The slenderness of the section is analysed to classify the section and understand whether the elastic or plastic material limits should be used. Sections with slender elements use an elastic approach to prevent buckling; sections with only compact elements are allowed to develop full plastic capacity (refer to Figure 3.4).

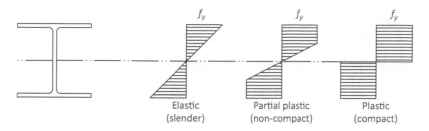

Figure 3.4 Section behaviour and slenderness.

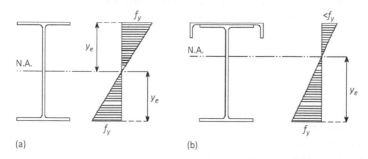

Figure 3.5 Example neutral axis and stress diagram for symmetrical (a) and non-symmetrical (b) sections.

$$M_s = f_y Z_e$$

$$\phi M_s = 0.9 f_y Z_e$$

3.4.1.1 Elastic section modulus

When using the elastic section modulus (Z), maximum stresses are found at the top and bottom of the section, reducing linearly to zero at the neutral axis (NA). As a section is loaded, symmetrical sections will reach the yield stress at the top and bottom of the section at the same time; however, non-symmetrical sections will reach yield at the part farthest from the neutral axis (y_e) first (refer to Figure 3.5).

3.4.1.1.1 Centroid and neutral axis

The neutral axis of a section is the location where no longitudinal bending stresses occur (refer to Figure 3.5). For elastic design, the axis occurs at the centroid of the shape. If you can imagine a slice of the cross-section, this point is the location where it could be balanced on one support.

The centroid is found by splitting the cross-section into simple shapes (such as rectangles), then multiplying the area of each shape by the distance from the centre of the shape to the base of the section, then dividing the resultant by the total area. The solution to this is the perpendicular distance from the base of the section to the neutral axis (y_c).

$$y_c = \frac{\sum_{1}^{N}(A_i y_{ci})}{\sum_{1}^{N}(A_i)}$$

where:
A_i = Area of each shape
y_{ci} = Distance from the base of the section to the centre of each shape

The neutral axis is located at the geometric centre of shapes which are symmetrical on the same axis.

3.4.1.1.2 Second moment of area

The second moment of area (I_x) for major axis bending $(I_y$ for minor axis bending) is calculated by splitting the section into rectangles, then summing the following calculation for each shape. Refer to Table 3.30 for common examples.

$$I_x = \frac{bh^3}{12} + bh\bar{y}^2$$

where:
b = Width of rectangle
h = Height of rectangle
$\bar{y}$ = Distance from the centroid of the rectangle to the neutral axis

3.4.1.1.3 Elastic modulus

The elastic section modulus (Z) is a useful tool for calculating the maximum elastic stress in a section with an applied moment (M).

The stress at any distance (y) from the neutral axis can be calculated:

Stress,
$$\sigma = \frac{M \times y}{I_x}$$

Therefore, at full elastic capacity, the stress at a distance of y_e is equal to the yield stress:

Elastic section modulus,
$$Z = \frac{I_x}{y_e}$$

Stress at yield,
$$\sigma_{max} = f_y = \frac{M \times y_e}{I_x} = \frac{M}{Z}$$

where:
I_x = Second moment of area
M = Applied moment
y_e = Maximum distance from neutral axis to extreme fibre of steel
(refer to Figure 3.5)

Therefore, the maximum applied moment within the elastic zone is calculated as

$$M_{max} = f_y Z$$

Note: This is a theory-based formula, the code requirement adopts an effective Z value (Z_e) and also factors down the yield strength ($\phi = 0.9$); refer to Section 3.4.1.3 for details.

Example 3.1: Elastic Section Modulus

Calculate the elastic section modulus (Z) for the section below:

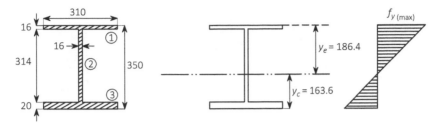

Centroid location,

$$y_c = \sum_{1}^{N}(A_i y_{ci}) / \sum_{1}^{N}(A_i) = \frac{(16 \times 310 \times 342) + (314 \times 16 \times 177) + (20 \times 310 \times 10)}{(16 \times 310) + (314 \times 16) + (20 \times 310)} = 163.6 \text{ mm}$$

Extreme fibre,

$$y_e = \text{MAX} \{y_c, (d - y_c)\} = \text{MAX} \{163.6, (350 - 163.6)\} = 184.6$$

Second moment of area,

$$I_x = \frac{bb^3}{12} + bb\bar{y}^2$$

$$= \left[\frac{310 \times 16^3}{12} + 310 \times 16 \times (186.4 - 16/2)^2 \right]$$

$$+ \left[\frac{16 \times 314^3}{12} + 16 \times 314 \times (177 - 163.6)^2 \right] + \left[\frac{310 \times 20^3}{12} + 310 \times 20 \times (163.6 - 20/2)^2 \right]$$

$$= 346.6 \times 10^6 \text{ mm}^4$$

Elastic section modulus,

$$Z = \frac{I_x}{y_e} = \frac{346.6 \times 10^6 \text{ mm}^4}{186.4 \text{ mm}} = 1.86 \times 10^6 \text{ mm}^3$$

3.4.1.2 Plastic section modulus

The plastic section modulus assumes that all steel has yielded and therefore occurs about a slightly different axis. Rather than the neutral axis adopted for elastic design, the plastic modulus uses a 'plastic neutral axis' (PNA). To achieve equilibrium, the axis occurs at a location where half of the area of steel is above the axis and half is below the axis (refer to Figure 3.6). The difference occurs because all of the steel is working equally hard, whereas for the elastic analysis, the steel near the neutral axis was not doing as much work.

No equation is provided to calculate the PNA location. Use simple geometry to equate the area above the axis with half of the total area (or seek the location where $A_c = A_t$).

The modulus is calculated by taking the moments about the PNA:

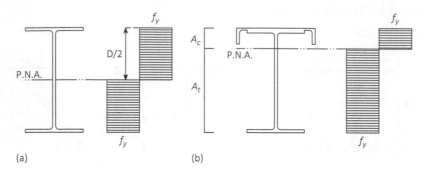

Figure 3.6 Example PNA and stress diagram for symmetrical (a) and non-symmetrical (b) sections.

$$S = A_c e_c + A_t e_t$$

where:

A_c = Area in compression
A_t = Area in tension
e_c = Lever arm from centroid of compression area to PNA
e_t = Lever arm from centroid of tension area to PNA

Centroids for the compression and tension areas are calculated separately by splitting the section into two at the PNA. The same calculation is used for centroids from the elastic chapter (refer to Section 3.4.1.1); however, the distance is measured to the PNA instead of the base.

$$e = \sum_1^N (A_i y_{pi}) \, / \sum_1^N (A_i)$$

where:

A_i = Area of each shape
y_{pi} = Distance from the PNA to the centre of each shape

Note: North American texts use the reverse notation for plastic and elastic section moduli (S and Z are swapped). In Australia, Z is always the elastic modulus, and S is the plastic modulus.

Example 3.2: Plastic Section Modulus

Calculate the plastic section modulus (S) for the section below:

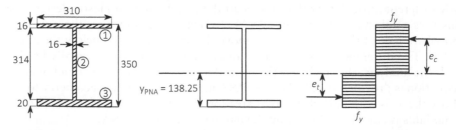

Area of steel, $A_s = 310 \times 16 + 314 \times 16 + 310 \times 20 = 16{,}184$ mm^2

Area in compression (or tension), $A_c = A_t = \dfrac{A_s}{2} = 8092$ mm^2

PNA location, $y_{PNA} = \left[\dfrac{A_s}{2} - 310 \times 20 + 20 \times 16\right] \div 16 = 138.25$ mm

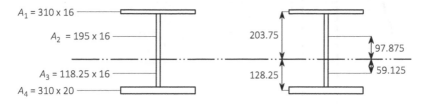

$A_1 = 310 \times 16$
$A_2 = 195 \times 16$
$A_3 = 118.25 \times 16$
$A_4 = 310 \times 20$
203.75
128.25
97.875
59.125

$$e = \sum_1^N (A_i y_{pi}) \Big/ \sum_1^N (A_i)$$

Compression lever arm,

$$e_c = [(310 \times 16) \times 203.75 + (195.75 \times 16) \times 97.875] \div 8092 = 162.77 \text{ mm}$$

Tension lever arm,

$$e_t = [(118.25 \times 16) \times 59.125 + (310 \times 20) \times 128.25] \div 8092 = 112.09 \text{ mm}$$

Plastic section modulus,

$$S = A_c e_c + A_t e_t$$
$$= 8092 \times (162.77 + 112.09) = 2.224 \times 10^6 \text{ mm}^3$$

3.4.1.3 Effective section modulus

To calculate the effective section modulus (Z_e), the cross-section is classified as compact, non-compact or slender. This is done by calculating the section slenderness (λ_S), which is taken as the slenderness value (λ_e) from the element in the section which has the highest value of (λ_e/λ_{ey}). The slenderness affects the design assumptions for the section failure modes (refer to Figure 3.7).

An individual plate element is classified as a plate which is supported on one or two ends. For example, a UC section (refer to Figure 3.8) consists of five elements (the web and the four flange outstands supported at the web intersection). Each of the flange outstands is identical, and therefore two elements should be checked: the web and the flange outstand.

Plate element slenderness, $\lambda_e = \left(\dfrac{b}{t}\right)\sqrt{\dfrac{f_y}{250}}$

where:
  b = Clear width of plate element (refer to Figure 3.8)
  f_y = Yield stress
  t = Thickness of plate element

For circular hollow sections, $\lambda_{e(CHS)} = \left(\dfrac{d_o}{t}\right)\sqrt{\dfrac{f_y}{250}}$

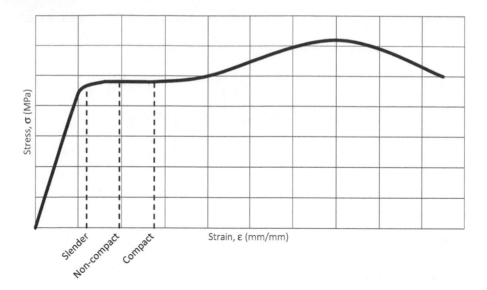

Figure 3.7 Effective modulus comparison.

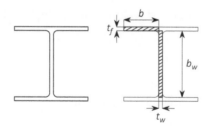

Figure 3.8 Example elements for slenderness calculation.

where: d_o = Outside diameter of section

Plate element slenderness limits are taken from Table 3.5.

For the example cross-section (refer to Figure 3.8), the steel is hot rolled; flanges are flat plate elements with one longitudinal support, under uniform compression for major axis bending (non-uniform compression for minor axis bending). The web is a flat plate element, supported at both ends, and under compression at one edge and tension at the other.

Typical values for UC and UB sections in major axis bending:

Flange: $\lambda_{ep} = 9, \lambda_{ey} = 16, \lambda_{ed} = 35$

Web: $\lambda_{ep} = 82, \lambda_{ey} = 115$

Typical values for UC and UB sections in minor axis bending:

Flange: $\lambda_{ep} = 9, \lambda_{ey} = 25$

Most open sections (UB, UC, PFC, EA and UA) are made by hot rolling (HR), and hollow structural sections (CHS, RHS, SHS) are generally cold formed (CF).

Table 3.5 Values of plate element slenderness limits

Plate element type	Longitudinal edges supported	Residual stresses (see notes)	Plasticity limit (λ_{ep})	Yield limit (λ_{ey})	Deformation limit (λ_{ed})
Flat	One	SR	10	16	35
(Uniform compression)		HR	9	16	35
		LW, CF	8	15	35
		HW	8	14	35
Flat	One	SR	10	25	–
(Maximum compression at		HR	9	25	–
unsupported edge, zero stress or		LW, CF	8	22	–
tension at supported edge)		HW	8	22	–
Flat	Both	SR	30	45	90
(Uniform compression)		HR	30	45	90
		LW, CF	30	40	90
		HW	30	35	90
Flat	Both	Any	82	115	–
(Compression at one edge, tension at the other)					
Circular hollow sections		SR	50	120	–
		HR, CF	50	120	–
		LW,	42	120	–
		HW	42	120	–

Source: AS 4100, Table 5.2. Copied by Mr L. Pack with the permission of Standards Australia under Licence 1607-c010.

Notes:
1. CF = cold formed; HR = hot rolled or hot finished; HW = heavily welded longitudinally; LW = lightly welded longitudinally; SR = stress relieved.
2. Welded members whose compressive residual stresses are less than 40 MPa may be considered to be lightly welded.

3.4.1.3.1 Compact sections: $\lambda_s \le \lambda_{sp}$

A section is classified as compact when it satisfies the following inequality: $\lambda_s \le \lambda_{sp}$, where λ_{sp} is equal to the value of λ_{ep} from the element which has the highest value of (λ_e/λ_{ey}).

Compact sections can reach full plastic capacity without failure due to local buckling effects. The effective section modulus (Z_e) is taken as the minimum of the plastic section modulus (S) and 1.5 times the elastic section modulus (Z).

$$Z_{e(compact)} = \text{MIN}\{S, 1.5 \times Z\}$$

3.4.1.3.2 Non-compact sections: $\lambda_{sp} < \lambda_s \le \lambda_{sy}$

A section is classified as non-compact when it satisfies the following inequality: $\lambda_{sp} < \lambda_s \le \lambda_{sy}$, where λ_{sy} is equal to the value of λ_{ey} from the element which has the highest value of (λ_e/λ_{ey}).

Non-compact sections are designed for a capacity between the yield limit and the plastic limit. The background behind the theory is essentially based on linear interpolation.

$$Z_{e(non\text{-}compact)} = Z + \left[\left(\frac{\lambda_{sy} - \lambda_s}{\lambda_{sy} - \lambda_{sp}} \right) (Z_c - Z) \right]$$

where: Z_c is the effective modulus if the section was classified as compact.

3.4.1.3.3 Slender sections: $\lambda_s > \lambda_{sy}$

A section is classified as slender when it satisfies the following inequality: $\lambda_s > \lambda_{sy}$.

Slender sections are designed for a capacity based on the yield limit. They are sections with elements that are too slender to undergo plastic yielding.

For sections with plate elements in uniform compression:

$$Z_{e(slender)} = Z\left(\frac{\lambda_{sy}}{\lambda_s}\right)$$

Alternatively, an effective cross-section can be made by reducing the width of slender elements until they become non-compact.

For sections with plate elements under maximum compression at the unsupported edge and zero stress at the supported edge:

$$Z_{e(slender)} = Z\left(\frac{\lambda_{sy}}{\lambda_s}\right)^2$$

For circular hollow sections:

$$Z_{e(slender)} = \text{MIN}\left[Z\left(\frac{\lambda_{sy}}{\lambda_s}\right), Z\left(\frac{2\lambda_{sy}}{\lambda_s}\right)^2\right]$$

3.4.1.3.4 Holes

Sections with holes that reduce the flange area by more than $100\{1 - [f_y/(0.85\, f_u)]\}\%$ shall have the plastic and elastic capacities reduced by (A_n/A_g) or have the capacity calculated using the net area.

Example 3.3: Section Capacity

Calculate the section capacity for the section used in the previous example for elastic and plastic modulus calculations. The section is hot rolled with $f_y = 250$ MPa.
Slenderness,
Flanges:

$$b_f = \frac{(310-16)}{2} = 147 \text{ mm} \quad t_f = 16 \text{ mm}$$

$$\lambda_e = (147/16)\sqrt{250/250} = 9.19 \quad \lambda_{ey} = 16$$

$$\lambda_e / \lambda_{ey} = 0.574$$

Web:

$$b_w = 314 \text{ mm} \quad t_w = 16 \text{ mm}$$

$$\lambda_e = (314/16)\sqrt{250/250} = 19.63 \quad \lambda_{ey} = 115$$

$$\lambda_e / \lambda_{ey} = 0.171$$

Section slenderness:
(Slenderness is governed by the flange because it has the larger λ_e/λ_{ey} value)

$$\lambda_{sp} = 9 \quad \lambda_s = 9.19 \quad \lambda_{sy} = 16$$

Slenderness classification:
The section is classified as 'non-compact',

$$\lambda_{sp} < \lambda_s \le \lambda_{sy}$$

Compact section modulus,

$$Z_c = \text{MIN}\{S, 1.5 \times Z\} = \text{MIN} \{2.224 \times 10^6, 1.5 \times 1.86 \times 10^6\} = 2.224 \times 10^6 \, \text{mm}^3$$

Non-compact section modulus,

$$
\begin{aligned}
Z_{e(\text{non-compact})} &= Z + \left[\left(\frac{\lambda_{sy} - \lambda_s}{\lambda_{sy} - \lambda_{sp}} \right)(Z_c - Z) \right] \\
&= 1.86 \times 10^6 + \left[\frac{16 - 9.19}{16 - 9}(2.224 \times 10^6 - 1.86 \times 10^6) \right] \\
&= 2.21 \times 10^6 \, \text{mm}^3
\end{aligned}
$$

Reduced section capacity,

$$\phi M_s = \phi f_y Z_e = 0.9 \times 250 \times 2.21 \times 10^6 \, \text{mm}^3 = 498.17 \text{ kNm}$$

3.4.2 Member capacity

The member capacity (M_b) is dependent on the section capacity, restraint type, moment distribution and length of the segment under consideration. The member can either be classified as having 'full lateral restraint' and therefore have a member capacity equal to the section capacity (M_s) or be classified as 'without full lateral restraint' and therefore have a member capacity which is smaller than the section capacity.

Members are often broken into segments by adding restraint points to achieve shorter effective lengths. A segment is defined as the distance between full or partial restraints, or the length from the last restraint to an end with no restraint.

3.4.2.1 Restraint types

Restraint types are required for the design of members with or without full lateral restraint. Definitions and examples are shown for each restraint type. The critical flange is denoted by 'C' in each case. AS 4100 shows diagrammatic representations in Clause 5.4.2 and defines the *critical flange* as the flange which would deflect farther under buckling in the absence of restraint (refer to AS 4100, Clause 5.5 for further details):

1. The compression flange for beams restrained at both ends
2. The tension flange for cantilevers

3.4.2.1.1 Continuous (C)

A restraint that provides full or partial 'lateral deflection' and 'twist rotation' restraints to the critical flange at intervals which achieve an equivalent effective length of zero (Figure 3.9).

3.4.2.1.2 Full restraint (F)

A restraint that provides either:
1. 'Lateral deflection' restraint to the critical flange and full or partial 'twist rotation' restraint to the section

   or
2. 'Lateral deflection' restraint anywhere on the section and full 'twist rotation' restraint (Figures 3.10 and 3.11)

   Note: Where shown in brackets, '(C)', the restraint is provided through type ii; others are type i.

3.4.2.1.3 Partial restraint (P)

A restraint that provides 'lateral deflection' fixity anywhere on the section and partial 'twist rotation' restraint to the section (Figures 3.12 and 3.13).

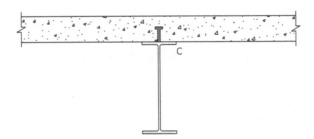

Figure 3.9 Continuous restraint example.

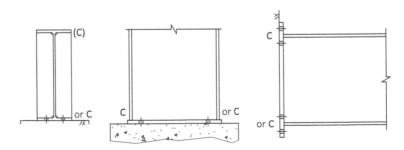

Figure 3.10 Full end restraint examples.

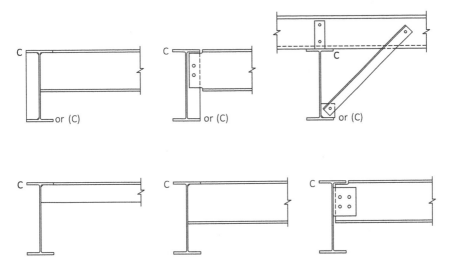

Figure 3.11 Full intermediate restraint examples.

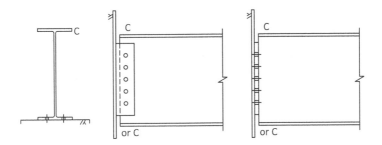

Figure 3.12 Partial end restraint examples.

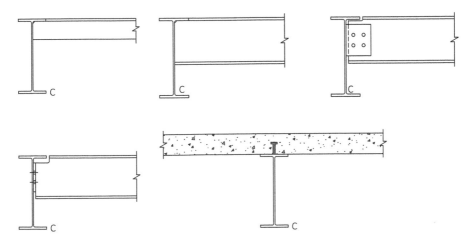

Figure 3.13 Partial intermediate restraint examples.

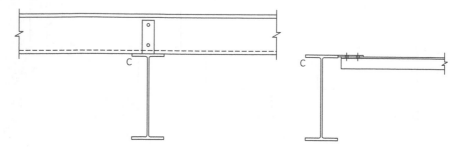

Figure 3.14 Lateral intermediate restraint examples.

3.4.2.1.4 Lateral restraint (L)

A restraint that provides only 'lateral deflection' restraint to a critical flange in a segment which has ends that are both either fully or partially restrained (Figure 3.14).

3.4.2.1.5 Unrestrained (U)

No restraint provided to the critical flange (e.g. the end of a cantilever).

3.4.2.2 Restraint element definitions

The following restraint element requirements need to be achieved to meet the requirements for each restraint type.

3.4.2.2.1 Lateral deflection

Lateral deflection restraints are required to transfer a minimum of 2.5% of the design flange force. A moment is also imparted if the centroid of the restraint is eccentric to the centroid of the flange. If the restraints are spaced closer than required for full lateral restraint, a smaller force may be adopted for each brace as long as the total force is equal to that which would be achieved using a maximum spacing (i.e. twice as many braces may have half the strength for each brace). Also, where parallel members are braced together, each brace is required to provide 2.5% of the immediate flange force in addition to 1.25% of the following flanges (considering a maximum of seven members).

3.4.2.2.2 Twist rotation

Twist rotation restraints are required to transfer a minimum of 2.5% of an unrestrained flange force to a restraint location without allowing any slip. This is in the torsional direction for the member. For a full twist rotation restraint, the restraining member should be stiff. For a partial twist rotation restraint, the restraining member may be flexible.

3.4.2.2.3 Lateral rotation

Lateral rotation restraints are designated to locations that are classified as fully (F) or partially (P) restrained by members which also restrain the critical flange about the section's minor axis with a comparable stiffness. This is generally achieved where the restraining member has full lateral restraint; refer to AS 4100, Clause 5.4.3.4 for further details.

3.4.2.3 Members with full lateral restraint

A member which has been provided with 'full lateral restraint' has a member capacity equal to the section capacity. 'Full lateral restraint' means that the length of each segment in the member is sufficiently short to prevent member lateral buckling by twisting out of the plane of loading.

$$M_b = M_s$$

'Full lateral restraint' is achieved by complying with at least one of the requirements listed in AS 4100, Clause 5.3.2:

1. Both ends are fully (F) or partially (P) restrained, and the critical flange is continuously (C) restrained.
2. Both ends are fully (F) or partially (P) restrained, and sufficient lateral (L) restraints are provided on the critical flange to reduce the segment length to L_s.
3. Both ends are fully (F) or partially (P) restrained, and the member length is less than (L_s).
4. The calculated member capacity is equal to or greater than the section capacity.

The maximum segment length for full lateral restraint, L_s, is defined in Table 3.6 for different section shapes.

Table 3.6 Maximum segment length for full lateral restraint

Section shape	Equation
Equal flanged I-sections (UC, UB, WC, WB)	$L_s = r_y(80 + 50\beta_m)\sqrt{\dfrac{250}{f_y}}$
Parallel flanged channels (PFC)	$L_s = r_y(60 + 40\beta_m)\sqrt{\dfrac{250}{f_y}}$
Unequal flanged I-sections	$L_s = r_y(80 + 50\beta_m)\left(\sqrt{\dfrac{2\rho A d_f}{2.5Z_e}}\right)\sqrt{\dfrac{250}{f_y}}$
Rectangular or square hollow sections (RHS, SHS)	$L_s = r_y(1800 + 1500\beta_m)\left(\dfrac{b_f}{b_w}\right)\left(\dfrac{250}{f_y}\right)$
Angles (EA, UA)	$L_s = t(210 + 175\beta_m)\left(\sqrt{\dfrac{b_2}{b_1}}\right)\left(\dfrac{250}{f_y}\right)$

where:
- A = Area of cross-section
- b_f = Width of flange b_w = Width of web
- b_1 = Longer leg length b_2 = Shorter leg length
- d_f = Distance between flange centroids
- I_{cy} = The second moment of area of the compression flange about the minor y-axis
- I_y = The second moment of area about the minor y-axis
- r_y = The radius of gyration about the principal y-axis (refer to section tables, Section 7.3)
- t = Thickness of an angle
- Z_e = Effective section modulus
- ρ = I_{cy}/I_y
- β_m = -1.0; or
  - -0.8 for segments with transverse loads; or
  - The ratio of smaller to larger end moments along the segment for segments without transverse loads (positive for reverse curvature, and negative when bent in single curvature).

3.4.2.4 Members without full lateral restraint

Where members do not have full lateral restraint, the element length reduces the member capacity to some value below that of the section capacity. For segments consisting of constant cross-sections, refer to the method below.

For segments made from varying cross-sections, either adopt the minimum cross-section or refer to AS 4100, Clause 5.6.1.1(b). Also, for I-sections with unequal flanges, refer to AS 4100, Clause 5.6.1.2.

3.4.2.4.1 Open sections with equal flanges, fully or partially restrained at both ends

For a segment (length between two points which are fully (F) or partially (P) restrained) or a sub-segment (length between two points of lateral restraint within a segment),

$$M_b = \alpha_m \alpha_s M_s \leq M_s$$

where:

α_m = Moment modification factor
α_s = Slenderness reduction factor
M_b = Member capacity

The moment modification factor (α_m) may be taken as 1, or it can be calculated using the following formula or selected from Tables 3.7 and 3.8. An additional method is also presented in AS 4100, Clause 5.6.4, which uses an elastic buckling analysis.

$$\alpha_m = \frac{1.7 M_m^*}{\sqrt{\left[(M_2^*)^2 + (M_3^*)^2 + (M_4^*)^2 \right]}} \leq 2.5$$

where:

M_2^*, M_3^*, M_4^* = Moments at quarter, half and three quarter points of segment

M_m^* = Maximum design moment in segment

$$\alpha_s = 0.6 \left[\sqrt{\left[\left(\frac{M_s}{M_o} \right)^2 + 3 \right]} - \left(\frac{M_s}{M_o} \right) \right]$$

where:

$$M_o = \sqrt{\left(\frac{\pi^2 E I_y}{l_e^2} \right) \left[GJ + \left(\frac{\pi^2 E I_w}{l_e^2} \right) \right]}$$

E = Modulus of elasticity = 200,000 MPa
G = Shear modulus $\approx$ 80,000 MPa
J = Section torsion constant (refer to Sections 3.9 and 7.3)
I_w = Section warping constant

Table 3.7 Moment modification factors (α_m) for segments fully or partially restrained at both ends

Beam segment	Moment distribution	Moment modification factor, α_m	Range
		$1.75 + 1.05\beta_m + 0.3\beta_m^2$ 2.5	$-1 \leq \beta_m \leq 0.6$ $0.6 < \beta_m \leq 1$
	$\dfrac{Fl}{2}\left(1 - \dfrac{2a}{l}\right)$	$1.0 + 0.35\left(1 - \dfrac{2a}{l}\right)^2$	$0 \leq \dfrac{2a}{l} \leq 1$
	$\dfrac{Fl}{4}\left[1 - \left(\dfrac{2a}{l}\right)^2\right]$	$1.35 + 0.4\left(\dfrac{2a}{l}\right)^2$	$0 \leq \dfrac{2a}{l} \leq 1$
	$\dfrac{Fl}{4}\left(1 - \dfrac{2\beta_m}{8}\right)$ $\dfrac{3\beta_m Fl}{16}$	$1.35 + 1.05\,\beta_m$ $-1.2 + 3.0\,\beta_m$	$0 \leq \beta_m < 0.9$ $0.9 \leq \beta_m \leq 1$

(Continued)

Table 3.7 (Continued) Moment modification factors (α_m) for segments fully or partially restrained at both ends

Beam segment	Moment distribution	Moment modification factor, α_m	Range
$\frac{\beta_m Fl}{8}$, F at midspan ($l/2$, $l/2$), $\frac{\beta_m Fl}{8}$	$\frac{Fl}{4}\left(1 - \frac{\beta_m}{2}\right)$; $\frac{\beta_m Fl}{8}$	$1.35 + 0.36\,\beta_m$	$0 \leq \beta_m < 1$
$\frac{\beta_m wl^2}{8}$, w, $\frac{\beta_m wl^2}{8}$	$\frac{wl^2}{8}\left(1 - \frac{\beta_m}{4}\right)^2$; $\frac{\beta_m wl^2}{8}$	$1.13 + 0.10\,\beta_m$ $-1.25 + 3.5\,\beta_m$	$0 \leq \beta_m \leq 0.7$ $0 \leq \beta_m \leq 1$
$\frac{\beta_m wl^2}{12}$, w, $\frac{\beta_m wl^2}{12}$	$\frac{wl^2}{8}\left(1 - \frac{2\beta_m}{3}\right)$; $\frac{\beta_m wl^2}{12}$	$1.13 + 0.12\,\beta_m$ $-2.38 + 4.8\,\beta_m$	$0 \leq \beta_m \leq 0.75$ $0.75 \leq \beta_m \leq 1$
M	M	1.00	N/A
F	Fl	1.75	N/A
w	$\frac{wl^2}{2}$	2.50	N/A

Source: AS 4100, Table 5.6.1. Copied by Mr L. Pack with the permission of Standards Australia under Licence 1607-c010.

Note: X = full or partial restraint.

Table 3.8 Moment modification factors (α_m) for segments unrestrained at one end

Member segment	Moment distribution	Moment modification factor, α_m
		0.25
		1.25
		2.25

Source: AS 4100, Table 5.6.2. Copied by Mr L. Pack with the permission of Standards Australia under Licence 1607-c010.

Note: X = full or partial restraint

I_y = Section second moment of area about y-axis
(refer to Sections 3.4.1.1 and 7.3)

l_e = Segment effective length = $k_t k_l k_r l$

where:

k_t = Twist restraint factor (refer to Table 3.9)
k_l = Load height factor (refer to Table 3.10)
k_r = Lateral rotation restraint factor (refer to Table 3.11)
l = Length

Table 3.9 Twist restraint factor (k_t)

Restraint arrangement	Twist restraint factor, k_t
FF, FL, LL, FU	1.0
FP, PL, PU	$1 + \dfrac{\left[\left(\dfrac{d_1}{l}\right)\left(\dfrac{t_f}{2t_w}\right)^3\right]}{n_w}$
PP	$1 + \dfrac{\left[2\left(\dfrac{d_1}{l}\right)\left(\dfrac{t_f}{2t_w}\right)^3\right]}{n_w}$

Source: AS 4100, Table 5.6.3(1). Copied by Mr L. Pack with the permission of Standards Australia under Licence 1607-c010.

Note: The restraint arrangement shows the restraint types at each end of the segment or sub-segment. F = Fixed, P = Partial, L = Lateral, U = Unrestrained.

where:
d_1 = Clear depth between flanges
n_w = Number of webs
t_f = Thickness of critical flange
t_w = Thickness of web

Table 3.10 Load height factors (k_l) for gravity loads

Longitudinal position of the load	Restraint arrangement	Load height position	
		Shear centre	Top flange
Within segment	FF, FP, FL, PP, PL, LL	1.0	1.4
	FU, PU	1.0	2.0
At segment end	FF, FP, FL, PP, PL, LL	1.0	1.0
	FU, PU	1.0	2.0

Source: AS 4100, Table 5.6.3(2). Copied by Mr L. Pack with the permission of Standards Australia under Licence 1607-c010.

Table 3.11 Lateral rotation restraint factor (k_r)

Restraint arrangement	Ends with lateral rotation restraints	Lateral rotation factor, k_r
FU, PU	Any	1.0
FF, FP, FL, PP, PL, LL	None	1.0
FF, FP, PP	One	0.85
FF, FP, PP	Both	0.70

Source: AS 4100, Table 5.6.3(3). Copied by Mr L. Pack with the permission of Standards Australia under Licence 1607-c010.

The length, l, is the distance between two points which are fully (F) or partially (P) restrained; or for a sub-segment, the distance between two points of lateral restraint.

The section warping constant (I_w) can be taken from tables (refer Section 7.3) or calculated as

$$I_w = \frac{I_y (d_f)^2}{4} \quad \text{(for doubly symmetric I-sections)}$$

$$I_w = I_{cy}(d_f)^2 \left(1 - \frac{I_{cy}}{I_y} \right) \quad \text{(for mono-symmetric I-sections)}$$

$$I_w = \frac{b_f{}^3 t_f b_w{}^2}{48} \left(8 - \frac{3 b_f t_f b_w{}^2}{I_x} \right) \quad \text{(for thin-walled channels)}$$

$$I_w \approx 0 \quad \text{(for angles, tee-sections, rectangular sections and hollow sections)}$$

where:
- b_f = Width of flange
- b_w = Depth of web
- t_f = Thickness of flange
- I_{cy} = The second moment of area of the compression flange about the minor y-axis
- I_x = The second moment of area about the major x-axis
- I_y = The second moment of area about the minor y-axis
- d_f = Distance between the centroid of each flange

3.4.2.4.2 Angle sections and hollow sections, fully or partially restrained at both ends

Angle sections and hollow sections are solved using the method in Section 3.4.2.4.1 (for open sections); however, the section warping constant (I_W) is set as zero.

3.4.2.4.3 Segments unrestrained at one end

Segments which are unrestrained at one end and fully or partially restrained at the other end (including either lateral continuity or restraint against 'lateral rotation') may be solved using the method in Section 3.4.2.4.1 (for segments restrained at both ends). However, Table 3.8 should be used to assign a value for α_m. Also, the value of α_m should not be simplified as equal to a value of 1, as it is not conservative for members with a uniform torsion that are restrained at only one end.

Example 3.4: Member Capacity

Calculate the member capacity for a 310UB40 (Grade 300) with a point load applied to the top flange at the centre of a 5.0 m span. Both ends of the beam are partially restrained without lateral rotation restraint.
Section capacity:

$$\phi M_s = 182 \text{ kNm} \qquad \text{(refer to Section 7.3 or calculated manually using Section 3.4.1)}$$

Member capacity:

$$M_b = \alpha_m \alpha_s M_s \leq M_s$$

A point load at the centre will result in maximum bending centrally, with half the bending at the quarter and three quarter points:

$$\alpha_m = \frac{1.7 M_m^*}{\sqrt{[(M_2^*)^2 + (M_3^*)^2 + (M_4^*)^2]}} \leq 2.5 = \frac{1.7 M_m^*}{\sqrt{[(0.5 M_m^*)^2 + (M_m^*)^2 + (0.5 M_m^*)^2]}}$$

$$= \frac{1.7}{\sqrt{[(0.5)^2 + (1)^2 + (0.5)^2]}} = 1.39$$

$$\alpha_s = 0.6 \left[\sqrt{\left[\left(\frac{M_s}{M_{oa}} \right)^2 + 3 \right]} - \left(\frac{M_s}{M_{oa}} \right) \right]$$

$$M_s = \phi M_s / \phi = 182 / 0.9 = 202.2 \text{ kNm}$$

$$M_{oa} = M_o = \sqrt{\left(\frac{\pi^2 E I_y}{l_e^2} \right) \left[GJ + \left(\frac{\pi^2 E I_w}{l_e^2} \right) \right]}$$

$E = 200{,}000$ MPa
$G = 80{,}000$ MPa
$J = 157 \times 10^3$ mm^4
$I_y = 7.65 \times 10^6$ mm^4
$I_w = 165 \times 10^9$ mm^6

Effective length:

$$l_e = k_t k_l k_r l = 1.066 \times 1.4 \times 1.0 \times 5000 \text{ mm} = 7462 \text{ mm}$$

$$k_t = 1 + \frac{\left[2\left(\dfrac{d_1}{l}\right)\left(\dfrac{t_f}{2t_w}\right)^3 \right]}{n_w} = 1 + \frac{\left[2\left(\dfrac{284}{5000}\right)\left(\dfrac{10.2}{2 \times 6.1}\right)^3 \right]}{1} = 1.066$$

$$k_l = 1.4$$
$$k_r = 1.0$$

$$M_{oa} = \sqrt{\left(\frac{\pi^2 \times 200,000 \times 7.65 \times 10^6}{7462^2} \right)\left[80,000 \times 157 \times 10^3 + \left(\frac{\pi^2 \times 200,000 \times 165 \times 10^9}{7462^2} \right) \right]}$$

$$= 70.7 \text{ kNm}$$

$$\alpha_s = 0.6\left[\sqrt{\left[\left(\frac{202.2}{70.7}\right)^2 + 3 \right]} - \left(\frac{202.2}{70.7}\right) \right] = 0.29$$

Member bending capacity:

$$\phi M_b = \alpha_m \alpha_s \phi M_s \leq \phi M_s = 1.39 \times 0.29 \times 182 \leq 182 = 73.4 \text{ kNm}$$

Alternatively, ϕM_b can be read from tables in Section 7.3 for the effective length calculated above, then multiplied by α_m (tables are based on $\alpha_m = 1$).

3.5 SHEAR

Members subject to shear are required to be assessed in accordance with AS 4100, Section 5. This section describes the design of unstiffened webs and web stiffeners. Web stiffeners are generally used either to provide increased global shear capacity to a member or to provide local load bearing capacity at a point load location. Shear capacities for common sections are tabulated in Section 7.3.

3.5.1 Unstiffened webs

The design of unstiffened webs is based on a minimum web thickness and a slenderness ratio for the web.

3.5.1.1 Minimum web thickness

The thickness of an unstiffened web (t_w) shall satisfy the following inequality:

$$t_w \geq \left(\frac{d_1}{180}\right)\sqrt{\frac{f_y}{250}} \qquad \text{(for webs with flanges on each side, e.g. I-sections)}$$

$$t_w \geq \left(\frac{d_1}{90}\right)\sqrt{\frac{f_y}{250}} \quad \text{(for webs with flanges on one side, e.g. channels)}$$

where:
  d_1 = Clear distance between flanges (depth of web)

3.5.1.2 Web capacity

The capacity of an unstiffened web depends on whether it is assumed to have an approximately uniform or a non-uniform shear stress distribution.

$$V^* \leq \phi V_v$$

$$\phi = 0.9 \quad \text{(for webs in shear)}$$

For a uniform stress distribution, $\quad V_v = V_u$

$$V_u = V_w = 0.6 f_y A_w \quad \left(\text{for webs where } t_w \geq \left(\frac{d_p}{82}\right)\sqrt{\frac{f_y}{250}}\right)$$

$$V_u = V_b = \alpha_v V_w \quad \left(\text{for webs where } t_w < \left(\frac{d_p}{82}\right)\sqrt{\frac{f_y}{250}}\right)$$

where:
  A_w = Gross cross-sectional area of web
      = $(d_1 - d_h)\, t_w$ (for a welded section)
      = $(d - d_h)\, t_w$ (for a hot-rolled section)
  d = Depth of entire section
  d_h = Total depth of holes in web cross-section (max 10% of d_1)
  d_p = Depth of web panel (= d_1 for an unstiffened web)

$$\alpha_v = \left[\frac{82}{\left(\dfrac{d_p}{t_w}\right)\sqrt{\dfrac{f_y}{250}}}\right]^2$$

A non-uniform stress distribution is used for cases such as members with unequal flanges, varying web thicknesses or holes (other than for fasteners),

$$V_v = \frac{2V_u}{0.9 + (f_{vm}^*/f_{va}^*)} \leq V_u$$

For circular hollow sections, the shear capacity is taken as $V_{w(CHS)} = 0.36\, f_y A_e$.
where:

  A_e = Effective area

(taken as gross area, provided that either there are no holes larger than those required for fasteners or the net area is greater than 0.9 times the gross area).

3.5.2 Combined bending and shear

If the bending moment can be resisted by only the flanges, then the full web capacity may be used for shear design (refer to AS 4100, Clause 5.12 for details). Otherwise, a reduction is made to the shear capacity, as shown:

$$V^* \leq \phi V_{vm}$$

where:

$$V_{vm} = V_v \qquad \text{(for } M^* \leq 0.75\phi M_s\text{)}$$

$$V_{vm} = V_v \left[2.2 - \left(\frac{1.6M^*}{\phi M_s} \right) \right] \qquad \text{(for } M^* > 0.75\phi M_s\text{)}$$

3.5.3 Globally stiffened webs

Globally stiffened webs are generally not economical in design, as the fabrication costs usually outweigh the reduction in mass. A method is provided for the design of stiffened webs in AS 4100, Clause 5.11.5.2.

3.5.4 Web bearing capacity

Web bearing capacity should be checked in areas of high shear transfer, such as bearing restraint locations and areas of high point loading. It should be checked under highly loaded pipe support locations where pipes rest on transverse beams. The unstiffened capacity (R_b) is calculated in this sub-section.

$$R^* \leq \phi R_b$$

$$\phi = 0.9 \quad \text{(for webs in shear or bearing)}$$

The method shown below is for I-sections and channels (refer to AS 4100, Clause 5.13 for hollow sections). The adopted value is the minimum of the bearing yield capacity and the bearing buckling capacity.

$$R_b = \text{MIN}\{R_{by}, R_{bb}\}$$

where:
Bearing yield, $R_{by} = 1.25 b_{bf} t_w f_y$
$\qquad\qquad b_{bf}$ = Width of stress distribution at inside of flange (refer to Figure 3.15)

Bearing buckling, R_{bb}, is calculated as shown for compression members $(d_1 \times b_b \times t_w)$, Section 3.7, with the following parameters:

$A \quad = t_w b_b$ (area of web)
$b_b \quad$ = Width of stress distribution at neutral axis (refer to Figure 3.15)
$k_f \quad = 1$
$l_e/r \quad = 2.5 d_1/t_w$ (where both flanges are restrained against lateral movement)

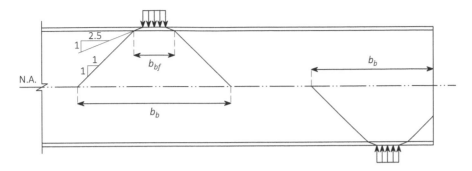

Figure 3.15 Bearing stress distribution. (AS 4100, Figure 5.13.1.1. Copied by Mr L. Pack with the permission of Standards Australia under Licence 1607-c010.)

l_e/r = $5.0d_1/t_w$ (where only one flanges is restrained against lateral movement)
α_b = 0.5

Values of $\phi R_{by}/b_{bf}$ and $\phi R_{bb}/b_b$ are tabulated for open sections in Section 7.3.

3.5.5 Web bearing stiffeners

Load bearing stiffeners are required at areas of high shear transfer, where the bearing capacity is exceeded. This commonly occurs at support locations for a beam and also at locations where high point loads are applied (e.g. a stiffener directly under a large pipe on a pipe rack). The stiffeners are required to be designed for yield and buckling:

$$R^* \leq \phi R_{sy} \quad \text{and} \quad R^* \leq \phi R_{sb}$$

$$R_{sy} = R_{by} + A_s f_{ys}$$

R_{sb} is calculated as shown for compression members, with the following parameters (refer to Section 3.7):

$$\alpha_b = 0.5 \quad k_f = 1 \quad \text{Area of web, } A = t_w b_b + A_{st}$$

The member length is d_1, with an effective length of $0.7d_1$ (where the flanges are restrained for rotation by other elements) or $1.0d_1$ (where the flanges are otherwise unrestrained). The effective section includes stiffeners on each side of the web, and the web to a maximum length each side of the stiffener equal to the minimum of $17.5t_w/\sqrt{f_y/250}$, or $s/2$, or dimensioned to the end of the beam (refer to Figure 3.16).

Refer to AS 4100, Section 5.14 for further details on stiffeners, such as maximum outstand dimensions, fitting requirements and design as torsional end restraints.

3.5.6 Openings in webs

Unstiffened webs may have openings with maximum dimensions of $0.1d_1$. This is increased to $0.33d_1$ for stiffened webs. Refer to AS 4100, Clause 5.10.7 for details.

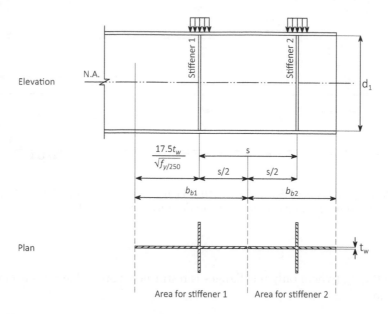

Figure 3.16 Effective bearing section for stiffened webs.

Example 3.5: Shear Capacity

Part 1 – Calculate the shear capacity of a 310UB40 (Grade 300) with a shear load applied over a width of 100 mm and located at 200 mm from the end of the unstiffened beam. Both flanges are restrained laterally.

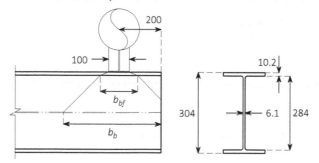

Web capacity:

$$\phi V_v = \phi V_w = 0.9 \times 0.6 f_y A_w \quad \left(\text{for webs where } t_w \geq \left(\frac{d_p}{82} \right) \sqrt{\frac{f_y}{250}} \right)$$

$$\phi V_v = 0.9 \times 0.6 \times 320 \times 6.1 \times 304 = 320 \text{ kN} \quad \left(\text{for } 6.1 \geq \left(\frac{284}{82} \right) \sqrt{\frac{320}{250}} = 3.92 \right)$$

Web bearing capacity (values selected using capacity tables):

$$\phi R_b = \text{MIN}[\phi R_{by}, \phi R_{bb}]$$

$\phi R_{by} / b_{bf} = 2.2$ kN/mm   (refer to Section 7.3)

$\phi R_{bb} / b_b = 0.588$ kN/mm

$b_{bf} = 100 + 2.5 \times t_f \times 2 = 151$ mm

$b_b = (b_{bf}/2 + d_1/2) + \text{MIN}[(b_{bf}/2 + d_1/2), 200]$

$\quad = (151/2 + 284/2) + \text{MIN}[(151/2 + 284/2), 200] = 417.5$ mm

$\phi R_{by} = \phi R_{by} / b_{bf} \times b_{bf} = 2.2 \times 151 = 332$ kN

$\phi R_{bb} = \phi R_{bb} / b_b \times b_b = 0.588 \times 417.5 = 245.5$ kN

$\phi R_b = \text{MIN}\{332, 245.5\} = 245.5$ kN

Web bearing capacity (values calculated manually):

$\phi R_b = \text{MIN}\left[\phi R_{by}, \phi R_{bb}\right]$

$\phi R_{by} = 0.9 \times 1.25 b_{bf} t_w f_y = 0.9 \times 1.25 \times 151 \times 6.1 \times 320 = 332$ kN

$\phi R_{bb} = \phi N_c$

$N_s = k_f A_n f_y = k_f t_w b_b f_y = 1 \times 6.1 \times 417.5 \times 320 = 815$ kN

$N_c = \alpha_c N_s$

$k_f = 1$

$l_e / r = 2.5 d_1 / t_w = 2.5 \times 284 \times 6.1 = 116$

$\lambda_n = \dfrac{l_e}{r}\left(\sqrt{k_f}\right)\sqrt{\dfrac{f_y}{250}} = 131.5$

$\alpha_a = \dfrac{2100\left(\lambda_n - 13.5\right)}{\lambda_n^2 - 15.3\lambda_n + 2050} = 14.3$

$\alpha_b = 0.5$

$\lambda = \lambda_n + \alpha_a \alpha_b = 139$

$\eta = 0.00326\left(\lambda - 13.5\right) \geq 0 = 0.408$

$\xi = \dfrac{\left(\dfrac{\lambda}{90}\right)^2 + 1 + \eta}{2\left(\dfrac{\lambda}{90}\right)^2} = 0.797$

$$\alpha_c = \xi \left[1 - \sqrt{\left[1 - \left(\frac{90}{\xi \lambda} \right)^2 \right]} \right] = 0.335$$

$$N_c = \alpha_c N_s = 0.335 \times 815 = 272.7$$

$$\phi R_{bb} = 0.9 \times 272.7 = 245.5 \text{ kN}$$

$$\phi R_b = \text{MIN}\{332, 245.5\} = 245.5 \text{kN}$$

The shear capacity is 245.5 kN, governed by web bearing buckling.

Part 2 – Recalculate the shear capacity with one stiffener added to each side of the beam under the applied load. The stiffeners are 70 mm wide, 10 mm thick and grade 250. Flanges are not restrained for rotation by other elements.

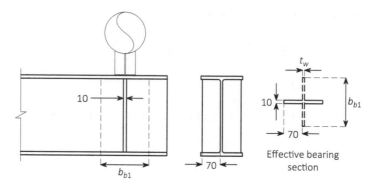

Web bearing capacity:

$$\phi R_b = \text{MIN}[\phi R_{sy}, \phi R_{sb}]$$

Stiffener yield:

$$\phi R_{sy} = \phi R_{by} + \phi A_s f_{ys} = 332,000 + 0.9 \times 2 \times 70 \times 10 \times 250 = 647 \text{ kN}$$

Stiffener buckling:

$$\phi R_{sb} = \phi N_c$$

Effective bearing section:

$$l_b = \frac{17.5 t_w}{\sqrt{f_y / 250}} = \frac{17.5 \times 6.1}{\sqrt{320/250}} = 94.4 \text{ mm}$$

$$b_{b1} = 94.4 + \text{MIN}[94.4, 200] = 188.8 \text{ mm}$$

Consider a cruciform section, comprised of a 188.8 mm segment of the web, intersected by the stiffeners at mid-point. The height of the bearing section is equal to the distance between flanges.

$$A_n = 188.8 \times 6.1 + 2 \times 70 \times 10 = 2551.7 \text{ mm}^2$$

$$N_s = k_f A_n f_y = 1 \times 188.8 \times 6.1 \times 320 + 1 \times 2 \times 70 \times 10 \times 250 = 718.5 \text{ kN}$$

$$N_c = \alpha_c N_s$$

$$l_e = k_e l = 1.0 \times 284 = 284 \text{ mm} \quad \text{(refer to Figure 3.19)}$$

$$I_x = 0.92 \times 10^6 \text{ mm}^4 \quad \text{(refer to Section 3.4.1.1)}$$

$$I_y = 0.624 \times 10^6 \text{ mm}^4$$

$$r_x = \sqrt{I_x / A} = \sqrt{0.92 \times 10^6 / 2551.7} = 19 \text{ mm}$$

$$r_y = \sqrt{I_y / A} = \sqrt{0.624 \times 10^6 / 2551.7} = 15.6 \text{ mm}$$

$$r = \text{MIN}[r_x, r_y] = 15.6 \text{ mm}$$

$$k_f = 1$$

$$\lambda_n = \frac{l_e}{r} \left(\sqrt{k_f} \right) \sqrt{\frac{f_y}{250}} = 19.24 \quad \text{(using the average yield stress)}$$

$$\alpha_a = \frac{2100(\lambda_n - 13.5)}{\lambda_n^2 - 15.3\lambda_n + 2050} = 5.7$$

$$\alpha_b = 0.5$$

$$\lambda = \lambda_n + \alpha_a \alpha_b = 22$$

$$\eta = 0.00326(\lambda - 13.5) \geq 0 = 0.028$$

$$\xi = \frac{\left(\dfrac{\lambda}{90} \right)^2 + 1 + \eta}{2 \left(\dfrac{\lambda}{90} \right)^2} = 9.042$$

$$\alpha_c = \xi \left[1 - \sqrt{1 - \left(\frac{90}{\xi \lambda} \right)^2} \right] = 0.97$$

$$N_c = \alpha_c N_s = 0.97 \times 718.5 = 698 \text{ kN}$$

$$\phi R_{sb} = \phi N_c = 0.9 \times 698 = 628 \text{ kN}$$

$$\phi R_b = \text{MIN}[\phi R_{sy}, \phi R_{sb}] = \text{MIN}[647, 628] = 628 \text{ kN}$$

The shear capacity of the stiffened section is limited to the stiffener yield and stiffener buckling. The stiffeners have increased the capacity from 245.5 to 628 kN, governed by stiffener buckling. AS 4100 does not give guidance on what yield stress to use when stiffeners are a different grade from the beam. The average has been adopted in the example above; however, the minimum value can be conservatively adopted.

3.6 TENSION

Members subject to tension are required to be assessed for their section capacity as shown. Tensile capacities for common sections are tabulated in Section 7.3.

$$N_t^* \leq \phi N_t$$

$$\phi = 0.9 \quad \text{(for members in tension)}$$

In accordance with AS 4100, Section 7, the tensile capacity of a member (ϕN_t) is taken as

$$\phi N_t = \text{MIN}[\phi A_g f_y, \phi 0.85 k_t A_n f_u]$$

where:

A_g = Gross cross-sectional area
A_n = Net cross-sectional area (gross area minus holes)
f_u = Tensile strength
f_y = Yield stress
k_t = Correction factor (Table 3.12)

Example 3.6: Tension Capacity

Calculate the tension capacity of a 75 x 6 EA (Grade 300) with M20 bolt holes penetrating the section at the connection.

$$\phi N_t = \text{MIN}[\phi A_g f_y, \phi 0.85 k_t A_n f_u]$$

A_g = 867 mm² (refer to Section 7.3)
A_n = A_g – 22 × 6 = 735 mm² (22 diameter holes for M20 bolts)
f_u = 440 MPa
f_y = 320 MPa
k_t = 0.85

$$\phi N_t = \text{MIN}[0.9 \times 867 \times 320, 0.9 \times 0.85 \times 0.85 \times 735 \times 440] = 210.3 \text{ kN}$$

The tensile capacity of the angle is calculated as 210.2 kN with M20 bolts connecting through one leg of the angle.

3.7 COMPRESSION

Members subject to compression are required to be assessed for section and member capacities in accordance with AS 4100, Section 6. The section capacity is the same for each axis; however, the member capacity varies. For I-sections, the member capacity is highest for

Table 3.12 Correction factor (k_t)

k_t	Description
1.0	Members with uniform tension, i.e. where the entire cross-section is uniformly restrained at the connection
0.9	Channels connected on the back of the web
0.85	Symmetrical, rolled or built-up, I-sections or channels, fully connected at both flanges; where the length of the connection (longitudinally between bolts or welds) is at least equal to the depth of the member; and each flange restraint is designed to take at least half the design load
0.85	Angles connected on the back of the long or equal leg
0.75	Unequal angles connected on the back of the short leg

Note: Refer to AS 4100, Clause 7.3 and Table 7.3.2 for additional cases, diagrams and details.

buckling in the major (x) axis. For this reason, the minor (y) axis is generally braced where an increased compressive capacity is required.

$$N^* \leq \phi N_s \quad \text{and} \quad N^* \leq \phi N_c$$

$$\phi = 0.9 \quad \text{(for members in compression)}$$

This section details the design of individual members in compression. For details on laced and battened compression members or back-to-back compression members, refer to AS 4100, Section 6.

3.7.1 Section compression capacity

The section capacity of a compression member is based on the factored net area of steel,

$$N_s = k_f A_n f_y$$

where: A_n = Net area of cross-section (gross area minus unfilled hole areas)

The gross area may be used instead of the net area where holes or unfilled penetrations reduce the area by less than $100\{1 - [f_y/(0.85f_u)]\}\%$

$$k_f = \text{Form factor} = A_e/A_g$$

where:

A_e = Effective cross-sectional area
A_g = Gross cross-sectional area (total area ignoring holes)

3.7.1.1 Effective cross-section

The effective area of a cross-section depends on the slenderness of each element which comprises the section. The width of each element is reduced to account for slenderness effects. Each element is measured using clear dimensions, as shown in Figure 3.17. Areas which

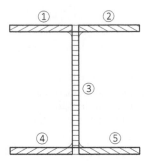

Figure 3.17 Effective cross-section elements.

are not enclosed in the diagram in this sub-section (the root radius and the area where the flanges meet) are assumed to be fully effective and may be added to the following equation without any reduction for effectiveness.

$$A_e = \sum_1^N \left(b_{ei} \, t_i \right)$$

where:

b_{ei} = Effective width of each element
t_i = Thickness of each element

3.7.1.1.1 Flat plate slenderness

Effective width of plate elements, $\qquad b_e = b \left(\dfrac{\lambda_{ey}}{\lambda_e} \right) \le b$

where:

$$\lambda_e = \frac{b}{t} \sqrt{f_y \big/ 250}$$

λ_{ey} = Yield slenderness (refer to Table 3.13)

3.7.1.1.2 Circular hollow sections

Effective outside diameter of section, $\qquad d_e = \mathrm{MIN}\left\{ \left[d_o \sqrt{\lambda_{ey}/\lambda_e} \right], \; \left[d_o \left(3\lambda_{ey}/\lambda_e \right)^2 \right], \; [d_o] \right\}$

where:

$$\lambda_e = \frac{d_o}{t} \sqrt{f_y/250}$$

λ_{ey} = Yield slenderness (refer to Table 3.13)

3.7.2 Member compression capacity

The compressive member capacity is calculated by reducing the section capacity by a slenderness reduction factor.

$$N_c = \alpha_c N_s$$

Table 3.13 Values of plate element yield slenderness limit

Plate element type	Longitudinal edges supported	Residual stresses (see notes)	Yield slenderness limit, λ_{ey}
Flat	One (outstand)	SR	16
		HR	16
		LW, CF	15
		HW	14
	Both	SR	45
		HR	45
		LW, CF	40
		HW	35
Circular hollow sections		SR	82
		HR, CF	82
		LW	82
		HW	82

Source: AS 4100, Table 6.2.4. Copied by Mr L. Pack with the permission of Standards Australia under Licence 1607-c010.

Notes:
1. SR = stress relieved; HR = hot rolled or hot finished; CF = cold formed; LW = lightly welded longitudinally; HW = heavily welded longitudinally.
2. Welded members whose compressive residual stresses are less than 40 MPa may be considered to be lightly welded.

where:

$$\alpha_c = \xi \left[1 - \sqrt{\left[1 - \left(\frac{90}{\xi \lambda} \right)^2 \right]} \right]$$

$$\xi = \frac{(\lambda/90)^2 + 1 + \eta}{2\left(\frac{\lambda}{90} \right)^2}$$

$\eta = 0.00326 \, (\lambda - 13.5) \geq 0$

$\lambda = \lambda_n + \alpha_a \alpha_b$

$$\lambda_n = \frac{l_e}{r} \left(\sqrt{k_f} \right) \sqrt{\frac{f_y}{250}}$$

$$r = r_x = \sqrt{\frac{I_x}{A_n}} \quad \text{(for capacity about } x\text{-axis)} \quad \text{or refer to Section 7.3}$$

$$r = r_y = \sqrt{\frac{I_y}{A_n}} \quad \text{(for capacity about } y\text{-axis)} \quad \text{or refer to Section 7.3}$$

$$\alpha_a = \frac{2100(\lambda_n - 13.5)}{\lambda_n^2 - 15.3\lambda_n + 2050}$$

α_b = Member section constant   (refer to Tables 3.14 and 3.15)

The capacity needs to be calculated about each axis (N_{cx} and N_{cy}), because the radius of gyration (r_x and r_y) and effective lengths (l_{ex} and l_{ey}) lead to different values for each case.

The guidelines presented in AS 4100 are based on flexural buckling. For members which may be subject to twisting (short cruciforms, tees and concentrically loaded angles), refer to AS 4100 Commentary, Clause C6.3.3.

Table 3.14 Values of member section constant (α_b) for $k_f = 1.0$

Compression member section constant, α_b	Section description
−1.0	Hot-formed RHS and CHS
	Cold-formed (stress relieved) RHS and CHS
−0.5	Cold-formed (non-stress relieved) RHS and CHS
	Welded H, I and box section fabricated from Grade 690 high-strength quenched and tempered plate
0	Hot-rolled UB and UC sections (flange thickness up to 40 mm)
	Welded H and I sections fabricated from flame-cut plates
	Welded box sections
0.5	Tees flame-cut from universal sections and angles
	Hot-rolled channels
	Welded H and I sections fabricated from as-rolled plates (flange thickness up to 40 mm)
	Other sections not listed in this table
1.0	Hot-rolled UB and UC sections (flange thickness over 40 mm)
	Welded H and I sections fabricated from as-rolled plates (flange thickness over 40 mm)

Source: AS 4100, Table 6.3.3(1). Copied by Mr L. Pack with the permission of Standards Australia under Licence 1607-c010.

Table 3.15 Values of member section constant (α_b) for $k_f < 1.0$

Compression member section constant, α_b	Section description
−0.5	Hot-formed RHS and CHS
	Cold-formed RHS and CHS (stress relieved)
	Cold-formed RHS and CHS (non-stress relieved)
0	Hot-rolled UB and UC sections (flange thickness up to 40 mm)
	Welded box sections
0.5	Welded H and I sections (flange thickness up to 40 mm)
1.0	Other sections not listed in this table

Source: AS 4100, Table 6.3.3(2). Copied by Mr L. Pack with the permission of Standards Australia under Licence 1607-c010.

3.7.2.1 Effective length

Programs such as Space Gass can nominate an effective length to members by using a buckling analysis. Alternatively, the length of the member can be factored by an effective length factor. Refer to Figure 3.18 for factors of members with idealised restraints; otherwise, refer to AS 4100, Clause 4.6.3 for a stiffness ratio method. Examples of major axis effective length factors are shown in Figure 3.19, using the idealised values.

Effective length, $l_e = k_e l$

where: k_e = Effective length factor (refer to Figure 3.18)

3.7.2.2 Braces

Where bracing is used to reduce the effective length for compression members, the brace should be designed to take the maximum of the design load and 2.5% of the column axial load (plus 1.25% of any parallel members beyond the connected member). This has been proved to conservatively provide the theoretical stiffness value required to reduce the effective length [28,29]. The minor axis is usually braced, because the buckling capacity is weakest about that axis; however, the major axis can be braced as well. Both flanges should be restrained at brace points to prevent twisting. Typical bracing configurations are shown in Figure 3.20. Portal frames are often braced with girts and purlins, with fly bracing to engage the bottom flange. Pipe racks are generally braced with either bolted or welded structural members. There are many ways to achieve this outcome.

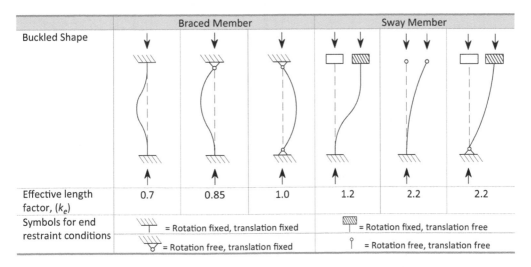

Figure 3.18 Effective length factors for members for idealised conditions of end restraint. (AS 4100, Figure 4.6.3.2. Copied by Mr L. Pack with the permission of Standards Australia under Licence 1607-c010.)

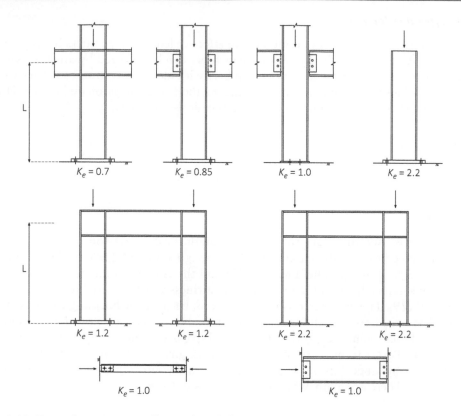

Figure 3.19 Example major axis effective length factors using idealised values.

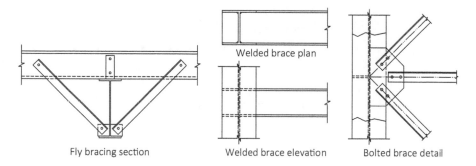

Figure 3.20 Typical minor axis brace configurations.

Example 3.7: Compression

Calculate the compression capacity of a 75 x 6 EA (Grade 300) with a length of 3 m and pinned connections at each end.

Section capacity:

$$\phi N_s = \phi k_f A_n f_y$$

A_g = 867 mm² (refer to Section 7.3)

$A_n = A_g$ = 867 mm²

f_u = 440 MPa

f_y = 320 MPa

Effective area of a cross-section:

$b = 75 - 6 = 69$ mm

$\lambda_{ey} = 16$ (refer to Table 3.13)

$$\lambda_e = \frac{b}{t}\sqrt{f_y/250} = \frac{69}{6}\sqrt{320/250} = 13$$

$$b_e = b\left(\lambda_{ey}/\lambda_e\right) \le b = 69\left(16/13\right) \le 69 = 69 \text{ mm}$$

A_e = 867 mm²

The effective width of the angle leg is not reduced; therefore, the effective area is equal to the gross area.

$$k_f = A_e/A_g = 1$$

$$\phi N_s = \phi k_f A_n f_y = 0.9 \times 1 \times 867 \times 320 = 250 \text{ kN}$$

Member capacity:

$$\phi N_c = \phi \alpha_c N_s$$

$$l_e = k_e l = 1.0 \times 3000 = 3000 \text{ mm}   \text{(refer to Figure 3.19)}$$

$$I_x = 0.722 \times 10^6 \text{mm}^4   \text{(refer Section 3.4.1.1 or Section 7.3)}$$

$$I_y = 0.187 \times 10^6 \text{mm}^4$$

$$r_x = \sqrt{I_x / A} = \sqrt{0.722 \times 10^6 / 867} = 28.9 \text{ mm}$$

$$r_y = \sqrt{I_y / A} = \sqrt{0.187 \times 10^6 / 867} = 14.7 \text{ mm}$$

$$r = \text{MIN}[r_x, r_y] = 14.7 \text{ mm}$$

$$\lambda_n = \frac{l_e}{r}\left(\sqrt{k_f}\right)\sqrt{\frac{f_y}{250}} = \frac{3000}{14.7}\left(\sqrt{1}\right)\sqrt{\frac{320}{250}} = 231$$

$$\alpha_a = \frac{2100\left(\lambda_n - 13.5\right)}{\lambda_n^2 - 15.3\lambda_n + 2050} = 8.8$$

$$\alpha_b = 0.5$$

$$\lambda = \lambda_n + \alpha_a \alpha_b = 231 + 88 \times 0.5 = 235.4$$

$$\eta = 0.00326(\lambda - 13.5) \ge 0 = 0.00326(235.4 - 13.5) \ge 0 = 0.72$$

$$\xi = \frac{\left(\dfrac{\lambda}{90}\right)^2 + 1 + \eta}{2\left(\dfrac{\lambda}{90}\right)^2} = 0.63$$

$$\alpha_c = \xi\left[1 - \sqrt{\left[1 - \left(\dfrac{90}{\xi\lambda}\right)^2\right]}\right] = 0.13$$

$$\phi N_c = \phi\alpha_c N_s = 0.13 \times 250 = 32.5 \text{ kN}$$

The compression capacity of the 3 m long 75 × 6 EA angle is calculated as 32.5 kN.

3.8 COMBINED ACTIONS

Members which are subject to more than one concurrent action (bending about an axis and axial load) are required to be checked for combined actions. The required check is dependent on the specific combination of loads. Checks need to be completed for combined section capacity and combined member capacity. Refer to AS 4100, Section 8, for additional details.

$$\phi = 0.9 \quad \text{(for combined actions)}$$

3.8.1 Combined section capacity

The design loads for a section capacity check are the loads which occur at one location, causing the worst combined effect. It is conservative, but not necessary, to adopt the maximum loads from the entire member. The calculations for combined section capacity are broken down into three categories:

1. Axial load with uniaxial bending about the major principal x-axis
2. Axial load with uniaxial bending about the minor principal y-axis
3. Axial load with biaxial bending

The axial load considered in this design should be chosen as either tension or compression, based on the load which is closest to failure: $(N^*/\phi N_s)$ or $(N^*_t/\phi N_t)$. For simplicity, the formulas in this section adopt the nomenclature for compressive loads.

Refer to AS 4100, Clause 8.3 for further details.
Refer to AS 4100, Clause 8.4.6 for angle members connected in trusses.

3.8.1.1 Axial load with uniaxial bending about the major principal x-axis

Sections shall satisfy

$$M_x^* \leq \phi M_{rx}$$

where: $M_{rx} = M_{sx}\left(1 - \dfrac{N^*}{\phi N_s}\right)$

This value can be conservative and may be increased for doubly symmetric I-sections, RHS and SHS, where the section is defined as compact:

$$M_{rx} = 1.18 M_{sx}\left(1 - \frac{N^*}{\phi N_s}\right) \le M_{sx} \quad \text{(for tension members, or for } k_f = 1\text{)}$$

$$M_{rx} = M_{sx}\left(1 - \frac{N^*}{\phi N_s}\right)\left[1 + 0.18\left(\frac{82 - \lambda_w}{82 - \lambda_{wy}}\right)\right] \le M_{sx} \quad \text{(for compression members, where } k_f < 1\text{)}$$

λ_w and λ_{wy} are the slenderness values (λ_e and λ_{ey}) for the web of the section (refer to Section 3.7.1.1)

3.8.1.2 Axial load with uniaxial bending about the minor principal y-axis

Sections shall satisfy

$$M_y^* \le \phi M_{ry}$$

where: $M_{ry} = M_{sy}\left(1 - \frac{N^*}{\phi N_s}\right)$

This value can be conservative and may be increased for doubly symmetric I-sections, where the section is defined as compact:

$$M_{ry} = 1.19 M_{sy}\left[1 - \left(\frac{N^*}{\phi N_s}\right)^2\right] \le M_{sy}$$

It can also be increased for RHS and SHS, where the section is defined as compact:

$$M_{ry} = 1.18 M_{sy}\left(1 - \frac{N^*}{\phi N_s}\right) \le M_{sy}$$

3.8.1.3 Axial load with biaxial bending

Sections shall satisfy

$$\frac{N^*}{\phi N_s} + \frac{M_x^*}{\phi M_{sx}} + \frac{M_y^*}{\phi M_{sy}} \le 1$$

This value can be conservative and may be increased for doubly symmetric I-sections, RHS and SHS, where the section is defined as compact:

$$\left(\frac{M_x^*}{\phi M_{rx}}\right)^\gamma + \left(\frac{M_y^*}{\phi M_{ry}}\right)^\gamma \le 1$$

where: $\gamma = 1.4 + \left(\dfrac{N^*}{\phi N_s}\right) \le 2$

3.8.2 Combined member capacity

The design loads for a member capacity check are the maximum loads along the entire member, regardless of whether they occur at the same exact location. The combined member capacity calculation is based on the load type and the analysis type:

1. Axial load with uniaxial bending: elastic analysis
2. Axial load with uniaxial bending: plastic analysis
3. Axial load with biaxial bending

Unless the governing load case is apparent, tension and compression loads should each be checked (if present) with their corresponding bending moments, as the capacity of the member varies differently for each case. Refer to AS 4100, Clause 8.3 for details.

3.8.2.1 Axial load with uniaxial bending: Elastic analysis

This section is mainly for compression members; tension members only need to be checked for out-of-plane capacity, and only where the member is in major axis bending and has insufficient restraint to prevent lateral buckling.

3.8.2.1.1 Compression members

For compression members, the following calculation can be completed with bending about either the major x-axis or the minor y-axis.

$$M^* \le \phi M_i$$

$$M_i = M_s\left(1 - \frac{N^*}{\phi N_c}\right)$$

where:
M_s is the section bending capacity
N_c is the member compression capacity

However, N_c may be recalculated with the effective length factor (k_e) reduced to 1.0 (for cases where k_e was originally greater than 1.0), although the member is still required to pass the check for compression only with the original k_e value.

This value can be conservative and may be increased for doubly symmetric I-sections, RHS and SHS, where the section is defined as compact and where the form factor (k_f) is equal to 1.0 (i.e. where the net area equals the gross area, and therefore, compression elements are not acting as slender):

$$M_i = M_s\left\{\left[1 - \left(\frac{1+\beta_m}{2}\right)^3\right]\left[1 - \frac{N^*}{\phi N_c}\right] + 1.18\left[\frac{1+\beta_m}{2}\right]^3\sqrt{\left(1 - \frac{N^*}{\phi N_c}\right)}\right\} \le M_{rx} \text{ or } M$$

where: β_m = Ratio of smaller to larger end bending moments

(Taken as positive for members with reverse curvature and no transverse load, otherwise taken conservatively as −1.0, or from values chosen from AS 4100, Clause 4.4.2.2 and AS 4100, Table 4.4.2.2.)

3.8.2.1.2 Additional check for out-of-plane capacity in compression members

Compression members which are bent about the major x-axis and have insufficient restraint to prevent lateral buckling require the following additional check:

$$M_x^* \leq \phi M_{ox}$$

$$M_{ox} = M_{bx}\left(1 - \frac{N^*}{\phi N_{cy}}\right)$$

where:

M_{bx} = Member bending capacity about x-axis
    (for a member without full lateral restraint, calculated with α_m chosen to represent the moment distribution along the entire member [rather than a segment])
N_{cy} = Member compression capacity for buckling about y-axis

This value can be conservative and may be increased for compact, doubly symmetric I-sections, fully or partially restrained at both ends, with no transverse loading, and where the form factor (k_f) is equal to 1.0 (i.e. where the net area equals the gross area, and therefore, compression elements are not acting as slender):

$$M_{ox} = \alpha_{bc}M_{bxo}\sqrt{\left[\left(1 - \frac{N^*}{\phi N_{cy}}\right)\left(1 - \frac{N^*}{\phi N_{oz}}\right)\right]} \leq M_{rx}$$

where:

$$\frac{1}{\alpha_{bc}} = \frac{1-\beta_m}{2} + \left(\frac{1+\beta_m}{2}\right)^3\left(0.4 - 0.23\frac{N^*}{\phi N_{cy}}\right)$$

M_{bxo} = Member bending capacity (without full lateral restraint), using $\alpha_m = 1$
N_{cy} = Member compression capacity for buckling about the y-axis

$$N_{oz} = \frac{GJ + (\pi^2 EI_w / l_z^2)}{(I_x + I_y) / A}$$

E = Modulus of elasticity = 200,000 MPa
G = Shear modulus ≈ 80,000 MPa
I_W = Section warping constant (refer to Sections 3.4.2.4 and 7.3)
I_x = Section second moment of area about x-axis (refer to Sections 3.4.1.1 and 7.3)
I_y = Section second moment of area about y-axis (refer to Sections 3.4.1.1 and 7.3)
J = Section torsion constant (refer to Section 3.9 and 7.3)
I_z = Distance between full or partial torsional restraints
β_m = Ratio of smaller to larger end bending moments
    (taken as positive when the member is bent in reverse curvature)

3.8.2.1.3 Check for out-of-plane capacity in tension members

Tension members which are bent about the major x-axis and have insufficient restraint to prevent lateral buckling require the following check:

$$M_x^* \leq \phi M_{ox}$$

$$M_{ox} = M_{bx}\left(1 + \frac{N^*}{\phi N_t}\right) \leq M_{rx}$$

where:
$\quad M_{bx}$ = Member bending capacity about x-axis
$\qquad$ (calculated with α_m chosen to represent the moment distribution along the entire member, rather than a segment)

$$M_{rx} = M_{sx}\left(1 - \frac{N^*}{\phi N_t}\right)$$

3.8.2.2 Axial load with uniaxial bending: Plastic analysis

The plastic analysis method is less common in design and is therefore not repeated in this text. For the combined capacity of doubly symmetric I-section members, analysed using a plastic analysis, refer to AS 4100, Clauses 4.5 and 8.4.3.

3.8.2.3 Axial load with biaxial bending

Members which are subject to axial load and biaxial bending are required to meet the following requirements.

3.8.2.3.1 Compression members

Members must satisfy

$$\left(\frac{M_x^*}{\phi M_{cx}}\right)^{1.4} + \left(\frac{M_y^*}{\phi M_{iy}}\right)^{1.4} \leq 1$$

where:
$\quad M_{cx}$ = MIN$\{M_{ix}, M_{ox}\}$ (Refer to Section 3.8.2.1)
$\quad M_{iy}$ = In-plane member moment capacity (Refer to Section 3.8.2.1)

3.8.2.3.2 Tension members

Members must satisfy

$$\left(\frac{M_x^*}{\phi M_{tx}}\right)^{1.4} + \left(\frac{M_y^*}{\phi M_{ry}}\right)^{1.4} \leq 1$$

where:

M_{tx} = MIN$\{M_{rx}, M_{ox}\}$ (Refer to Section 3.8.2.1)

M_{ry} = In-plane member moment capacity (Refer to Section 3.8.2.1)

3.9 TORSION

Design for torsion is not covered in AS 4100; however, a detailed explanation and some information on the theory are presented in AS 4100, Supplement 1, Clause C8.5.

Torsion is essentially grouped into two forms: uniform torsion and warping torsion. Uniform torsion is where the rate of change of the angle of twist in a member is constant along the length, and longitudinal warping deflections are also constant. Warping torsion is where the rate of change of the angle of twist in a member varies along the length.

3.9.1 Uniform torsion

Uniform torsion resists load through a single set of shear stresses, which act around the entire cross-section. Uniform torsion is generally seen in cross-sections which have negligible warping constants (refer to Section 3.4.2.4) and high torsional stiffness (hollow sections, angles, cruciforms and tees), as they are not prone to warping. It is for this reason that hollow sections perform better than open sections in torsion. This can be seen in the circular hollow section shown in Figure 3.21.

Uniform torsion can also be experienced in members with high warping constants (or low torsional stiffness) if they are torsionally unrestrained at both ends, or at sections in a member which are a sufficient distance (a) away from a restraint (refer to Figure 3.21, I-section at point B).

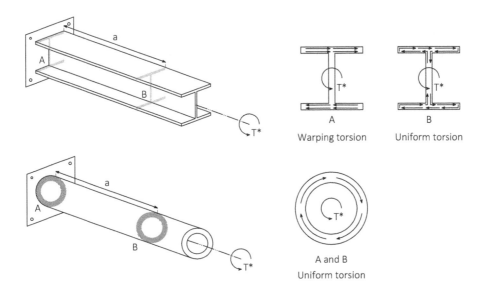

Figure 3.21 Torsion diagrams.

3.9.2 Warping torsion

Warping torsion resists load by forming a couple, acting at the flanges, which push in opposing directions. Warping torsion is generally seen in cross-sections which have high warping constants (refer to Section 3.4.2.4) and low torsional stiffness (I-sections, PFC), as they are prone to warping. It occurs in locations where the section is restrained, such as at a base plate or welded connection. It can also occur in the location where torsion is applied for members in which torsion varies along the length. Essentially, the flanges develop opposing loads which form a couple and therefore can resist torsion. The effects of warping torsion are less pronounced in compact sections compared with non-compact sections.

Warping torsion is often ignored in members which have negligible warping constants, which is generally acceptable but may lead to high local stresses at areas of load application or restraint.

3.9.3 Non-uniform torsion

A cantilever comprised of a section which is prone to warping (such as an I-section), experiences warping torsion at the restrained end and gradually transitions into uniform torsion at some distance (a) away from the restraint (refer to Figure 3.21). Conversely, thin-walled hollow sections experience uniform torsion in both locations. Where a section is subject to a reasonable amount of each type of torsion (uniform and warping), it is described as *non-uniform torsion*. The resultant solution is a combination of the two methods.

3.9.4 Finite element analysis of torsion

Finite element analysis (FEA) is a recommended method for designing structures which are subject to torsion (especially warping torsion). This is primarily because the design guidelines are vague, and an FEA can be conducted quickly and provide a good depiction of the behaviour. Figure 3.22 shows a linear static analysis of a cantilever with torsion (similar to that in Figure 3.21): reactions are shown on the section to the left (warping torsion) and the stress is shown on the section to the right (uniform torsion). The analysis confirms the example diagram in Figure 3.21.

Figure 3.23 shows the stress in the beam, calculated using a linear static analysis, and buckling factors and displacement using a linear buckling analysis. The buckling check is important, because the flange can have a large compressive force where the warping torsion

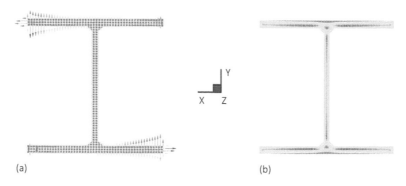

(a)

(b)

Figure 3.22 FEA analysis showing linear buckling analysis sections. (a) Warping torsion reactions, (b) uniform torsion stress.

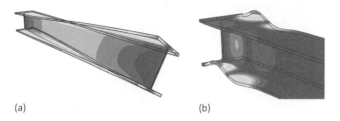

(a)   (b)

Figure 3.23 FEA analysis showing linear stress and linear buckling deflections. (a) Linear analysis of beam, (b) buckling analysis at restraint.

occurs, near the restraint. Refer to Section 6.11 for further details on FEA modelling. The solution can be found by modelling the beam using plate or brick elements.

3.9.5 Torsion calculations

Torsion calculations can be performed manually; however, care needs to be taken to ensure the correct form of torsion is considered. Uniform torsion calculations are straightforward; however, warping torsion calculations are more complex, and many varied methods are available.

$$T^* \leq \phi T_u$$

$\phi = 0.9$ (adopted from members in bending)

3.9.5.1 Uniform torsion calculations

The method in this sub-section can be used for members which are under uniform torsion and are free to warp. Twisting angle and shear stress can be calculated as shown. Shear stresses should be limited to $0.6\ f_y$. Values of ϕT_u are tabulated for hollow sections in Section 7.3.

Ultimate torsion capacity

$$\phi T_u = 0.9 \times 0.6 \times f_y \times (T/\tau) = 0.9 \times 0.6 \times f_y \times C$$

Twisting angle, $$\theta = \frac{T^* L}{KG}$$

where:

    C = Torsion modulus = T/τ
    G $\approx$ 80,000 MPa
    K = Torsion variable (refer to Table 3.16)
    L = Length of member
    T = Torsion
    τ = Shear stress (refer to Table 3.16)

The torsion variable (K) is equal to the torsion constant (J) for circular cross-sections, and is equal to a smaller amount for other sections. Refer to Table 3.16 for a list of torsion variables and formulas for shear stress, τ.

Table 3.16 Uniform torsion formulas

Section type	Torsion variable, K	Shear stress, τ (MPa)	
Solid circle	$0.5\pi r^4$	$2T/\pi r^3$	At boundary
Solid square	$2.25a^4$	$2T/\pi r^3$	At midpoint on side
Circular hollow section	$0.5\pi({r_o}^4 - {r_i}^4)$	$2Tr_o/\pi({r_o}^4 - {r_i}^4)$	At outer boundary
Rectangular hollow section	$\dfrac{2t\,t_1(a-t)^2(b-t_1)^2}{at+bt_1-t^2-t_1^2}$	$\dfrac{T}{2t(a-t)(b-t_1)}$	Near midpoint on short side (b)
		$\dfrac{T}{2t_1(a-t)(b-t_1)}$	Near midpoint on long side (a)

Source: These equations are adopted from *Roark's Formulas for Stress and Strain* (Young, W.C. et al., 8th ed., McGraw-Hill, USA, 2012). Refer to text for advanced calculations and additional sections.

3.9.5.1.1 Torsion constant

The torsion constant (J) is best taken from tables (refer to Section 7.3); however, it can be closely approximated using the formulas

$$J \approx \sum \left(\frac{bt^3}{3} \right) \quad \text{(for open sections)}$$

$$J \approx \frac{4A_e^{\,2}}{\sum \left(\dfrac{b}{t} \right)} \quad \text{(for hollow sections)}$$

where: A_e = Area enclosed by the hollow section

3.9.5.2 Warping torsion calculations

Accurate warping torsion calculations can be extremely detailed and complex. The following is a simplified and conservative approach to warping torsion based on the twin beam method. FEA is recommended as a more convenient approach if a greater capacity is required. Detailed formulas can be found in *Roark's Formulas for Stress and Strain* [18] if calculations are preferred.

The twin beam method essentially ignores any contribution from the web and resists the torsion by treating the two flanges as individual beams which are bent about their major axes (refer to Figure 3.24). The elastic bending capacity is shown for flange bending. This may be increased to a plastic capacity (×1.5) especially for thick flanges and short lengths; however, the flange should be checked for slenderness and member effects (refer to Sections 3.4.2 and 3.10.6.6).

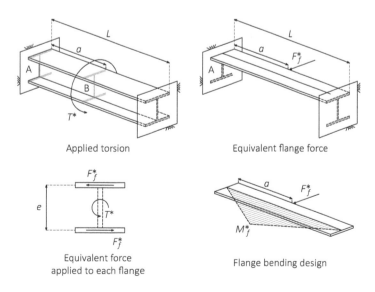

Figure 3.24 Warping torsion twin beam diagram (beam fixed at both ends).

$$M_f^* \le \phi M_f$$

$$\phi = 0.9 \quad \text{(members in bending)}$$

Flange force, $\qquad\qquad\qquad\qquad F_f^* = T^*/e$

Flange moment, $\qquad\qquad\qquad\quad M_f^* = F_f^* \times a \times (L-a)/L \quad$ (beam fixed at both ends)

$$M_f^* = F_f^* \times a \quad \text{(cantilever fixed at } A\text{)}$$

Flange bending capacity, $\qquad\quad \phi M_f = 0.9 \times f_{yf} t_f b_f^2 / 6 \quad$ (elastic capacity)

where:
 a = Distance from support to applied torsion
 b_f = Width of flange
 e = Distance between flanges $(d - t_f)$
 f_{yf} = Yield strength of flange
 t_f = Thickness of flange

Example 3.8: Uniform Torsion

Calculate the torsion capacity of a 168.3 × 6.4 CHS (Grade C350L0). Calculate the twist angle with a torsion of 20 kNm and a length of 3.0 m.

Torsion capacity (uniform torsion):

$$\phi T_u = 0.9 \times 0.6 \times f_y \times (T/\tau) = 0.9 \times 0.6 \times f_y \times C$$

$$f_y = 350 \text{ MPa}$$

$$r_o = D/2 = 168.3/2 = 84.15 \text{ mm}$$

$$r_i = r_o - t = 84.15 - 6.4 = 77.75 \text{ mm}$$

$$\tau = 2Tr_o / \pi(r_o^4 - r_i^4)$$

$$C = T/\tau = \pi(r_o^4 - r_i^4)/2r_o = (2 \times 84.15)/\pi(84.15^4 - 77.75^4) = 254 \times 10^3 \text{ mm}^3$$

$$\phi T_u = 0.9 \times 0.6 \times 350 \times 254 \times 10^3 = 48 \text{ kNm}$$

Twisting angle:

$$\theta = T^* L / KG$$

$$G \approx 80{,}000 \text{ MPa}$$

$$K = 0.5\pi(r_o^4 - r_i^4) = 0.5\pi(84.15^4 - 77.75^4) = 21.4 \times 10^6 \text{ mm}^4$$

$$L = 3000 \text{ mm}$$

$$\theta = (20 \times 10^6 \times 3{,}000)/(21.4 \times 10^6 \times 80{,}000) = 0.035 \text{ radians}$$

$$\theta = 0.035 \times 180/\pi = 2°$$

Example 3.9: Warping Torsion

Check a 150UC37 (Grade 300) with the torsion of 2.5 kNm, applied centrally on a 1 m long beam, fixed at both ends.

Section details:

$b_f = 154$ mm

$d = 162$ mm

$f_{yf} = 300$ MPa

$t_f = 11.5$ mm

$e = d - t_f = 162 - 11.5 = 150.5$ mm

Torsion check:

$$F_f^* = T^*/e = 2.5 \times 10^6/150.5 = 16.6 \text{ kN}$$

$$M_f^* = F_f^* \times a \times (L-a)/L = 33.2 \times 10^3 \times 500 \times (1000 - 500)/1000 = 4.2 \text{ kNm}$$

$$\phi M_f = 0.9 \times f_{yf} t_f b_f^2/6 = 0.9 \times 300 \times 11.5 \times 154^2/6 = 4.6 \text{ kNm}$$

The beam is sufficient to transfer the torsion load using only the elastic capacity of the flanges.

3.10 CONNECTIONS

It is necessary to define connections as either simple (pinned) or rigid (fixed) for the purpose of a structural analysis. Simple connections transfer shear and axial loads while allowing rotation. Rigid connections transfer shear, axial and bending loads. A semi-rigid connection may also be used, whereby a predictable level of rotation is determined; however, this is less common. These definitions are a simplification of the true behaviour in connections; however, this is a common and necessary step in the economic design of structures. The design of connections essentially uses the loads on a connection to determine forces in bolts, welds and plates. Each of these items is then checked individually. The Australian Steel Institute (ASI) produces a comprehensive set of connection design guides which provide standard connection geometry, theory and capacity tables [6,7]. Connections are more commonly designed either using simplified hand calculations or a software package such as Limcon.

3.10.1 Minimum actions

Connections are required to be designed for a minimum value of the calculated design load or a minimum design action. AS 4100, Clause 9.1.4 specifies minimum actions for different types of construction. The purpose is generally to provide structures which have a minimum capacity, even when lightly loaded. The minimum actions are calculated as fractions of the smallest member size required for strength limit states. Therefore, members which are chosen for other reasons (such as deflection limits) may be designed to smaller minimum actions. Also, they are in reference to member capacities, not section capacities; therefore, length effects may be taken into account (Table 3.17).

Table 3.17 Minimum actions

Type of construction	*Minimum action requirement*
Rigid (fixed moment connections)	50% of member design bending capacity
Simple (pinned connections)	Minimum of 15% shear capacity or 40 kN
Tension or compression members (braces or struts)	30% of member design axial capacity (100% for threaded rods with turnbuckles)
Splices (tension members)	30% of member design axial tension capacity
Splices (compression members, with ends prepared for full contact)	15% of member design axial compression capacity through fasteners (in addition to calculated design moment for second-order effects)
Splices (compression members, with ends not prepared for full contact)	30% of member design axial compression capacity (in addition to calculated design moment for second-order effects)
Splices (flexural members)	30% of member design bending capacity
Splices (shear)	Design shear force only (including any bending due to eccentricity of fasteners)
Splices (combined actions)	Each item above simultaneously

Source: Refer to AS 4100, Clause 9.1.4 for details on calculating design moments for splice connections.

3.10.2 Bolting

Australia uses two categories of bolts and three tensioning specifications. Category 4.6 is a low carbon steel bolt and should only be tensioned using the snug-tightened procedure. The supplement to AS 4100 defines the snug-tight condition as a bolt being tightened to the full capacity of a person on a podger spanner (ranging from 400 to 800 mm) or until plies are effectively contacted using a pneumatic impact wrench (at which point a change in pitch is heard). Category 8.8 bolts are made from a high-strength, heat-treated, medium carbon steel and therefore should not be welded without specialised procedures. Category 8.8 bolts may be tensioned via the snug-tightened procedure or fully tensioned using either the bearing (TB) or friction (TF) types. Fully tensioned bolts are tightened either by using the part turn method or by using direct tension indication (proprietary items). The only difference between bearing and friction types is that the contact surface for friction types is specified to assure a serviceability slip capacity. The surface should be 'as-rolled' to achieve standard values; testing should be completed for other surface conditions. Details, applications and capacities for each of the bolt types are outlined in Tables 3.18 through 3.21 and 3.24.

3.10.2.1 Bolt capacities

Bolts are required to be specified in terms of diameter, class and tensioning. Grade 8.8 bolts are high strength and low ductility; they are most commonly used for structural connections. Grade 4.6 bolts are comparatively lower strength and higher ductility, and they are used for base plates and minor connections such as handrails. Snug-tightened tensioning is the cheapest and most common bolt torque and is used for standard connections. Fully tensioned bearing types are commonly used as standard on mine sites, for structures with vibrations and for rigid connections (splices and bolted moment connections). Fully tensioned friction types are used when the friction capacity of a bolt needs to be ensured to prevent slippage at serviceability limit states.

Table 3.18 Bolt categories

Category	Class	Tensioning	Yield strength (MPa)	Tensile strength (MPa)	Application
4.6/S	4.6	Snug-tightened	400	240	Hold-down bolts for baseplates Attachments for girts/purlins, handrails, etc.
8.8/S	8.8	Snug-tightened	830	660	Simple (pinned) structural connections Web side plate, flexible end-plate, cleat connections
8.8/TB	8.8	Fully tensioned Bearing type	830	660	Rigid moment connections (splices, bolted end-plates) or equipment with vibrations
8.8/TF	8.8	Fully tensioned Friction type	830	660	Connections where bolt slip needs to be limited, such as movement-sensitive installations or equipment with vibrations

Table 3.19 4.6/S Strength bolting capacities

Bolt size	Axial tension ϕN_{tf} (kN)	Shear 4.6N/S (standard – threads included) ϕV_{fn} (kN)	Shear, 4.6X/S (non-standard –threads excluded) ϕV_{fx} (kN)
M12	27.0	15.1	22.4
M16	50.2	28.6	39.9
M20	78.4	44.6	62.3
M24	113	64.3	89.7
M30	180	103	140
M36	261	151	202

Table 3.20 8.8/S, 8.8/TB, 8.8/TF strength bolting capacities

Bolt size	Axial tension ϕN_{tf} (kN)	Shear 8.8N/S (standard – threads included) ϕV_{fn} (kN)	Shear, 8.8X/S (non-standard –threads excluded) ϕV_{fx} (kN)
M16	104	59.3	82.7
M20	163	92.6	129
M24	234	133	186
M30	373	214	291

Table 3.21 8.8/TF serviceability bolting capacities (slip capacity)

Bolt size	Axial tension ϕN_{ti} (kN)	Shear (standard holes) ϕV_{sf} (kN)	Shear (oversized holes) ϕV_{sf} (kN)	Shear (slotted holes) ϕV_{sf} (kN)
M16	66.5	23.3	19.8	16.3
M20	101	35.5	30.2	24.9
M24	147	51.5	43.7	36.0
M30	234	82.1	69.8	57.5

Note: Capacities are based on a friction slip factor of 0.35, requiring as-rolled contact between bolt and steel.

Bolts are required to be checked individually for shear and tensile loads, as well as with a combined action check. Loads on individual bolts are calculated by using a bolt group analysis. Design actions on bolts need to satisfy the following inequalities:

Shear, $\qquad V_f^* \le \phi V_f$

Tension, $\qquad N_{tf}^* \le \phi N_{tf}$

Combined, $\qquad \left[\dfrac{V_f^*}{\phi V_f}\right]^2 + \left[\dfrac{N_{tf}^*}{\phi N_{tf}}\right]^2 \le 1.0$

$\phi = 0.8$ (for bolts in shear and tension)

Bolt capacity is calculated in accordance with AS 4100:

Bolt shear capacity, $\qquad V_f = 0.62\, f_{uf} k_r (n_n A_c + n_x A_o)$

Bolt tension capacity, $\qquad N_{tf} = A_s f_{uf}$

where:
- A_c = Minor diameter area of bolt (AS 1275)
- A_o = Nominal plain shank area of bolt
- A_s = Tensile stress area of bolt (AS 1275)
- f_{uf} = Tensile strength of bolts
- k_r = Bolted lap reduction factor = 1.0 except as shown in Figure 3.25
- n_n = Number of shear planes with threads intersecting
- n_x = Number of shear planes without threads intersecting

The number of shear planes is one when two plates are bolted to each other as shown in Figure 3.25. It is two when there are two plates on one side sandwiching the inside plate. Tables 3.19 through 3.21 provide bolt capacities based on a single shear plane. The capacity of the attaching plate to transfer the load also needs to be checked (including ply in bearing for shear loads); refer to Section 3.10.6.

3.10.2.1.1 Lapped connections

Bolted lap connections (bracing cleats and bolted flange splices) with joint lengths (l_j) greater than 300 mm are required to have the bolt shear capacity reduced by k_r. This is to account for uneven shear distribution, which occurs in connections where plates behave in a more flexible manner.

$k_r = \text{MIN}[1.075 - l_j/4000,\quad 0.75]$

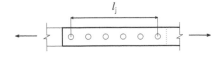

Figure 3.25 Bolted lap connections.

3.10.2.2 Bolt group analysis

To determine whether a bolted connection is capable of transferring a design load, the force on each individual bolt should be calculated. The loads on a bolt group may consist of in-plane actions (shear loads) and out-of-plate actions (axial loads).

3.10.2.2.1 Shear loading on bolt groups

If a shear force is applied to the centroid of a bolt group, the shear load is distributed evenly between the bolts. This is a common scenario for simple connections, especially in bracing members or simply supported beams. Shear forces in vertical and horizontal directions are then calculated, and the resultant is taken as the design force in each bolt.

Design shear in a single bolt, $\qquad V_f^* = \sqrt{V_y^{*2} + V_x^{*2}}$ (for shear applied at the centroid)

where:

Vertical shear in a single bolt, $\qquad V_y^* = V_{by}^*/n_b$

Horizontal shear in a single bolt, $\qquad V_x^* = V_{bx}^*/n_b$

n_b = Number of bolts in group
V_{bx}^* = Horizontal shear load applied to bolt group
V_{by}^* = Vertical shear load applied to bolt group

For cases where a shear load is applied with a torsion load (or the shear load is applied at an eccentricity to the centroid of the bolt group), the load will vary in each of the bolts. The load is dependent on the distance of the bolt to the centroid of the bolt group. The torsion in the group is equal to the shear load multiplied by the eccentricity (Figure 3.26).

Design shear in a single bolt, $\qquad V_f^* = \sqrt{V_y^{*2} + V_x^{*2}}$ (for shear applied at an eccentricity)

where:
Vertical shear in a single bolt, $\qquad V_y^* = V_{by}^*/n_b + T_b^* x_n/I_{bp}$
Horizontal shear in a single bolt, $\qquad V_x^* = V_{bx}^*/n_b + T_b^* y_n/I_{bp}$

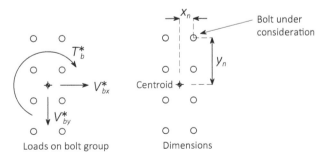

Figure 3.26 Shear on bolt groups.

$$I_{bp} = \sum (x_n^2 + y_n^2)$$

T_b^* = Torsion applied to bolt group
x_n = Horizontal distance from bolt to centroid of bolt group
y_n = Vertical distance from bolt to centroid of bolt group

For shear loading on base plates, it is commonly recommended that the load is spread between half of the bolts. This is because bolt holes are typically 6 mm oversized, and therefore, some bolts may not be engaged. Many clients specify that the full shear load should be able to be carried by only two bolts, and some clients even limit the load to one bolt (refer to Section 3.10.3.3 for details).

3.10.2.2.2 Axial loading on bolt groups

Pure tensile loading on connections is assumed to be distributed evenly between bolts (in the same way as shear). Bending-induced axial forces are calculated using a plastic distribution for bolts located symmetrically about a section (Figure 3.27). Therefore, all bolts above the neutral axis are assumed to take the same load. An additional allowance on each of the calculated loads should be made to account for prying forces, especially when using thin plates.

Tension in a single bolt, $\qquad N_{tf}^* = p \, (N_{ta}^* + N_{tb}^*)$

where:

Tension due to axial loading, $\quad N_{ta}^* = N_{tb}^*/n_b$

Tension due to bending, $\qquad N_{tb}^* = M_b^*/(n_t y_m)$

M_b^* = Bending moment applied to bolt group
N_{tb}^* = Tensile load applied to bolt group
n_t = Number of bolts in tensile zone (bolts on neutral axis are excluded)
p = Prying allowance (see following paragraph)
y_m = Lever arm (between centre of tension and compression zones)
$\quad = (n_p + 1)s_p/2$ (for an odd number of bolt rows)
$\quad = (n_p s_p)/2$ (for an even number of bolt rows)
n_p = Number of bolt rows
s_p = Spacing of bolt rows

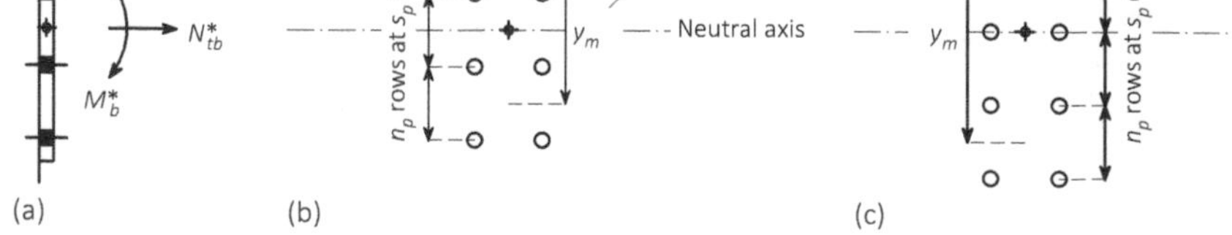

Figure 3.27 Tension on bolt groups. (a) Loads on bolt group, (b) even rows, (c) odd rows.

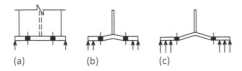

Figure 3.28 Prying examples. (a) Rigid, (b) thick and (c) thin prying.

Prying is caused by the flexural displacement of the connecting plate, resulting in compression at the edges (refer to Figure 3.28). The compressive force on the plate is balanced by increased tension in the bolts. Prying is therefore dependent on the flexural stiffness of the connecting plate. It can be limited by increasing the plate thickness (typically 1.25 times the bolt diameter) and by adopting minimum edge distances for bolts. A prying allowance of 10% is recommended for thick plates, or stiff arrangements; 20%–40% is recommended for thinner, more flexible arrangements. AS 4100 provides no guidance on calculating prying forces.

$p = 110\%$ (rigid connections or thick plates)

$p = 120\%$ to 140% (flexible connections or thin plates)

FEA (refer to Section 6.11) can be used to calculate prying forces using 'compression-only' elements under the plate. The bolt and plate calculations should both include the prying allowance.

A plastic approach may be adopted instead of a typical bolt group analysis. The tension may be distributed among the bolts on the tensile side of the neutral axis, and the shear forces are distributing among the bolts on the compression side.

3.10.2.3 General bolting requirements

General bolting requirements are concerned with hole size, edge distance and bolt spacing to allow bolts to be installed and remain effective. Bolting requirements are set out in AS 4100, Clauses 9.6 and 14.3.

3.10.2.3.1 Holing

Hole sizes are typically 2 mm larger than the bolt for steel to steel connections (M12 to M24) and 6 mm larger for base plates. Oversized or slotted holes may be used if required; however, large washers are required to be installed, so that the holes are covered. Refer to Table 3.22 for a full list of standard hole sizes.

3.10.2.3.2 Edge distances

Minimum edge distances for bolts are dependent on the diameter of the bolt and the construction method of the edge. For normal bolt holes, the distance is measured from the centre of the bolt hole. For oversized or slotted holes, the distance is measured from the centre of the bolt when located as close as possible to the edge. Minimum edge distances are provided in Table 3.23.

Maximum edge distance is limited to 12 times the thickness of the thinnest outer connected ply, or 150 mm, whichever is smaller.

Table 3.22 Standard hole sizes

Bolt size	Normal hole size	Base plate hole size	Short slotted hole length	Long slotted hole length	Oversized hole diameter
M12	14	18	22	30	20
M16	18	22	26	40	24
M20	22	26	30	50	28
M24	26	30	34	60	32
M30	33	36	40	75	38
M36	39	42	48	90	45

Notes:
1. Slotted holes are the same width as normal holes; however, the length is increased.
2. Plate washers shall be used under the nut for holes 3 mm or greater in diameter than the bolt size. The minimum distance from the edge of the washer to the edge of the hole is 50% of the hole diameter. The minimum thickness of plate washers is 4 mm for oversized holes and 8 mm for slotted holes.
3. Refer to AS 4100, Clause 14.3.5 for further details.

Table 3.23 Minimum edge distance

Sheared or hand flame-cut edge	Rolled plate, flat bar or section: Machine-cut, sawn or planed edge	Rolled edge of a rolled flat bar or section
$1.75\ d_f$	$1.50\ d_f$	$1.25\ d_f$

Source: AS 4100, Table 9.6.2.

Table 3.24 Bolt dimensions

Bolt size		Height of head	Height of nut		Width of nut across corners		Washer diameter		Washer thickness	
Class	Thread pitch		4.6	8.8	4.6	8.8	4.6	8.8	4.6	8.8
M12	1.75	8	11	11	21	21	24	24	2.5	2.5
M16	2	11	15	17	28	31	30	34	3	4
M20	2.5	13	18	21	35	39	37	42	3	4
M24	3	16	21	25	42	47	44	50	4	4
M30	3.5	20	26	31	53	58	56	60	4	4
M36	4	24	31	27	63	69	66	72	5	4

3.10.2.3.4 Bolt spacing

The minimum pitch (spacing) of bolts is typically 2.5 times the fastener diameter for the purposes of installing the bolts. The maximum pitch is 12 times the thickness of the thinnest connected ply, or 200 mm, whichever is smaller. Where the fastener does not carry design actions and is not subject to corrosion, the maximum pitch may be increased to the minimum of 32 times the plate thickness or 300 mm. For fasteners on the outside of members, in the direction of the design load, the maximum pitch may be increased to four times the plate thickness plus 100 mm, or 200 mm, whichever is smaller.

Gauge lines are typically used to detail bolt centres. Refer to Table 3.24 for bolt dimensions and Section 7.3 for recommended gauge lines based on section size (Figure 3.29).

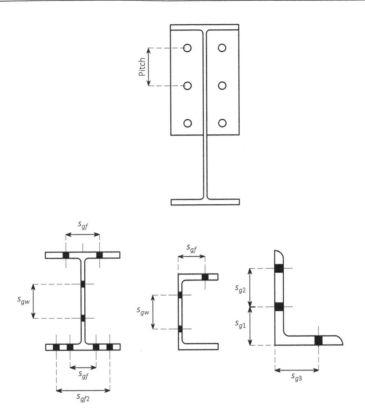

Figure 3.29 Gauge line diagram.

Example 3.10: Bolt Group

Calculate bolt forces for the connection below with a rigid prying assumed, then specify an appropriate bolt:

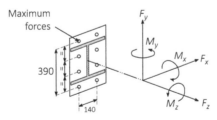

$F_x^* = -5$ kN

$F_y^* = -40$ kN

$F_z^* = 50$ kN

$M_x^* = 150$ kNm

$M_y^* = 10$ kNm

$M_z^* = 5$ kNm

Bolt group properties:

$$n_b = 8$$

$$x_n = -140/2 = -70 \text{ mm} \quad \text{(for bolt in top left corner)}$$

$$y_n = -360/2 = -195 \text{ mm} \quad \text{(for bolt in top left corner)}$$

$$p = 110\% \quad \text{(for thick prying)}$$

$$I_{bp} = \sum (x_n^2 + y_n^2)$$

$$= 4 \times \left(\frac{140}{2}\right)^2 + 4 \times \left(\frac{-140}{2}\right)^2 + 2 \times \left(\frac{130}{2}\right)^2 + 2 \times \left(\frac{-130}{2}\right)^2 + 2 \times \left(\frac{390}{2}\right)^2 + 2 \times \left(\frac{-390}{2}\right)^2$$

$$= 208,200 \text{ mm}^2$$

Shear loading on bolt group:

$$V_y^* = V_{by}^*/n_b + T_b^* x_n/I_{bp} = 40,000/8 + (-5 \times 10^6) \times (-70)/208,200 = 6.68 \text{ kN}$$

$$V_x^* = V_{bx}^*/n_b + T_b^* y_n/I_{bp} = (-5,000)/8 + (-5 \times 10^6) \times (195)/208,200 = -5.31 \text{ kN}$$

$$V_f^* = \sqrt{V_y^{*2} + V_x^{*2}} = \sqrt{6.68^2 + (-5.31)^2} = 8.53 \text{ kN}$$

Axial loading on bolt group:

Bending about major axis,

$$n_p = 4$$

$$s_p = 130 \text{ mm}$$

$$y_m = (n_p s_p)/2 = (4 \times 130)/2 = 260 \text{ mm}$$

$$n_t = 4$$

$$N_{tb(\text{major})}^* = M_b^*/(n_t y_m) = 150 \times 10^6/(4 \times 260) = 144.2 \text{ kN}$$

Bending about minor axis,

$$n_p = 2$$

$$s_p = 140 \text{ mm}$$

$$y_m = (n_p s_p)/2 = (2 \times 140)/2 = 140 \text{ mm}$$

$$n_t = 4$$

$$N_{tb(\text{minor})}^* = M_b^*/(n_t y_m) = 10 \times 10^6/(4 \times 140) = 17.9 \text{ kN}$$

Total axial force due to bending moments,

$$N_{tb}^* = 144.2 + 17.9 = 162.1 \text{ kN}$$

Axial tension due to pure tension load,

$$N_{ta}^* = N_{tb}^*/n_b = 50/8 = 6.25 \text{ kN}$$

Axial force in bolt,

$$N_{tf}^* = p \ (N_{ta}^* + N_{tb}^*) = 110\% \ (6.25 + 104.4) = 185.2 \text{ kN}$$

Selecting an M24, Grade 8.8/S,

Shear,

$$V_f^* \leq \phi V_f \qquad 8.53 \text{ kN} \leq 133 \text{ kN} \quad \text{Pass (ratio} = 15.6)$$

Tension,

$$N_{tf}^* \leq \phi N_{tf} \quad 185.2 \text{ kN} \leq 234 \text{ kN} \quad \text{Pass (ratio} = 1.26)$$

Combined,

$$\left[\frac{V_f^*}{\phi V_f}\right]^2 + \left[\frac{N_{tf}^*}{\phi N_{tf}}\right]^2 \leq 1.0 \quad \left[\frac{8.53}{133}\right]^2 + \left[\frac{185.2}{234}\right]^2 = 0.63 \leq 1.0 \quad \text{Pass (ratio} = 1.59)$$

The M24 Gr 8.8/S is sufficient and is governed by tensile loading, with a ratio of 1.26. Note: Plate thickness should be designed for 111% of the bolt capacity to ensure prying assumption.

3.10.3 Anchor bolts

The ASI adopts the American Institute of Steel Construction (AISC) design principles for the calculation of anchor bolt capacities [4,6]. Anchor bolts are typically constructed using category 4.6/S bolts because of the higher ductility; however, threaded rod (Grade 250) may also be used.

Cast-in anchors are typically used for new construction, because they are cheaper. Post-installed anchors are generally only used for existing concrete surfaces or for areas where bolt layouts are complex. Cast-in anchors are designed in accordance with the details presented within this section; refer to Table 3.25 for typical values. Post-installed anchors, such as chemical anchors and mechanical anchors, are designed by following vendor guides (such as Hilti; refer to Chapter 8). Post-installed Hilti anchors are typically grade 5.8.

3.10.3.1 Grout

Grout is used under most base plates to ensure that steelwork is installed at the correct level. Construction tolerances for footings are typically governed by a construction specification (15 or 20 mm is a commonly used limit). Grout thickness is designed to ensure a minimum vendor specified thickness is achieved when tolerances are at a maximum (values typically range from 30 to 50 mm; refer to Chapter 8 for grout suppliers). Cementitious grout is generally used, because it is cheaper. However, epoxy grout is more flexible, durable and waterproof, and can be used for greater thicknesses and temperatures. Epoxy grout is commonly used for locations with dynamic loading and has been adopted for standard applications by some clients within the oil and gas industry.

Table 3.25 Typical anchor bolt and plate details

Bolt diameter	Embedment, h_{ef}	Plate width, B	Plate thickness, t	Weld size	Min spacing, s	Min edge, c_1	Protrusion, P	ϕN_{tf} (kN)	ϕV_f (kN)
12	150	50	10	6	100	130	120	27	15.1
16	225	50	12	6	150	175	130	50.2	28.6
20	325	50	16	6	150	225	135	78.4	44.6
24	425	50	16	8	150	275	155	113	64.3
30	550	75	20	10	200	350	185	180	103
36	635	75	25	12	400	390	200	261	151

Notes:
1. Refer to Anchor (Plate) details shown in Figure 3.30.
2. Values are selected to provide full anchor capacity for single bolts and groups of up to four bolts; with bolt grade 4.6, weld electrode $f_{uf} = 490$ MPa, concrete $f'_c = 32$ MPa and plate $f_y = 250$ MPa.
3. Minimum bolt spacing of s is applicable for groups of four bolts spaced in each direction to provide full tension or shear capacity.
4. Minimum edge distance of c_1 is applicable for shear capacity only.
5. Combined shear and tension should be checked for anchors with high combined loading.
6. Shear capacity should be reduced to 80% where a grout pad is used.
7. Units are in millimetres unless noted otherwise.

The specification of grout may lead to bending being induced in anchor bolts. Tension calculations should exclude any contribution from grout, and shear capacities should be reduced to 80% when using grout. For cases where steelwork levels are insignificant, and water will not pool at the base plate, columns may be installed directly onto concrete surfaces without grout.

The protrusion (P) for anchor bolts (refer to Figure 3.30) should be selected by adding the construction tolerances to the grout thickness, plate thickness, washer thickness, nut height and two threads, and then rounding up to the nearest 5 or 10 mm.

3.10.3.2 Tension in anchor bolts

Typical anchor bolt details should include a bolt head, nut or end plate to ensure bolt slip does not occur (refer to Figure 3.30). The tensile strength is then governed by either the bolt capacity or the break-out and pull-out capacities. Standard details are typically used for anchor bolt designs; therefore, the bolt size is usually selected (refer to Section 3.10.2.1), and the embedment depth is detailed for the full fastener capacity. Further checks are required when adjacent fastener bolts are in tension, because the break-out zones may overlap.

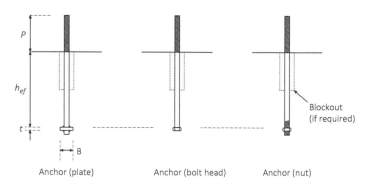

Figure 3.30 Typical anchor bolt details.

The capacity of a group of anchor bolts in tension shall satisfy

$$N_t^* \leq \text{MIN}[\phi N_{tf} \times n_b, \phi N_{cb}, \phi N_p]$$

Prying is often ignored in anchor bolt design, because plates are typically thicker, and the bolts have increased ductility resulting from the Grade 4.6 selection and the typically long embedment lengths. For cases where this may not be applicable, prying may be calculated in accordance with Section 3.10.2.2 for bolt sizing only (not necessary for embedment calculations).

3.10.3.2.1 Tension break-out capacity

Break-out capacity is calculated by using a pyramid failure shape (this has superseded the cone shape due to simplicity). The surface area of the break-out zone is the critical element for design (refer to Figure 3.31). It is calculated using an angle of 1:1.5 and is limited by adjacent edges and overlapping of adjacent bolts. This method should not be used for bolts which exceed 50 mm in diameter or embedment depths which exceed 635 mm, or where edge distances are less than six times the bolt diameter.

Break-out capacity,

$$\phi N_{cb} = \phi \psi_3 10.1 \sqrt{f_c'} h_{ef}^{1.5} \frac{A_N}{A_{N0}} \times 10^{-3} \qquad \text{where:} \quad h_{ef} < 280 \text{ mm}$$

$$\phi N_{cb} = \phi \psi_3 3.9 \sqrt{f_c'} h_{ef}^{5/3} \frac{A_N}{A_{N0}} \times 10^{-3} \qquad \text{where:} \quad 280 \text{ mm} \leq h_{ef} \leq 635 \text{ mm}$$

where:
ϕ = 0.7
ψ_3 = 1.0 (for cracked concrete)
      = 1.25 (for uncracked concrete)
A_N = Concrete break-out surface area for anchor group
A_{N0} = Concrete break-out surface area for single anchor = $3h_{ef} \times 3h_{ef} = 9h_{ef}^2$
f_c' = Concrete compressive strength
h_{ef} = Effective embedment depth

The value for A_N is calculated by using the same break-out angle and by incorporating the effects of edge distances and bolt spacing (refer to Figure 3.32). For a single bolt with no concrete edges, $A_N/A_{N0} = 1$.

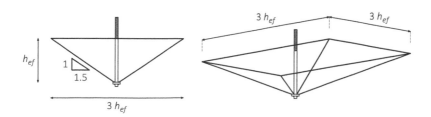

Figure 3.31 Anchor break-out capacity.

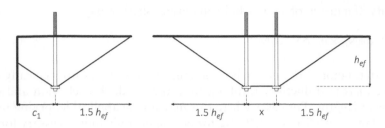

Figure 3.32 Geometry for anchor pull-out.

The tensile capacity due to concrete break-out may be increased by lapping of reinforcement into the cone area. The capacity is increased by $\phi = 0.9$ times the steel area; however, development lap lengths are required as per Section 4.1.5.

3.10.3.2.2 Tension pull-out capacity

Pull-out is where the concrete fails immediately around the bolt by crushing around the bolt plate, head or nut.

Pull-out capacity, $\qquad \phi N_p = n_b \times (\phi \psi_4 A_{brg} 8 f_c' \times 10^{-3})$

where:

$\qquad \phi \quad = 0.7$
$\qquad \psi_4 \quad = 1.0$ (for cracked concrete)
$\qquad \qquad = 1.4$ (for uncracked concrete)
$\qquad A_{brg} =$ The bearing area of the bolt plate, head or nut
$\qquad n_b \quad =$ Number of anchor bolts in a group

3.10.3.3 Shear in anchor bolts

Shear forces may be restrained by plate friction, shear keys or direct anchor bolt loading. Plate friction can only be used where a compressive force is consistently present. Shear keys may be welded to the underside of the base plate; however, careful consideration needs to be given to the reinforcement detailing. Direct anchor bolt loading is the most common method of shear transfer; however, the load is typically assumed to act on only half of the anchor bolts due to the size of the bolt holes. Many clients specify that the shear load should be carried by only two bolts, and some even limit the load to one bolt. This author recommends either two bolts or an alternate method of engaging bolts.

The number of engaged bolts may be increased by welding smaller plates with holes only 1 mm larger than the bolt on top of each anchor bolt during erection (refer to Figure 3.58); however, this leads to induced bending in the anchor bolts and should be checked with a double curvature assumed (i.e. a moment equal to the shear force multiplied by half the distance from the plate to the concrete). AISC recommends a further reduction to 80% of the bolt capacity in shear when grout is used to account for bending effects in the bolt.

The shear strength is governed by either the bolt capacity or the break-out capacity. The bolt capacity should be selected using the 'thread included' values. The capacity of a group of anchor bolts in shear shall therefore satisfy

$$V_f^* \leq \text{MIN}[\phi V_f \times n_{bv} k_g, \phi V_{cb}]$$

k_g = 0.8 (when using a grout pad) = 1.0 (when not using grout)

n_{bv} = Number of bolts assumed to carry shear (recommended limit of 2)

3.10.3.3.1 Shear break-out capacity

Break-out capacity,

$$\phi V_{cb} = \phi \times 1.23 \times 10^{-3} (d_f)^{0.5} (c)^{1.5} \sqrt{f'_c} \left(\frac{A_v}{A_{vo}} \right) \psi_6$$

where:

ϕ = 0.7

Ψ_6 = Edge distance reduction factor

= 1.0 (when $c_1 \geq \frac{2}{3} \times$ concrete thickness)

(no guidance provided for cases where edge distance limits break-out cone size)

c = Edge distance in direction of shear force

= c_1 (for near edge bolt) = $c_1 + s$ (for far edge bolt)

A_v = Projected area for group of bolts on side of concrete face in direction of shear force

(refer to Figure 3.34)

A_{vo} = Projected area for single bolt on side of concrete face in direction of shear force

= $4.5 \, c_1^2$ (where edge distance is not limited by depth; refer to Figures 3.33 and 3.34)

d_f = Diameter of bolt (using a maximum value of 30 mm)

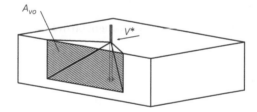

Figure 3.33 Single bolt shear break-out failure surface.

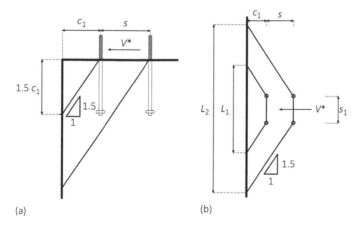

Figure 3.34 Bolt group break-out failure surfaces. (a) Elevation of shear failure plane. (b) Plan of shear failure plane.

Break-out capacity should be checked for two failure planes:

1. Half the total load on the failure plane caused by the bolts closest to the concrete face

$$A_v = L_1 \times 1.5c_1 \qquad L_1 = s_1 + 3c_1$$

2. Full load on the failure plane caused by the bolts farthest from the concrete face

$$A_v = L_2 \times 1.5(c_1 + s) \qquad L_2 = s_1 + 3(c_1 + s)$$

This method should not be used for embedment lengths which are less than eight times the bolt diameter. The concrete should either be uncracked or have adequate supplementary reinforcement.

3.10.3.4 Combined tension and shear in anchor bolts

Anchor bolts which are subject to combined axial tension and shear should be checked for each load separately as well as undergoing a combined check on the bolt. There is no need to check for a combined concrete failure capacity.

$$\left[\frac{V_f^*}{\phi V_f \times n_{bv} k_g} \right]^2 + \left[\frac{N_t^*}{\phi N_t \times n_b} \right]^2 \leq 1.0$$

Refer to Sections 3.10.3.2 and 3.10.3.3 for nomenclature.

3.10.4 Pin connections

Pin connections (such as a clevis or a locking pin) are checked in a similar manner to bolts, although pin bending and pin bearing capacities are also calculated. Bending is relevant only for pin connections where the pin has potential to fail in bending (or double-bending). Pin bearing needs to be checked to ensure that the pin does not fail locally under bearing loads. The capacity of the attaching plate to transfer the load also needs to be checked (including ply in bearing for shear loads); refer to Section 3.10.6. Calculations vary from bolts as shown; note that the shear capacity is based on the yield strength rather than the tensile strength for pins.

Shear, $V_f^* \leq \phi V_f$

Bearing, $V_b^* \leq \phi V_b$

Bending, $M^* \leq \phi M_p$

$\phi = 0.8$ (for pins in shear, bearing and bending)

Pin capacity is calculated in accordance with AS 4100,

Pin shear capacity, $\qquad V_f = 0.62 \, f_{yp} n_s A_p$

Pin bearing capacity, $\qquad V_b = 1.4 \, f_{yp} d_f t_p k_p$

Pin tension capacity, $\qquad N_{tf} = f_{yp} S$

where:
$\quad A_p \quad$ = Cross-sectional area of pin
$\quad d_f \quad$ = Diameter of pin
$\quad f_{yp} \quad$ = Yield strength of pin
$\quad k_p \quad$ = 1 (for pins without rotation), 0.5 (for pins with rotation)
$\quad n_s \quad$ = Number of shear planes
$\quad S \quad$ = Plastic section modulus (refer to Table 7.2)
$\qquad$ = $4r^3/3 = d_f^3/6 \quad$ (circle)

$\qquad$ = $(bh^2)/4$ (rectangle)
$\quad t_p \quad$ = Thickness of connecting plate(s)

A standard clevis arrangement has two shear planes (as shown in Figure 3.35). This means that the pin capacity is effectively doubled ($n_s = 2$) and compared with the full applied load(V*). The moment in the pin is calculated using force diagrams (refer to Section 7.2) and is equal to $V^* e/4$. Refer to Section 6.3 for details on plate calculations required for a lug, which are similar for a clevis. Pin bearing needs to be checked for the full shear load (V*) applied to the central plate as well as half the shear load applied to each of the external plates (assuming the central plate is located at an equal distance from each of the external plates).

3.10.5 Welding

Two types of welds are typically used in structural engineering: fillet welds and butt welds. Fillet welds are cheaper, because no edge preparation is required to the mating surfaces;

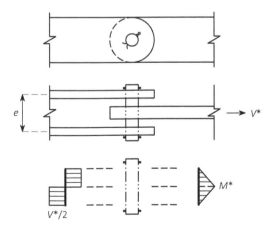

Figure 3.35 Clevis pin forces.

however, they can only be used where the connecting elements are at an angle to each other. The weld can be created in a single pass for sizes up to 8 mm, therefore resulting in an increased cost for 10 and 12 mm fillets. Butt welds cost more, primarily because of the surface preparation; however, increased inspection requirements also lead to increased cost. The butt weld is a much more versatile weld and can be used to connect more complex elements, such as straight edges.

3.10.5.1 Weld capacities

General purpose (GP) fillet welds are cheaper than structural purpose (SP) fillet welds; however, they are not commonly used for structural applications. GP welds have a lower inspection rate and a higher percentage of permitted imperfections. The capacity reduction factor for a GP fillet weld is 0.6; however, it increases to 0.8 for SP fillet welds.

The tensile strength of weld metals increased in the first amendment to AS 4100, 2012. New structures should comply with the increased values. Both the new and the old values are provided for the purpose of assessing old structures. Refer to Table 3.26 for GP and SP fillet weld capacities.

E48XX/W50X were previously the most commonly used values; therefore, it is expected that the 490 MPa designation will be the common choice when complying with the new amendment (the value to remember is that a 6 mm SP fillet weld can resist 1 kN/mm).

The tabulated values can be calculated by hand using the following equation. The same procedure can be used for calculating capacities for incomplete penetration butt welds; however, the throat thickness should be calculated in accordance with AS 4100, Clause 9.7.2.3.

$$\phi v_w = \phi 0.6 f_{uf} t_t k_r$$

where:

f_{uf} = Tensile strength of weld metal

k_r = Reduction factor for lap splice connections

    = 1.0 (typically for all welds, except as below)

    = $1.10 - 0.06 l_w$ (for lap splices, 1.7 m $\leq l_w \leq$ 8.0 m)

    = 0.62 (for lap splices, l_w > 8.0 m)

l_w = Welded lap splice length

t_t = Throat thickness (refer to Figure 3.36)

    = $t_w / \sqrt{2}$ (for equal leg fillet welds)

t_w = Leg length

ϕ = 0.8 (for SP fillet welds)

    = 0.6 (for GP fillet welds)

Incomplete penetration butt welds are designed with the cross-section reduced by 3 mm when the angle of preparation is less than 60°. Complete penetration (CP) butt welds are required to have a minimum strength equal to that of the smaller of the connecting sections. A capacity reduction factor of 0.9 is relevant for (SP) butt welds. The design is typically ignored for CP butt welds, as it does not govern.

Table 3.26 Fillet weld design capacities

Weld size (Leg length) (mm)	Weld capacity, ϕV_w (kN/mm)				
	New structures AS 4100 Amendment 1 (2014)			Previous structures AS 4100 (1990 and 1998 editions)	
Electrode	Refer to Notes 2 and 3			E41XX/W40X	E48XX/W50X
Tensile strength, f_{uf}	430 MPa	490 MPa	550 MPa	410 MPa	480 MPa
General purpose welds (GP)					
3	0.329	0.374	0.420	0.313	0.367
4	0.438	0.499	0.560	0.417	0.489
5	0.547	0.624	0.700	0.522	0.611
6	0.657	0.746	0.840	0.626	0.733
8	0.876	0.998	1.12	0.835	0.978
10	1.09	1.25	1.40	1.04	1.22
12	1.31	1.50	1.68	1.25	1.47
Structural purpose welds (SP)					
3	0.438	0.499	0.560	0.417	0.489
4	0.584	0.665	0.747	0.557	0.652
5	0.730	0.832	0.933	0.696	0.815
6	0.876	0.998	1.12	0.835	0.978
8	1.17	1.33	1.49	1.11	1.30
10	1.46	1.66	1.87	1.39	1.63
12	1.75	2.00	2.24	1.67	1.96

Notes:
1. Check project specifications to ensure new electrodes or tensile strengths have been updated to comply with AS 4100, Amendment 1. Adopting previous values is conservative.
2. A wide range of electrodes are now available; refer to AS 4100, Amendment 1, Table 9.7.3.10(1) for a complete list.
3. It is now recommended that the weld is specified in relation to the tensile strength of the weld metal, rather than the electrode.
4. Refer to ASI Technical Note TN008 V1, February 2012 for further information on changes in welding electrodes.
5. Capacities are required to be reduced for welded lap splice connections with lengths greater than 1.7 m; refer to the calculations method (Section 3.10.5.1).

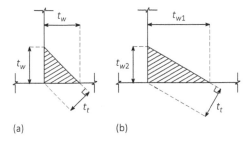

Figure 3.36 Fillet weld dimensions. (a) Equal leg fillet weld. (b) Unequal leg fillet weld.

3.10.5.2 Weld group analysis

A weld group analysis is used to calculate the force per unit length on a welded connection. The theory is similar to an elastic analysis of a section in bending; however, the thickness of each weld is taken as unity. The calculated force is then compared with the capacities shown in Table 3.26 or calculated manually.

$$V_{res}^* \leq \phi V_w$$

Weld group requirements are set out in AS 4100, Clause 9.8, and are specific to fillet welds. The force is calculated in each direction, and then the magnitude is resolved.

In-plane force in x-direction,
$$V_x^* = \frac{F_x^*}{L_w} - \frac{M_z^* y_s}{I_{wp}}$$

In-plane force in y-direction,
$$V_y^* = \frac{F_y^*}{L_w} + \frac{M_z^* x_s}{I_{wp}}$$

Out-of-plane force in z-direction,
$$V_z^* = \frac{F_z^*}{L_w} + \frac{M_x^* y_s}{I_{wx}} - \frac{M_y^* x_s}{I_{wy}}$$

Resultant,
$$V_{res}^* = \sqrt{V_x^{*2} + V_y^{*2} + V_z^{*2}}$$

where:

$F_{x,y,z}^*$ = Force applied to weld group in x, y and z directions (respectively)
I_{wp} = Polar moment of area = $I_{wx} + I_{wy}$
I_{wx} = Second moment of area about x-axis
I_{wy} = Second moment of area about y-axis
L_w = Total length of weld
$M_{x,y,z}^*$ = Moment applied to weld group about x, y and z axis (respectively)
$V_{x,y,z}^*$ = Design force in weld for x, y and z directions (respectively)
x_s = Horizontal distance from centroid of group to weld under consideration
y_s = Vertical distance from centroid of group to weld under consideration

The second moment of area needs to be calculated using the weld group geometry. Refer to Section 3.4.1.1 for the elastic theory; the only variation is that a weld is provided on each side of the plate (however, measurements are generally to the centre line) and the thickness is taken as unity. Typical values for fully welded shapes are provided in Table 3.27, and an example calculation is provided. The typical values are simplified to remove components which have insignificant effects. Maximum stresses occur at the maximum distance from the centroid (X_{max} and Y_{max}). Alternatively, a plastic approach may be adopted, distributing the moment between flanges and the shear to the web. Compression should be assumed to be transferred through welds unless connections are prepared for full contact.

Table 3.27 Weld group properties

Weld group	I_{wx}	I_{wy}	x dimensions	y dimensions
	$\dfrac{d^3}{12}$	0	$\bar{x} = x_{max} = 0$	$\bar{y} = y_{max} = \dfrac{d}{2}$
	$\dfrac{d^3}{6}$	$\dfrac{db^2}{2}$	$\bar{x} = x_{max} = \dfrac{b}{2}$	$\bar{y} = y_{max} = \dfrac{d}{2}$
	$\dfrac{d^3}{6} + \dfrac{bd^2}{2}$	$\dfrac{b^3}{6} + \dfrac{db^2}{2}$	$\bar{x} = x_{max} = \dfrac{b}{2}$	$\bar{y} = y_{max} = \dfrac{d}{2}$
	$\dfrac{\pi d^3}{8}$	$\dfrac{\pi d^3}{8}$	$\bar{x} = x_{max} = \dfrac{d}{2}$	$\bar{y} = y_{max} = \dfrac{d}{2}$

(Continued)

Table 3.27 (Continued) Weld group properties

Weld group	I_{wx}	I_{wy}	x dimensions	y dimensions
I-section diagram	$bd^2 + \dfrac{d^3}{6}$	$\dfrac{b^3}{3}$	$\bar{x} = x_{max} = \dfrac{b}{2}$	$\bar{y} = y_{max} = \dfrac{d}{2}$
C-section diagram	$bd^2 + \dfrac{d^3}{6}$	$\dfrac{2b^4 + 4db^3}{6b + 3d}$	$\bar{x} = \dfrac{b^2}{2b + d}$ $x_{max} = \dfrac{b^2 + db}{2b + d}$	$\bar{y} = y_{max} = \dfrac{d}{2}$
channel diagram	$2b\bar{y}^2 + \dfrac{d^3}{6}$ $+ 2d\left(\dfrac{d}{2} - \bar{y}\right)^2$	$2d\bar{x}^2 + \dfrac{b^3}{6}$ $+ 2b\left(\dfrac{b}{2} - \bar{x}\right)^2$	$\bar{x} = \dfrac{b^2}{2(b + d)}$ $x_{max} = b - \bar{x}$	$\bar{y} = \dfrac{d^2}{2(b + d)}$ $y_{max} = d - \bar{y}$

Note: Values are approximated using centre lines for each plate.

Example 3.11: Weld Group

Calculate weld forces for the fully welded section below, then specify an appropriate fillet weld:

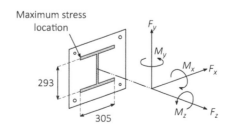

$$F_x^* = 16 \text{ kN}$$

$$F_y^* = 40 \text{ kN}$$

$$F_z^* = 20 \text{ kN}$$

$$M_x^* = 100 \text{ kNm}$$

$$M_y^* = 40 \text{ kNm}$$

$$M_z^* = -20 \text{ kNma}$$

Weld group properties:

$$I_{wx} = bd^2 + \frac{d^3}{6} = 305 \times 293^2 + \frac{293^3}{6} = 30,376,238 \text{ mm}^3$$

$$I_{wy} = \frac{b^3}{3} = \frac{305^3}{3} = 9,457,542 \text{ mm}^3$$

$$I_{wp} = I_{wx} + I_{wy} = 39,833,780 \text{ mm}^3$$

$$L_w = 4b + 2d = 4 \times 305 + 2 \times 293 = 1806 \text{ mm}$$

Location of maximum stress:

$$x_s = -x_{\max} = \frac{-b}{2} = \frac{-305}{2} = -152.5 \text{ mm}$$

$$y_s = y_{\max} = \frac{d}{2} = \frac{293}{2} = 146.5 \text{ mm}$$

Force on weld:

$$V_x^* = \frac{F_x^*}{L_w} - \frac{M_z^* y_s}{I_{wp}} = \frac{16,000}{1,806} - \frac{(-20 \times 10^6) \times 146.5}{39,833,780} = 82 \text{ N/mm}$$

$$V_y^* = \frac{F_y^*}{L_w} + \frac{M_z^* x_s}{I_{wp}} = \frac{40,000}{1,806} + \frac{(-20 \times 10^6) \times -152.5}{39,833,780} = 99 \text{ N/mm}$$

$$V_z^* = \frac{F_z^*}{L_w} + \frac{M_x^* y_s}{I_{wx}} - \frac{M_y^* x_s}{I_{wy}} = \frac{20,000}{1,806} + \frac{100,000 \times 146.5}{30,376,238} - \frac{40,000 \times -152.5}{9,457,542} = 1138 \text{ N/mm}$$

$$V_{res}^* = \sqrt{V_x^{*2} + V_y^{*2} + V_z^{*2}} = \sqrt{0.082^2 + 0.099^2 + 1.138^2} = 1.15 \text{ kN/mm}$$

Weld selection:
An 8 mm SP fillet weld is chosen, with a tensile strength of 490 MPa.

$$V_{res}^* \le \phi V_w$$

$$\phi V_w = 1.33 \text{ kN/mm}$$

3.10.5.3 Weld symbols

Welding symbols are defined in AS 1101.3. A simplified summary of weld symbol construction is shown in Figure 3.37. Basic weld symbols are given in Figure 3.38 (shown on the 'arrow side'); these are used to designate the weld type. Supplementary symbols (refer to Figure 3.39) are used on the outside of the weld symbol to give additional requirements for welds. Figure 3.40 shows commonly used examples of basic welds. Fillet welds are the cheapest and most commonly used structural welds. The CP specification is the simplest way to achieve maximum capacity of a weld (equal to the capacity of the connecting element).

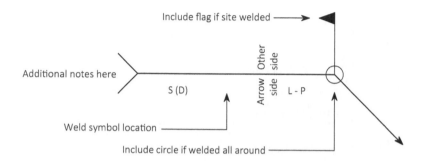

Figure 3.37 Standard location of elements of a welding symbol S, depth of preparation or size of weld; D, design throat thickness; L, length of weld (full length if not specified); P, spacing of weld (if stitch welding). (Modified from AS 1101.3, figure 2.3.)

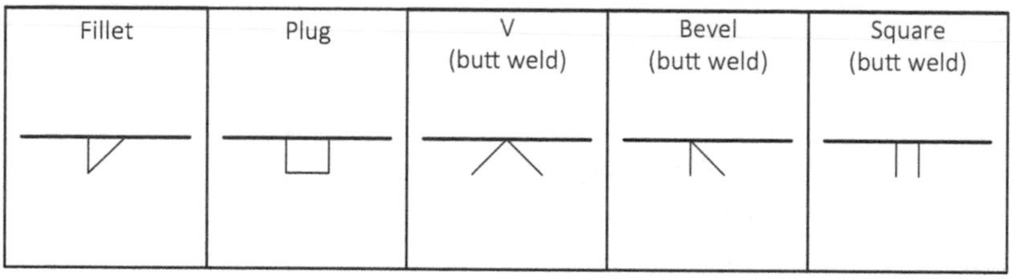

Fillet	Plug	V (butt weld)	Bevel (butt weld)	Square (butt weld)

Figure 3.38 Basic weld symbols. (Modified from AS 1101.3, figure 2.1.)

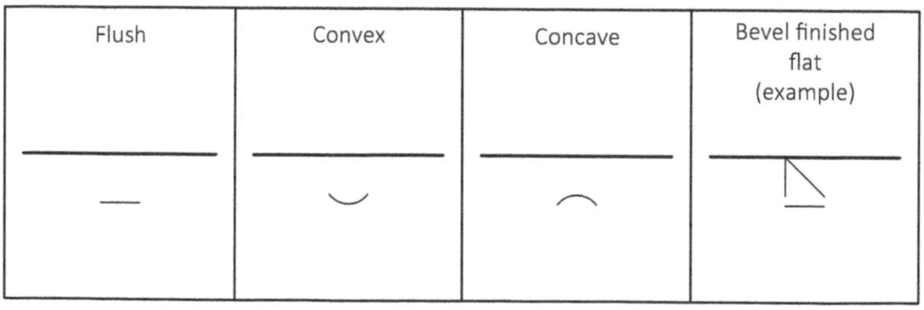

Flush	Convex	Concave	Bevel finished flat (example)

Figure 3.39 Supplementary symbols. (Modified from AS 1101.3, figure 2.2.)

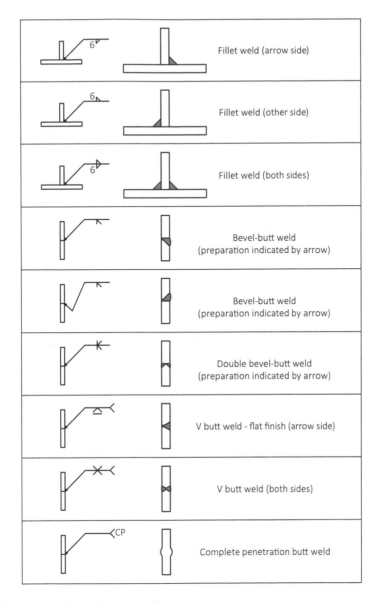

Figure 3.40 Common weld symbol examples.

3.10.5.4 General weld requirements

Maximum and minimum weld sizes are specified to ensure that a weld can be installed properly and correctly transfer force. The maximum allowable size of a weld is equal to the minimum thickness of a connecting element. The minimum size is equal to the thinner connecting element or the value taken from Table 3.28, whichever is smaller.

Access for the welding rod also needs to be considered (refer to Figure 3.41). The weld should have a minimum of 30° clearance for the welding rod, and the welder should be able to see the weld while welding. A minimum angle of 45° is often required for access on the acute side between inclined surfaces.

Table 3.28 Minimum fillet weld size

Thickness of thickest part, t (mm)	*Minimum size of fillet weld*
$t \leq 3$	$2t/3$
$3 > t \leq 7$	3
$7 < t \leq 10$	4
$10 < t \leq 15$	5
$t > 15$	6

Note: Minimum weld size is taken as the smaller value of the thinner connecting plate thickness and the value tabulated above.

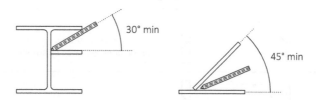

Figure 3.41 Welding rod clearances.

3.10.6 Plate analysis

The analysis of connecting plates is complex and varied. Different methods are recommended for connections based on the type of applied load. The following summary is based on the general theory presented in AS 4100 and recommendations by the ASI and AISC.

3.10.6.1 Tension

The tensile capacity of a connecting plate element can be governed by either the yield or the tensile failure.

Tensile stress failure, $N_t^* \leq \phi N_t = \phi \times 0.85 k_t A_n f_u$

Yielding stress failure, $N_t^* \leq \phi N_t = \phi \times A_g f_y$

where:

A_g = Gross cross-sectional area = d × t
A_n = Net cross-sectional area (gross area minus hole areas)
f_u = Tensile strength
f_y = Yield strength
k_t = Distribution correction factor
= 1.0 (for connections made at each part of the section, placed symmetrically about the centroid, with each part sized for the force in that component of the section)
=0.75 (for unequal angles connected on the short leg only)
= 0.85 (for unequal angles connected on the long leg or equal angles on either leg)
= 0.75 to 1.0 (for other connections depending on geometry; refer to AS 4100, Cl 7.3)
ϕ = 0.9 (plate tension)

3.10.6.2 Ply in bearing

Bolted plate connections need to be checked to ensure that bolts do not tear through the plate connection or cause a bearing failure on the hole (Figure 3.42).

Bearing failure, $V_f^* \leq \phi V_b = \phi \times 3.2 d_f t_p f_u$

Tear-out failure, $V_f^* \leq \phi V_b = \phi \times a_e t_p f_u$

where:

  a_e = Minimum clearance (minimum of a_{e1} and a_{e2})
        (measured from centre of bolt to nearest edge in direction of bolt reaction)
  d_f = Bolt diameter
  f_u = Tensile strength
  t_p = Plate thickness
  ϕ = 0.9 (ply in bearing)

3.10.6.3 Block shear

The design of block shear is a new requirement of AS 4100, added in Amendment 1. The following method is for connection components. The equation combines the previous methods for calculating shear and tensile failure. It assumes that the tension plane is governed by the tensile capacity; however, the shear plane can be governed by either tensile or yielding capacity (Figure 3.43).

Block shear failure, $R_{bs}^* \leq \phi R_{bs} = \phi[0.6 f_u A_{nv} + k_{bs} f_u A_{nt}] \leq \phi[0.6 f_y A_{gv} + k_{bs} f_u A_{nt}]$

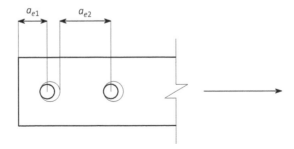

Figure 3.42 Ply in bearing.

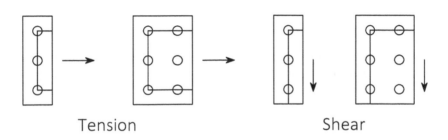

Figure 3.43 Block failure diagrams.

where:

A_{gv} = Gross cross-sectional area in shear
A_{nt} = Net cross-sectional area in tension (gross area minus hole areas)
A_{nv} = Net cross-sectional area in shear (gross area minus hole areas)
f_u = Tensile strength
f_y = Yield strength
k_{bs} = 1 (for uniform tensile stress distribution)
= 0.5 (for non-uniform tensile stress distribution)
ϕ = 0.75 (block shear)

3.10.6.4 Compression

The following procedure can be followed for typical connections, which are not subject to member buckling effects and do not have empty penetrations (unused bolt holes). For other components, such as long and slender plate connections, refer to Section 3.7.

$$N_c^* \leq \phi N_t = \phi \times k_f A_n f_y$$

where:

A_n = Net area of cross-section (gross area minus hole areas)
Note: The gross area may be used instead of the net area where bolt holes reduce the area by less than $100\{1-[f_y/(0.85f_u)]\}\%$.
f_y = Yield strength
k_f = 1 (where not subject to local buckling and all bolt holes are filled)
ϕ = 0.9 (plate compression)

3.10.6.5 Shear

The shear capacity of connections can be calculated using Section 3.5. For a rectangular plate, this can be simplified to

$$V^* \leq \phi V_v = \phi \times 0.6 dt f_y$$

where:

d = Depth of plate
f_y = Yield strength
t = Thickness of plate
ϕ = 0.9 (plate shear)

3.10.6.6 Bending

The bending capacities of connecting plate elements are typically calculated using plastic limits. The following information provides a background explanation and recommendations for design (Figure 3.44).

3.10.6.6.1 Major axis bending

The major axis bending capacity of a rectangular plate is dependent on the selection of elastic or plastic methods. If the plate slenderness is greater than the appropriate slenderness

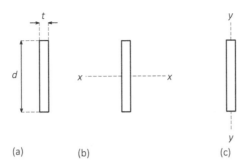

Figure 3.44 Rectangular elements in bending. (a) Geometry, (b) major axis bending, (c) minor axis bending.

limit, and the plate is therefore classified as non-compact or slender, then an elastic approach must be used (partial elastic for non-compact); refer to Section 3.4.1.3. Alternatively, if the plate is classified as compact, a plastic capacity may be adopted, therefore giving a higher capacity.

AS 4100 does not provide guidance on slenderness limits for plates with no longitudinal edges supported (refer to AS 4100, Table 5.2). If values were provided, they would likely be most similar to the values provided for an element with one edge supported, with maximum compression at the unsupported edge, and with zero stress or tension at the supported edge (refer to Table 3.5); however, the value may be lower. This would result in a limit of $(d/t) < 9$ for hot-rolled steel to be classified as compact. The ASI design guide states that most connections can be treated as compact for major axis bending of rectangular plate components and provides no further explanation [6].

It is therefore recommended that capacities are calculated using elastic methods to be conservative. Plastic methods may be used where the plate slenderness is less than the value obtained from Table 3.5 (AS 4100, Table 5.2); however, caution should be used when approaching the value. A non-linear FEA is recommended for cases where a plastic capacity is required for elements with slenderness values near or above the tabulated values.

Elastic capacity, $M_x^* \leq \phi M_{sx} = \phi \times f_y t d^2 / 6$

Plastic capacity, $M_x^* \leq \phi M_{sx} = \phi \times f_y t d^2 / 4$

3.10.6.6.2 Minor axis bending

The minor axis bending capacity of a rectangular plate is always classified as compact, and therefore a plastic capacity is appropriate.

Plastic capacity, $M_y^* \leq \phi M_{sy} = \phi \times f_y d t^2 / 4$

where:
- d = Depth of plate
- f_y = Yield strength
- t = Thickness of plate
- ϕ = 0.9 (plate bending)

3.10.6.7 Yield line analysis

Plate elements which undergo bending in complex patterns can be solved by using yield line analysis or by using FEA (refer to Section 6.11). Yield line analysis is frequently adopted for base plate calculations, with the Murray method [27] being widely accepted. The ASI has extrapolated on the Murray method [6] with the inclusion of a reduction factor ($\phi = 0.9$) and by conservatively removing the plate capacity across the bolt hole. Most available references provide formulas for common yield patterns on standard connections; however, in a design office, typical connections are generally done using software packages such as Limcon or FEA. This section presents the theory, as hand calculations are not likely to be completed for standard connections.

The concept of yield line analysis is based on calculating and equating internal and external work. The external work is equal to the force in each bolt multiplied by the distance which is deflected. The internal work is equal to the bending capacity of the plate multiplied by the angle of deformation. Yield lines are assumed, and then the two formulas are solved to find the bolt force. The process is then repeated with different yield patterns to find the minimum bolt force which causes failure. A worked example is provided using a UC section, as per Murray's original test data; however, the theory is frequently applied to various shapes (refer to Figure 3.48). The best way to decide on the appropriate yield pattern is to imagine the plate failing and draw yield lines across the bent sections of the plate.

3.10.6.7.1 Basic example

The first example shows a simple proof using yield theory to calculate the capacity of a rectangular section in minor axis bending (refer to Figure 3.45).

External work,	$W_e = F \times 1$ unit
Internal work,	$W_i = (\phi f_y\ t^2/4) \times b \times (1\ \text{unit}/L)$
Equating W_e with W_i and simplifying,	$F \times L = \phi f_y\ bt^2/4$
Bending capacity,	$\phi M = 0.9\ f_y b\ t^2/4$

The resultant equation is the same as that shown in Section 3.10.6.6.

3.10.6.7.2 Base plate example

The same theory can now be applied to more complex shapes (refer to Figure 3.46). A base plate loaded in tension is considered by calculating the capacity of each side of the plate separately and then doubling the value. Many yield shapes are possible; however, the diagram shown is the most commonly used assumption.

The angle of the yield lines is the only variable for this scenario. The minimum angle is limited by the size of the bolt holes, and the maximum angle is limited by the depth of the section.

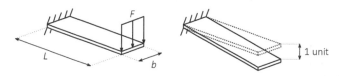

Figure 3.45 Plate bending example.

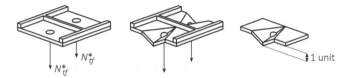

Figure 3.46 Base plate in tension.

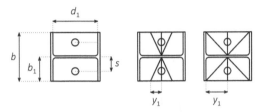

Figure 3.47 Base plate geometry and possible yield lines.

The external work is reduced because the bolt is located at a location which deflects by less than 1 unit.

External work, $W_e = N_{tf}^* \times 1 \, (s/b_1)$

The internal work is most easily calculated by taking the orthogonal components of each failure segment. The formula below is created by multiplying the bending capacity per unit length by the sum of the breadth of each yield line multiplied by the yield line rotations. The breadth and rotations are found by taking the horizontal (y_1) component of the diagonal yield lines (as shown in Figure 3.47) multiplied by the rotation in that plane $(1/b_1)$, then the vertical (b_1) component of the diagonal yield lines multiplied by rotation $(1/y_1)$, and finally the central failure plane (b_1) multiplied by rotation $(2/y_1)$. The components for the diagonal yield lines are multiplied by two, because there are two diagonal yield lines on each side of the plate.

Internal work, $W_i = (\phi f_y t^2 /4) \times \left[2\left(y_1 \dfrac{1}{b_1} \right) + 2\left(b_1 \dfrac{1}{y_1} \right) + \left(b_1 \dfrac{2}{y_1} \right) \right]$

Equating W_e with W_i and simplifying to solve for the maximum bolt force,

$$N_{tf}^* = (\phi f_y t^2 /4) \times \left[2\left(y_1 \dfrac{1}{b_1} \right) + 2\left(b_1 \dfrac{1}{y_1} \right) + \left(b_1 \dfrac{2}{y_1} \right) \right](b_1/s)$$

The solution is dependent on the value of y_1. The value which results in the smallest bolt force is needed; therefore, a value can be found by differentiating the formula for internal work, with respect to y_1, and solving for zero.

$$\frac{dW_i}{dy_1} = (\phi f_y t^2 /4) \times \left[2\left(\dfrac{1}{b_1} \right) + 2\left(b_1 \dfrac{-1}{y_1^2} \right) + \left(b_1 \dfrac{-2}{y_1^2} \right) \right] = 0$$

$$\frac{1}{b_1} + b_1 \frac{-1}{y_1^2} + b_1 \frac{-1}{y_1^2} = 0$$

$$b_1 \frac{2}{y_1^2} = \frac{1}{b_1}$$

$$y_1 = \sqrt{2}b_1 \quad \left(y_1 \text{ is also limited to } \frac{d_1}{2} \text{ for geometric reasons} \right)$$

therefore,

$$y_1 = \text{MIN}\left[\sqrt{2}b_1, \frac{d_1}{2} \right]$$

As discussed previously, the capacity of the connection is equal to double the capacity of each single bolt.

$$N_{tf}^* = 2(\phi f_y t^2 / 4) \times \left[2\left(y_1 \frac{1}{b_1} \right) + 2\left(b_1 \frac{1}{y_1} \right) + \left(b_1 \frac{2}{y_1} \right) \right](b_1/s)$$

Note: The ASI design guide also removed the bolt hole from the capacity; therefore, the internal work would be calculated as

$$W_i = (\phi f_y t^2 / 4) \times \left[2\left(y_1 \frac{1}{b_1} \right) + 2\left(b_1 \frac{1}{y_1} \right) + \left((b_1 - d_h) \frac{2}{y_1} \right) \right];$$

however, this was not done in the original work by Murray [27]. Judgement should be used based on the number and size of holes.

3.10.6.7.3 Further yield line examples

The following yield line examples can be used with the theory as previously explained (Section 3.10.6.7.2) to solve more complex connection geometries. The same patterns shown for the UB profile can be adopted for UCs or PFCs (Figure 3.48).

3.10.6.7.4 Alternate cantilever method

The cantilever method is often used as a quick and simple check for plate capacity near tension bolts. The method assumes a 45° load distribution and checks the plastic capacity at the support point. Where there are bolts on both sides of the supporting plate, the moment may be calculated by equating it to a simply supported beam (supported by the bolts) with a central point load equal to the total tension in both bolts ($M^* = N^*L/4 = 2N_{tf}^*L/4$). For the case of single bolts, the moment is calculated by equating it to a cantilever ($M^* = N_{tf}^*L$). The plate capacity is calculated using a plastic capacity with the effective width (b) shown in Figure 3.49.

Plate capacity, $\qquad \phi M_{sy} = 0.9 \times f_y bt^2/4 \qquad$ (refer to Section 3.10.6.6 for details)

3.10.6.8 Base plates in compression

Plates which bear against concrete surfaces are checked for plate capacity and concrete bearing capacity. The concrete bearing check is in accordance with AS 3600, Cl 12.6.

$$\phi N_{cb} = \phi A_1 f_b$$

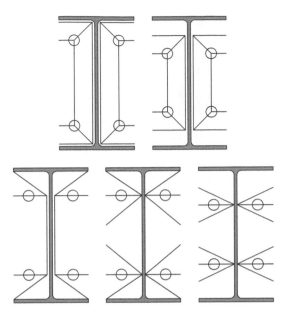

Figure 3.48 Common yield line patterns.

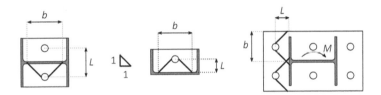

Figure 3.49 Cantilever plate bending model.

where:

$$\phi = 0.6$$

$$f_b = \text{MIN}\left[f_c' \times 0.9\sqrt{\frac{A_2}{A_1}} , f_c' \times 1.8 \right]$$

A_1 = bd = Area of base plate
A_2 = Largest area of supporting surface (concrete or grout) that is geometrically simi-
lar to and concentric with A_1
b = Width of base plate
d = Depth of base plate

There are many plate bending theories for bearing surfaces. The current recommended model is the Thornton method [32], modified to suit Australian Standards by Ranzi and Kneen [30] and updated to the latest revisions by Hogan [6]. It focuses on both the cantilever (equivalent area using a_1, a_2) and pressure (using a_3 to derive λa_4) methods of analysis; refer to Figure 3.50 for a diagrammatic representation.

$$\phi N_{sb} = \frac{\phi f_y bdt^2}{2a_0^{\,2}}$$

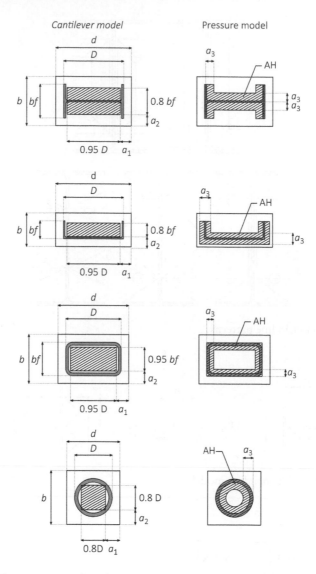

Figure 3.50 Base plate pressure distributions. (From Hogan, T.J. and Munter, S.A., *Structural Steel Connections Series Simple Connections Suite*, 1st ed., Australian Steel Institute, Australia, 2007.)

where:

t = Thickness of base plate

a_0 = MAX $[a_1, a_2, \lambda a_4]$

ϕ = 0.9 (plate bending)

$$\lambda = \text{MIN} \left[\frac{k_x \sqrt{X}}{1 + \sqrt{1 - X}}, 1.0 \right] = 1 \text{ (when } X > 1)$$

Refer to Figure 3.50 and Table 3.29 for remaining variables.

Table 3.29 Variables for base plates in compression

Section shape	UC, UB, WB, WC	PFC	RHS	SHS	CHS (rectangular plate)	CHS (circular plate)
a_1	$0.5(d-0.95D)$	$0.5(d-0.95D)$	$0.5i(d-0.95D)$	$0.5(d-0.95D)$	$0.5(d-0.8D)$	$0.5(d-0.8D)$
a_2	$0.5(b-0.8b_f)$	$0.5(b-0.8b_f)$	$0.5i(b-0.95b_f)$	$0.5(b-0.95b_f)$	$0.5(b-0.8D)$	a_1
a_3	$\dfrac{a_5-\sqrt{a_5^2-4A_H}}{4}$	$\dfrac{a_5-\sqrt{a_5^2-8A_H}}{4}$	$\dfrac{a_5-\sqrt{a_5^2-4A_H}}{4}$	$\dfrac{a_5-\sqrt{a_5^2-4A_H}}{4}$	$\dfrac{D-\sqrt{D^2-4A_H/\pi}}{2}$	
a_4	$0.25\sqrt{Db_f}$	$0.33\sqrt{2Db_f}$	$0.295\sqrt{Db_f}$	$0.306b_f$	$0.29D$	$0.29D$
a_5	$D+b_f$	$2b_f+D$	b_f+D	$2b_f$	$-$	$-$
k_x	$2\sqrt{\dfrac{db}{Db_f}}$	$1.5\sqrt{\dfrac{db}{Db_f}}$	$1.7\sqrt{\dfrac{db}{Db_f}}$	$1.65\dfrac{\sqrt{db}}{b_f}$	$1.95\dfrac{\sqrt{db}}{D}$	$1.7\dfrac{d}{D}$
X	$\dfrac{4N^*}{\phi f_b a_5^2}$	$\dfrac{8N^*}{\phi f_b a_5^2}$	$\dfrac{4N^*}{\phi f_b a_5^2}$	$\dfrac{4N^*}{\phi f_b a_5^2}$	$\dfrac{4N^*}{\phi f_b \pi D^2}$	$\dfrac{4N^*}{\phi f_b \pi D^2}$
A_H	$2b_f a_3+2a_3(D-2a_3)$	$2b_f a_3+a_3(D-2a_3)$	$Db_f-(D-2a_3)(b_f-2a_3)$		$\pi(Da_3-a_3^2)$	

3.11 ELASTIC STRESS ANALYSIS

A combined stress analysis can be completed for sections which undergo concurrent stresses caused by different load types. This section summarises the elastic analysis of axial, bending, shearing and torsion loading, as well as the von Mises stress combination. For items which adopt an elastic analysis, the stress should be limited as shown. This is a sectional analysis method and does not consider member effects, such as buckling.

$$f_{\max}^* \leq \phi f_y = 0.9 \times f_y$$

3.11.1 Principal stresses

Principal stresses $(\sigma_{XX}, \sigma_{YY}, \sigma_{ZZ})$ are the tensile and compressive forces on an area of the section under consideration. Longitudinal axial stress (σ_{ZZ}) is caused by pure tension or compression, and also by bending moments. A positive bending moment about the major axis causes tension in the bottom flange and compression in the top flange. Therefore, when a positive moment is combined with a tensile load, the maximum stress will be a tension in the bottom flange. When further combined with a minor axis moment, the maximum stress will be a tension at the corner of the bottom flange.

3.11.2 Shear stresses

Shear stresses are caused by shear loads in the primary (σ_{YZ}) and secondary (σ_{ZX}) directions, as well as by torsion (σ_{XY}). Figure 3.51 shows the coordinate system used for stresses. This chapter deals only with beam shear; for torsion shear stresses, refer to Section 3.9. Shear stresses are largest at the centre of sections and on thin elements within the cross-section.

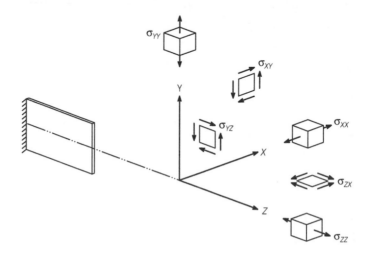

Figure 3.51 Stress coordinate system.

3.11.3 Typical beam stresses

The most common stresses in design are those caused by axial, bending and primary shear. General formulas for these spanning-induced stresses are

Axial stress from tension or compression, $\quad \sigma_{ZZ} = N^*/A$

Axial stress from major axis bending, $\quad \sigma_{ZZ} = M_x^* y/I_x$

Axial stress from minor axis bending, $\quad \sigma_{ZZ} = M_y^* x/I_y$

Shear stress from major axis shear, $\quad \sigma_{YZ} = V_y^* Q/I_x t$

Shear stress from minor axis shear, $\quad \sigma_{ZX} = V_x^* Q/I_y t$

where:

$\quad A \quad$ = Cross-sectional area of section

$\quad I_x, I_y$ = Second moment of area $\left[= bb^3/12 + bb\bar{y}^2 \right]$ (refer to Section 3.4.1.1 and Table 3.30)

$\quad t \quad$ = Thickness of section at point under consideration, transverse to shear load direction

$\quad Q \quad$ = Area of cross-section above (in direction of shear) point under consideration multiplied by the distance between the neutral axis of the entire section and the neutral axis of the above cross-section.

$\quad x,y \quad$ = Distance from neutral axis to point under consideration, transverse to bending axis

3.11.4 Combined stress

The von Mises stress criterion uses one formula to calculate the maximum stress magnitude based on three principal stresses and three shear stresses.

$$f_{\max}^* = \sqrt{\frac{1}{2}[(\sigma_{XX} - \sigma_{YY})^2 + (\sigma_{YY} - \sigma_{ZZ})^2 + (\sigma_{ZZ} - \sigma_{XX})^2 + 6(\sigma_{XY}^2 + \sigma_{YZ}^2 + \sigma_{ZX}^2)]}$$

The formula can be modified to account for stress states with fewer inputs by removing any zero values.

Principal stresses only:

$$f_{\max}^* = \sqrt{\frac{1}{2}[(\sigma_{XX} - \sigma_{YY})^2 + (\sigma_{YY} - \sigma_{ZZ})^2 + (\sigma_{ZZ} - \sigma_{XX})^2]}$$

Typical beam stresses only:

$$f_{\max}^* = \sqrt{\sigma_{ZZ}^2 + 3(\sigma_{YZ}^2 + \sigma_{ZX}^2)}$$

Table 3.30 Second moment of areas

Section type	I_x	I_y
Circle	$\dfrac{\pi}{4}r^4$	
Rectangle	$\dfrac{bh^3}{12}$	$\dfrac{hb^3}{12}$
CHS	$\dfrac{\pi}{4}(r_o^4 - r_i^4)$	
RHS	$\dfrac{bh^3}{12} - \dfrac{b_i h_i^3}{12}$	$\dfrac{hb^3}{12} - \dfrac{h_i b_i^3}{12}$
Equal flanged I-sections (UC, UB, WC, WB)	$\dfrac{b_f D^3 - (b_f - t_w)d_1^3}{12}$	$\dfrac{2t_f b_f^3 + d_1 t_w^3}{12}$

(*Continued*)

Table 3.30 (Continued) Second moment of areas

Section type	I_x	I_y
Parallel flanged channels (PFC)	$$\dfrac{b_f D^3 - (b_f - t_w)d_1^3}{12}$$	$$\dfrac{2t_f b_f^3 + d_1 t_w^3}{3} - AC_x^2$$ where: $A = 2b_f t_f + d_1 t_w$ $$C_x = \dfrac{2t_f b_f^2 + d_1 t_w^2}{2A}$$
Equal angle (EA)	$\dfrac{1}{3}\left(ty^3 + b(b-y)^3 - (b-t)(b-y-t)^3\right)$ where: $y = b - C_y$ $C_y = (b^2 + bt - t^2)/(2(2b-t))$	
Unequal angle (UA)	$\dfrac{1}{3}\left\{\begin{array}{l} ty^3 + b(h-y)^3 \dots \\ -(b-t)(h-y-t)^3 \end{array}\right\}$ where: $y = h - C_y$	$\dfrac{1}{3}\left\{\begin{array}{l} tx^3 + h(b-x)^3 \dots \\ -(h-t)(b-x-t)^3 \end{array}\right\}$ where: $x = b - C_x$

$$C_x = [t(2(b-t)+h)+(b-t)^2]/(2(b-t+h))$$
$$C_y = [t(2(h-t)+b)+(h-t)^2]/(2(h-t+b))$$

Note: For all diagrams above, the x-axis is horizontal and the y-axis is vertical.

Example 3.12: Stress Analysis

Calculate the maximum stresses on the section shown, subject to the following design actions:

$$N_t^* = 50 \text{ kN} \quad M_x^* = 3 \text{ kNm} \quad M_y^* = 0.2 \text{ kNm} \quad V_y^* = 110 \text{ kN} \quad V_x^* = 20 \text{ kN}$$

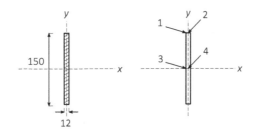

Shear stress is highest at the centre of sections and at minimum thicknesses; bending is highest at the edges of a section, and axial stresses are uniform. Therefore, the maximum stress on the section will occur at either the corner (1), the centre of an edge (2, 3) or the geometric centre (4).

$$I_x = \frac{bh^3}{12} = \frac{12 \text{ mm} \times (150 \text{ mm})^3}{12} = 3.375 \times 10^6 \text{ mm}^4$$

$$I_y = \frac{hb^3}{12} = \frac{150 \text{ mm} \times (12 \text{ mm})^3}{12} = 21.6 \times 10^3 \text{ mm}^4$$

Stress at Point 1:

Axial stress from tension or compression, $\quad \sigma_{ZZ} = \dfrac{N_t^*}{A} = \dfrac{50 \text{ kN}}{150 \text{ mm} \times 12 \text{ mm}} = 27.8 \text{ MPa}$

Axial stress from major axis bending, $\quad \sigma_{ZZ} = M_x^* y/I_x = 3 \text{ kNm} \times \left(\dfrac{150 \text{ mm}}{2} \right)/I_x = 66.7 \text{ MPa}$

Axial stress from minor axis bending, $\quad \sigma_{ZZ} = M_y^* x/I_y = 0.2 \text{ kNm} \times \left(\dfrac{12 \text{ mm}}{2} \right)/I_y = 55.6 \text{ MPa}$

Shear stress from major axis shear, $\qquad \sigma_{YZ} = 0$ (edge of section)

Shear stress from minor axis shear, $\qquad \sigma_{ZX} = 0$

Von Mises stress,

$$f_{\max}^* = \sqrt{\sigma_{ZZ}^2 + 3(\sigma_{YZ}^2 + \sigma_{ZX}^2)} = \sqrt{(27.8 \text{ MPa} + 66.7 \text{ MPa} + 55.6 \text{ MPa})^2 + 3(0^2 + 0^2)} = 150 \text{ MPa}$$

Stress at Point 2:

Axial stress from tension or compression, $\quad \sigma_{ZZ} = 27.8 \text{ MPa}$

Axial stress from major axis bending, $\qquad \sigma_{ZZ} = 66.7 \text{ MPa}$

Axial stress from minor axis bending, $\qquad \sigma_{ZZ} = M_y^* x/I_y = 0.2 \text{ kNm} \times (0)/I_y = 0$

Shear stress from major axis shear, $\qquad \sigma_{YZ} = 0$

Shear stress from minor axis shear,

$$\sigma_{ZX} = V_x^* Q/I_y t = 20 \text{ kN} \times Q/(I_y t) = 16.7 \text{ MPa}$$

$$\{Q = (h \times b/2)(b/4) = 2700 \text{ mm}^3\}$$

$$\{t = 150 \text{ mm}\}$$

Von Mises stress,

$$f_{\max}^* = \sqrt{\sigma_{ZZ}^2 + 3(\sigma_{YZ}^2 + \sigma_{ZX}^2)}$$

$$= \sqrt{(27.8 \text{ MPa} + 66.7 \text{ MPa})^2 + 3[0^2 + (16.7 \text{ MPa})^2]} = 98.8 \text{ MPa}$$

Stress at Point 3:

Axial stress from tension or compression, $\quad \sigma_{ZZ} = 27.8 \text{ MPa}$

Axial stress from major axis bending, $\qquad \sigma_{ZZ} = M_x^* y/I_x = 3 \text{ kNm} \times (0)/I_x = 0$

Axial stress from minor axis bending, $\sigma_{ZZ} = 55.6 \text{ MPa}$

Shear stress from major axis shear, $\sigma_{YZ} = V_y^* Q / I_x t = 110 \text{ kN} \times Q / I_x t = 91.7 \text{ MPa}$

$$\{Q = (b \times h/2)(h/4) = 33,750 \text{ mm}^3\}$$

$$\{t = 12 \text{ mm}\}$$

Shear stress from minor axis shear, $\sigma_{ZX} = 0$

Von Mises stress,

$$f_{\text{max}}^* = \sqrt{\sigma_{ZZ}^2 + 3(\sigma_{YZ}^2 + \sigma_{ZX}^2)}$$

$$= \sqrt{(27.8 \text{ MPa} + 55.6 \text{ MPa})^2 + 3[(91.7 \text{ MPa})^2 + 0^2]} = 179.3 \text{ MPa}$$

Stress at Point 4:

Axial stress from tension or compression, $\sigma_{ZZ} = 27.8 \text{ MPa}$

Axial stress from major axis bending, $\sigma_{ZZ} = 0$

Axial stress from minor axis bending, $\sigma_{ZZ} = 0$

Shear stress from major axis shear, $\sigma_{YZ} = 91.7 \text{ MPa}$

Shear stress from minor axis shear, $\sigma_{ZX} = 16.7 \text{ MPa}$

Von Mises stress,

$$f_{\text{max}}^* = \sqrt{\sigma_{ZZ}^2 + 3(\sigma_{YZ}^2 + \sigma_{ZX}^2)}$$

$$= \sqrt{(27.8 \text{ MPa})^2 + 3[(91.7 \text{ MPa})^2 + (16.7 \text{ MPa})^2]} = 163.7 \text{ MPa}$$

3.12 STEEL DETAILING

This section includes typical steel details which are used for standard connections. Individual connections should always be checked by the engineer.

3.12.1 Steel notes

Notes should be included on standard design drawings. The following notes are suggested for a typical project. Project-specific details such as design loads and wind or seismic parameters should be indicated on drawings. The construction category should also be stated on drawings, where relevant (refer to Section 1.7.2).

3.12.1.1 General

All works shall be completed in accordance with Australian Standards and legislation as well as project specifications.

All dimensions are expressed in millimetres and levels expressed in metres unless noted otherwise.

All material test certificates shall be generated through testing by an independent laboratory accredited by ILAC or APLAC.

3.12.1.2 Steel

All steel shall comply with the following grades:

- Plates – Grade 350 to AS/NZS 3678
- UB, UC, PFC, TFB, EA, UA, Flats – Grade 300 to AS/NZS 3679.1
- WB and WC – Grade 300 to AS 3679.2
- SHS and CHS – Grade 450 to AS 1397 and AS/NZS 4600

3.12.1.3 Welding

- All welds and welding procedures shall be qualified in accordance with AS/NZS 1554.
- All welds shall be category SP except handrails, ladders and floor plates.
- Complete penetration (CP) welds shall be qualified butt welds that develop the full strength of the steel.
- Welds shall be a minimum of 6 mm fillet weld for steel thickness ≤12 mm, 8 mm fillet weld for steel thickness ≤20 and 10 mm fillet weld for steel thickness >20 mm.

3.12.1.4 Bolting

- All bolts, nuts and washers shall be hot dipped galvanised in accordance with AS 1214.
- All bolts shall be M20 Grade 8.8/S unless noted otherwise.
- All bolts for handrails shall be M16 Grade 4.6/S unless noted otherwise.
- Load indicating washers shall be used for all fully tensioned bolts.
- Bolts shall be installed on standard gauge lines unless noted otherwise.
- Standard bolt holes shall be 2 mm larger than bolt diameters. Base plate holes shall be 6 mm larger than bolt diameters.
- Edge distances shall be a minimum of 1.75 bolt diameters for sheared or hand flame-cut edges; 1.50 bolt diameters for rolled plate, flat bar or section machine cut, sawn or planed edges; 1.25 bolt diameters for rolled edges of a rolled flat bar or section.

3.12.1.5 Surface protection

Steel shall be coated in accordance with the project specifications.

3.12.1.6 Grouting

All structural steel to concrete interfaces shall be grouted in accordance with project specifications. Nominal grout thickness shall be 40 mm.

3.12.2 Additional steel details

Steel details for the following items are shown in other sections of this text.

Table 3.31 Additional steel details

Design item	Reference
Lifting lugs	Figure 6.16
Stairs	Figure 6.23
Ladders	Figure 6.24

3.12.3 Coping

Example coping details are provided below, including maximum allowable cope dimensions. Coping is required for certain connections, such as a web side plate on a beam to beam connection (Figure 3.52).

3.12.4 Bracing cleat

It is not necessary to detail every dimension of a typical bracing cleat. Minimum clearances and bolt spacing are usually sufficient. A table is often used to standardise the dimensions as shown. Details are shown for an equal angle, minor axis bracing, and a circular hollow section with cheek plates (Table 3.32 and Figure 3.53).

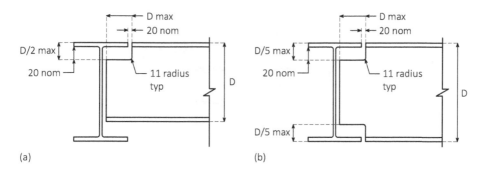

Figure 3.52 Example coping details. (a) Single cope, (b) double cope.

Table 3.32 Bracing table

Member designation	Cleat and tongue plate thickness	Weld size 'W'	Cheek plate thickness	Bolts	
				'N' rows	Size GR 8.8/S
50x5EA to 65x10EA	10	6	–	2	M16
75x5EA to 90x10EA	10	6	–	2	M20
10x6EA to 150x12EA	12	8	–	3	M20
150x16EA to 200x26EA	16	8	–	3	M20
65x50x5UA to 65x50x8UA	10	6	–	2	M16
75x50x5UA to 75x50x8UA	10	6	–	2	M20
100x75x6UA to 150x100x12UA	12	8	–	3	M20
76.1x3.2CHS to 101.6x3.2CHS	16	6	–	1	M20
114.3x3.2CHS to 139.7x3.5CHS	16	6	–	2	M20
165.1x5.4CHS to 168.3x7.1CHS	20	8	12	2	M20
219.1x4.8CHS to 273.1x9.3CHS	25	8	12	2	M24

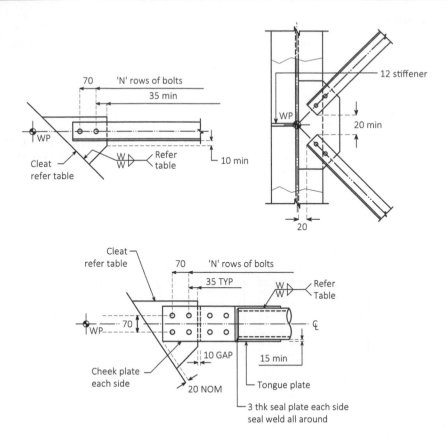

Figure 3.53 Example bracing cleat details.

3.12.5 Web side plate

Web side plates are most commonly used for beam to beam connections. They behave as simple (pinned) connections. The double column web side plate should be used for high shear or axial loads. The weld size is recommended to be a minimum of 75% of the plate thickness [6] (Table 3.33 and Figure 3.54).

Table 3.33 Web side plate and end plate table

Member designation	'N' Bolt rows M20 GR 8.8/S
200PFC to 250PFC, 200UC to 250UC, 200UB to 250UB	2
300PFC, 310UC, 310UB, 350WB	3
380PFC, 360UB to 410UB, 400WC	4
460UB, 500WC	5
530UB	6
610UB	7
700WB	8
800WB to 1200WB	9

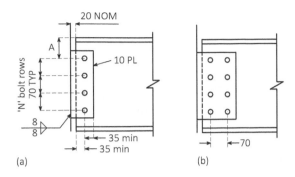

Figure 3.54 Example web side plate (a) and double column web side plate (b) details.

3.12.6 End plates

End plates are most commonly used for beam to column connections. The flexible end plate behaves as a simple (pinned) connection. The full depth end plate is used when a semi-rigid connection is required, as it can transfer some moment. Thickness and number of bolts can be specified depending on design requirements. Plate width is equal to the attaching member bolt gauge plus two times the minimum edge distance (2 × 35). Standard gauges (refer to Section 7.3) can be tabularised on the design drawings for commonly used sections (Figure 3.55).

3.12.7 Bolted moment connections

Bolted moment connections are shown for a beam to column connection and splices. The column splice has the ends of each section prepared for full contact in accordance with AS 4100 to transfer high axial loads without engaging bolts in shear. Plate dimensions can be calculated based on bolt dimensions and edge distances (Table 3.34 and Figure 3.56).

3.12.8 Welded moment connections

The welded moment connection is used for rigid welded connections. The web doubler plate or diagonal stiffeners are specified to strengthen the web for high loads (Figure 3.57).

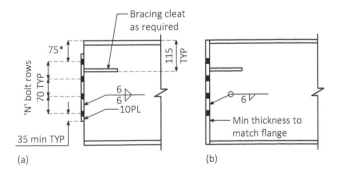

Figure 3.55 Example end plate details. (a) Flexible end plate, (b) full debth end plate.

Table 3.34 Bolted moment dimensions

Member	Gauge	Bolted moment connection					Splice connection						
		Bolt (8.8/ TB)	A	N	T	Web doubler	Bolt	E	B	C	N	X	Y
610UB	140	M24	75	4	28	10	M24	25	N/A	6	5	6	2
530UB	140	M20	75	4	25	10	M24	20	N/A	6	4	5	1
460UB	90	M20	75	4	25	N/A	M24	20	N/A	6	3	5	1
410UB	90	M20	75	4	25	N/A	M24	16	N/A	6	3	4	1
360UB	90	M20	75	4	25	N/A	M20	16	N/A	6	4	4	1
310UB	70	M20	75	2	25	N/A	M20	16	N/A	6	3	3	1
250UB	70	M20	75	2	25	N/A	M20	12	N/A	6	2	2	1
200UB	70*	M20	65	2	20	N/A	M20	12	N/A	6	2	2	1
310UC	140	M24	75	2	25	10	M24	16	100x16	10	3	3	1
250UC	140	M24	75	2	25	10	M24	12	100x12	8	2	2	1
200UC	140	M20	75	2	20	N/A	M24	10	75x10	10	2	2	1
150UC	90	N/A					M20	12	N/A	10	2	2	1

Note: W based on stiffener thickness, *t*: 6 mm ($t \leq 12$), 8 mm ($t \leq 20$), 10 mm ($t > 20$). Width of plate E to match flange width. Gauge for 200UB18 is 50 mm.

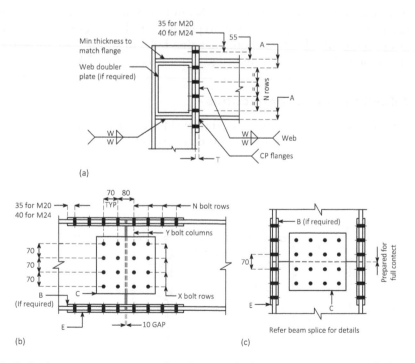

Figure 3.56 Bolted moment connection details. (a) Bolted moment connection, (b) beam splice, (c) column splice.

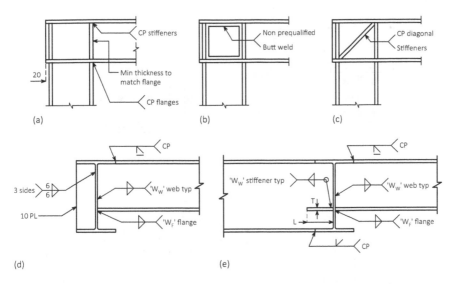

Figure 3.57 Welded moment connection detail. (a) Welded moment connection, (b) web doubler plate, (c) diagonal stiffener, (d) welded beam intersection, (e) welded beam continuous.

3.12.9 Base plates

Base plate details are provided for UB and UC sections. Shear keys should be specified for connections with high shear loads (ensure pockets are detailed on concrete drawings). Alternatively, steel washers can be welded (to the base plate) over the anchor bolts to reduce the hole size and ensure all bolts are engaged despite the oversized holes (Figure 3.58; Tables 3.36 and 3.37).

Table 3.35 Base plate table

Member designation	Pinned base plate							Moment base plate						
	X	Y	B	L	T	W	Bolt grade 4.6/S	X	Y	B	L	T	W	Bolt grade 4.6/S
610UB	180	450	300	650	32	8	M30	–	–	–	–	–	–	–
530UB	180	400	300	600	25	8	M30	–	–	–	–	–	–	–
460UB	180	300	300	500	25	6	M30	–	–	–	–	–	–	–
410UB	120	250	250	450	25	6	M24	–	–	–	–	–	–	–
360UB	120	200	250	400	25	6	M24	–	–	–	–	–	–	–
310UB	120	150	200	350	20	6	M20	–	–	–	–	–	–	–
250UB	120	130	200	300	20	6	M20	–	–	–	–	–	–	–
200UB	100	100	200	250	16	6	M20	210	270	300	360	25	6	M20
180UB	100	–	180	200	16	6	M20	150	240	230	320	25	6	M20
150UB	100	–	180	200	16	6	M20	150	210	230	290	20	6	M20
310UC	180	180	350	350	25	6	M30	–	–	–	–	–	–	–
250UC	150	150	280	280	20	6	M24	330	330	410	410	32	8	M24
200UC	120	120	230	230	20	6	M20	300	300	400	400	32	8	M24
150UC	120	–	200	200	16	6	M20	220	220	300	300	25	6	M20
100UC	100	–	180	120	12	6	M20	170	170	250	250	20	6	M20

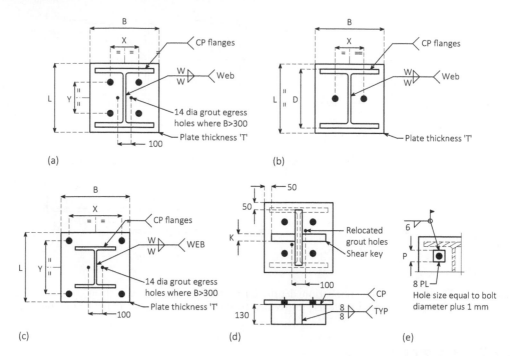

Figure 3.58 Example base plate details. (a) Pinned base plate, (b) pinned base plate (D<190), (c) moment base plate, (d) shear keys, (e) anchor bolt plate.

Table 3.36   Shear key dimension

Column designation	K
150UC, 200UB, 250UB,	25
200UC, 250UC, 310UC, 310UB, 360UB, 410UB	32
460UB, 530UB, 610UB	36

Table 3.37   Anchor bolt plate dimension

Bolt (grade 4.6/S)	P
20	70
24	80
30	100

Chapter 4

Concrete design

This section summarises the design of concrete structures in accordance with AS 3600, current at the time of writing (AS 3600-2009, incorporating Amendments 1 and 2). The status of standards should be checked at the beginning of a project or calculation. Only specific clauses are outlined in this section, and the full standard should be followed to ensure that all requirements are met.

4.1 MATERIAL

The material properties presented in this section are based on those presented in AS 3600, Section 3. The specification and supply of concrete should be in accordance with AS 1379.

4.1.1 Concrete

The most commonly used concrete strengths for industrial designs are 32 and 40 MPa. Higher-strength grades are common for precast design and for commercial designs, where weight and thickness are of higher importance. These higher-strength concrete grades are often unavailable at regional locations due to timing of delivery and capacity of batching plants.

AS 3600 is applicable to concrete with characteristic compressive strength between 20 and 100 MPa and with density between 1800 and 2800 kg/m^3. Other restrictions are also placed on reinforcing bars; refer to AS 3600, Section 1 for details.

The characteristic compressive strength of concrete, f'_c, is the compressive cylinder strength measured 28 days after placement. The cylinders may be 150 mm dia $\times$ 300 mm high or 100 mm dia $\times$ 200 mm high. If international codes are used, the compressive strength is not always measured in the same way, and therefore may have to be converted. For example, the British Standards use a 150 mm cube; however, the Eurocode and American codes use cylinders (150 mm dia $\times$ 300 mm high). Cube tests result in measured compressive strengths in the order of 25% higher than cylinder tests.

AS 3600 provides other concrete properties in relation to the compressive strength; refer to Table 4.1. Equations for values are also presented in AS 3600, Section 3.

Table 4.1 Concrete properties at 28 days

f_c' (MPa)	20	25	32	40	50	65	80	100
F_{cmi}(MPa)	22	28	35	43	53	68	82	99
E_c(MPa)	24,000	26,700	30,100	32,800	34,800	37,400	39,600	42,200

Source: AS 3600, Table 3.1.2. Copied by Mr L. Pack with the permission of Standards Australia under Licence 1607-c010.

Density, $\rho = 2400 \text{ kg}/\text{m}^3$ (for normal-strength concrete)

Density may also be calculated by testing in accordance with AS 1012.12.1 and AS 1012.12.2.

Poisson's ratio, $\upsilon = 0.2$

Poisson's ratio may also be calculated by testing in accordance with AS 1012.17.

Characteristic uni-axial tensible strength, $f_{ct}' = 0.36\sqrt{f_c'}$

Characteristic flexural tensible strength, $f_{ct.f}' = 0.6\sqrt{f_c'}$

Note: The characteristic uni-axial and flexural tensile strengths can be converted to mean and upper values by multiplying the characteristic values by 1.4 and 1.8, respectively.

The coefficient of thermal expansion is $10 \times 10^{-6}/°C$, with a range of $\pm 20\%$, and may be calculated from test data (refer to Table 6.18 for additional values based on aggregate selection).

Special (S) concrete grades such as high-early strength and lightweight concrete are also available. Normal (N) concrete is typically lightly loaded at 7 days and reaches design strength at 28 days. These periods can be reduced to 3 and 7 days, respectively, for high-early strength concrete. Details pertaining to special concrete grades should be discussed with suppliers prior to completion of the design phase.

4.1.2 Reinforcement

D500N is the most commonly used reinforcement for general reinforcing bars. It is a deformed bar with a yield strength of 500 MPa and a normal ductility class. Ligatures are also commonly made from D500N, so that only one type of reinforcement is required; however, R250N is also common and is used when smaller bars or lower yield strengths are acceptable. D500L is used where smaller diameter bars are required. The properties for reinforcement presented are from AS 3600, Clause 3.2. Yield strengths and uniform strains are available in Table 4.2. Available bar diameters and areas are detailed in Table 4.3.

Modulus of elasticity, $E_s = 200,000$ MPa

Reinforcing mesh is commonly used in the construction of slabs. The mesh is an automatically machine-welded grid of bars in each direction. Areas are commonly less than that of bars spaced by hand; however, the sizes go up to an approximate equivalent of N12 at 180 centres for square mesh. Care should be taken when specifying mesh due to the reduced capacity reduction factor and strain limit. For further detail on available reinforcement and mesh sizes, refer to Section 7.3 and Chapter 8.

4.1.3 Cost

An approximate cost should be considered during the design phase. The relative prices of items vary between different locations (refer to Table 4.4). Costs generally increase by

Table 4.2 Yield strength and ductility class of reinforcement

Reinforcement		Characteristic yield strength, f_{sy} (MPa)	Uniform strain, ε_{su}	Ductility class
Type	Designation grade			
Bar plain to AS/ NZS 4671	R250N	250	0.05	N
Bar deformed to AS/NZS 4671	D500L (fitments only)	500	0.015	L
	D500N	500	0.05	N
Welded wire mesh, plain, deformed or indented to AS/ NZS 4671	D500L	500	0.015	L
	D500N	500	0.05	N

Source: AS 3600, Table 3.2.1. Copied by Mr L. Pack with the permission of Standards Australia under Licence 1607-c010.

Note: Refer to AS/NZS 4671 for details on steel reinforcing materials.

Table 4.3 Bar diameters, areas and masses

Bar diameter (mm)	Area (mm²)	Mass (kg/m)
D500N		
10	78.5	0.617
12	113	0.888
16	201	1.58
20	314	2.47
24	452	3.55
28	616	4.83
32	804	6.31
36	1020	7.99
40	1257	9.86
D500L		
6.00	28.3	0.222
6.75	35.8	0.281
7.60	45.4	0.356
8.60	58.1	0.456
9.50	70.9	0.556
10.7	89.9	0.706
11.9	111.2	0.873
R250N		
6.5	33.2	0.2605
10	78.5	0.6165

Notes:
1. Areas and masses do not include any tolerances or allowances; some manufactures allow an additional 2.5%.
2. Designations shown are for Australia only.
3. Refer to AS/NZS 4671 for details.

Table 4.4 Approximate costs for supply of concrete

Sections	Adelaide	Brisbane	Melbourne	Perth	Sydney
General					
20 MPa plain concrete ($/m³)	150	160	135	160	182
25 MPa plain concrete ($/m³)	156	165	142	165	186
32 MPa plain concrete ($/m³)	163	170	151	172	191
40 MPa plain concrete ($/m³)	172	180	165	179	202
50 MPa plain concrete ($/m³)	190	198	180	199	240
60 MPa plain concrete ($/m³)	205	212	200	262	272
Reinforcement cut and supply ($/t)	1350	1145	1200	1250	1440
Reinforcement placing and fixing ($/t)	750	950	900	800	850
SL92 mesh ($/m²)	5.96	6.37	7.45	5.92	6.50
SL72 mesh ($/m²)	3.71	3.98	4.73	4.08	4.97
Specific items (32 MPa)					
Blinding – 25 MPa ($/m³)	228	252	286	318	368
Beam ($/m³)	224	283	221	249	286
Column/pier ($/m³)	228	287	221	250	287
Strip footing ($/m³)	239	287	221	250	287
Raft foundation ($/m³)	212	283	211	239	274
Slab ($/m³)	232	258	243	273	314
Additional costs					
Pumping rate ($/h)	–	–	80	200	220
Minimum pumping ($)	500	–	320	600	500
Pneumatic placing (additional $)	29	42	25	25	32
Testing (site test and three cylinders)	205	173	252	184	221
Curing ($/m²)	2.00	1.70	1.95	1.70	2.05
Formwork (Class 3) ($/m²) approx.	140	125	200	131	154
Excavation 1 m trench (sand/soft rock) ($/m³)	74/94	37/89	46/95	29/93	49/214

Source: Rawlinsons Quantity Surveyors and Construction Cost Consultants, *Rawlinsons Australian Construction Handbook,* 35th ed., Rawlinsons Publishing, Australia, 2017. Reproduced with permission from Rawlinsons.

Notes:
1. Listed prices are approximate and vary over time. Values are taken from 2017.
2. Some costs are modified from those found in Rawlinsons to provide simple values.
3. Unlisted values are not currently available.
4. Refer to the current edition of the Australian Construction Handbook, Rawlinsons, for further details including waffle slabs, waterproofing, light weight concrete and regional price indices.

10%–20% for remote locations. Additional fees may also be applicable for site access and training. A very rough number to use is $1000–$2000/m³ for supply and installation of concrete (including sundries) on Mining or Oil & Gas sites. Refer to the current edition of the *Australian Construction Handbook*, Rawlinsons [13], for a more detailed estimate. Contact a local supplier when a more accurate estimate is required.

4.1.4 Cover

The cover to reinforcement is the clear distance from outside a reinforcing bar or tendon to the edge of a concrete surface. The required value is dependent on the climate, exposure and concrete

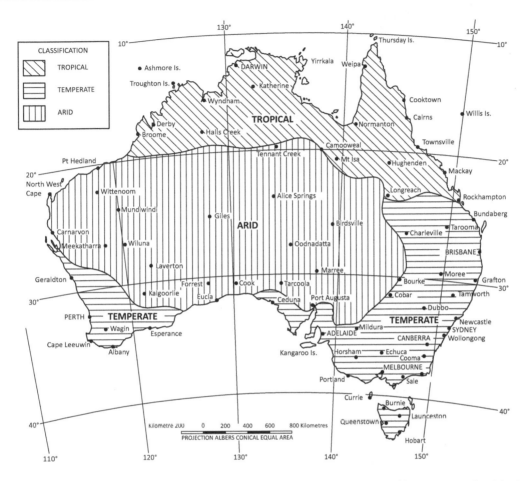

Figure 4.1 Climatic zones in Australia. (From AS 3600, figure 4.3. Copied by Mr L. Pack with the permission of Standards Australia under Licence 1607-c010.)

strength. Typically, the structural team will create a table for each specific project, detailing the required cover for individual items such as beams, slabs, columns and so on (refer to Table 4.8).

Initially, the climatic zone of the site is selected from Figure 4.1. The climatic zone is then used to select the correct exposure classification; refer to Table 4.5. The selected values should be increased by 10 mm when installed against ground with a damp proof membrane, and 20 mm when installed directly against ground. Refer to AS 3600, Section 4 for specific details in relation to cases where abrasion, freezing, thawing, aggressive soils or non-standard concrete chemical content are applicable. The exposure classification and concrete strength are then used to select the appropriate cover from Tables 4.6 and 4.7.

4.1.5 Bar development

The calculations presented for strength operate under the assumption that the bars in the cross-section are fully anchored. This means that the bars have a hook, cog or straight length sufficiently long to ensure that strength can be developed.

Development length calculations are based on information presented in AS 3600, Section 13. The details provided in this section are simplified, and savings may be made for some specific cases when following the full requirements of AS 3600. Many projects create

Table 4.5 Exposure classifications

Surface and exposure environment	Exposure classification reinforced or prestressed concrete members[a]
1. Surface of members in contact with the ground:	
a. Members protected by a damp proof membrane	A1
b. Residential footings in non-aggressive soils	A1
c. Other members in non-aggressive soils	A2
d. Members in aggressive soils:	
i. Sulfate bearing (magnesium content <1 g/L)	Refer to AS 3600, T 4.8.1
ii. Sulfate bearing (magnesium content ≥1 g/L)[b]	U
iii. Other	U
e. Salt-rich soils and soils in areas affected by salinity	Refer to AS 3600, T 4.8.2
2. Surface of members in interior environments:	
a. Fully enclosed within a building except for a brief period of weather exposure during construction:	
i. Residential	A1
ii. Non-residential	A2
b. In industrial buildings, the member being subject to repeated wetting and drying	B1
3. Surface of members in above-ground exterior environments in areas that are	
a. Inland (>50 km from coastline), environment being	
i. Non-industrial and arid climatic zone[c]	A1
ii. Non-industrial and temperate climatic zone	A2
iii. Non-industrial and tropical climatic zone	B1
iv. Industrial[d] and any climatic zone	B1
b. Near-coastal (1 to 50 km from coastline), any climatic zone	B1
c. Coastal[e] and any climatic zone	B2
4. Surfaces of members in water:	
a. In freshwater	B1
b. In soft or running water	U
5. Surfaces of maritime structures in sea water:	
a. Permanently submerged	B2
b. In spray zone[f]	C1
c. In tidal/splash zone[g]	C2
6. Surfaces of members in other environments, i.e. any exposure environment not specified in Items 1 to 5[h]	U

Source: AS 3600, table 4.3. Copied by Mr L. Pack with the permission of Standards Australia under Licence 1607–c010.

Note: In this table, classifications A1, A2, B1, B2, C1 and C2 represent increasing degrees of severity of exposure, while classification U represents an exposure environment not specified in this table but for which a degree of severity of exposure should be appropriately assessed. Protective surface coatings may be taken into account in the assessment of the exposure classification.

[a] In this context, reinforced concrete includes any concrete containing metals that rely on the concrete for protection against environmental degradation. Plain concrete members containing metallic embedments should be treated as reinforced members when considering durability.

[b] Severity of sulfate attack depends on the type of sulfate. For example, magnesium sulfate is more aggressive than sodium sulfate. The use of sulfate-resisting cement and concrete would be adequate for sodium sulfate conditions. For the magnesium sulfate conditions, specific consideration should be given to the cement and concrete that are likely to resist this type of sulfate.

[c] The climatic zones referred to are those given in Figure 4.1, which is based on the Bureau of Meteorology map, Major seasonal rainfall zones of Australia, Commonwealth of Australia, 2005.

[d] *Industrial* refers to areas that are within 3 km of industries that discharge atmospheric pollutants.

[e] For the purpose of this table, the coastal zone includes locations within 1 km of the shoreline of large expanses of saltwater. Where there are strong prevailing winds or vigorous surf, the distance should be increased beyond 1 km, and higher levels of protection should be considered.

[f] The spray zone is the zone from 1 m above wave crest level.

[g] The tidal/splash zone is the zone 1 m below lowest astronomical tide (LAT) and up to 1 m above highest astronomical tide (HAT) on vertical structures and all exposed soffits of horizontal structures over the sea.

[h] Further guidance on measures appropriate in exposure classification U may be obtained from AS 3735.

Table 4.6 Required cover where standard formwork and compaction are used

Exposure classification	Required cover (mm)				
	Characteristic strength, (f'c) (MPa)				
	20	25	32	40	≥50
A1	20	20	20	20	20
A2	(50)	30	25	20	20
B1	–	(60)	40	30	25
B2	–	–	(65)	45	35
C1	–	–	–	(70)	50
C2	–	–	–	–	65

Source: AS 3600, Table 4.10.3.2. Copied by Mr L. Pack with the permission of Standards Australia under Licence 1607-c010.

Note: Bracketed figures are the appropriate covers when the concession given in AS 3600, Clause 4.3.2, relating to the strength grade permitted for a particular exposure classification, is applied.

Table 4.7 Required cover where repetitive procedures and intense compaction or self-compacting concrete are used in rigid formwork

Exposure classification	Required cover (mm)				
	Characteristic strength (f'c) (MPa)				
	20	25	32	40	≥50
A1	20	20	20	20	20
A2	(45)	30	20	20	20
B1	–	(45)	30	25	20
B2	–	–	(50)	35	25
C1	–	–	–	(60)	45
C2	–	–	–	–	60

Source: AS 3600, Table 4.10.3.3. Copied by Mr L. Pack with the permission of Standards Australia under Licence 1607-c010.

Note: Bracketed figures are the appropriate covers when the concession given in AS 3600, Clause 4.3.2, relating to the strength grade permitted for a particular exposure classification, is applied.

a table of development lengths to be used to prevent excessive repetition of these calculations (refer to Table 4.9) and to avoid confusion.

4.1.5.1 Bars in tension

The basic development length for tension is the straight development length that is used in general cases.

Basic development length, $L_{sy.tb} = \dfrac{0.5 k_1 k_3 f_{sy} d_b}{k_2 \sqrt{f'_{c*}}} \geq 29 k_1 d_b$

where:

f'_{c*} = Compressive concrete strength using a maximum value of 65
d_b = bar diameter (mm)
k_1 = 1.3 (for horizontal bars with more than 300 mm of concrete below the bar)
   = 1.0 (otherwise)

Table 4.8 Example table for standard cover on a project

Item	Cover (mm)
Concrete deposited directly against ground	75
Concrete deposited directly against a damp proof membrane on ground	65
Concrete deposited against a damp proof membrane on formwork and backfilled against ground	50
Top surfaces of slabs	40
Bottom surfaces of slabs	50
Bored piers	65
Other formed exposed surfaces	50

Note: This table uses selected values for a hypothetical project with non-aggressive soil in a non-coastal area.

Table 4.9 Example table for standard bar development lengths

Bar size	Lap length (mm)		Cog length (mm)
	Horizontal bars with >300 mm concrete cast below bar	*All other bars*	
N10	500	400	160
N12	600	450	170
N16	800	600	210
N20	1000	800	250
N24	1400	1100	300
N28	1800	1400	350

Notes:
1. Reinforcement splices should use 125% of the value obtained.
2. Splices should be staggered where possible.
3. Where less than 50% of the reinforcement is spliced at one section, lap lengths may be 80% of the values calculated using Note 1.
4. This table uses selected values for a hypothetical project.

$$k_2 = (132 - d_b)/100$$
$$k_3 = 1.0 - 0.15 (c_d - d_b)/d_b \quad [0.7 \leq k_3 \leq 1.0]$$

$$c_d = \text{MIN}[a/2, c_1, c] \quad \text{(straight bars) (refer to Figure 4.2)}$$
$$c_d = \text{MIN}[a/2, c_1] \quad \text{(cogged or hooked bars) (refer to Figure 4.2)}$$
$$c_d = c \text{ (looped bars)} \quad \text{(refer to Figure 4.2)}$$
$$c_d = \text{MIN}[(S_L - d_b)/2, c] \quad \text{(for lapped splices) (refer to Figure 4.4)}$$
$$c_d = \text{MIN}[S_L, c] \quad \text{(for staggered lapped splices) (refer to Figure 4.5)}$$

The edge distance variable (c_1) can be removed from equations in wide cross-sections such as slabs and walls (where side cover is not relevant). For staggered laps, the clear spacing (a) can be taken as the distance between bars being developed (rather than the distance between adjacent bars in the section).

This 'basic' length should be increased by the following factors where applicable:

$$L_{sy.t} = 1.0 \times L_{sy.tb} \quad \text{(for normal deformed bars)}$$
$$L_{sy.t} = 1.5 \times L_{sy.tb} \quad \text{(for epoxy-coated bars)}$$
$$L_{sy.t} = 1.3 \times L_{sy.tb} \quad \text{(for lightweight concrete)}$$
$$L_{sy.t} = 1.3 \times L_{sy.tb} \quad \text{(for structural elements built with slip forms)}$$
$$L_{sy.t} = 1.5 \times L_{sy.tb} \quad \text{(for plain bars)} \ L_{sy} \geq 300 \text{ mm}$$
$$L_{sy.t} = 0.4 \times L_{sy.tb} \quad \text{(for bars with end plates complying with AS 3600, Cl 13.1.4)}$$

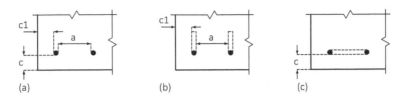

Figure 4.2 Values of c_d. (a) Straight bars, (b) cogged or hooked, (c) looped bars. (From AS 3600, figure 13.1.2.3(A). Copied by Mr L. Pack with the permission of Standards Australia under Licence 1607-c010.)

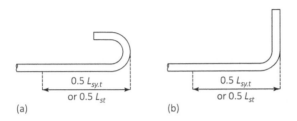

Figure 4.3 Hook and cog lengths. (a) Standard hook, (b) standard cog. (From AS 3600, figure 13.1.2.6. Copied by Mr L. Pack with the permission of Standards Australia under Licence 1607-c010.)

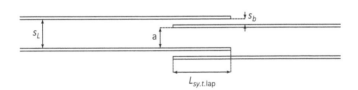

Figure 4.4 Value of c_d for lapped splices. (From AS 3600, figure 13.2.2. Copied by Mr L. Pack with the permission of Standards Australia under Licence 1607-c010.)

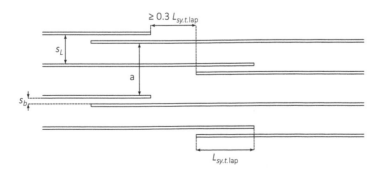

Figure 4.5 Value of c_d for staggered lapped splices. (From AS 3600, figure 13.2.2. Copied by Mr L. Pack with the permission of Standards Australia under Licence 1607-c010.)

Hooks and cogs are generally used at the ends of beams and columns to provide sufficient anchorage. The length required for development in a hook or cog is half of that required for a straight bar, as shown in Figure 4.3.

A hook is required to complete a 135° or 180° turn on a diameter of $4d_b$ (for D500N bars), followed by a straight extension of $4d_b$ or 70 mm, whichever is greater. A cog is required to bend 90° with an internal diameter equal to the smaller of $4d_b$ (for D500N bars) and $8d_b$, and with the same total length required for a hook.

If the required lengths cannot be achieved, the appropriate percentage of the strength can be adopted (i.e. a bar which only has 75% of the required development length can be assumed to reach 75% of its yield strength), although the length cannot be less than 12 times the bar diameter, and additional rules are supplied for slabs.

Reduced development length, $L_{st} = L_{sy.t}\left(\sigma_{st}/f_{sy}\right)$

where: σ_{st} = Developed tensile stress

Methods for calculating lengths for mesh and bundled bars are presented in AS 3600, Cl 13.1.

4.1.5.2 Lapped splices in tension

Laps should not be used for bars over 40 mm in diameter. The lap dimension required for normal bars in tension is calculated as follows:

Lap length, $L_{sy.t.lap} = k_7 L_{sy.t} \geq 29k_1 d_b$

where:
  k_7 = 1.25 (for general cases)
  k_7 = 1.0 (where A_s is at least twice that required and no more than half the reinforcement in the section is being spliced)

For narrow elements (beam webs and columns), $L_{sy.t.lap}$ shall also be not less than $L_{sy.t} + 1.5s_b$, where s_b is taken from Figures 4.4 and 4.5; however, a value of zero is used if $s_b < 3d_b$.

4.1.5.3 Bars in compression

The basic development length for bars in compression may be calculated as follows:

$$L_{sy.cb} = \frac{0.22f_{sy}}{\sqrt{f_c'}} d_b \geq 0.0435 f_{sy} d_b \geq 200 \text{ mm}$$

This value may also be factored down where less than yield strength is required; however, it must remain above 200 mm. Bends and hooks should not be used in length calculations for bars in compression. Plain bars require double the development length for compression.

4.1.5.4 Lapped splices in compression

Laps should not be used for bars over 40 mm in diameter. The lap dimension required for normal bars in compression is calculated as follows:

Lap length, $L_{sy.c.lap} = L_{sy.c} \geq 40\ d_b \geq 300 \text{ mm}$

4.2 BEAMS

Beams are designed for bending, shear and torsion loads. If a beam is subject to axial loads, it should be checked as a column. The bending and shear capacities of beams should be determined in accordance with Section 8 of AS 3600. Capacities for typical sections are provided in Section 7.3.2.

4.2.1 Reinforcement requirements

Reinforcing bars should be spaced at minimum centres which allow proper placement of concrete (i.e. larger than 1.5 times the aggregate size is recommended and at least one bar diameter). Maximum clearances provided are for crack control: a maximum of 100 mm from the side or soffit of a beam to the centre of the nearest bar, and a maximum of 300 mm centres on tension faces. Bars with diameters equal to less than half of the largest diameter in the section should be ignored.

Shear reinforcement is required when $V^* > 0.5\phi V_{uc}$. The ligatures should be spaced longitudinally at maximum centres of 0.5 D or 300 mm, whichever is less. This value may be increased to 0.75 D or 500 mm when $V^* \leq \phi V_{uo.min}$. The maximum transverse spacing of shear ligatures should be less than the minimum of 600 mm and D.

4.2.2 Crack control

Cracking in reinforced concrete beams is deemed to be controlled where the above reinforcement requirements are met, minimum strength is provided (refer to Section 4.2.3) and stress in the bars is controlled.

Where beams are fully enclosed within a building (except for a brief period during construction) and crack control is not required, the bar stress limits can be ignored.

Stress under service loading shall not exceed 80% of the reinforcement yield strength and shall also be limited to the values shown in Tables 4.10 and 4.11. This stress can be calculated using the method shown in Section 4.5.

Table 4.10 Maximum steel stress for tension or flexure in reinforced beams

Nominal bar diameter (d_b) (mm)	Maximum steel stress (MPa)
10	360
12	330
16	280
20	240
24	210
28	185
32	160
36	140
40	120

Source: AS 3600, table 8.6.1(A). Copied by Mr L. Pack with the permission of Standards Australia under Licence 1607-c010.

Note: Values for other bar diameters may be calculated using the equation
Max stress $= -173 \log_e (d_b) + 760$ MPa.

Table 4.11 Maximum steel stress for flexure in reinforced beams

Centre-to-centre spacing (mm)	Maximum steel stress (MPa)
50	360
100	320
150	280
200	240
250	200
300	160

Source: AS 3600, table 8.6.1(B). Copied by Mr L. Pack with the permission of Standards Australia under Licence 1607-c010.

Note: Intermediate values may be calculated using the equation Max stress = –0.8 x centre-to-centre spacing + 400 MPa.

4.2.3 Beams in bending

The strength of a beam in bending is calculated by resolving the tension and compressive forces about the neutral axis. The bending capacity is the sum of each compressive and tensile force multiplied by its distance from the neutral axis.

The tensile force in the steel is calculated based on the strain, which may or may not reach yield. For class 500N reinforcing bars, a strain of 0.0025 will achieve the full yield strength of the bar. Any smaller value will result in a fraction of the bar strength being used. A strain greater than 0.05 would result in failure of the bar. Calculations in this section assume a normal ductility; for low ductility reinforcement, the limiting strain value of 0.05 should be replaced with 0.015 (refer to Table 4.2) and the reduction factor is decreased.

$$\sigma = E_s \varepsilon_s \leq f_{sy}$$

$$E_s = 200,000 \text{ MPa}$$

$$\varepsilon_s = \text{Strain in steel} \leq 0.05 \quad \text{(Normal ductility – standard reinforcement)}$$
$$\leq 0.015 \quad \text{(Low ductility – mesh)}$$

Non-linear behaviour of the compression block is converted into an equivalent compression block. For the purpose of this text, the compression zone is defined as the concrete on the compression side of the neutral axis; the compression block is defined as the equivalent uniform stress block (equal to the size of the compression zone multiplied by γ). Figure 4.6 shows how the actual stress is converted into an equivalent stress block.

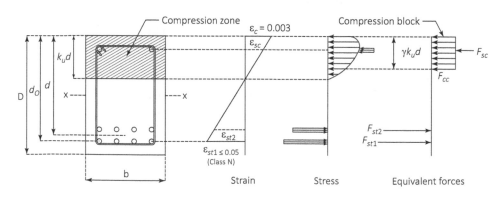

Figure 4.6 Concrete bending behaviour.

Beams should generally be designed to be ductile to provide early visual warnings of failure. If a beam is 'over-reinforced', it is governed by compressive failure of the concrete; however, an 'under-reinforced' beam is governed by the failure of steelwork. The amount of steel reinforcement in a beam should ideally be proportioned to ensure its failure prior to concrete crushing failure. An 'under-reinforced' classification is therefore preferable.

Both bending and torsion loads cause tension in the longitudinal reinforcement of a beam. If torsion is present, additional tension steel is often required in the section. This affects the ductility of the beam, and therefore, calculations may have to be repeated. Beams with compressive loads should be designed as columns (refer to Section 4.4).

4.2.3.1 Minimum strength requirements

The minimum bending moment required for a section, $(M_{uo})_{min}$, is based on 1.2 times the cracking moment. This is to prevent beams from being designed in which the section fails prior to concrete cracking, as this failure mode is sudden and can be catastrophic. It is also because strength calculations for beams in bending are based on the cracked section. This clause becomes a limiting requirement when very small percentages of reinforcement are used.

$$(M_{uo})_{min} = 1.2\left[Z\left(f'_{ct.f}\right) \right]$$

where:

$$f'_{ct.f} = 0.6\sqrt{f'_c}$$

$$Z = bd^2 / 6$$

Note: Calculations for prestressed beams are excluded; refer to AS 3600, Section 8 for details.

The minimum strength requirement is deemed to be satisfied for rectangular sections where

$$A_{st} \geq \left[\alpha_b \left(D/d \right)^2 f'_{ct.f} / f_{sy} \right] bd$$

where: $\alpha_b = 0.2$ (for rectangles; for other shapes, refer to AS 3600 Cl 8.1.6)

Sections in which $k_{uo} > 0.36$ and in which $M^* > 0.6 M_{uo}$ should only be used under certain conditions; refer to AS 3600, Cl 8.1.5 for details.

4.2.3.2 Ultimate strength in bending

The ultimate strength of a beam in bending, ϕM_{uo}, is calculated with the restrictions stipulated in AS 3600, Clause 8.1. Concrete is assumed to have no tensile strength and a maximum compressive strain of 0.003. The neutral axis is at a depth of $k_u d$, and a uniform compressive stress of $\alpha_2 f'_c$ is bounded by an area of $b \times \gamma k_u d$ (refer to Figures 4.7 and 4.8) for rectangular cross-sections.

$$M^* \leq \phi M_{uo}$$

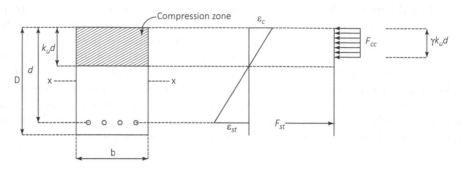

Figure 4.7 Singly reinforced beam in bending.

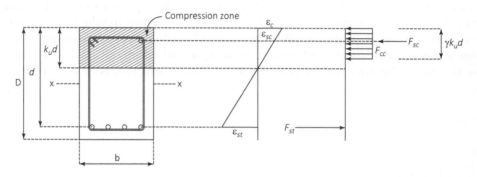

Figure 4.8 Doubly reinforced beam in bending.

b = Width of section (refer to Figure 4.8)
d = Depth to the centroid of tension reinforcement
d_o = Depth to the outermost tension reinforcement
k_u = Ratio of depth of the compressive zone to d
$\alpha_2 = 1 - 0.003\,f_c'$ where: $[0.67 \le \alpha_2 \le 0.85]$

$\gamma = 1.05 - 0.007\,f_c'$ where: $[0.67 \le \gamma \le 0.85]$

Reduction factor for pure bending (no compression),
$\phi = 0.6 \le (1.19 - 13k_{uo}/12) \le 0.8$ (Class N reinforcement)
$\phi = 0.6 \le (1.19 - 13k_{uo}/12) \le 0.64$ (Class L reinforcement)

k_{uo} is equal to the ratio of the depth of the compression zone to the depth of the outermost tension bar (d_o) from the extreme compressive fibre under pure bending. For cases where there is only one layer of tension reinforcement, $d = d_o$, and therefore $k_u = k_{uo}$. If there are two or more layers of reinforcement in the tensile zone, d is the depth to the centroid of the reinforcement, and d_o is the depth to the centre of the outer layer.

Ratio of compressive zone to d_o, $k_{uo} = k_u d/d_o$

4.2.3.3 Singly reinforced beam

Beams which only have reinforcement on the tension face can be solved using the follow-ing method. Beams which have an insignificant amount of compression reinforcement may also be solved using the same method and by ignoring the contribution of compression steel. If there is more than one layer of tension reinforcement, the same method can be used (ensuring that d is taken as the depth to the centroid of the tension reinforcement; refer to Figure 4.6); however, strain should be checked using the depth to each bar to confirm yield and check for rupture.

Assuming tension steel has yielded,

Tension in steel, $\qquad\qquad F_{st} = -A_{st}f_{sy}$

Compression in concrete, $\quad F_{cc} = \alpha_2 f'_c b \gamma k_u d$

Equilibrium can be used to equate tension and compression and solve for the only unknown (k_u).

Equilibrium equation, $\quad F_{st} + F_{cc} = 0$

Neutral axis ratio, $\qquad k_u = A_{st}f_{sy} / \left(\alpha_2 f'_c b \gamma d\right)$

The bending moment capacity is then calculated by multiplying the tension force by the distance from the centroid of the tension bars to the centre of the compression zone.

Moment capacity, $\quad M_{uo} = A_{st}f_{sy} \times \left(d - 0.5\gamma k_u d\right)$

Strain in the steel can then be checked to ensure that the tension steel can yield with the maximum compressive strain of 0.003 in the concrete. This is done by using similar triangles (refer to Figure 4.7). The tensile strain should be between −0.0025 (to ensure yield) and −0.05 (ductility class N; refer to Table 4.2). A negative sign is used for tension.

Strain in concrete, $\qquad \varepsilon_c = 0.003$

Strain in tensile steel, $\quad \varepsilon_{st} = \dfrac{\varepsilon_c}{k_u d}\left(k_u d - d\right)$

If the strain is less than yield, the previous calculations need to be repeated with $A_{st}f_{sy}$ replaced by $A_{st}\varepsilon_{st}E_s$. This replaces the full tensile capacity of each bar with the tension that can be achieved prior to concrete failure. This will result in an over-reinforced design. Alternatively, the amount of reinforcement can be reduced to achieve a more ductile (under-reinforced) design.

Strain can also be checked at any depth, y, measured from the extreme compressive (top) fibre (used for multiple layers of reinforcement).

Strain at y, $\quad \varepsilon_{sy} = \dfrac{\varepsilon_c}{k_u d}\left(k_u d - y\right)$

4.2.3.4 Doubly reinforced beam

A doubly reinforced beam uses tension and compression reinforcement. Strain in the concrete is limited to 0.003, and the steel strains are solved by using similar triangles and balancing the tension and compression forces. For a typical rectangular section, strain is calculated using similar triangles, as shown in Figure 4.8.

Strain in concrete, $\qquad \varepsilon_c = 0.003$

Strain in tensile steel, $\qquad \varepsilon_{st} = \left(k_u d - d\right)\dfrac{\varepsilon_c}{k_u d}$

Strain in compression steel, $\quad \varepsilon_{sc} = \varepsilon_c - \dfrac{\varepsilon_c\left(c + d_b/2\right)}{k_u d}$

where:

$\quad c \qquad$ = Cover to compression bars

$\quad d_b \qquad$ = Diameter of compression bars

$\quad c + \dfrac{d_b}{2} \quad$ = Depth from top of section to centre of compression bars

The steel tension and compression strains should be smaller in magnitude than 0.05 (ductility class N; refer to Table 4.2).

Tension in steel, $\quad F_{st} = A_{st}\varepsilon_{st}E_s \geq -A_{st}f_{sy}$

Note: The absolute value of $A_{st}\varepsilon_{st}E_s$ is less than $A_{st}f_{sy}$; however, a negative sign is used for tension. The value with the lowest magnitude is required.

Compression in concrete, $\quad F_{cc} = \alpha_2 f'_c b\gamma k_u d$

Compression in steel, $\qquad F_{sc} = A_{sc}\varepsilon_{sc}E_s \leq A_{sc}f_{sy}$

The equilibrium equation requires the sum of forces to equal zero. This can be used to solve for the unknown value of k_u by using an iterative approach (such as goal seek in Excel or guess and check) or by making assumptions (presented in the next section) and then solving for k_u and checking the assumptions.

Equilibrium equation, $\quad F_{st} + F_{cc} + F_{sc} = 0$

Once k_u is calculated, the moment capacity is found by taking moments about any point. It is convenient to take the moments about the centre of tensile steel:

Moment capacity, $\quad M_{uo} = F_{cc}\times\left[d - 0.5\gamma k_u d\right] + F_{sc}\times\left[d - \left(c + d_b/2\right)\right]$

4.2.3.5 Assumption method

The following equations are only required when the assumption method is used to solve for k_u. The purpose of them is to take account of the $\leq$ and $\geq$ operators in the force equations for steel. Start at assumption 1, and move down the list if the assumptions are proved incorrect.

1. Assume tension steel has yielded and compression steel has yielded:

    therefore, $\qquad F_{st} = -A_{st}f_{sy} \quad$ and $\quad F_{sc} = A_{sc}f_{sy}$

    using equilibrium, $\quad -A_{st}f_{sy} + \alpha_2 f'_c b\gamma k_u d + A_{sc}f_{sy} = 0$

    $$k_u = \dfrac{\left(A_{st} - A_{sc}\right)f_{sy}}{\alpha_2 f'_c b\gamma d}$$

Check the assumptions, $\varepsilon_{st} \leq -0.0025$ and $\varepsilon_{sc} \geq 0.0025$ (Negative sign for tension)
Check the steel has not ruptured, $\varepsilon_{st} \geq -0.05$ and $\varepsilon_{sc} \leq 0.05$

2. Assume tension steel has yielded and compression steel has not yielded:

using equilibrium, $\quad -A_{st}f_{sy} + \alpha_2 f_c' b \gamma k_u d + A_{sc}\varepsilon_{sc}E_s = 0$

$$-A_{st}f_{sy} + \alpha_2 f_c' b \gamma k_u d + A_{sc}\left[\varepsilon_c - \frac{\varepsilon_c(c + d_b/2)}{k_u d}\right]E_s = 0$$

This can be solved using the quadratic formula (after multiplying both sides by k_u and expanding):

$$k_u = \frac{-B \pm \sqrt{B^2 - 4AC}}{2A}$$

$$A = \left[\alpha_2 f_c' b \gamma d\right]$$

$$B = \left[-A_{st}f_{sy} + A_{sc}\varepsilon_c E_s\right]$$

$$C = \left[-A_{sc}\frac{\varepsilon_c(c + d_b/2)}{d}E_s\right]$$

Check the assumptions, $\varepsilon_{st} \leq -0.0025$ and $\varepsilon_{sc} < 0.0025$
Check the tension steel has not ruptured, $\varepsilon_{st} \geq -0.05$

3. Assume tension steel has not yielded and compression steel has yielded:

using equilibrium, $\quad A_{st}\varepsilon_{st}E_s + \alpha_2 f_c' b \gamma k_u d + A_{sc}f_{sy} = 0$

$$A_{st}\left[(k_u d - d)\frac{\varepsilon_c}{k_u d}\right]E_s + \alpha_2 f_c' b \gamma k_u d + A_{sc}f_{sy} = 0$$

This can be solved using the quadratic formula:

$$k_u = \frac{-B \pm \sqrt{B^2 - 4AC}}{2A}$$

$$A = \left[\alpha_2 f_c' b \gamma d\right]$$

$$B = \left[A_{st}\varepsilon_c E_s + A_{sc}f_{sy}\right]$$

$$C = \left[-A_{st}\varepsilon_c E_s\right]$$

Check the assumptions, $\varepsilon_{st} > -0.0025$ and $\varepsilon_{sc} \geq 0.0025$
Check the compression steel has not ruptured, $\varepsilon_{sc} \leq 0.05$

4. Assume tension steel has not yielded and compression steel has not yielded:

using equilibrium, $\quad A_{st}\varepsilon_{st}E_s + \alpha_2 f_c' b \gamma k_u d + A_{sc}\varepsilon_{sc}E_s = 0$

$$A_{st}\left[(k_u d - d)\frac{\varepsilon_c}{k_u d}\right]E_s + \alpha_2 f_c' b \gamma k_u d + A_{sc}\left[\varepsilon_c - \frac{\varepsilon_c(c + d_b/2)}{k_u d}\right]E_s = 0$$

This can be solved using the quadratic formula:

$$k_u = \frac{-B \pm \sqrt{B^2 - 4AC}}{2A}$$

$$A = [\alpha_2 f_c' b \gamma d]$$

$$B = [A_{st}\varepsilon_c E_s + A_{sc}\varepsilon_c E_s]$$

$$C = \left[-A_{st}\varepsilon_c E_s - A_{sc}\frac{\varepsilon_c(c + d_b/2)}{d}E_s\right]$$

Check the assumptions, $\varepsilon_{st} > -0.0025$ and $\varepsilon_{sc} < 0.0025$

Example 4.1: Bending Capacity

Calculate the bending capacity of a 600 mm deep × 400 mm wide concrete beam. 4-N16 reinforcing bars are provided on the tension face. Shear ligatures are R10-200 with a cover of 50 mm. Concrete is grade N40.

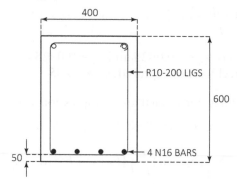

Beam details:

$$A_{st} = 4 \times \pi \times 16^2/4 = 804.2 \text{ mm}^2$$

$$d = 600 - 50 - 10 - 16/2 = 532 \text{ mm}$$

$$f_{sy} = 500\,\text{MPa}$$

$$f_c' = 40\,\text{MPa}$$

$$\alpha_2 = 1 - 0.003 f_c' = 0.88 \le 0.85 = 0.85 \qquad [0.67 \le \alpha_2 \le 0.85]$$

$$\gamma = 1.05 - 0.007 f_c' = 0.77 \qquad [0.67 \le \gamma \le 0.85]$$

Bending capacity:

$$F_{st} = -A_{st}f_{sy} = -804.2 \times 500 = -402 \text{ kN}$$

$$k_u = A_{st}f_{sy}/(\alpha_2 f_c' b \gamma d) = 804.2 \times 500/(0.85 \times 40 \times 400 \times 0.77 \times 532) = 0.0722$$

$$F_{cc} = \alpha_2 f_c' b \gamma k_u d = 0.85 \times 40 \times 400 \times 0.77 \times 0.0722 \times 532 = 402 \text{ kN}$$

$$F_{st} + F_{cc} = 0 \quad \text{PASS}$$

$$M_{uo} = A_{st}f_{sy} \times (d - 0.5\gamma k_u d) = 804.2 \times 500 \times (532 - 0.5 \times 0.77 \times 0.0722 \times 532) = 208 \text{ kNm}$$

Reduction factor for pure bending (no compression) (Class N reinforcement)

$$\phi = 0.6 \le (1.19 - 13k_{uo} / 12) \le 0.8$$

$$= 0.6 \le (1.19 - 13 \times 0.0708 / 12) \le 0.8 = 0.8$$

$$\phi M_{uo} = 0.8 \times 208 = 166.4 \text{ kNm}$$

Check strain:

$$\varepsilon_c = 0.003$$

$$\varepsilon_{st} = \frac{\varepsilon_c}{k_u d}(k_u d - d) = \frac{0.003}{0.0722 \times 532}(0.0722 \times 532 - 532) = -0.039$$

$-0.05 \le \varepsilon_{st} \le -0.0025$ PASS – The steel has yielded, but not failed.

The bending capacity of the beam is 166.4 kNm.

4.2.4 Beams in shear

The shear capacity of a beam (or column) is calculated in accordance with AS 3600, Clause 8.2. The total shear capacity is a combination of the shear capacity of the concrete and the shear capacity of the reinforcement. Effects of prestressing are excluded from this section; refer to AS 3600 for details.

4.2.4.1 Ultimate shear strength

The ultimate shear strength of a beam is calculated in accordance with the following equation; however, it may be increased to $\phi V_{u.min}$ in beams which are provided with the required minimum area of shear reinforcement.

$$\phi V_u = \phi(V_{uc} + V_{us}) \le \phi V_{u.max}$$

where:

ϕ = 0.7 (for shear)

V_{uc} should be excluded if significant load reversal or torsion may occur and cause cracking of the zone usually used in compression.

4.2.4.2 Area of shear reinforcement

The area of shear reinforcement, A_{sv}, is the cross-sectional area of the ligatures which are tied in the same direction as the applied load. They may also be inclined longitudinally so that they would run perpendicular to the shear failure plane. The number of bars which are included for shear capacity in each direction is shown diagrammatically in Figure 4.9.

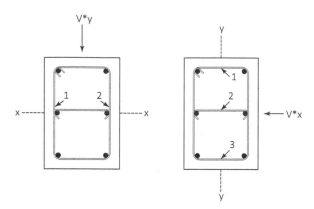

Figure 4.9 Number of shear bars.

4.2.4.3 Minimum shear strength

If a minimum area of shear reinforcement, $A_{sv.\min}$, is provided in a beam, it is deemed to achieve a minimum shear strength of $V_{u.\min}$. There is no need for further calculations if $\phi V_{u.\min}$ is greater than the design shear load.

$$A_{sv.\min} = 0.06\sqrt{f_c'}b_v s/f_{sy.f} \geq 0.35 b_v s/f_{sy.f}$$

where:

b_v $= b_w - 0.5\sum d_d$
b_v $= b$ for rectangular cross-sections with no ducts
b_w $=$ width of web
$\sum d_d$ $=$ sum of diameter of grouted ducts in any horizontal plane across the web
$f_{sy.f}$ $=$ Yield strength of shear bars, commonly 500 MPa (500 MPa for D500L or D500N and 250 MPa for R250N bars)
s $=$ Centre to centre spacing of shear bars

$$V_{u.\min} = V_{uc} + 0.10\sqrt{f_c'}b_v d_o \geq V_{uc} + 0.6\,b_v d_o$$

4.2.4.4 Concrete shear strength

Shear capacity is usually checked at the support face and at $2d_o$ from the support face, as they are typically the governing locations, due to the β_3 coefficient.

Shear strength of concrete, $\quad V_{uc} = \beta_1\beta_2\beta_3 b_v d_o f_{cv}\left[\dfrac{A_{st}}{b_v d_o}\right]^{1/3}$

where:

A_{st} $=$ Area of longitudinal steel, fully anchored in the tensile zone

$f_{cv} = f_c'^{1/3} \leq 4MPa$

where shear reinforcement is greater than $A_{sv.\min}$, $\beta_1 = 1.1(1.6 - d_o/1000) \geq 1.1$

otherwise, $\beta_1 = 1.1(1.6 - d_o/1000) \geq 0.8$

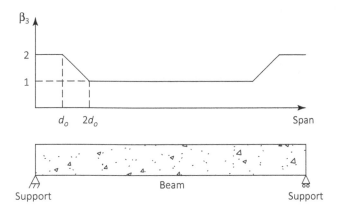

Figure 4.10 β_3 coefficient diagram.

and where:

β_2 = 1, (for members subject to pure bending)

β_2 = $1 - (N^*/3.5A_g) \geq 0$, (for members subject to axial tension)

β_2 = $1 + (N^*/14\ A_g) \geq 0$, (for members subject to axial compression)

β_3 = 1,

β_3 can be increased linearly from 1 at $2d_o$ away from a support up to 2 at d_o from the support; and remains at a maximum of 2 between the face of the support and d_o from the support (refer to Figure 4.10)

4.2.4.5 Shear strength of ligatures

Shear strength of steel,

$$V_{us} = \left(A_{sv}f_{sy.f}d_o/s\right)\cot\theta_v \qquad \text{(perpendicular shear reinforcement)}$$

$$V_{us} = \left(A_{sv}f_{sy.f}d_o/s\right)\left(\sin\alpha_v \cot\theta_v + \cos\alpha_v\right) \qquad \text{(inclined shear reinforcement)}$$

where:

$\cot\theta_v = 1/\tan\theta_v$

α_v = Angle between inclined shear reo and longitudinal tensile reo

θ_v = 45°; or

θ_v = chosen in the range of 30°–60°

(except that the minimum value shall be taken as varying linearly from 30° when $V^* = \phi V_{u.\min}$ to 45° when $V^* = \phi V_{u.\max}$)

4.2.4.6 Maximum shear strength

The maximum shear strength of a beam is limited by web crushing and shall not be taken as greater than $V_{u.\max}$ irrespective of any other calculations.

$$V_{u.\max} = 0.2f_c'b_v d_o$$

Example 4.2: Beam shear

Calculate the shear capacity of a 600 mm deep x 400 mm wide concrete beam at the support face and at $2d_o$ from the support. 4-N16 reinforcing bars are provided on the tension face. Shear ligatures are a single closed loop of R10-200 with a cover of 50 mm. Concrete is grade N40. The beam is not loaded axially.

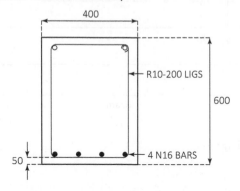

Beam details:

$$s = 200 \text{ mm}$$
$$f_{sy.f} = 250 \text{ MPa}$$
$$f'_c = 40 \text{ MPa}$$
$$f_{cv} = f'^{1/3}_c \le 4 \text{ MPa} = 40^{\frac{1}{3}} = 3.42 \text{ MPa}$$

$$A_{st} = 4 \times \pi \times 16^2 / 4 = 804.2 \text{ mm}^2$$
$$A_{sv} = 2 \times \pi \times 10^2 / 4 = 157.1 \text{ mm}^2$$

$$b_v = 400 \text{ mm}$$
$$d_o = 600 - 50 - 10 - 16/2 = 532 \text{ mm}$$

$$\beta_1 = 1.1\left(1.6 - d_o / 1000\right) \ge 1.1$$
$$= 1.1\left(1.6 - 532 / 1000\right) \ge 1.1$$
$$= 1.175 \ge 1.1 = 1.175$$

$$\beta_2 = 1, \quad \text{(for members subject to pure bending)}$$
$$\beta_3 = 2, \quad \text{(at the support face)}$$
$$\beta_3 = 1, \quad \text{(at } 2d_o \text{ from the support face)}$$

Minimum shear reinforcement:

$$A_{sv.min} = 0.06\sqrt{f'_c}\,b_v s / f_{sy.f} \ge 0.35\,b_v s / f_{sy.f}$$

$$0.06\sqrt{40}\times400\times200/250 \geq 0.35\times400\times200/250$$

$$121.4 \geq 112 = 121.4 \text{ mm}$$

The area of shear reinforcement provided in the beam is greater than $A_{sv.\min}$; therefore, the minimum shear strength clause is relevant.

Shear capacity (using subscript 1 for the support face and 2 for $2d_o$ from the support face):

Concrete capacity,

$$V_{uc(1)} = \beta_1\beta_2\beta_3 b_v d_o f_{cv}\left[\frac{A_{st}}{b_v d_o}\right]^{1/3} = 1.175\times1\times2\times400\times532\times3.42\left[\frac{804.2}{400\times532}\right]^{1/3} = 266.4 \text{ kN}$$

$$V_{uc(2)} = \beta_1\beta_2\beta_3 b_v d_o f_{cv}\left[\frac{A_{st}}{b_v d_o}\right]^{1/3} = 1.175\times1\times1\times400\times532\times3.42\left[\frac{804.2}{400\times532}\right]^{1/3} = 133.2 \text{ kN}$$

Steel capacity,

$$V_{us} = \left(A_{sv}f_{sy.f}d_o/s\right)\cot\theta_v = (157.1\times250\times532/200)\cot45 = 104.5 \text{ kN}$$

Minimum shear strength,

$$V_{u.\min(1)} = V_{uc(1)} + 0.10\sqrt{f_c'}b_v d_o \geq V_{uc(1)} + 0.6b_v d_o$$

$$= 266.4 + 0.10\sqrt{40}\times400\times532 \geq 266.4 + 0.6\times400\times532$$

$$= 401 \geq 394.1 = 401 \text{ kN}$$

$$V_{u.\min(2)} = V_{uc(2)} + 0.10\sqrt{f_c'}b_v d_o \geq V_{uc(2)} + 0.6b_v d_o = 267.8 \text{ kN}$$

Maximum shear strength,

$$V_{u.\max} = 0.2f_c'b_v d_o = 0.2\times40\times400\times532 = 1702.4 \text{ kN}$$

Ultimate shear strength,

$$V_{u(1)} = V_{uc(1)} + V_{us} \leq V_{u.\max}$$

$$= 266.4 + 104.5 \leq 1702.4 = 370.9 \text{ kN}$$

Minimum shear reinforcement has been provided; therefore, the capacity can be increased,

$$V_{u(1)} \geq V_{u.\min(1)} = 401 \text{ kN}$$

$$\phi V_{u(1)} = 0.7\times401 = 280.7 \text{ kN}$$

$$V_{u(2)} = V_{uc(2)} + V_{us} \leq V_{u.\max}$$

$$= 133.2 + 104.5 \leq 1702.4 = 237.7 \text{ kN}$$

Minimum shear reinforcement has been provided; therefore, the capacity can be increased,

$$V_{u(2)} \geq V_{u.\min(2)} = 267.8 \text{ kN}$$

$$\phi V_{u(2)} = 0.7 \times 267.8 = 187.4 \text{ kN}$$

The shear capacity of the beam is 280.7 kN at the support face and 187.4 kN at $2d_o$ from the support.

4.2.5 Beams in Torsion

The calculation for torsion in a beam is generally completed after the bending and shear calculations. The reason for this is that the torsion calculations require the shear capacity and may also affect the bending design.

The torsional capacity of the concrete (ϕT_{uc}) is calculated first; however, beams without closed fitments are only permitted to carry a maximum torsion of $0.25\phi T_{uc}$. Where this is exceeded, closed fitments should be provided, and ϕT_{us} becomes the torsional capacity of the beam. T_{uc} and T_{us} are not added together in the way that shear calculations are completed.

4.2.5.1 Torsion in beams without closed fitments

Beams which do not have closed fitments (ligatures) are required to meet the following inequalities:

$$T^* \leq 0.25\phi T_{uc}; \qquad \text{and,}$$

$$\frac{T^*}{\phi T_{uc}} + \frac{V^*}{\phi V_{uc}} \leq 1.0 \quad \text{for beams where:} \quad D \leq \text{MAX}[b_w/2, 250 \text{ mm}]$$

$$\frac{T^*}{\phi T_{uc}} + \frac{V^*}{\phi V_{uc}} \leq 0.5 \quad \text{otherwise}$$

where:

T_{uc}	= Torsional strength of concrete
V_{uc}	= Shear strength of concrete (refer to Section 4.2.4)
ϕ	= 0.7 (for torsion and shear)
b_w	= Width of web

Torsional reinforcement is required, in the form of closed fitments, where any of the above requirements are not achieved. The torsional capacity of concrete without closed fitments is calculated as follows. Prestressing components are not included; refer to AS 3600, Cl 8.3 for details.

$$T_{uc} = J_t \left(0.3\sqrt{f_c'} \right)$$

where:

J_t	= $0.33x^2y$ (for a rectangle)
J_t	= $0.33\sum x^2y$ (for a T-shape, L-shape or I-shape)
J_t	= $2A_m b_w$ (for a thin-walled hollow section)
A_m	= Area enclosed by the median lines of the walls of single cell
b_w	= Minimum wall thickness of a hollow section
x	= Shorter dimension of a rectangle
y	= Longer dimension of rectangle

4.2.5.2 Torsion in beams with closed fitments

For beams which require closed fitments, the ultimate strength in torsion is calculated as follows. The minimum longitudinal and closed fitment reinforcement requirements are also specified as shown.

$$T^* \le \phi T_{us}$$

$$\phi = 0.7 \qquad \text{(for torsion and shear)}$$

$$T_{us} = f_{sy.f}\left(A_{sw}/s\right)2A_t \cot\theta_v$$

where:

$\quad A_{sw}\quad$ = Cross-sectional area of torsional bar (forming a closed fitment)

$\quad A_t\quad$ = Area of a polygon with vertices at the centre of longitudinal bars at the corners of the cross-section

$\quad f_{sy.f}\quad$ = Yield strength of torsional bars (commonly 500 MPa) (500 MPa for D500L or D500N and 250 MPa for R250N bars)

$\quad s\quad$ = Spacing of torsional reinforcement

$\quad \theta_v\quad$ = Compression strut angle (refer to Section 4.2.4)

$\cot\theta_v\quad$ = $1/\tan\theta_v$

4.2.5.2.1 Closed fitment requirements

The minimum area of closed fitments for torsion design should comply with $A_{sv.min}$ from the shear calculations; however, it must be in the form of closed fitments. It should also provide a minimum capacity of at least $0.25T_{uc}$.

Note: $A_{sv.min}$ uses the diameter of each vertical leg which forms part of a closed fitment; however, A_{sw} uses only the area of one bar forming a closed loop (Figure 4.11).

4.2.5.2.2 Longitudinal torsion reinforcement requirements

Torsional loading causes longitudinal tension on reinforcing bars in addition to that caused by bending. Where they occur in the same load cases, the provided area needs to include bending-induced tension and torsion-induced tension. This can be done by designing the section for bending, then adding the appropriate amount of longitudinal steel for torsion. In this case, the bending design should then be rechecked, as it may affect the ductility and strength design.

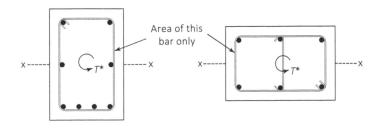

Figure 4.11 Torsional bar diagram for A_{sw}.

Alternatively, the load can be added as an equivalent moment. It is likely that the design will be governed by the tensile steel, and therefore, the calculated tension induced by torsion can be multiplied by the lever arm from the compression zone to the tensile steel and added as a design moment.

Tensile force in the flexural tensile zone,

$$F^*_{t.\text{tor}} = 0.5 f_{sy.f} \frac{A_{sw}}{s} u_t \cot^2 \theta_v$$

where: u_t = perimeter of a polygon defined for A_t

Although not stated in the standards, it is assumed that this formula should be factored to account for the actual applied torsion (i.e. if only half the capacity is used, then only half of the load should be added). The equation would then be modified to

$$F^*_{t.\text{tor}} = 0.5 f_{sy.f} \frac{A_{sw}}{s} u_t \cot^2 \theta_v \times (T^*/\phi T_{us}) \qquad \text{(if reducing based on percentage of load)}$$

Required area of torsional steelwork in tensile zone,

$$A_{st.\text{tor}} = F^*_{t.\text{tor}} / f_{sy}$$

Tensile force in the flexural compressive zone,

$$F^*_{c.\text{tor}} = 0.5 f_{sy.f} \frac{A_{sw}}{s} u_t \cot^2 \theta_v - F^*_c \geq 0$$

$$F^*_{c.\text{tor}} = 0.5 f_{sy.f} \frac{A_{sw}}{s} u_t \cot^2 \theta_v \times (T^*/\phi T_{us}) - F^*_c \geq 0$$

(if reduced based on percentage of load)

where: F^*_c = Absolute value of design force in compressive zone due to bending

Required area of torsional steelwork in compressive zone,

$$A_{sc.\text{tor}} = F^*_{c.\text{tor}} / f_{sy}$$

The alternative method of equivalent additional bending moment (force multiplied by eccentricity for beams governed by tension reinforcement) is shown, as it is often more practicable in design,

Equivalent additional bending moment, $\qquad M^*_{\text{tor}} = F^*_{t.\text{tor}} \times (d - \gamma k_u d/2)$

Revised design bending moment in beam, $\qquad M^*_{\text{revised}} = M^* + M^*_{\text{tor}}$

4.2.5.2.3 Web crushing check

The following requirement is to prevent crushing of the web in combined shear and torsion:

$$\frac{T^*}{\phi T_{u.\max}} + \frac{V^*}{\phi V_{u.\max}} \leq 1$$

where:

$$T_{u.\max} = 0.2f_c'J_t$$

$V_{u.\max}$ = Maximum shear strength (refer to Section 4.2.4)
$\phi = 0.7$ (for torsion and shear)

4.2.5.2.4 Torsional detailing

The reinforcement required for torsion includes closed fitments as well as longitudinal bars at each corner of the fitments. The closed fitments should be capable of developing full yield strength over each leg of the bars, unless calculations show that it is not required. The maximum spacing of the fitments should be limited to the smaller of $0.12u_t$ and 300 mm.

Example 4.3: Torsion

Check the torsional capacity of a 600 mm deep×400 mm wide concrete beam at $2d_o$ from the support face. 4-N16 reinforcing bars are provided on the tension face. Shear ligatures are a single closed loop of R10-200 with a cover of 50 mm. Concrete is grade N40. N16 reinforcement is provided in the top corners; however, it is excluded from the calculations for simplicity.

ULS design loads: (previous examples)

$M^* = 50$ kNm $\phi M_{uo} = 166.4$ kNm

$V^* = 75$ kNm $\phi V_{u(2)} = 0.7 \times 267.8 = 187.4$ kN

$T^* = 15$ kNm

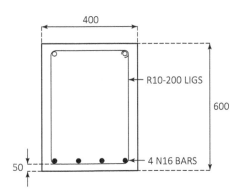

Beam details:

$s = 200$ mm

$f_{sy.f} = 250$ MPa

$f_c' = 40$ MPa

$\gamma = 1.05 - 0.007f_c' = 0.77$ $\left[0.67 \leq \gamma \leq 0.85\right]$

Torsion capacity of beam without closed fitments (calculated for reference only):

$$T_{uc} = J_t \left(0.3\sqrt{f_c'}\right)$$

$$J_t = 0.33x^2y = 0.33 \times 400^2 \times 600 = 31.68 \times 10^6 \, \text{mm}^3$$

$$T_{uc} = 31.68 \times 10^6 \left(0.3\sqrt{40}\right) = 60.1 \, \text{kNm}$$

$$0.25 \, T_{uc} = 0.25 \times 60.1 = 15 \, \text{kNm}$$

$$0.25 \, \phi T_{uc} = 0.7 \times 15 = 10.5 \, \text{kNm}$$

Capacity ratios,

$$\frac{T^*}{\phi T_{uc}} + \frac{V^*}{\phi V_{uc}}$$

$$\frac{15}{0.7 \times 60.1} + \frac{75}{187.4} = 0.76$$

Closed fitments are required for the beam, both because T^* is greater than 10.5 kNm and because the sum of the torsion and shear capacity ratios is greater than 0.5.

Torsion capacity of beam with closed fitments:

$$T_{us} = f_{sy.f} \left(A_{sw}/s\right) 2A_t \cot\theta_v$$

$$A_{sw} = 1 \times \pi \times 10^2/4 = 78.5 \, \text{mm}^2$$

$$A_t = \left(600 - 2(50+10+16/2)\right) \times \left(400 - 2(50+10+16/2)\right) = 122\,496 \, \text{mm}^2$$

$$\theta_v = 45 \, \text{deg}$$

$$T_{us} = 250 \times (78.5/200) \times 2 \times 122\,496 \cot 45 = 24.1 \, \text{kNm}$$

$$\phi T_{us} = 0.7 \times 24 = 16.8 \, \text{kNm} \qquad \geq 10.5 \, \text{kNm} \quad \text{Pass}$$

Equivalent additional moment from torsion load:

$$F^*_{t.tor} = 0.5 f_{sy.f} \frac{A_{sw}}{s} u_t \cot^2\theta_v \times \left(T^*/\phi T_{us}\right)$$

$$\theta_v = 45 \, \text{deg} \qquad \text{(conservative)}$$

$$u_t = 2\left(600 - 2(50+10+16/2)\right) + 2\left(400 - 2(50+10+16/2)\right) = 1456 \, \text{mm}$$

$$F^*_{t.tor} = 0.5 \times 250 \times \frac{78.5}{200} \times 1456 \times \cot^2 45 \times (15/16.8) = 63.8 \, \text{kN}$$

$$d = 600 - 50 - 10 - 16/2 = 532 \, \text{mm}$$

$$k_u = 0.0722 \qquad \text{(previous bending example)}$$

$$k_u d = 0.0722 \times 532 = 41.1 \text{ mm}$$

$$M_{\text{tor}}^* = F_{t.\text{tor}}^* \times (d - \gamma k_u d/2) = 32.6 \text{ kNm}$$

$$63.8 \times (0.532 - 0.77 \times 0.0411/2) = 32.9 \text{ kNm}$$

$$M_{\text{revised}}^* = M^* + M_{\text{tor}}^* = 50 + 32.9 = 82.9 \text{ kNm} \qquad \leq \phi M_{uo} = 166.4 \text{ kNm} \quad \text{Pass}$$

Web crushing check:

$$\frac{T^*}{\phi T_{u.\max}} + \frac{V^*}{\phi V_{u.\max}} \leq 1$$

$$T_{u.\max} = 0.2 f_c' J_t = 0.2 \times 40 \times 31.68 \times 10^6 = 253.4 \text{ kNm}$$

$$V_{u.\max} = 0.2 f_c' b_v d_o = 0.2 \times 40 \times 400 \times 532 = 1702.4 \text{ kN}$$

$$\frac{15}{0.7 \times 253.4} + \frac{75}{0.7 \times 1702.4} = 0.15 \qquad \leq 1 \qquad \text{Pass}$$

Detailing:
   Maximum spacing of fitments,

$$s_{max} = \text{MIN}[0.12 u_t, 300] = 175 \text{ mm} \qquad > 200 \qquad \text{Fail}$$

The torsional capacity of the beam is 16.8 kNm. Combined shear and torsion is acceptable. The bending capacity of the beam remains sufficient with the equivalent torsional moment added to the design moment. The maximum spacing of fitments (ligatures) is exceeded; the spacing should be revised from 200 to 175 mm.

4.3 SLABS

The design of slabs in engineering offices increasingly involves finite element analysis (FEA) packages (refer to Section 6.11), which have the capability to analyse plates. However, the guidelines presented in AS3600 are generally based on a simplified beam approach with moment redistribution. This chapter presents only a simple hand analysis method for slabs. Refer to AS 3600, Section 6 for various acceptable analysis methods.

Slabs are generally designed for bending and shear in the same way as beams (refer to Sections 4.2.3 and 4.2.4); however, an additional check is added for punching shear. It is also more convenient to remove the width dimensions from equations and calculate a resultant capacity in kilonewton-metres per metre and kilonewtons per metre. Strain should be checked in the reinforcing bars, as they may not reach full yield strength prior to strain failure in the concrete. Capacities for typical sections are provided in Section 7.3.2.

Care should be taken as to which way the reinforcement is placed, as the placing sequence affects the bending strength in each direction. Figure 4.12 shows two opposing cross-sections, where the bending strength in the primary direction (about the x-axis) is stronger than in the secondary direction (about the y-axis). This is because the bars are farther from the neutral axis and will therefore have a longer lever arm.

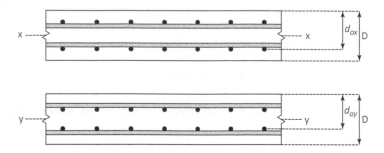

Figure 4.12 Typical slab cross-section.

4.3.1 Reinforcement requirements

For two-way reinforced slabs, the minimum reinforcement provided should meet the following equations to achieve minimum strength:

$$A_{st}/bd \geq 0.24\left(\frac{D}{d}\right)^2 f'_{ct.f}/f_{sy} \quad \text{(for slabs supported by columns at their corners)}$$

$$A_{st}/bd \geq 0.19\left(\frac{D}{d}\right)^2 f'_{ct.f}/f_{sy} \quad \text{(for slabs supported by beams or walls on four sides)}$$

4.3.1.1 Tensile reinforcement detailing

AS 3600 provides the following requirements for detailing in slabs, followed by deemed-to-comply arrangements for one-way and two-way slabs which meet specific provisions (refer AS 3600, Cl 9.1.3 for complete, detailed list):

1. The distribution of reinforcement should be based on a hypothetical moment envelope formed by displacing the calculated positive and negative values by a distance of D in each direction along the slab (refer to Figure 4.13). If an envelope is not calculated, the deemed-to-comply rules should be followed.
2. At least one-third of the required top reinforcement at supports should extend the greater of $12d_b$ or D beyond the point of contraflexure.
3. For simply supported discontinuous ends of slabs, at least half of the bottom reinforcement required at mid-span should continue for a distance equal to the greater of $12d_b$ and D past the face of supports. Where there is no requirement for shear reinforcement, this length may be reduced to $8d_b$ where half the bars are extended or $4d_b$ where all bars are extended.

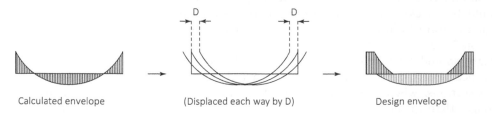

Calculated envelope   (Displaced each way by D)   Design envelope

Figure 4.13 Slab design envelope.

4. For continuous or flexurally restrained slabs, at least one-quarter of the bottom bars required at mid-span should continue past the face of the support.
5. For frames that incorporate slabs and are intended to carry lateral loads, the effects of the lateral load on slab reinforcement should be taken into account, and the reinforcement should not be less than the relevant deemed-to-comply solution.

Depending on construction preference, 100% of the top and bottom reinforcement is often continued through the entire slab. This is especially common for small slabs, where the excess reinforcement is not significant.

4.3.2 Crack control

The following crack control measures are aimed at preventing cracking due to shrinkage and temperature effects. Effects of prestressing are not shown here; refer to AS 3600, Section 9 for details. Refer to Section 5.4.5 for slabs on grade.

4.3.2.1 Reinforcement in the primary direction

No additional reinforcement is required for one-way slabs or two-way slabs where the provided area for each spanning direction is greater than or equal to the area required for minimum strength and 75% of the value required in the secondary direction. This reduction is made because part of the section will be in compression.

4.3.2.2 Reinforcement in the secondary direction

For slabs that are free to expand or contract in the secondary direction, the minimum area of reinforcement in that direction is $(1.75)bd \times 10^{-3}$.

Note that for slabs less than or equal to 500 mm thick, the reinforcement on the top and bottom surface counts towards the total area. For slabs that are thicker than 500 mm, each face should be calculated separately, using only the reinforcement on the face under consideration and using a thickness of 250 mm.

For slabs that are restrained from expanding or contracting in the secondary direction, the minimum area of reinforcement is as follows:

For slabs which are fully enclosed within buildings (except for a brief period during construction):

Minor degree of control, $(1.75)bd \times 10^{-3}$

Moderate degree of control, $(3.5)bd \times 10^{-3}$

Strong degree of control, $(6)bd \times 10^{-3}$

For other A1 and A2 surface exposures:

Moderate degree of control, $(3.5)bd \times 10^{-3}$

Strong degree of control, $(6)bd \times 10^{-3}$

For B1, B2, C1 and C2:

Strong degree of control, $(6)bd \times 10^{-3}$

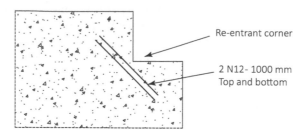

Figure 4.14 Example plan for re-entrant corner reinforcing.

Minor = Cracks are aesthetically inconsequential (unseen)
Moderate = Cracks can be seen but tolerated
Strong = Appearance of cracks is unacceptable

Special consideration should be paid to detailing of reinforcement at restraints, openings and discontinuities, as these are common areas of cracking. Trimmer bars should be used at re-entrant corners, as shown in Figure 4.14. Depending on slab thickness, these can be used in the centre of the slab or on both faces.

4.3.3 Analysis

AS 3600 provides guidance on one-way and two-way calculations for slab design. Only the one-way calculations are summarised here due to the prevalence of FEA for two-way slab designs.

This method (refer to AS 3600, Clause 6.10.2) is applicable for one-way slabs where:

1. The ratio of longer to shorter adjacent spans does not vary by more than 1.2.
2. Loads are essentially uniformly distributed.
3. The live load (Q) does not exceed twice the dead load (G).
4. Cross-sections are uniform.
5. There are deemed-to-comply reinforcement arrangements (AS 3600, Cl 9.1.3.2).
6. Bending moments at supports are caused only by the actions applied to the beam or slab.

4.3.3.1 Negative moment calculation

1. At the first interior support:

   a. Two spans only for ductility class N $\qquad M^*_{neg} = F_d L_n^2 /9$

   Two spans only for ductility class L $\qquad M^*_{neg} = F_d L_n^2 /8$

   b. More than two spans $\qquad M^*_{neg} = F_d L_n^2 /10$

2. At other interior supports, $\qquad M^*_{neg} = F_d L_n^2 /11$

3. At interior faces of exterior supports for members built integrally with their supports:

   a. For beams where the support is a column $\qquad M^*_{neg} = F_d L_n^2 /16$

   b. For slabs and beams where the support is a beam $\qquad M^*_{neg} = F_d L_n^2 /24$

4.3.3.2 Positive moment calculation

1. For an end span $\qquad\qquad\qquad\qquad$ $M^*_{pos} = F_d L_n^2 /11$

2. For an interior span with ductility class N $\quad$ $M^*_{pos} = F_d L_n^2 /16$

$\quad$ For an interior span with ductility class L $\quad$ $M^*_{pos} = F_d L_n^2 /14$

where:

$\quad$ F_d $\quad$ = Uniformly distributed design load, factored for strength (force per unit length)
$\quad$ L_n $\quad$ = Clear length of span being considered

The denominator (x) for these equations is shown graphically in Figure 4.15 for slabs with different numbers of spans, using ductility class N reinforcement,

$\quad$ $M^* = F_d L_n^2 /x$ $\qquad$ (refer to Figure 4.15 for x value)

4.3.3.3 Transverse shear calculation

1. For an end span:

   a. At the face of the interior support $\quad$ $V^* = 1.15\ F_d L_n /2$

   b. At mid-span $\qquad\qquad\qquad\qquad$ $V^* = F_d L_n /7$

   c. At the face of the end support $\qquad$ $V^* = F_d L_n /2$

2. For an interior span:

   a. At the face of supports $\qquad\qquad$ $V^* = F_d L_n /2$

   b. At mid-span $\qquad\qquad\qquad\qquad$ $V^* = F_d L_n /8$

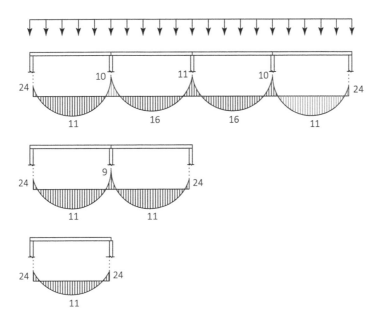

Figure 4.15 Denominators (x) for one-way slab moment calculations (class N reo).

4.3.4 Bending

The bending capacity of a slab is calculated in the same way as that of a beam (refer to Section 4.2.3). A width of 1000 mm is typically adopted to calculate the bending capacity per metre width. If using reinforcing mesh instead of normal bars, the ductility becomes low, and therefore the steel strain should be limited to 0.015 instead of 0.05 (refer to Table 4.2). Capacities for typical sections are provided in Section 7.3.2.

4.3.5 Shear

The shear design of a slab needs to consider two separate failure mechanisms: standard shear and punching shear. Standard shear capacity is calculated in the same way as shear for beams and is used where shear failure is possible across the width of a slab. Punching shear is the local failure of a slab in shear around the perimeter of a highly loaded area. It is common where high loads are transferred to columns, or where high point loads are applied to slabs. Punching shear is often the governing shear load for slab design, especially for flat slabs supported on columns. Capacities for shear and punching shear are provided in Section 7.3.2 for typical slab cross-sections.

4.3.5.1 Standard shear

Where shear failure is possible across the width of a slab, refer to Section 4.2.4; however, adopt a width (b) of 1000 mm and calculate the strength in units of kilonewtons per metre.

4.3.5.2 Punching shear

Punching shear is dependent on the assumed failure length, also known as the *critical shear perimeter*. This is the perimeter of a shape formed by outlining the loaded area with a clear gap of $d_{om}/2$ (refer to Figure 4.16).

$$V^* \leq \phi\, V_u$$
$$\phi = 0.7 \text{ (for punching shear)}$$

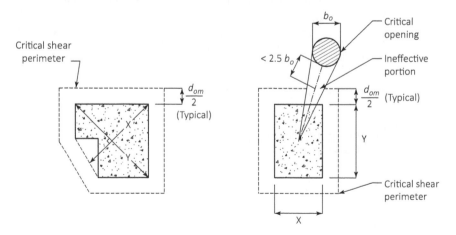

Figure 4.16 Punching shear perimeter (u). (From AS 3600, figure 9.2.1(A). Copied by Mr L. Pack with the permission of Standards Australia under Licence 1607-c010.)

The calculation method adopted for punching shear design depends on whether any moment (M_v^*) is being transferred from the slab to the support in the direction being considered.

Note: The pre-stressing component has been removed; refer to AS 3600, Cl 9.2.3 for details.

1. Where M_v^* is zero, V_u is taken as equal to V_{uo},

$$V_{uo} = ud_{om}\left(f_{cv}\right)$$

where:

d_{om} = Mean value of d_o around the critical shear perimeter

$$f_{cv} = 0.17\left(1+\frac{2}{\beta_h}\right)\sqrt{f_c'} \leq 0.34\sqrt{f_c'}$$

u = Length of critical shear perimeter (Refer to Figure 4.16)

β_h = Ratio of longest dimension, Y, to perpendicular dimension, X

The calculated capacity can be increased through the use of shear-heads or shear-studs:

$$V_{uo\,(\text{with shear reinforcing})} = ud_{om}\left(0.5\sqrt{f_c'}\right) \leq 0.2ud_{om}f_c'$$

2. Where M_v^* is not zero, V_u is calculated as shown,
   Minimum area of closed fitments,

$$A_{sw}/s \geq 0.2y_1/f_{sy.f}$$

   a. Where there are no closed fitments in the torsion strip or spandrel beam,

$$V_u = V_{uo}/\left[1 + uM_v^*/\left(8V^*ad_{om}\right)\right]$$

   b. Where the torsion strip contains the minimum quantity of closed fitments, V_u is taken as equal to $V_{u.\,min}$,

$$V_{u.min} = 1.2\,V_{uo}/\left[1 + u\,M_v^*/\left(2V^*a^2\right)\right]$$

   c. Where there are spandrel beams perpendicular to the direction of M_v^*, which contain the minimum quantity of closed fitments, V_u is taken as equal to $V_{u.\,min}$,

$$V_{u.min} = 1.2\,V_{uo}\left(D_b/D_s\right)/\left[1 + uM_v^*/\left(2V^*ab_w\right)\right]$$

   d. If the torsion strip or spandrel beam contains more than the minimum quantity of closed fitments,

$$V_u = V_{u.min}\sqrt{\left[\left(A_{sw}/s\right)/\left(0.2/f_{sy.f}\right)\right]}$$

   where: $V_{u.\,min}$ is calculated using (b) or (c) as appropriate.

In no case shall the adopted value be taken as greater than

$$V_{u.max} = 3V_{u.min}\sqrt{(x/y)}$$

where:

a = Dimension of critical shear perimeter measured parallel to direction of M_v^*

A_{sw} = Cross-sectional area of closed fitment

b_w = Web width of spandrel beam

D_b = Depth of spandrel beam

D_s = Overall depth of slab or drop panel

$f_{sy.f}$ = Yield strength of fitments (500 MPa for D500L or D500N and 250 MPa for R250N bars)

M_v^* = Design moment being transferred from slab into support

s = Spacing of fitments

x = Shorter dimension of cross-section for torsion strip or spandrel beam

y = Longer dimension of cross-section for torsion strip or spandrel beam

y_1 = Longer cross-sectional dimension of closed fitments

4.3.6 Deflection check

Deflection may be solved by using a range of refined methods, such as non-linear analysis or FEA: however, these methods should include

1. Two-way action
2. Shrinkage and creep
3. Load history
4. Construction procedure
5. Deflection of formwork

Alternatively, the deflection can be calculated using the simplified calculation (AS 3600, Cl 9.3.3) or the deemed-to-comply span/depth ratio (AS 3600, Cl 9.3.4). The deemed-to-comply method is outlined in the next sub-section.

4.3.6.1 Deemed-to-comply span/depth procedure

The deflection requirements of AS 3600, Clause 2.3.2, shall be deemed to comply where the span/depth ratio outlined in this sub-section is achieved.

This procedure is applicable for one-way slabs or multiple-span two-way slabs where

1. The slab is flat and essentially uniform in depth.
2. The slab is fully propped during construction.
3. Loads are essentially uniformly distributed.
4. The live load (Q) does not exceed twice the dead load (G).

$$L_{ef}/d \le k_3 k_4 \left[\frac{(\Delta/L_{ef})1000E_c}{F_{d.ef}} \right]^{1/3}$$

where:

d = Effective depth of slab (depth to centre of tensile reinforcement)

E_c = Young's modulus of concrete (MPa) (Section 4.1.1)

$F_{d.ef}$ = Effective design service load (kPa)

L_{ef} = Effective span

k_3 = 1.0 for a one-way slab

= 0.95 for a two-way flat slab without drop panels

= 1.05 for a two-way flat slab with drop panels (drop panels must extend at least L/6 in each direction on each side of the support centre line and have an overall depth of 1.3D)

k_4 = 1.4 for simply supported slabs

= 1.75 for continuous end spans

= 2.1 for continuous interior spans (for continuous spans, the ratio of longer to shorter span must not exceed 1.2, and end spans must not be longer than adjacent interior span)

Δ/L_{ef} = Deflection limitation (Refer to Table 2.24)

Example 4.4: Slab shear

Calculate the shear capacity (per metre) of a 200 mm deep slab at the support face and at $2d_o$ from the support. N16-200 reinforcing bars are provided each way on the tension face with a cover of 50 mm. SL82 mesh is provided on the top face for crack control. Concrete is grade N32.

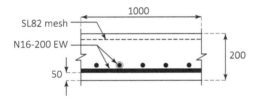

Slab details:

$$f_c' = 32\,\text{MPa}$$

$$f_{cv} = f_c'^{1/3} \leq 4\ \text{MPa} = 32^{\frac{1}{3}} = 3.175\ \text{MPa}$$

$$A_{st} = (1000/200) \times \pi \times 16^2/4 = 1005.3\ \text{mm}^2$$

$$b_v = 1000\,\text{mm} \qquad (\text{using 1000 mm to calculate capacity per metre})$$

$$d_o = 200 - 50 - 16 - 16/2 = 126\ \text{mm} \quad (\text{conservatively using depth to inner layer of bars})$$

$$\beta_1 = 1.1(1.6 - d_o/1000) \geq 1.1$$

$$= 1.1(1.6 - 126/1000) \geq 1.1$$

$$= 1.62 \geq 1.1 = 1.62$$

$$\beta_2 = 1, \quad (\text{for members subject to pure bending})$$

$$\beta_3 = 2, \quad (\text{at the support face})$$

$$\beta_3 = 1, \quad (\text{at } 2d_o \text{ from the support face})$$

The mesh is excluded, as it is on the compression face.

Shear capacity: (using subscript 1 for the support face and 2 for $2d_o$ from the support face):

Concrete capacity,

$$V_{uc(1)} = \beta_1\beta_2\beta_3 b_v d_o f_{cv}\left[\frac{A_{st}}{b_v d_o}\right]^{1/3} = 1.62\times1\times2\times1000\times126\times3.175\left[\frac{1005.3}{1000\times126}\right]^{1/3} = 259.2\text{ kN}$$

$$V_{uc(2)} = \beta_1\beta_2\beta_3 b_v d_o f_{cv}\left[\frac{A_{st}}{b_v d_o}\right]^{1/3} = 1.62\times1\times1\times1000\times126\times3.175\left[\frac{1005.3}{1000\times126}\right]^{1/3} = 129.6\text{ kN}$$

Maximum shear strength,

$$V_{u.\max} = 0.2f_c'b_v d_o = 0.2\times32\times1000\times126 = 806.4\text{ kN}$$

Ultimate shear strength,

$$V_{u(1)} = V_{uc(1)} + V_{us} \le V_{u.\max}$$

$$= 259.2 + 0 \le 806.4 = 259.2\text{ kN}$$

$$\phi V_{u(1)} = 0.7\times259.2 = 181.4\text{ kN}$$

$$V_{u(2)} = V_{uc(2)} + V_{us} \le V_{u.\max}$$

$$= 129.6 + 0 \le 806.4 = 129.6\text{ kN}$$

$$\phi V_{u(2)} = 0.7\times129.6 = 90.7\text{ kN}$$

The shear capacity of the slab is 181.4 kN (per metre width) at the support face and 90.7 kN (per metre width) at $2d_o$ from the support.

Example 4.5: Punching shear

Calculate the punching shear capacity of the slab from the previous example. The applied load is from a wheel which has a contact area of 400 mm wide by 250 mm long. The wheel is located at 300 mm from a free edge.

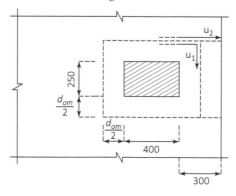

Slab details:

$$f'_c = 32 \, \text{MPa}$$

Two possible shear perimeters are considered. Shear perimeter 1 is for a failure of the slab around the wheel. Shear perimeter 2 is when the failure perimeter has three edges and is free at the unsupported edge.

$$\beta_h = 400/250 = 1.6$$

$$f_{cv} = 0.17\left(1 + \frac{2}{\beta_h}\right)\sqrt{f'_c} \le 0.34\sqrt{f'_c}$$

$$= 0.17\left(1 + \frac{2}{1.6}\right)\sqrt{32} \le 0.34\sqrt{32} = 1.92 \, \text{MPa}$$

Punching shear can be roughly estimated using $d_{om(av)}$, the average value of d_o. however, for this calculation, the exact value is calculated.

$$d_{om(av)} \approx 200 - 50 - 16 = 134 \, \text{mm} \qquad \text{(using average depth)}$$

The direction of bars is not specified; therefore, the shallow dimension is assumed for the longer length:

$$d_{o1} = 200 - 50 - 16 - 16/2 = 126 \, \text{mm} \quad \text{(shallow direction)}$$

$$d_{o2} = 200 - 50 - 16/2 = 142 \, \text{mm} \quad \text{(deep direction)}$$

The governing shear perimeter is not always obvious, because the value of d_{om} is different for each failure plane. The mean value of d_o can be calculated by adding the depth multiplied by the length for each direction, then dividing by the total length of the shear perimeter. The resultant can be solved using the quadratic formula (although adopting the previously estimated value is often sufficiently accurate).

Shear perimeter 1:

$$u_1 = \left(800 + 2d_{om}\right) + \left(500 + 2d_{om}\right)$$

$$d_{om} = \frac{d_{o1}\left(800 + 2d_{om}\right) + d_{o2}\left(500 + 2d_{om}\right)}{\left(800 + 2d_{om}\right) + \left(500 + 2d_{om}\right)}$$

$$0 = -4\left(d_{om}\right)^2 + \left(2d_{o1} + 2d_{o2} - 1300\right)\left(d_{om}\right) + \left(800 d_{o1} + 500 d_{o2}\right)$$

Solving using the quadratic formula,

$$d_{om} = \frac{-B \pm \sqrt{B^2 - 4AC}}{2A}$$

$$A = -4$$

$$B = \left(2d_{o1} + 2d_{o2} - 1300\right)$$

$$C = \left(800 d_{o1} + 500 d_{o2}\right)$$

$$d_{om} = 132.7\,\text{mm}$$

$$u_1 = (800 + 2 \times 132.7) + (500 + 2 \times 132.7) = 1831\,\text{mm}$$

$$V_{uo1} = u_1 d_{om}(f_{cv}) = 1831 \times 132.7 \times 1.92 = 467\,\text{kN}$$

Shear perimeter 2:

$$u_2 = (1400 + d_{om}) + (250 + d_{om})$$

$$d_{om} = \frac{d_{o1}(1400 + d_{om}) + d_{o2}(250 + d_{om})}{(1400 + d_{om}) + (250 + d_{om})}$$

$$0 = -2(d_{om})^2 + (d_{o1} + d_{o2} - 1650)(d_{om}) + (1400 d_{o1} + 250 d_{o2})$$

$$d_{om} = 129.2\,\text{mm}$$

(Shear perimeter 2 has a lower d_{om}, because it has a longer failure length in the shallow direction)

$$u_2 = (1400 + 129.2) + (250 + 129.2) = 1908\,\text{mm}$$

$$V_{uo2} = u_2 d_{om}(f_{cv}) = 1908 \times 129.2 \times 1.92 = 473\,\text{kN}$$

Punching shear capacity:
Punching shear is governed by shear perimeter 1,

$$V_{uo} = \text{MIN}\left[V_{uo1}, V_{uo2}\right] = 467\,\text{kN}$$

$$\phi V_{uo} = 0.7 \times 467 = 327\,\text{kN}$$

The punching shear capacity for the specified wheel load and location is 327 kN.

4.4 COLUMNS

Columns are characterised as being different from beams because of the presence of axial loading. They can be designed to transmit bending, shear and axial loads. Notably, the bending capacity is affected by the axial load in the column. Columns should be designed in accordance with Section 10 of AS 3600. Capacities for typical sections are provided in Section 7.3.2.

All columns should be designed for a minimum bending moment of $0.05\,D \times N^*$ to allow for a minimum eccentricity of $0.05\,D$. This should be checked in each principal axis.

4.4.1 Reinforcement requirements

Columns should generally be designed with longitudinal reinforcing bars taking between 1% and 4% of the total cross-sectional area. In large columns, the area of steel may be less than 1% provided that $A_{sc}f_{sy}$ is greater than $0.15N^*$ and all other design requirements are satisfied.

Table 4.12 Bar diameters for fitments and helices

Longitudinal bar diameter (mm)	Minimum bar diameter for fitment and helix (mm)
Single bars up to 20	6
Single bars 24 to 28	10
Single bars 28 to 36	12
Single bars 40	16
Bundled bars	12

Source: AS 3600, Table 10.7.4.3. Copied by Mr L. Pack with the permission of Standards Australia under Licence 1607-c010.

Confinement is provided by shear bars (fitments/ligatures) with closed hooks (for rectangular cross-sections) or by helices (in circular cross-sections). Corner bars are required to be restrained by the fitments. All other bars spaced at over 150 mm centres should be restrained, and alternate bars spaced at less than 150 mm centres should also be restrained. A bar is considered confined if it is restrained by a transverse bar with a minimum bend of 135°, or by a circular fitment when longitudinal bars are equally spaced. Refer to AS 3600, Cl 10.7.4 for additional details.

The spacing of the fitments should not exceed the minimum value of the shortest cross-sectional dimension and 15 times the smallest bar diameter. When bundled bars are used, this is changed to half the shortest cross-sectional dimension and 7.5 times the smallest bar diameter. AS 5100.5 further limits the spacing to a maximum of 300 mm.

Bar diameters for fitments and helices should not be lower than the values shown in Table 4.12. The first and last fitments should be within 50 mm from each end of the column. Helices should have an additional 1.5 flat turns at the end of the run to provide sufficient anchorage.

Additional requirements are provided in AS 3600 for splicing of reinforcement within columns. Further confinement requirements are also specified for columns with a compressive strength greater than 50 MPa (including special confinement rules for locations where $N^* \geq \phi N_{uo}$, or where both $N^* \geq \phi 0.3 f_c' A_g$ and $M^* \geq 0.6 \phi M_u$); refer to AS 3600, Clause 10.7.3.

4.4.2 Effective length

The effective length of a column (L_e) is calculated by multiplying the unsupported length of a column (L_u) by an effective length factor (k).

$$L_e = k L_u$$

The effective length factor can be based on simple end restraint conditions, as shown in Figure 4.17. Alternatively, it can be calculated by using the stiffness of restraining elements, as shown in AS 3600, Clause 10.5. Column slenderness should be checked in both the x- and y-axes with appropriate end fixity in each direction.

4.4.3 Short columns

The design of short columns does not need to consider the moment magnifier (δ), caused by slenderness effects, and can be further simplified when axial or bending forces are small.

Where the axial force in a short column is less than $0.1 f_c' A_g$, the cross-section may be designed for bending only. For a short braced column with a small moment, refer to AS 3600, Cl 10.3.3.

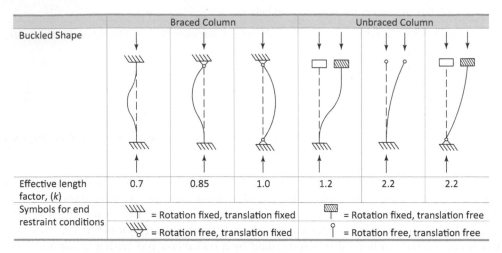

Figure 4.17 Effective length factors for columns. (From AS 3600, figure 10.5.3(A). Copied by Mr L. Pack with the permission of Standards Australia under Licence 1607-c010.)

An unbraced column is considered short when

$$L_e/r \leq 22$$

A braced column is considered short when either

$$L_e/r \leq 25$$

or,

$$L_e/r \leq \alpha_c \, (38 - f_c'/15)(1 + M_1^*/M_2^*)$$

where:

$$\alpha_c = \sqrt{2.25 - 2.5N^*/0.6N_{uo}} \quad \text{(for } N^*/0.6N_{uo} \geq 0.15) \text{ or,}$$

$$\alpha_c = \sqrt{1/(3.5N^*/0.6N_{uo})} \quad \text{(for } N^*/0.6N_{uo} < 0.15)$$

L_e = slenderness ratio   (refer to Section 4.4.4)

M_1^*/M_2^* = Ratio of smaller to larger end moments   (see note below)

r = Radius of gyration = 0.3D (for rectangular), 0.25D (for circular)
D = Cross-sectional dimension for direction being considered

M_1^*/M_2^* is taken as negative for a single curvature and positive for a double curvature. When M_1^* is less than or equal to $0.05 \, D \times N^*$, the moment should be treated as a single curvature despite the moment sign. When the absolute value of M_2^* is less than or equal to $0.05 \, D \times N^*$, the ratio shall be taken as -1.0.

4.4.4 Slender columns

Slender columns may be defined as columns which are not short. The design bending moment for slender columns is increased by a moment magnifier, δ.

$M^*_{\text{(Design Moment)}} = \delta M^*_{\text{(Calculated Moment)}}$

For a braced column,

$$\delta_b = k_m/(1 - N^*/N_c) \geq 1$$

where: $k_m = (0.6 - 0.4\ M^*_1/M^*_2)$

k_m should not be taken as less than 0.4 unless there is significant transverse loading between ends and in the absence of more exact calculations, shall be taken as 1.0.
buckling load,

$$N_c = (\pi^2/L_e^2)[182d_o(\phi M_c)/(1 + \beta_d)]$$

where:
$\phi = 0.6$
$M_c = M_{ub}$ with $k_u = 0.\overline{54}$

For an unbraced column,
The moment magnifier shall be taken as the larger value of δ_b and δ_s.

$$\delta_s = 1/\left(1 - \sum N^*/\sum N_c\right)$$ (for each column in a storey)

If the forces and moments were calculated using a linear elastic analysis (refer to AS 3600, Cl 6.2), the design of slender columns should include the effects of the radius of gyration and effective length. The slenderness ratio (L_e/r) should not exceed 120 unless a rigorous analysis is completed in accordance with AS 3600, Clause 10.5.

4.4.5 Columns in compression and bending

Columns which are subject to a combined compression and bending load should be checked using an interaction diagram. The diagram shows the increase in bending capacity (from pure bending) caused by an increase in axial load until the balanced point is reached; the bending capacity then reduces to zero at the squash load. The interaction diagram is created to allow the actual combination of bending and axial compression to be plotted on the graph. The plotted value should be within the area encapsulated by the interaction diagram. The interaction diagram is defined by four key points, as shown in Figure 4.18.

The column design capacities are calculated using strain compatibility across the section (refer to Figure 4.21). The capacity reduction factor, however, requires k_u to be defined in terms of d. For the purposes of column design, calculations are based on d being the depth to the outermost tension bar (previously denoted as d_o). This is because the number of bars in tension changes as the axial load is increased, and it is more convenient to work with the ratio of the depth to the outer bar. k_u is therefore redefined as the ratio of the depth to the neutral axis (from the extreme compressive fibre) to the depth to the outermost bar (d). This is defined as k_{uo} for the case of pure bending.

4.4.5.1 Squash load point

The squash load is the point on the diagram where a column fails in pure compression. The capacity is based on the cross-sectional area of steel and concrete with a capacity reduction factor, ϕ, of 0.6. The axial load needs to be applied at the plastic centroid for the section to

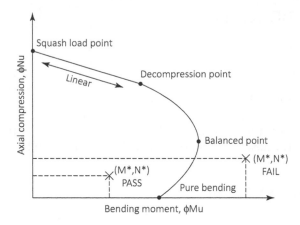

Figure 4.18 Column interaction diagram.

remain in compression without bending being caused by eccentric loading. Since the concrete is at an ultimate strain of 0.003, the steel will have exceeded its yield strain of 0.0025 and will be at yield strength.

$$\phi N_{uo} = \phi[\text{Area of concrete} \times \alpha_1 f_c' + \text{area of steel} \times f_{sy}]$$

$$= \phi[(A_g - A_s) \times \alpha_1 f_c' + A_s \times f_{sy}]$$

where:
$$\alpha_1 = 1 - 0.003\, f_c' \quad [0.72 \leq \alpha_1 \leq 0.85]$$

A_g = Gross area of cross-section
A_s = Area of steel

4.4.5.2 Squash load point through to decompression point

Linear interpolation is used between the squash load point and the decompression point. The capacity reduction factor, ϕ, is equal to 0.6 in this region.

4.4.5.3 Decompression point through to pure bending

The same theory is used for calculating the capacities at all points from the decompression point through to the pure bending point. The easiest way to manually calculate values is to create a solution based on an input parameter of k_u. Once k_u is selected, the tension and compression forces in the section can be calculated. The axial load on the section (applied at the plastic centroid) is equal to the sum of tension and compression forces, and the bending moment can be calculated by resolving the forces about the neutral axis.

The capacity reduction factor varies in certain areas of this section of the graph. The equation for ϕ ensures that the resultant reduction factor remains at 0.6 from the decompression point down to the balanced point and then increases linearly (with N_u) up to ϕ_{pb} at pure bending.

4.4.5.4 Decompression point

The decompression point is where the bending and axial loads result in a maximum concrete strain equal to 0.003, and there is zero strain in the outermost tension steel. This is the point where $k_u = 1$. The capacity reduction factor, ϕ, at decompression is 0.6.

4.4.5.5 Balanced point

The balanced point is the location where the concrete strain is at its limit and the outer steel strain reaches yield. The balanced point occurs when the concrete strain is equal to 0.003 and the strain in the outermost tension steel is equal to 0.0025. This is the point where $k_u = \dfrac{\varepsilon_c}{\varepsilon_c + f_{sy}/E_s} = 0.\overline{54}$. The capacity reduction factor, ϕ, at the balanced point is 0.6.

4.4.5.6 Pure bending point

The pure bending point is where the external axial force is zero, and therefore, the internal tension force is equal to the compression force. This capacity is calculated in the same manner as for a beam, with the capacity reduction factor (defined as ϕ_{pb}) between 0.6 and 0.8, depending on the value of k_{uo}.

4.4.5.7 Rectangular cross-sections

The interaction diagram for a rectangular cross-section can be calculated using the following steps, with calculations shown for combined compression and bending about the major (x) axis (Figure 4.19).

4.4.5.7.1 Squash load point

$$\phi N_{uo} = \phi[(A_g - A_s) \times \alpha_1 f_c' + A_s \times f_{sy}]$$

where:

$\alpha_1 = 1 - 0.003\, f_c'$ $[0.72 \le \alpha_1 \le 0.85]$

A_g = Gross area of cross-section = bD
A_s = Area of steel

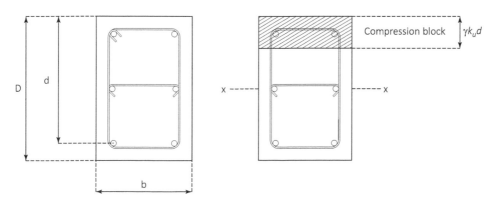

Figure 4.19 Rectangular cross-section.

b = Width of column
D = Overall depth of column
f_{sy} = Yield strength of steel (typically 500 MPa)
f_c' = Compressive strength of concrete
ϕ = 0.6 for pure compression

The plastic centroid can now be calculated. The depth to the plastic centroid is the distance (measured from the extreme compressive fibre) to the applied axial load which results in pure compression. This value is also used later to calculate moment capacities. It is calculated by multiplying the compressive force of the concrete by the depth to the concrete centroid, then adding the force in each bar multiplied by the depth to the bar, and dividing the entire value by the squash load:

$$\text{Plastic centroid,} \quad d_q = \left[([bD - A_s] \times \alpha_1 f_c') \frac{D}{2} + \sum_{i=1}^{n} A_{bi} f_{sy} d_{yi} \right] / N_{uo}$$

where:

A_{bi} = Area of each bar in cross-section
d_{yi} = Depth to each bar in cross-section

Note: $d_q = \dfrac{D}{2}$ for symmetrical sections (about x-axis) with symmetrical reinforcement.

4.4.5.7.2 Decompression point through to pure bending

The most convenient way to create a solution is to solve the formula for k_u values starting at $k_u = 1$ (decompression), decreasing until the point of pure bending is reached. This can be done quickly using software such as Excel.

For a given k_u,

Compression block area, $\quad A_c = b\gamma k_u d$

$$\gamma = 1.05 - 0.007 f_c' \quad [0.67 \le \gamma \le 0.85]$$

Compressive force, $\quad F_{cc} = \alpha_2 f_c' A_c$

$$\alpha_2 = 1 - 0.003 f_c' \quad [0.67 \le \alpha_2 \le 0.85]$$

Compressive force lever arm, $\quad y_c = \gamma k_u d / 2$

For a rectangular cross-section with n bars, the force in each bar needs to be calculated separately based on the strain at the bar position (Figure 4.20).

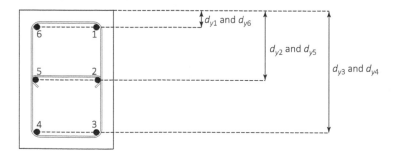

Figure 4.20 Bar numbers and identification.

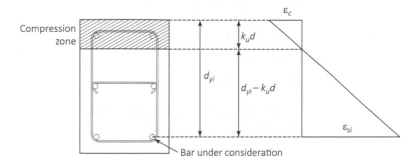

Figure 4.21 Rectangular column strain diagram.

Considering an individual bar, i,

Strain in bar, $\varepsilon_{si} = \dfrac{\varepsilon_c}{k_u d} \times (k_u d - d_{yi})$

where:

d_{yi} = Depth to the bar being considered (refer to Figure 4.21)

ε_c = 0.003

if $\varepsilon_{si} < 0$, the bar is in tension; if $\varepsilon_{si} > 0$, the bar is in compression.

Steel strains should always be checked to ensure that the magnitude is smaller than 0.05 (max strain for ductility class N; refer to Table 4.2) to prevent rupture.

Stress in bar, $\sigma_{si} = \text{MIN}[\varepsilon_{si}\, F_s, f_{sy}] = \text{MIN}[\varepsilon_{si} \times 200\,000, 500]$ (compression)

$\sigma_{si} = \text{MAX}[\varepsilon_{si}\, E_s, -f_{sy}] = \text{MAX}[\varepsilon_{si} \times 200\,000, -500]$ (tension)

Force in tension bars, $F_{si} = \sigma_{si} \times A_{bi}$

Force in compression bars, $F_{si} = (\sigma_{si} - \alpha_2 f_c') \times A_{bi} \geq 0$

Note: $\alpha_2\, f_c'$ is subtracted in the compression bars because this area is already accounted for in the concrete compression calculation.

Moment contributed by bar, $M_i = F_{si} \times d_{yi}$

Section calculations:

Compression force applied to section, $\phi N_u = \phi \left[F_{cc} + \sum_{i=1}^{n} F_{si} \right]$

Bending capacity, $\phi M_u = \phi \left[N_u d_q - F_{cc} y_c - \sum_{i=1}^{n} M_i \right]$

Note: $N_u d_q$ is subtracted to remove any applied axial load from the moment calculation, adopting the lever arm for the squash load. The calculation for ϕ is shown for class N reinforcement.

Reduction factor for bending with compression,

for, $N_u \geq N_{ub}$ $\phi = 0.6$

for, $N_u < N_{ub}$ $\phi = 0.6 + (\phi_{pb} - 0.6)(1 - N_u/N_{ub})$ (AS 3600, Table 2.2.2)

N_{ub} = Axial capacity at balanced point $(k_u = 0.\overline{54})$

Reduction factor for pure bending (no compression, Class N reinforcement)

$\phi_{pb} = 0.6 \leq (1.19 - 13\,k_{uo}/12) \leq 0.8$ (AS 3600, Table 2.2.2)

k_{uo} is the ratio of the depth to the neutral axis (from the extreme compressive fibre) to the depth to the outermost bar while subject to pure bending.

4.4.5.8 Circular cross-sections

The interaction for a circular cross-section is similar to that for a rectangle; however, the maths varies because of the shape, and the bars are all at different depths.

4.4.5.8.1 Squash load point

$$\phi N_{uo} = \phi[(A_g - A_s) \times \alpha_1 f_c' + A_s \times f_{sy}]$$

where:
$\alpha_1 = 1 - 0.003 f_c'$ $[0.72 \leq \alpha_1 \leq 0.85]$

A_g = Gross area of cross-section = $\pi D^2/4$
A_s = Area of steel
f_{sy} = Yield strength of steel (typically 500 MPa)
f_c' = Compressive strength of concrete
ϕ = 0.6 for pure compression

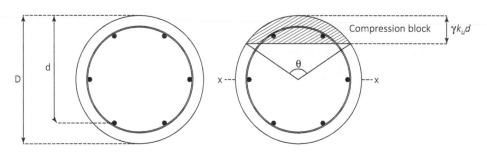

Figure 4.22 Circular cross-section.

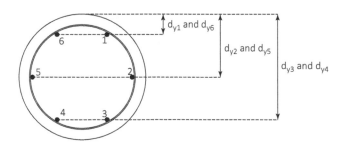

Figure 4.23 Bar numbers and identification.

Plastic centroid, $d_q = \left[([\pi D^2/4 - A_s] \times \alpha_1 f_c') \dfrac{D}{2} + \sum_{i=1}^{n} A_{bi} f_{sy} d_{yi}\right] / N_u$

where:

A_{bi} = Area of each bar in cross-section

d_{yi} = Depth to each bar in cross-section

Note: $d_q = \dfrac{D}{2}$ for symmetrical sections (about x-axis) with symmetrical reinforcement.

4.4.5.8.2 Decompression point through to pure bending

For a given k_u,

Angle to compressive block,

$$\frac{\theta}{2} = \cos^{-1}\left[\left(\frac{D}{2} - \gamma k_u d\right) \Big/ \left(\frac{D}{2}\right)\right] \quad \text{(refer to Figure 4.22 for representation of } \theta)$$

$$\gamma = 1.05 - 0.007 f_c' \quad [0.67 \le \gamma \le 0.85]$$

Compression block area, $A_c = (\theta - \sin\theta)\left(\dfrac{D}{2}\right)^2 / 12$

Compressive force, $F_{cc} = \alpha_2 f_c' A_c$

$$\alpha_2 = 1 - 0.003 f_c' \quad [0.67 \le \alpha_2 \le 0.85]$$

Compressive force lever arm, $y_c = \dfrac{D}{2} - 4\left(\dfrac{D}{2}\right)\sin\left(\dfrac{\theta}{2}\right)^3 / [3(\theta - \sin\theta)]$ (Figure 4.23)

For a circular cross-section with n bars (Figure 4.23), the force in each bar needs to be calculated separately. Considering an individual bar, i,

Strain in bar, $\varepsilon_{si} = \dfrac{\varepsilon_c}{k_u d} \times (k_u d - d_{yi})$

where: d_{yi} is the depth to the bar being considered (refer to Figure 4.24)

$\varepsilon_c = 0.003$

If $\varepsilon_{si} < 0$, the bar is in tension; if $\varepsilon_{si} > 0$, the bar is in compression.

Stress in bar, $\sigma_{si} = \text{MIN}[\varepsilon_{si} E_s, f_{sy}] = \text{MIN}[\varepsilon_{si} \times 200\,000, 500]$ (compression)

$\sigma_{si} = \text{MAX}[\varepsilon_{si} E_s, -f_{sy}] = \text{MAX}[\varepsilon_{si} \times 200\,000, -500]$ (tension)

Force in tension bars, $F_{si} = \sigma_{si} \times A_{bi}$

Force in compression bars, $F_{si} = (\sigma_{si} - \alpha_2 f_c') \times A_{bi} \ge 0$

Moment contributed by bar, $M_i = F_{si} \times d_{yi}$

Section calculations:

Compression force applied to section, $\phi N_u = \phi\left[F_{cc} + \sum_{i=1}^{n} F_{si}\right]$

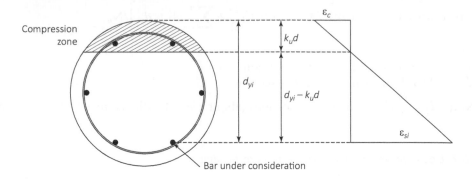

Figure 4.24 Circular column strain diagram.

Bending capacity, $\phi M_u = \phi \left[N_u d_q - F_{cc} y_c - \sum_{i=1}^{n} M_i \right]$

Reduction factor for bending with compression,

for, $N_u \geq N_{ub}$ $\phi = 0.6$

for, $N_u < N_{ub}$ $\phi = 0.6 + (\phi_{pb} - 0.6)(1 - N_u/N_{ub})$ (AS3600, Table 2.2.2)

N_{ub} = Axial capacity at balanced point $(k_u = 0.\overline{54})$

Reduction factor for pure bending (no compression, Class N reinforcement)

$\phi_{pb} = 0.6 \leq (1.19 - 13k_{uo}/12) \leq 0.8$ (AS3600, Table 2.2.2)

Example 4.6: Interaction diagram

Create an interaction diagram for a 600 mm deep × 400 mm wide concrete column about the major axis. 4-N20 reinforcing bars are placed on the top and bottom faces and 1-N20 at the midpoint on each side. Shear ligatures are a single closed loop of R10-200 with a cover of 50 mm. Concrete is grade N40.

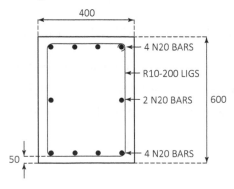

$b = 400$ mm

$D = 600$ mm

$f_{sy} = 500 \text{ MPa}$

$f'_c = 40 \text{ MPa}$

$d = 600 - 50 - 10 - 20/2 = 530 \text{ mm}$

$d_q = \dfrac{D}{2} = \dfrac{600}{2} = 300 \text{ mm}$

$\alpha_2 = 1 - 0.003 f'_c = 1 - 0.003 \times 40 = 0.85 \quad [0.67 \le \alpha_2 \le 0.85]$

$\gamma = 1.05 - 0.007 f'_c = 1.05 - 0.007 \times 40 = 0.77 \quad [0.67 \le \gamma \le 0.85]$

SQUASH LOAD POINT:

$\phi N_{uo} = \phi[(bD - A_s) \times \alpha_1 f'_c + A_s \times f_{sy}]$

$A_s = 10 \times \pi (20)^2/4 = 3142 \text{ mm}^2$

$\alpha_1 = 1 - 0.003 f'_c = 1 - 0.003 \times 40 = 0.85 \quad [0.72 \le \alpha_1 \le 0.85]$

$N_{uo} = (400 \times 600 - 3142) \times 0.85 \times 40 + 3142 \times 500 = 9624 \text{ kN}$

$\phi M_u = 0 \text{ kN}$

$\phi N_{uo} = 0.6 \times 9624 = 5774 \text{ kN}$

DECOMPRESSION POINT:

$k_u = 1$

$A_c = b\gamma k_u d = 400 \times 0.77 \times 1 \times 530 = 163{,}240 \text{ mm}^2$

$F_{cc} = \alpha_2 f'_c A_c = 0.85 \times 40 \times 163{,}240 = 5{,}550 \text{ kN}$

$y_c = \gamma k_u d/2 = 0.77 \times 1 \times 530/2 = 204.1 \text{ mm}$

Top bars,

$d_{y(\text{top})} = 50 + 10 + 20/2 = 70 \text{ mm}$

$\varepsilon_{s(\text{top})} = \dfrac{\varepsilon_c}{k_u d} \times (k_u d - d_{y(\text{top})}) = \dfrac{0.003}{1 \times 530} \times (1 \times 530 - 70) = 0.0026$

$\varepsilon_{s(\text{top})} > 0 \quad \text{(top bars are in compression)}$

$\sigma_{s(\text{top})} = \text{MIN}[\varepsilon_{s(\text{top})} E_s, f_{sy}] = \text{MIN}[0.0026 \times 200\ 000,\ 500] = 500 \text{ MPa}$

$F_{s(\text{top})} = (\sigma_{s(\text{top})} - \alpha_2 f'_c) \times A_{b(\text{top})} = (500 - 0.85 \times 40) \times 4 \times \pi \times (20)^2/4 = 586 \text{ kN}$

Middle (side) bars,

$d_{y(\text{mid})} = 600/2 = 300 \text{ mm}$

$$\varepsilon_{s(mid)} = \frac{\varepsilon_c}{k_u d} \times (k_u d - d_{y(mid)}) = \frac{0.003}{1 \times 530} \times (1 \times 530 - 300) = 0.0013$$

$\varepsilon_{s(mid)} > 0$ (middle bars are in compression)

$$\sigma_{s(mid)} = MIN[\varepsilon_{s(mid)} E_s, f_{sy}] = MIN[0.0013 \times 200\,000, 500] = 260 \text{ MPa}$$

$$F_{s(mid)} = (\sigma_{s(mid)} - \alpha_2 f_c') \times A_{b(mid)} = (260 - 0.85 \times 40) \times 2 \times \pi \times (20)^2 / 4 = 142 \text{ kN}$$

Bottom bars,

$$d_{y(bot)} = 600 - 50 - 10 - 20/2 = 530 \text{ mm}$$

$$\varepsilon_{s(bot)} = \frac{\varepsilon_c}{k_u d} \times (k_u d - d_{y(bot)}) = \frac{0.003}{1 \times 530} \times (1 \times 530 - 530) = 0.0$$

$\varepsilon_{s(bot)} = 0$ (bottom bars are not loaded)

$$\sigma_{s(bot)} = MIN[\varepsilon_{s(bot)} E_s, f_{sy}] = MIN[0 \times 200\,000, 500] = 0 \text{ MPa}$$

$$F_{s(bot)} = \sigma_{s(bot)} \times A_{b(bot)} = 0 \times 4 \times \pi \times (20)^2 / 4 = 0 \text{ kN}$$

Compression capacity,

$$N_u = F_{cc} + \sum_{i=1}^{n} F_{si} = 5550 + 586 + 142 + 0 = 6278 \text{ kN}$$

Bending capacity,

$$M_u = N_u d_q - F_{cc} y_c - \sum_{i=1}^{n} M_i$$

$$M_{(top)} = F_{s(top)} \times d_{y(top)} = 586 \times 0.070 = 41 \text{ kNm}$$

$$M_{(mid)} = F_{s(mid)} \times d_{y(mid)} = 91.5 \times 0.300 = 42.7 \text{ kNm}$$

$$M_{(bot)} = F_{s(bot)} \times d_{y(bot)} = 0 \times 0.530 = 0 \text{ kNm}$$

$$M_u = 6278 \times 0.3 - 5550 \times 0.2041 - 41 - 42.7 - 0 = 667.2 \text{ kNm}$$

$$\phi = 0.6 \qquad N_u \geq N_{ub}$$

(Decompression point has a higher axial force than balanced point)

$$\phi M_u = 0.6 \times 667.2 = 400 \text{ kN}$$

$$\phi N_u = 6278 \text{ kN}$$

BALANCED POINT:

$$k_u = 0.\overline{54}$$

$$A_c = b \gamma k_u d = 400 \times 0.77 \times 0.\overline{54} \times 530 = 89,040 \text{ mm}^2$$

$$F_{cc} = \alpha_2 f_c' A_c = 0.85 \times 40 \times 89{,}040 = 3{,}027 \text{ kN}$$

$$y_c = \gamma k_u d/2 = 0.77 \times 0.\overline{54} \times 530/2 = 111.3 \text{ mm}$$

Top bars,

$$d_{y(\text{top})} = 50 + 10 + 20/2 = 70 \text{ mm}$$

$$\varepsilon_{s(\text{top})} = \frac{\varepsilon_c}{k_u d} \times (k_u d - d_{y(\text{top})}) = \frac{0.003}{0.\overline{54} \times 530} \times (0.\overline{54} \times 530 - 70) = 0.0023$$

$$\varepsilon_{s(\text{top})} > 0 \quad (\text{top bars are in compression})$$

$$\sigma_{s(\text{top})} = \text{MIN}[\varepsilon_{s(\text{top})} E_s, f_{sy}] = \text{MIN}[0.0023 \times 200\,000, 500] = 455 \text{ MPa}$$

$$F_{s(\text{top})} = (\sigma_{s(\text{top})} - \alpha_2 f_c') \times A_{b(\text{top})} = (455 - 0.85 \times 40) \times 4 \times \pi \times (20)^2/4 = 528.7 \text{ kN}$$

Middle (side) bars,

$$d_{y(\text{mid})} = 600/2 = 300 \text{ mm}$$

$$\varepsilon_{s(\text{mid})} = \frac{\varepsilon_c}{k_u d} \times (k_u d - d_{y(\text{mid})}) = \frac{0.003}{0.\overline{54} \times 530} \times (0.\overline{54} \times 530 - 300) = -0.0001$$

$$\varepsilon_{s(\text{mid})} < 0 \quad (\text{middle bars are in tension})$$

$$\sigma_{s(\text{mid})} = \text{MAX}[\varepsilon_{s(\text{mid})} E_s, -f_{sy}] = \text{MAX}[-0.0001 \times 200\,000, -500] = -22.6 \text{ MPa}$$

$$F_{s(\text{mid})} = \sigma_{s(\text{mid})} \times A_{b(\text{mid})} = -22.6 \times 2 \times \pi \times (20)^2/4 = -14.2 \text{ kN}$$

Bottom bars,

$$d_{y(\text{bot})} = 600 - 50 - 10 - 20/2 = 530 \text{ mm}$$

$$\varepsilon_{s(\text{bot})} = \frac{\varepsilon_c}{k_u d} \times (k_u d - d_{y(\text{bot})}) = \frac{0.003}{0.\overline{54} \times 530} \times (0.\overline{54} \times 530 - 530) = -0.0025$$

$$\varepsilon_{s(\text{bot})} < 0 \quad (\text{bottom bars are in tension})$$

$$\sigma_{s(\text{bot})} = \text{MAX}[\varepsilon_{s(\text{mid})} E_s, -f_{sy}] = \text{MAX}[-0.0025 \times 200\,000, -500] = -500 \text{ MPa}$$

$$F_{s(\text{bot})} = \sigma_{s(\text{bot})} \times A_{b(\text{bot})} = -500 \times 4 \times \pi \times (20)^2/4 = -628.3 \text{ kN}$$

Compression capacity,

$$N_u = F_{cc} + \sum_{i=1}^{n} F_{si} = 3027 + 528.7 - 14.2 - 628.3 = 2913.5 \text{ kN}$$

Bending capacity,

$$M_u = N_u d_q - F_{cc} y_c - \sum_{i=1}^{n} M_i$$

$$M_{(\text{top})} = F_{s(\text{top})} \times d_{y(\text{top})} = 528.7 \times 0.070 = 37 \text{ kNm}$$

$$M_{(\text{mid})} = F_{s(\text{mid})} \times d_{y(\text{mid})} = -14.2 \times 0.300 = -4.3 \text{ kNm}$$

$$M_{(bot)} = F_{s(bot)} \times d_{y(bot)} = -628.3 \times 0.530 = -333 \text{ kNm}$$

$$M_u = 2913.5 \times 0.3 - 3027 \times 0.1113 - 37 + 4.3 + 333 = 837.4 \text{ kNm}$$

$$N_u = N_{ub} \qquad \phi = 0.6$$

$$\phi M_u = 0.6 \times 837.4 = 502 \text{ kNm}$$

$$\phi N_u = 1748 \text{ kN}$$

PURE BENDING POINT:

The method from the previous two sets is usually repeated with numerous values of k_u decreasing incrementally from 1 down to the point where compression capacity is calculated as zero. Alternatively, trial and error or the quadratic method shown in Section 4.2.3.4 can be used to find k_u at pure bending.

$$k_u = k_{uo} = 0.1564$$

$$A_c = b\gamma k_u d = 400 \times 0.77 \times 0.1564 \times 530 = 25,529 \text{ mm}^2$$

$$F_{cc} = \alpha_2 f_c' A_c = 0.85 \times 40 \times 25,529 = 868 \text{ kN}$$

$$y_c = \gamma k_u d/2 = 0.77 \times 0.1564 \times 530/2 = 31.9 \text{ mm}$$

Top bars,

$$d_{y(top)} = 50 + 10 + 20/2 = 70 \text{ mm}$$

$$\varepsilon_{s(top)} = \frac{\varepsilon_c}{k_u d} \times (k_u d - d_{y(top)}) = \frac{0.003}{0.1564 \times 530} \times (0.1564 \times 530 - 70) = 0.0005$$

$$\varepsilon_{s(top)} < 0 \quad \text{(top bars are in tension)}$$

$$\sigma_{s(top)} = \text{MIN}[\varepsilon_{s(top)} E_s, f_{sy}] = \text{MIN}[-0.0005 \times 200\,000, 500] = 93.3 \text{ MPa}$$

$$F_{s(top)} = \sigma_{s(top)} \times A_{b(top)} = 93.3 \times 4 \times \pi \times \frac{(20)^2}{4} = 74.5 \text{ kN}$$

Middle (side) bars,

$$d_{y(mid)} = 600/2 = 300 \text{ mm}$$

$$\varepsilon_{s(mid)} = \frac{\varepsilon_c}{k_u d} \times (k_u d - d_{y(mid)}) = \frac{0.003}{0.1564 \times 530} \times (0.1564 \times 530 - 300) = -0.0079$$

$$\varepsilon_{s(mid)} < 0 \quad \text{(middle bars are in tension)}$$

$$\sigma_{s(mid)} = \text{MAX}[\varepsilon_{s(mid)} E_s, -f_{sy}] = \text{MAX}[-0.0079 \times 200\,000, -500] = -500 \text{ MPa}$$

$$F_{s(mid)} = \sigma_{s(mid)} \times A_{b(mid)} = -500 \times 2 \times \pi \times (20)^2/4 = -314.2 \text{ kN}$$

Bottom bars,

$$d_{y(bot)} = 600 - 50 - 10 - 20/2 = 530 \text{ mm}$$

$$\varepsilon_{s(bot)} = \frac{\varepsilon_c}{k_u d} \times (k_u d - d_{y(bot)}) = \frac{0.003}{0.1564 \times 530} \times (0.1564 \times 530 - 530) = -0.0162$$

$\varepsilon_{s(bot)} < 0$ (bottom bars are in tension)

$\sigma_{s(bot)} = \text{MAX}[\varepsilon_{s(mid)} E_s, -f_{sy}] = \text{MAX}[-0.0162 \times 200\,000, -500] = -500 \text{ MPa}$

$F_{s(bot)} = \sigma_{s(bot)} \times A_{b(bot)} = -500 \times 4 \times \pi \times (20)^2/4 = -628.3 \text{ kN}$

Compression capacity,

$$N_u = F_{cc} + \sum_{i=1}^{n} F_{si} = 868 + 74.5 - 314.2 - 628.3 = 0 \text{ kN} \quad \text{(confirmed pure bending)}$$

Bending capacity,

$$M_u = N_u d_q - F_{cc} y_c - \sum_{i=1}^{n} M_i$$

$M_{(top)} = F_{s(top)} \times d_{y(top)} = 74.5 \times 0.070 = 5.2 \text{ kNm}$

$M_{(mid)} = F_{s(mid)} \times d_{y(mid)} = -314.2 \times 0.300 = -94.3 \text{ kNm}$

$M_{(bot)} = F_{s(bot)} \times d_{y(bot)} = -628.3 \times 0.530 = -333 \text{ kNm}$

$M_u = 0 \times 0.3 - 868 \times 0.0319 - 5.2 + 94.3 + 333 = 394.3 \text{ kNm}$

$\phi_{pb} = 0.6 \leq (1.19 - 13k_{uo}/12) \leq 0.8$

$0.6 \leq \left(1.19 - 13 \times \dfrac{0.1564}{12}\right) \leq 0.8 = 0.8$

$N_u < N_{ub} \qquad \phi = 0.6 + (\phi_{pb} - 0.6)(1 - N_u/N_{ub}) = 0.6 + (0.8 - 0.6)(1 - 0/2913.5) = 0.8$

$\phi M_u = 0.8 \times 394.3 = 315 \text{ kNm}$

$\phi N_u = 0 \text{ kN}$

The interaction diagram is created by plotting all four points (dashed line). Additional detail can be added by solving more values of k_u (solid line). The column loading is acceptable if the bending and axial loading plot within the graph boundary.

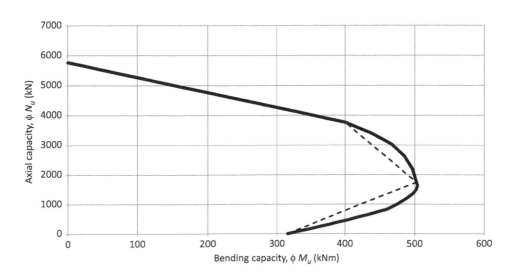

4.5 ELASTIC ANALYSIS

Elastic analysis is generally used for calculating stresses in bars and in concrete under service loading. This is usually done for crack control (refer to Section 4.2.2) or for liquid-retaining structures (refer to Section 6.9).

The concept is to create an equivalent section by factoring up the area of steel by the ratio (n) of the Young's modulus of steel to concrete (refer to Table 4.1). The section is then analysed as if it were one material. This works because both materials strain at the same rate; however, the steel stress is n times greater at the same strain as the concrete.

Transformation ratio, $\quad n = E_s / E_c$

for 32 MPa concrete, $\quad n_{32} = 200,000/30,100 = 6.64$

for 40 MPa concrete, $\quad n_{40} = 200,000/32,800 = 6.10$

4.5.1 Calculate depth to neutral axis

The first step is to assume that the neutral axis is above the top bars; that is, all bars are in tension. This is often the case for slabs, walls and shallow beams. Refer to Figure 4.25 for details.

The depth to the neutral axis needs to be solved using the quadratic formula by equating the tensile and compressive moments:

Tensile moment, $\qquad M_t = n A_{st}(d - kd) + n A_{sc}(d_{tb} - kd)$

Compressive moment, $\quad M_c = bkd\ (kd/2)$

where:
$\quad A_{sc}$ = Area of top bars
$\quad A_{st}$ = Area of bottom bars
$\quad d_{tb}$ = Depth to top bars
$\quad kd$ = Depth to neutral axis

Equating moments, $\quad M_t = M_c$

$$0 = \frac{b}{2}(kd)^2 + nA_{st}(kd) + nA_{sc}(kd) - nA_{st}d - nA_{sc}d_{tb}$$

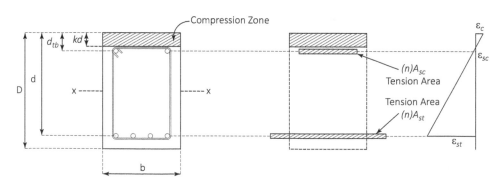

Figure 4.25 Top and bottom bars in tension.

This can be solved using the quadratic formula,

$$kd = \frac{-B \pm \sqrt{B^2 - 4AC}}{2A}$$

$$A = \frac{b}{2}$$

$$B = n(A_{st} + A_{sc})$$

$$C = n(-A_{st}d - A_{sc}d_{tb})$$

Check the assumption $k_d \leq d_{tb}$; if this is not the case, the assumption needs to be changed so that the neutral axis is below the top bars; that is, the top bars are in compression and the bottom bars are in tension. This is often the case for beams and columns. Refer to Figure 4.26 for details.

The depth to the neutral axis needs to be solved again using the quadratic formula by equating the tensile and compressive moments:

Tensile moment, $M_t = n A_{st}(d - kd)$

Compressive moment, $M_c = bkd\ (kd/2) + (n-1)A_{sc}(kd - d_{tb})$

Note: $(n - 1)$ is used because the area is already accounted for once in the concrete side of the calculation.

Equating moments, $M_t = M_c$

$$0 = \frac{b}{2}(kd)^2 + (n-1)A_{sc}(kd) + nA_{st}(kd) - (n-1)A_{sc}d_{tb} - nA_{st}d$$

This can be solved using the quadratic formula,

$$kd = \frac{-B \pm \sqrt{B^2 - 4AC}}{2A}$$

$$A = \frac{b}{2}$$

$$B = (n-1)A_{sc} + nA_{st}$$

$$C = -(n-1)A_{sc}d_{tb} - nA_{st}d$$

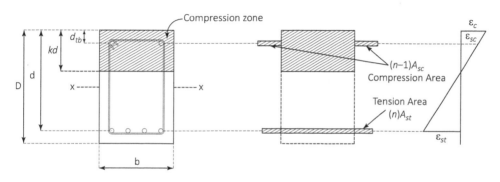

Figure 4.26 Top bars in compression and bottom bars in tension.

4.5.2 Calculate moment of inertia

The moment of inertia is calculated using the standard equation, $I_x = \dfrac{bh^3}{12} + bh(d_{NA})^2$, where d_{NA} is the depth from the centre of the shape to the neutral axis. This can be simplified further by combining terms and ignoring the small terms (Figure 4.27).

All bars in tension: $I_x = \dfrac{b(kd)^3}{3} + nA_{sc}(kd - d_{tb})^2 + nA_{st}(d - kd)^2$ (Assumption 1)

Top bars in compression: $I_x = \dfrac{b(kd)^3}{3} + (n-1)A_{sc}(kd - d_{tb})^2 + nA_{st}(d - kd)^2$ (Assumption 2)

4.5.3 Calculate stress

The stress can now be calculated using the following equation and accounting for the modulus conversion. The transformation ratio is included in the calculation of the steel stresses because it would undergo significantly more stress to strain the same distance as the concrete equivalent.

$$\sigma = My/I_x$$

where:

M	= Moment
y	= Distance from neutral axis to point being considered
σ	= Stress

Maximum concrete stress, $\quad \sigma_{cc} = M(-kd)/I_x$

Top steel stress, $\quad \sigma_{sc} = n\,M(d_{tb} - kd)/I_x$

Bottom steel stress, $\quad \sigma_{st} = n\,M(d - kd)/I_x$

Note: Negative values are calculated as compression.

This method can be used for beams in other shapes; however, additional assumptions need to be added for changes in geometry. For example, in a t-beam, an assumption would be made for the neutral axis being in the top flange, and also for it being in the web.

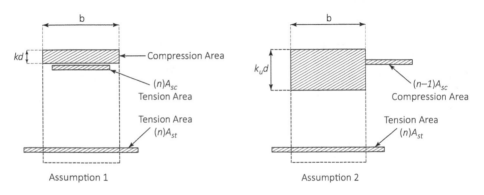

Figure 4.27 Each case for moment of inertia.

4.5.4 Calculate strain

Strain can be checked using the following formula. For the elastic method to be relevant, the magnitude of concrete strain should be less than 0.003, and steel strain should be less than f_{sy}/E_s.

Concrete compression strain, $\varepsilon_{cc} = \sigma_{cc}/E_c$

Top steel strain, $\varepsilon_{sc} = \sigma_{sc}/E_s$

Bottom steel strain, $\varepsilon_{st} = \sigma_{st}/E_s$

Example 4.7: Elastic transformed analysis

Calculate the elastic stresses for a 600 mm deep × 400 mm wide concrete column with 50 kNm about the major axis. 4-N16 reinforcing bars are placed on the tension face and 2-N16 are on the top face. Shear ligatures are a single closed loop of R10-200 with a cover of 50 mm. Concrete is grade N40.

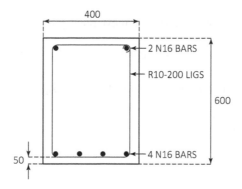

$M = 50 \text{ kNm}$

$b = 400 \text{ mm}$

$D = 600 \text{ mm}$

$f_{sy} = 500 \text{ MPa}$

$f'_c = 40 \text{ MPa}$

$E_c = 32,800 \text{ MPa}$

$E_s = 200,000 \text{ MPa}$

$n_{40} = 200,000/32,800 = 6.10$

Elastic section:
  Assuming the neutral axis is above the top bars (all bars in tension),

$A_{sc} = 2 \times \pi \times 16^2/4 = 402 \text{ mm}^2$

$A_{st} = 4 \times \pi \times 16^2/4 = 804 \text{ mm}^2$

$$d = 600 - 50 - 10 - 16/2 = 532 \text{ mm}$$

$$d_{tb} = 50 + 10 + 16/2 = 68 \text{ mm}$$

$$0 = \frac{b}{2}(kd)^2 + nA_{st}(kd) + nA_{sc}(kd) - nA_{st}d - nA_{sc}d_{tb}$$

$$kd = \frac{-B \pm \sqrt{B^2 - 4AC}}{2A}$$

$$A = \frac{b}{2} = \frac{400}{2} = 200$$

$$B = n(A_{st} + A_{sc}) = 6.10(804 + 402) = 7357$$

$$C = n(-A_{st}d - A_{sc}d_{tb}) = 6.10(-804 \times 532 - 402 \times 68) = -2.776 \times 10^6$$

$$kd = 100.8 \text{ mm}$$

$k_d > d_{tb}$; therefore, the neutral axis is not above the top bars, and the assumption needs to be modified.

Taking the neutral axis below the top bars,

$$0 = \frac{b}{2}(kd)^2 + (n-1)A_{sc}(kd) + nA_{st}(kd) - (n-1)A_{sc}d_{tb} - nA_{st}d$$

$$kd = \frac{-B \pm \sqrt{B^2 - 4AC}}{2A}$$

$$A = \frac{b}{2} = \frac{400}{2} = 200$$

$$B = (n-1)A_{sc} + nA_{st} = (6.10 - 1)402 + 6.10 \times 804 = 6954$$

$$C = -(n-1)A_{sc}d_{tb} - nA_{st}d = -(6.10 - 1)402 \times 68 - 6.10 \times 804 \times 532$$

$$= -2.749 \times 10^6$$

$$kd = 101.1 \text{ mm}$$

$$I_x = \frac{b(kd)^3}{3} + (n-1)A_{sc}(kd - d_{tb})^2 + nA_{st}(d - kd)^2$$

$$= \frac{400(101.1)^3}{3} + (6.10 - 1) \times 402 \times (101.1 - 68)^2 + 6.10 \times 804(532 - 101.1)^2 = 1051 \times 10^6 \text{ mm}^4$$

Elastic stress:

$$\sigma_{cc} = M(-kd)/I_x = 50 \times 10^6 \times (-101.1)/(1051 \times 10^6) = -4.8 \text{ MPa}$$

$$\sigma_{sc} = nM(d_{tb} - kd)/I_x = 6.10 \times 50 \times 10^6 \times (68 - 101.1)/(1051 \times 10^6) = -9.6 \text{ MPa}$$

$$\sigma_{st} = nM(d - kd)/I_x = 6.10 \times 50 \times 10^6 \times (532 - 101.1)/(1051 \times 10^6) = 125 \text{ MPa}$$

Check strain,

$$\varepsilon_{cc} = \sigma_{cc}/E_c = -4.8/32,800 = -0.00015 \qquad > -0.003 \qquad \text{Pass}$$

$$\varepsilon_{sc} = \sigma_{sc}/E_s = -9.6/200,000 = -0.00005 \qquad > -0.0025 \qquad \text{Pass}$$

$$\varepsilon_{st} = \sigma_{st}/E_s = 125/200,000 = 0.00063 \qquad < 0.0025 \qquad \text{Pass}$$

A moment of 50 kNm results in an elastic stress of 4.8 MPa (compression) at the top of the concrete, 9.6 MPa (compression) in the top layer of steel and 125 MPa (tension) in the bottom layer of steel.

4.6 STRUT AND TIE

Strut and tie modelling is a means of analysing concrete in zones which behave in a non-flexural manner. This includes beams which have a small span-to-depth ratio and also non-flexural zones of flexural members (e.g. supports). Strut and tie modelling requirements are provided in AS 3600, Chapters 7 and 12. Non-flexural members are approximated as those with a span-to-depth ratio of less than 1.5 for cantilevers, 3 for simply supported beams, and 4 for continuous beams. The design approach is based on a lower bound theorem, which means that if you can find a way for the structure to work, it will find a better way. Strut and tie modelling involves creating an imaginary truss within a concrete element (refer to Figure 4.28). Chords which take compression, called struts, rely only on the concrete strength. Chords which take tension, called ties, rely on the reinforcing strength. Nodes are the points at which struts and/or ties intersect.

The force in each element can be solved using the laws of statics. AS 3600 requires that loads be applied only at the nodes and that struts and ties carry only axial loading. Ties are permitted to cross struts; however, struts cannot cross other struts and must intersect at nodes. Angles between struts and/or ties must not be less than 30°, or 20° for prestressed concrete. Prestressing requirements are not shown in this text.

4.6.1 Ties

Ties are sized to carry the tensile load between nodes. AS 3600 requires that reinforcing steel is at least 50% developed before entering the node (refer to Section 4.1.5). The full tensile load is also required to be developed prior to exiting the node. Furthermore, if the

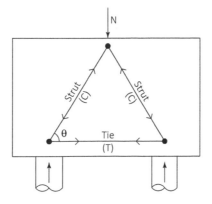

Figure 4.28 Strut and tie concept.

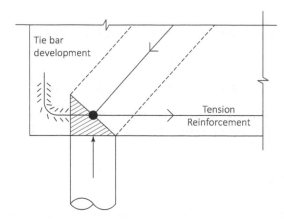

Figure 4.29 Tie development.

full load is developed before entering the node, the node stress from the back face can be accepted without further calculations [20] (refer to Figure 4.29).

$$T^* \leq \phi_{st(s)} A_{st} f_{sy}$$

$$\phi_{st(s)} = 0.8 \quad \text{(for ties)}$$

where:

    A_{st} = Area of steel in tension
    f_{sy} = Yield strength of steel (typically 500 MPa)

4.6.2 Struts

Struts are sized to carry the compressive loads between nodes. AS 3600 defines struts as fan-shaped, prismatic or bottle-shaped. Fan-shaped struts are ideal for distributed loads. Prismatic struts require specific confinement and are therefore not commonly used. This text focuses on the more common bottle-shaped struts. Bottle-shaped struts are the standard selection for transferring compressive forces through a concrete medium between two nodes. The strut is often drawn as prismatic for simplicity; however, bursting reinforcement is provided to spread the load while preventing cracking.

The capacity of the strut is governed at the smallest cross-section, which occurs at the interface to the nodes (Figure 4.30).

$$C^* \leq \phi_{st(c)} \, \beta_s \, 0.9 f_c' A_c$$

$$\phi_{st(c)} = 0.6 \quad \text{(for struts)}$$

where:

    A_c = Smallest cross-sectional area of strut (perpendicular to load) = $d_s b$
    b = Width of strut (measured into page)
    d_s = Depth of strut at node face (perpendicular to strut)
    f_c' = Compressive strength of concrete (typically 32 or 40 MPa)
    β_s = $1/[1 + 0.66 \cot^2 \theta]$ within the limits $0.3 \leq \beta_s \leq 1$
    θ = Smallest angle between a strut and a tie (considering both ends) (refer to Figure 4.28)

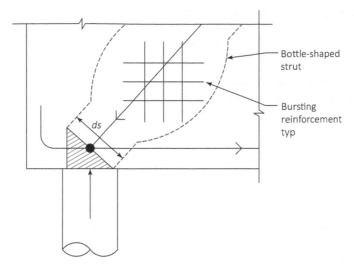

Figure 4.30 Strut geometry.

4.6.3 Bursting reinforcement

Bursting reinforcement is provided in bottle-shaped struts to prevent cracking. For serviceability limit states, the stress in bursting reinforcement should be limited depending on the required degree of crack control. For strength limit states, the stress is limited to the factored yield strength.

Bursting reinforcement is not required for struts where the following inequality is met, showing that concrete strength is sufficient:

$$C^* < T_{b.cr}$$

where:

$\quad$ C^* $\quad$ = ULS compression in strut
$\quad$ $T_{b.cr}$ $\quad$ = $0.7\,bl_b f'_{ct}$
$\quad$ b $\quad$ = Width of strut (measured into page)
$\quad$ l_b $\quad$ = $1 - d_c$
$\quad$ d_c $\quad$ = Length of node face
$\quad$ l $\quad$ = Length of strut (between node faces)

$$f'_{ct} = 0.36\sqrt{f'_c}$$

The bursting force is calculated for struts which require bursting reinforcement. Strength limit states and serviceability limit states both need to be checked for these struts. The assumed bursting angle is dependent on the limit state being considered.

$$T_b^* \le \phi_{st(s)} T_{bu} \qquad (ULS)$$

$$MAX[T_{b.s}^*, T_{b.cr}] \le T_{bc} \qquad (SLS)$$

$$\phi_{st(s)} = 0.8 \quad \text{(for bursting reinforcement)}$$

ULS bursting force, $T_b^* = C^* \tan \alpha = 0.2\,C^*$

SLS bursting force, $T_{b.s}^* = C \tan \alpha = 0.5\,C$

where:

  C^* = ULS compression in strut
  C = SLS compression in strut
  $\tan \alpha$ = 0.2 (ULS)
      = 0.5 (SLS)

The capacity of the bursting reinforcement is calculated using the component of the bar strength in the axis which is perpendicular to the strut. Only reinforcement with $\gamma_i > 40°$ is included in the following calculations. Reinforcement which is steeper is excluded, often resulting in the inclusion of reinforcement in only one direction (horizontal or vertical) when completing calculations. Reinforcing bars on both sides of the strut can be included in the areas of steel. The reinforcing bars shall be evenly distributed along l_b at maximum centres of 300 mm (Figure 4.31).

Bursting capacity, $T_{bu} = A_{sh} f_{sy} \sin \gamma_h + A_{sv} f_{sy} \sin \gamma_v$

Burst cracking capacity, $T_{bc} = A_{sh} f_{yc} \sin \gamma_h + A_{sv} f_{yc} \sin \gamma_v$

where:

  A_{sh} = Area of horizontal bursting reinforcement which crosses l_b
  A_{sv} = Area of vertical bursting reinforcement which crosses l_b
  f_{sy} = Yield strength of steel (typically 500 MPa)
  f_{yc} = 150 MPa (strong degree of crack control)
      = 200 MPa (moderate degree of crack control)
      = 250 MPa (minor degree of crack control)
  γ_h = Angle between strut and horizontal reinforcement
  γ_v = Angle between strut and vertical reinforcement

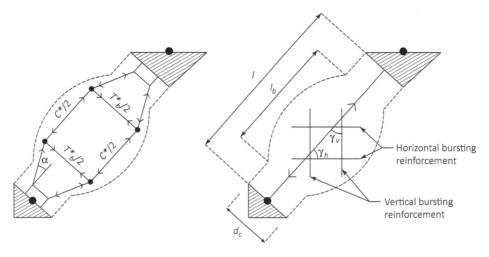

Figure 4.31 Bursting reinforcement.

4.6.4 Nodes

Nodes are the points at which struts and/or ties intersect. The combined stress in a node is checked using a Mohr's circle to calculate the maximum principal stress. Nodes are typically accepted without further calculation (based on the strut having been checked) where all forces applied to the node are perpendicular to the node face, or when concrete stresses are low and end forces are close to perpendicular. Node stresses need to be checked when concrete stresses are high or where angles are not square.

Nodes are classified into one of three categories (refer to Figure 4.32). All nodes will have at least three chords intersecting at the same point. CCC nodes have only compression struts. CCT nodes have two or more compressive struts and only one tension tie. CTT nodes have only one compression strut and two or more tension ties. The principal stress applied to the face of each node is checked against the design capacity using the combined normal and shear stress.

Nodes are dimensioned by intersecting parallel lines between the struts and ties. The support width is governed by the geometry of the applied load or the support width (circular supports are reduced to an equivalent 70% width). The width of a tie face is dimensioned by locating the centroid of the tension bars centrally on the face (refer to Figure 4.33). It has been shown that if ties are fully developed before entering a node [20], the node stress from the back face can be accepted without further calculations. If a plate is used to engage the tie, it creates an effective compressive strut on the back face of the node. The node stress should be checked at the node for this scenario. Nodes which are over supports can be split centrally into two CCT nodes.

The formula for node capacity is similar to that for strut capacity, as both are checking the strength of concrete in compression. The maximum compressive stress is limited as shown:

$$\sigma_{MAX} \leq \phi_{st(n)} \, \beta_n \, 0.9f'_c$$

$$\phi_{st(n)} = 0.6 \quad \text{(for nodes)}$$

where:
$$\beta_n \quad = 1 \text{ (CCC nodes)}$$
$$= 0.8 \text{ (CCT nodes)}$$
$$= 0.6 \text{ (CTT nodes)}$$

For nodes where the strut is perpendicular to the node face, the stress at the outside face of the node is equal to the strut force divided by the area of the face. This stress is the same as that used to check the strut capacity. The principal and shear stresses which

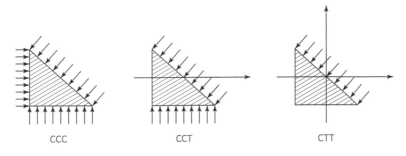

CCC   CCT   CTT

Figure 4.32 Node types.

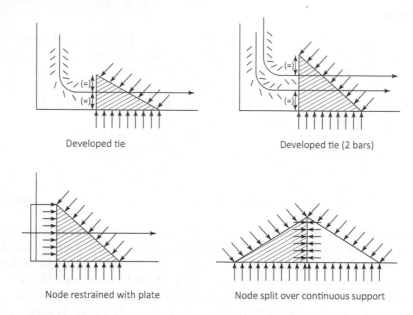

Developed tie   Developed tie (2 bars)

Node restrained with plate   Node split over continuous support

Figure 4.33 Node examples.

are perpendicular and parallel (respectively) to the node face should be calculated where the chord force is not perpendicular to the node face. The stress on the strut axis is pure compression; therefore, a Mohr's circle can be drawn with one end based at (0,0) and the other end at $(\sigma_{pi},0)$. The compressive stress (σ_i) and shear stress (τ_i) in the direction of the face of the node are found by rotating around the Mohr's circle by two times the angle between the strut and a line normal to the node face $(2\theta_i)$. Mohr's circles are a graphical representation of stress conditions in which the point on the circle is always rotated by double the angle by which the element is rotated to find the stress in a different direction (Figure 4.34).

Stress on strut axis i, $\qquad\qquad \sigma_{pi} = C^*/A_{ci}$ (as per strut design)

Mohr's circle radius for face i, $\qquad R_i = \sigma_{pi}/2$

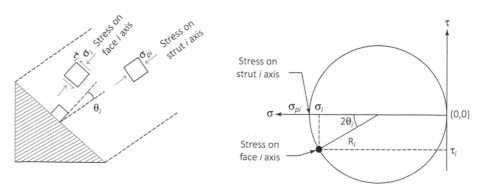

Figure 4.34 Stress at the face of a node (i).

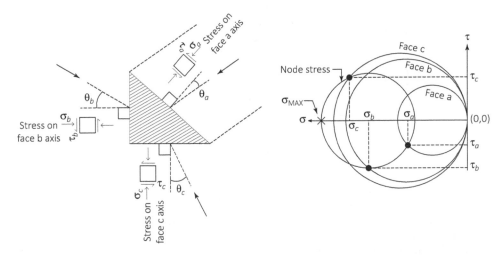

Figure 4.35 Total stress in node.

Compression normal to face i, $\sigma_i = R_i(1 + \cos(2\theta_i))$

Shear parallel to face i, $\tau_i = R_i \sin(2\theta_i)$

where:

A_{ci} = Cross-sectional area of strut i (measured perpendicular to strut)

θ_i = Angle between chord and line normal to face of node i

The combined stress within the node is found by drawing circles for each of the node faces, then drawing a circle to represent the combined stress. The combined stress circle is centred on the x-axis and intersects each of the stress points that are calculated for the node faces. Technically, only two of the points need to be calculated to create the circle, because the node is in equilibrium. The third point (for a three-sided node) is typically completed as a check of the result (Figure 4.35).

Equations are shown for the calculation of the resultant node stress based on faces a and b.

Centre of Mohr's circle, $\sigma_m = \dfrac{\sigma_a + \sigma_b}{2} + \dfrac{\tau_b^2 - \tau_a^2}{2(\sigma_b - \sigma_a)}$

Mohr's circle radius, $R = \sqrt{(\sigma_a - \sigma_m)^2 + \tau_a^2}$

Minimum compressive stress, $\sigma_{MIN} = \sigma_m - R$

Maximum compressive stress, $\sigma_{MAX} = \sigma_m + R$

Example 4.8: Strut and Tie

Design a strut and tie system for a 2000 kN ULS load applied to the concrete beam shown below.

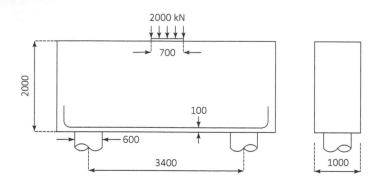

STRUT AND TIE MODEL:

A strut and tie system is modelled using a right-angled triangle for simplicity. The side of the node is dimensioned as twice the distance to the centre of the tie. The base of the node is 70% the width of the piers/columns (for circular supports). The remaining geometry is driven from these dimensions.

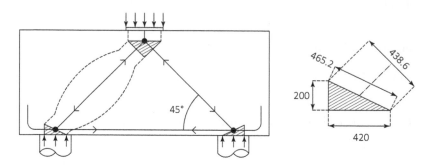

The supports are assumed to take 50% of the load each (1000 kN) due to the model symmetry.

Strut compression, $C^* = 1000\,\text{kN} \div \sin 45 = 1414\,\text{kN}$

Tie tension, $T^* = 1000\,\text{kN} \div \tan 45 = 1000\,\text{kN}$

TIE:

Try 6N24 reinforcing bars,

$$T^* \leq \phi_{st(s)} A_{st} f_{sy}$$

$$A_{st} = 6 \times \pi (24\,\text{mm})^2 / 4 = 2714\,\text{mm}^2$$

$$f_{sy} = 500\,\text{MPa}$$

$$\phi_{st(s)} A_{st} f_{sy} = 0.8 \times 2714\,\text{mm}^2 \times 500\,\text{MPa} = 1086 \quad > 1000\,\text{kN} \quad \text{Pass}$$

STRUTS:

The strut is governed at the cross-section as it enters the bottom node.

$$C^* \leq \phi_{st(c)} \beta_s 0.9 f_c' A_c$$

$$A_c = d_s b = 438.6\,\text{mm} \times 420\,\text{mm} = 184{,}212\,\text{mm}^2$$

$$b = 0.7 \times 600\,\text{mm} = 420\,\text{mm}$$

$d_s = 438.6$ mm

$\beta_s = 1/[1 + 0.66 \cot^2 \theta] = 1/[1 + 0.66 \cot^2 45] = 0.602$ $[0.3 \le \beta_s \le 1]$

$\phi_{st(c)} \beta_s 0.9 f_c' A_c = 0.6 \times 0.602 \times 40$ MPa $\times 184,212$ mm$^2 = 2397$ kN > 1414 kN   Pass

BURSTING REINFORCEMENT:

$T_{b.cr} = 0.7 b l_b f_{ct}'$

Average $d_c = (465.2$ mm $+ 438.6$ mm$)/2 = 451.9$ mm

$l_b = l - d_c = 2183$ mm $- 451.9 = 1731$ mm

$f_{ct}' = 0.36\sqrt[3]{f_{ct}'} = 0.36\sqrt[3]{40}$ MPa $= 2.28$ MPε

$T_{b.cr} = 0.7 \times 420$ mm $\times 1730.9$ mm $\times 2.28$ MPa $= 1159 < C^*$ Bursting reo required

Due to the 45° angle, vertical and horizontal reinforcement (on both sides) can be included.

Using 8N20 bars on each side, distributed at 150 mm centres along l_b (16 total each way),

$A_{sh} = 16 \times \pi (20$ mm$)^2 / 4 = 5027$ mm^2

$A_{sv} = 5027$ mm^2

ULS,

$T_b^* = C^* \tan \alpha = 0.2 C^* = 0.2 \times 1414$ kN $= 283$ kN

$\phi_{st(s)} T_{bu} = 0.8[A_{sh} f_{sy} \sin \gamma_h + A_{sv} f_{sy} \sin \gamma_v]$

$= 0.8\left[5027 \text{ mm}^2 \times 500 \text{ MPa} \sin 45 + 5027 \text{ mm}^2 \times 500 \text{MPa} \sin 45 \right]$

$= 2843$ kN > 283 kN   Pass

SLS,

Assuming ULS is governed by the 1.35 G combination case,

$C = C^* / 1.35 = 1414 / 1.35 = 1048$ kN

$T_{b.s}^* = C \tan \alpha = 0.5 C = 0.5 \times 1048 \text{kN} = 524$ kN

$\text{MAX}[T_{b.s}^*, T_{b.cr}] = \text{MAX}[524 \text{ kN}, 1159 \text{ kN}] = 1159$ kN

For a moderate degree of crack control,

$f_{yc} = 200$ MPa

$T_{bc} = A_{sh} f_{yc} \sin \gamma_h + A_{sv} f_{yc} \sin \gamma_v$

$= 5027 \text{ mm}^2 \times 200 \text{ MPa} \sin 45 + 5027 \text{ mm}^2 \times 200 \text{ MPa} \sin 45$

$= 1422$ kN > 1159 kN   Pass

The geometry of the beam governs the bursting reinforcement requirements. The horizontal bars should be continuous around the back of the node (or can be lapped with U-bars) to prevent splitting. Vertical splitting is prevented by the compressive load.

NODES:

The top node is CCC with all loading perpendicular to the node face; therefore, the node is accepted based on the strut capacity.

The bottom node is CCT and has higher stresses due to the inclination of the compression strut to the node face.

Maximum node stress,

$$\phi_{st(n)}\beta_n 0.9 f_c' = 0.6 \times 0.8 \times 0.9 \times 40 \text{ MPa} = 17.3 \text{ MPa}$$

Stress on outside face of node (a) at diagonal strut,

$$A_{ca} = 438.6 \text{ mm} \times 420 \text{ mm} = 184{,}212 \text{ mm}^2$$

$$\sigma_{pa} = C^*/A_{ca} = 1{,}414 \text{ kN}/184{,}212 \text{ mm}^2 = 7.68 \text{ MPa}$$

$$R_a = \sigma_{pa}/2 = 7.68/2 = 3.84 \text{ MPa}$$

$$\sigma_a = R_a(1 + \cos(2\theta_a)) = 3.84 \text{ MPa}(1 + \cos(2 \times 19.5)) = 6.82 \text{ MPa}$$

$$\tau_a = R_a \sin(2\theta_i) = 3.84 \text{ MPa} \sin(2 \times 19.5) = 2.42 \text{ MPa}$$

θ_a can be scaled from the drawing or calculated based on the geometry at the node

$$\theta_a = 90 - [45 - [90 - \tan^{-1}(420/200)]] = 19.5°$$

Stress on outside face of node (b) at tie,

$$A_{cb} = 200 \text{ mm} \times (0.7 \times 600 \text{ mm}) = 84{,}000 \text{ mm}^2$$

$$\sigma_{pb} = \sigma_b = C^*/A_{cb} = 1{,}000 \text{ kN}/84{,}000 \text{ mm}^2 = 11.9 \text{ MPa}$$

Stress on outside face of node (c) at support,

$$A_{cc} = (0.7 \times 600 \text{ mm}) \times (0.7 \times 600 \text{ mm}) = 176{,}400 \text{ mm}^2$$

$$\sigma_{pc} = \sigma_c = C^*/A_{cc} = 1{,}000 \text{ kN}/176{,}400 \text{ mm}^2 = 5.67 \text{ MPa}$$

Stress inside node,

$$\sigma_m = \frac{\sigma_a + \sigma_b}{2} + \frac{\tau_b^2 - \tau_a^2}{2(\sigma_b - \sigma_a)} = \frac{6.82 + 11.9}{2} + \frac{0^2 - 2.42^2}{2(11.9 - 6.82)} = 8.79 \text{ MPa}$$

$$R = \sqrt{(\sigma_a - \sigma_m)^2 + \tau_a^2} = \sqrt{(6.82 - 8.79)^2 + 2.42^2} = 3.12 \text{ MPa}$$

$$\sigma_{MAX} = \sigma_m + R = 8.79 \text{ MPa} + 3.12 \text{ MPa} = 11.9 \text{ MPa} \quad < 17.3 \text{ MPa} \quad \text{Pass}$$

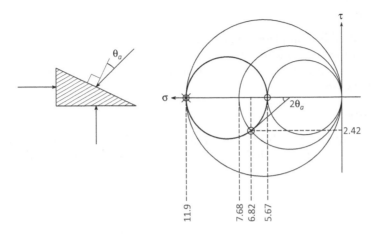

Node stresses are found to pass, and the model is accepted.

SUMMARY:

The beam has been designed as a non-flexural member. 6N24 reinforcing bars are required on the bottom (tie) face. N20 reinforcing bars are required at 150 mm centres, each way and on each face to prevent bursting at the sides. The bursting bars should be lapped with U-bars (in plan) around the short edges of the beam to prevent splitting. N20-150 bars are nominally adopted as standard for the remaining portion of the beam.

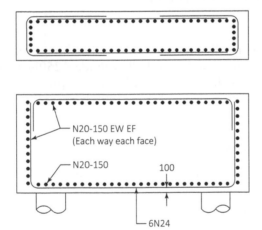

4.7 CONCRETE DETAILING

This section includes typical concrete details which are used on standard drawings. Individual details should always be checked by the engineer.

4.7.1 Concrete notes

Notes should be included on standard design drawings. The following notes are suggested for a typical project. Project-specific details such as design loads and wind or seismic parameters should be indicated on drawings. The construction category should also be stated on drawings where relevant (refer to Section 1.7.2).

4.7.1.1 General

- All works shall be completed in accordance with Australian Standards and legislation as well as project specifications.
- All dimensions are expressed in millimetres and levels expressed in metres unless noted otherwise.

4.7.1.2 Concrete

- Concrete shall be grade N32 unless noted otherwise. Slump shall be 80 mm and the maximum aggregate size shall be 20 mm. Concrete materials shall comply with the requirements of AS 1379. All concrete shall use type 'GP' or 'GB' cement in accordance with AS 3972.
- Concrete cover shall be 40 mm unless noted otherwise on drawings. Concrete cast directly against ground shall be 65 mm.
- Formwork shall be class 4 in accordance with AS 3610 unless noted otherwise. Concrete below ground and obstructed from view shall be class 3.
- High-frequency immersion type vibrators shall be used to compact all concrete.
- Free dropping of concrete is not permitted from heights greater than 1500 mm.
- All exposed surfaces shall be cured for a minimum of 7 days after pouring.
- A 20 × 20 mm chamfer shall be provided on all exposed edges of concrete.
- A damp proof membrane of 0.2 mm thick polythene shall be provided under all concrete cast against ground except for piles.

4.7.1.3 Reinforcement

Reinforcement shall be normally deformed D500N bars in accordance with AS/NZS 4671 unless noted otherwise. Plain round bars shall be grade R250N when specified.

Reinforcement lap lengths shall be in accordance with project standards.

All re-entrant corners shall include 2 N12 – 1000 bars on the top and bottom reinforcement layers.

4.7.2 Additional concrete details

Concrete details for the following items (Table 4.13) are shown in other sections of this text.

4.7.3 Miscellaneous details

Standard details are provided for various miscellaneous items not covered elsewhere in the text (Figures 4.36 through 4.39).

Table 4.13 Additional concrete details

Design item	Reference
Cover requirements	Table 4.8
Development lengths	Table 4.9
Anchor bolts	Figure 3.30
Pad footings	Section 5.1.3
Piles	Section 5.2.5
Retaining walls	Figure 5.18
Slabs	Section 5.4.6
Machine foundations	Section 6.4.6
Concrete structures for retaining liquids	Figure 6.46

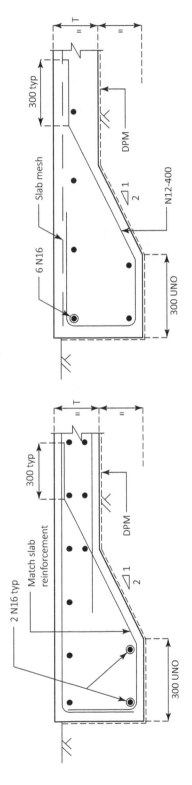

Figure 4.36 Slab thickening details.

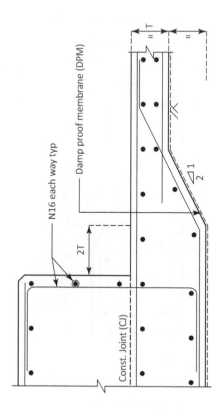

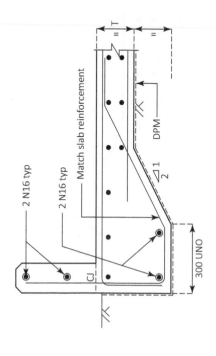

Figure 4.37 Upstand and plinth details.

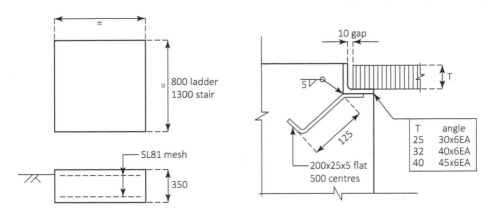

Figure 4.38 Landing and cast-in edge details.

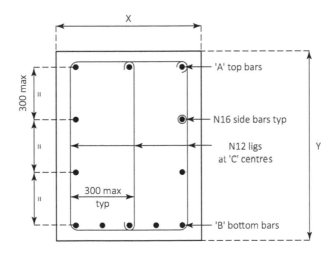

Figure 4.39 Beam detail.

Chapter 5

Geotechnical

Geotechnical engineering is often performed by structural engineers for simple scenarios, and designs of low significance or failure consequence. The scope of this section is to present simple methods which can be used in the design of typical pad footings, piled footings, slabs and retaining walls. Engineering judgement should be used on the applicability of each method in relation to specific designs. Information is presented with the assumption that a geotechnical report is available with the required information and recommendations.

5.1 PAD FOOTINGS

The design of pad footings requires analysis of stability (uplift, sliding and overturning) and bearing capacity. Concrete strength should also be checked in accordance with Sections 4.2.3, 4.3.5 and 4.4.

This author recommends that the resultant geotechnical capacities for pad footings are factored by a geotechnical reduction factor (Table 5.1) in accordance with the Australian Standard for Bridge Design – Foundations and Soil-Supporting Structures (AS 5100.3) for ultimate limit states design. Characteristic soil parameters should be adopted for calculations, which are defined as a cautious estimate of the mean value for a material property. Selection of a reduction factor should consider the structure type and redundancy such that the chance of a lower value governing the design should not exceed 5%. Structures with a low redundancy should therefore adopt more conservative values. A typical value has been suggested as 0.5 standard deviations below the mean.

5.1.1 Stability

The stability of pad footings is checked using a combination of self-weight, soil weight, friction, cohesion and passive soil pressure. Increased pad dimensions and depth can be used to increase the sliding and overturning capacities. Any mass which causes a stabilising effect should be factored at 0.9 (AS/NZS 1170.0, Clause 4.2.1). Sliding and overturning are typically checked in each of the orthogonal directions (Figure 5.1); however, if one direction clearly governs it is generally sufficient to provide calculations for the governing direction.

Table 5.1 Geotechnical strength reduction factor, ϕ_g

Method of soil testing	Geotechnical strength reduction factor, ϕ_g
Standard penetration test (SPT)	0.35–0.4
Cone penetration test (CPT)	0.4–0.5
Advanced laboratory testing	0.45–0.6
Advanced in situ tests	0.5–0.65

Source: From AS 5100, table 10.3.3(A).

Note: Judgement should be used when adopting values. Lower values should be selected when investigations are limited, calculations are simple, construction control is limited, consequence is high, loading is cyclical, foundations are permanent, and published correlations are used.

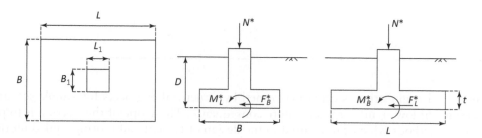

Figure 5.1 Footing parameters.

Uplift requirement, $\quad\quad E_{d,stb} \geq E_{d,dst}$

Sliding requirement, $\quad\quad F_{d,stb} \geq F_{d,dst} \quad\quad (F_{d,stb(B)} \geq F_B^*,\ F_{d,stb(L)} \geq F_L^*)$

Overturning requirement, $\quad M_{d,stb} \geq M_{d,dst} \quad\quad (M_{d,stb(L)} \geq M_L^*, M_{d,stb(B)} \geq M_B^*)$

where:

$E_{d,dst}$ = Destabilising (uplift) force

$F_{d,dst}$ = Destabilising sliding force

$M_{d,ds}$ = Destabilising overturning moment about base

The sliding capacity of a footing is calculated using the vertical reaction multiplied by the coefficient of friction, along with cohesion multiplied by the base area and the passive resistance (refer to Section 5.3.2). Passive resistance can also be included for the pedestal; however, the contribution is typically minor.

Stabilising uplift force, $\quad\quad\quad\quad\quad\quad\quad E_{d,stb} = 0.9(F_c + F_s)$

Stabilising sliding force in B direction, $\quad\quad F_{d,stb(B)} = \phi_g(F_f + F_{coh} + F_{p(L)})$

Stabilising sliding force in L direction, $\quad\quad F_{d,stb(L)} = \phi_g(F_f + F_{coh} + F_{p(B)})$

Frictional resistance, $\quad\quad\quad\quad\quad\quad\quad F_f = \mu\,[N^* + F_c + F_s]$

$\quad$ Coefficient of friction, $\quad \mu = \tan(\delta)$

$\quad$ External friction, $\quad\quad\quad \delta \approx 0.667\phi \quad$ (for smooth concrete)

$\quad\quad\quad\quad\quad\quad\quad\quad\quad\quad \delta \approx 1.0\phi \quad\quad$ (for no-fines concrete)

Cohesion resistance, $\qquad F_{coh} = BLc$

Passive resistance on L side, $\qquad F_{p(L)} = LK_p\gamma D^2/2 - LK_p\gamma(D-t)^2/2$

$$= LK_p\gamma(2Dt - t^2)/2$$

Passive resistance on B side, $\qquad F_{p(B)} = BK_p\gamma(2Dt - t^2)/2$

Passive pressure coefficient, $\qquad K_p = (1 + \sin\phi)/(1 - \sin\phi)$

where:

- B = Width of footing
- c = Characteristic cohesion
- D = Depth to base of footing
- F_c = Weight of concrete
- F_s = Weight of soil = $(BL - B_1L_1)(D - t)\gamma$
- L = Length of footing
- N^* = Ultimate limit state axial load applied to footing
- t = Thickness of base
- ϕ = Characteristic internal friction angle
- γ = Characteristic soil density

The overturning capacity is checked about each axis and uses the vertical reaction multiplied by the eccentricity from the applied load to the edge of the footing. Passive resistance is shown in the next set of equations; however, it is often ignored due to low contributions. The following equations assume that the footing pedestal is central to the pad (Figure 5.2).

Stabilising overturning moment about L, $\qquad M_{d,stb(L)} = M_{e(L)} + \phi_g M_{p(L)}$

Stabilising overturning moment about B, $\qquad M_{d,stb(B)} = M_{e(B)} + \phi_g M_{p(B)}$

Eccentric overturning resistance about L, $\qquad M_{e(L)} = N^*(B/2) + 0.9(F_c + F_s)(B/2)$

Eccentric overturning resistance about B, $\qquad M_{e(B)} = N^*(L/2) + 0.9(F_c + F_s)(L/2)$

Passive overturning resistance about L, $\qquad M_{p(L)} = F_{p(L)}e_p$

Passive overturning resistance about B, $\qquad M_{p(B)} = F_{p(B)}e_p$

Eccentricity from centroid of passive pressure to base,

$e_p = t/3 \times (2a+b)/(a+b)$ $\quad$ (centroid of a trapezoid)

for, $\qquad a = K_p\gamma(D-t) \quad b = K_p\gamma D$

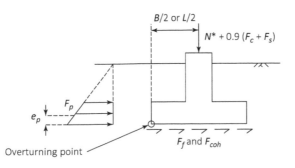

Figure 5.2 Footing stability.

5.1.2 Bearing capacity

Bearing capacity is a variable which is specific to an individual foundation. A site cannot be classified to accept a particular bearing capacity, because it is dependent on numerous input parameters such as soil properties, foundation size, geometry and depth. Significant care also needs to be taken when communicating the bearing capacity of a footing, because definitions have started to overlap and become confused. Geotechnical engineers define the ultimate bearing capacity of a foundation as the pressure which would cause it to fail: this will be referred to as the *ultimate (geotechnical) bearing capacity*, P_{ug}. Conversely, structural engineers often refer to the ultimate bearing capacity of a foundation as the failure capacity reduced by a factor of ϕ_g: this will be referred to as the *factored ultimate bearing capacity*, $\phi_g P_{ug}$. The *allowable bearing capacity* is an additional term which describes the bearing pressure that should not be exceeded under working (unfactored) loads. The allowable bearing capacity is typically equal to the ultimate (geotechnical) bearing capacity divided by a factor of safety (FOS), typically 3. Technically, serviceability loads should be compared to expected settlements to evaluate the performance of the structure; however, for the purpose of simplicity, they will be compared to the allowable bearing capacity in this chapter.

The details presented in this section are drawn from the Brinch–Hansen method [23], commonly used in Europe and adopted to some extent within Eurocode 7 and the German foundation code (DIN 4017). Characteristic soil parameters and the geotechnical strength reduction factor are recommended as described in Section 5.1. Typical soil parameters are provided in AS 4678, Appendix D, and repeated in Tables 5.12 through 5.14. The values are intended to be used in the absence of more reliable information and are useful for the concept phase of a project.

5.1.2.1 Linear elastic bearing pressures

A linear elastic bearing pressure is often adopted for serviceability limit states and for pressures under non-yielding materials, such as rock (refer to Figures 5.3 and 5.4). The analysis assumes that the soil behaves in a perfectly elastic manner.

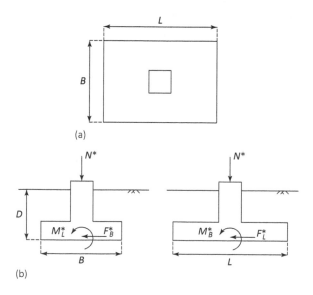

Figure 5.3 Pad footing plan and elevations. (a) Footing plan, (b) footing elevations.

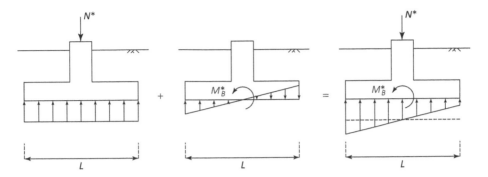

Figure 5.4 Linear soil pressure distribution.

The same stress analysis theory from the steel section of this text applies:

Stress, $\qquad\qquad\sigma = F / A \pm My / I$

For a footing with no bending moments, this simplifies to

Bearing pressure, $\qquad P = N / (BL)$

For axial load and a bending moment about one axis,

Bearing pressure, $\qquad P = N/A + 6M_L/(B^2 L)$

For axial load and bending moments about both axes,

Bearing pressure, $\qquad P = N/A + 6M_L/(B^2 L) + 6M_B/(L^2 B)$

where:
- A = Base area of pad
- B = Width of footing
- L = Length of footing
- M_L = SLS moment about L
- M_B = SLS moment about B
- N = SLS axial compression

Care needs to be taken to check that negative stresses are not seen. This is done by changing the plus signs in the above equations to minus signs and ensuring a positive result. If a negative value is calculated, the pad dimensions need to be increased to ensure a fully elastic response under the applied loads.

5.1.2.2 Plastic bearing pressures

Yielding materials are assumed to adopt a plastic pressure distribution under ultimate load conditions. This is done by calculating an equivalent pad area which gains uniform bearing pressure. Applied loads are resolved to an equivalent vertical reaction at an eccentricity. The equivalent pad dimensions are equal to twice the distance from the reaction to the nearest edge. The maximum vertical load is equal to the effective footing area multiplied by the ultimate bearing capacity (Figure 5.5).

Eccentricity of load in L direction, $\qquad e_L = M_B^* / N_b^*$

Eccentricity of load in B direction, $\qquad e_B = M_L^* / N_b^*$

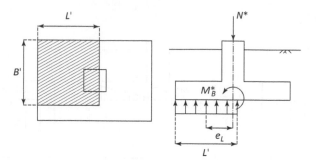

Figure 5.5 Equivalent pad size.

Effective footing length, $L' = L - 2e_L$

Effective footing width, $B' = B - 2e_B$

Effective footing area, $A' = L' \times B'$

ULS pressure at base of pad, $P^* = [N^* + 1.2(F_c + F_s)] \div A'$ (if governed by live load)

$$P^* = [N^* + 1.35(F_c + F_s)] \div A'$$ (if governed by dead load)

All further bearing calculations are completed using the effective dimensions.

5.1.2.3 Brinch–Hansen design method

The ultimate (geotechnical) bearing capacity is calculated by adopting the minimum bearing capacity for failure along the long (L) and short (B) edges of a foundation. The Brinch–Hansen method is presented in this section for the common case of a horizontal ground surface and a horizontal footing base.

Base and ground inclination has been assumed to be flat, as it is unusual to use sloping footings and surfaces for typical designs. For information on the factors with sloping levels refer to the original work, available online at the Geo website [23].

ULS requirement, $P^* \le \phi_g P_{ug}$

SLS requirement, $P \le P_{ug}/FOS$

   $\phi_g = 0.35$ to 0.65 (refer to Table 5.1)

   $FOS \approx 3.0$ (refer to Section 5.1.2)

Ultimate (geotechnical) bearing capacity, $P_{ug} = Q/A' = MIN[Q_{(L)}/A', Q_{(B)}/A']$

Bearing failure along long edge,

   $$Q_{(L)}/A' = 0.5\gamma N_\gamma B' s_{\gamma B} i_{\gamma B} b_\gamma + (\bar{q} + c\cot\phi)N_q d_{qB} s_{qB} i_{qB} b_q - c\cot\phi$$

Bearing failure along short edge,

   $$Q_{(B)}/A' = 0.5\gamma N_\gamma L' s_{\gamma L} i_{\gamma L} b_\gamma + (\bar{q} + c\cot\phi)N_q d_{qL} s_{qL} i_{qL} b_q - c\cot\phi$$

   Bearing capacity factors, $N_q = e^{\pi\tan\phi}\tan^2(45 + \phi/2)$

   $$N_\gamma = 1.5(N_q - 1)\tan\phi$$

Load inclination factors,
$$i_{qB} = [1 - 0.5F_B^*/(V + A'c\cot\phi)]^5$$
$$i_{qL} = [1 - 0.5F_L^*/(V + A'c\cot\phi)]^5$$
$$i_{\gamma B} = [1 - 0.7F_B^*/(V + A'c\cot\phi)]^5$$
$$i_{\gamma L} = [1 - 0.7F_L^*/(V + A'c\cot\phi)]^5$$

Shape factors,
$$s_{qB} = 1 + \sin(\phi)B'i_{qB}/L'$$
$$s_{qL} = 1 + \sin(\phi)L'i_{qL}/B'$$
$$s_{\gamma B} = 1 - 0.4B'i_{\gamma B}/(L'i_{\gamma L})$$
$$s_{\gamma L} = 1 - 0.4L'i_{\gamma L}/(B'i_{\gamma B})$$

Depth factors,
$$d_{qB} = 1 + 2\tan(\phi(1 - \sin\phi)^2)D/B' \quad \text{(for } D \leq B')$$
$$= 1 + 2\tan(\phi(1 - \sin\phi)^2)\tan^{-1}(D/B') \quad \text{(for } D > B')$$
$$d_{qL} = 1 + 2\tan(\phi(1 - \sin\phi)^2)D/L' \quad \text{(for } D \leq L')$$
$$= 1 + 2\tan(\phi(1 - \sin\phi)^2)\tan^{-1}(D/L') \quad \text{(for } D > L')$$

Base and ground inclination factors, $b_q = 1.0$ (for flat foundation bases)
$$b_\gamma = 1.0 \quad \text{(for flat soil surfaces)}$$

Soil weight, $\bar{q} = \gamma D$

where:

c = Characteristic cohesion
H_B^* = Ultimate horizontal load in B direction
H_L^* = Ultimate horizontal load in L direction
ϕ = Characteristic internal friction
γ = Characteristic soil density

5.1.3 Pad footing detailing

The drawings in this section are provided as recommended details. Specific pads are often called up using tables to provide dimensions and reinforcement requirements. Anchor bolts (refer to Table 3.25 and Figure 3.30) and shear keys (Figure 3.58) should also be shown as required. The selection of shear ligature geometry should consider the geometry of cast-in items to prevent potential clashes (Figure 5.6).

Example 5.1: Pad Footing

Check the stability and ultimate bearing capacity for the footing shown. The soil is a drained in situ material.

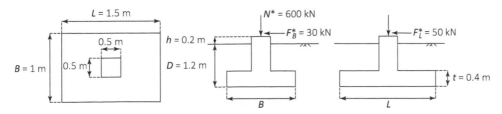

SOIL PROPERTIES

$\gamma = 18 \text{ kN/m}^3 \quad \phi = 35° \quad c = 5 \text{ kN/m}^2 \quad \phi_g = 0.5$

FOOTING PROPERTIES

Concrete footing mass, $\quad F_c = (1.5 \times 1 \times 0.4 + 1 \times 0.5 \times 0.5) \, 24 = 20.4 \text{ kN}$

Soil mass, $\quad\quad\quad F_s = (BL - B_1 L_1)(D - t)\gamma = (1 \times 1.5 - 0.5 \times 0.5)(1.2 - 0.4)18 = 18 \text{ kN}$

$F_B^* = 30 \text{ kN} \quad\quad M_L^* = F_B^*(D + h) = 30(1.2 + 0.2) = 42 \text{ kNm}$

$F_L^* = 50 \text{ kN} \quad\quad M_B^* = F_L^*(D + h) = 50(1.2 + 0.2) = 70 \text{ kNm}$

STABILITY

Sliding,

Frictional resistance,

$\delta \approx 0.667\phi = 0.667 \times 35 = 23.35° \quad$ (for smooth concrete)

$\mu = \tan(\delta) = \tan 23.35 = 0.43$

$F_f = \mu[N^* + F_c + F_s] = 0.43[600 + 20.4 + 18] = 275.5 \text{ kN}$

$F_{coh} = BLc = 1 \times 1.5 \times 5 = 7.5 \text{ kN}$

$K_p = (1 + \sin\phi)/(1 - \sin\phi) = (1 + \sin 35)/(1 - \sin 35) = 3.7$

$F_{p(L)} = LK_p\gamma(2Dt - t^2)/2 = 1.5 \times 3.7 \times 18 \times (2 \times 1.2 \times 0.4 - 0.4^2)/2 = 39.9 \text{ kN}$

$F_{p(B)} = BK_p\gamma(2Dt - t^2)/2 = 1 \times 3.7 \times 18 \times (2 \times 1.2 \times 0.4 - 0.4^2)/2 = 26.6 \text{ kN}$

$F_{d,stb(B)} = \phi_g(F_f + F_{coh} + F_{p(L)}) = 0.5(275.5 + 7.5 + 39.9) = 161.5 \text{ kN}$

$F_{d,stb(L)} = F_f + F_{coh} + F_{p(B)} = 0.5(275.5 + 7.5 + 26.6) = 154.8 \text{ kN}$

Pass (161.4 > 30)

Pass (154.8 > 50)

Overturning,

$M_{e(L)} = N^*(B/2) + 0.9(F_c + F_s)(B/2) = 600(1/2) + 0.9(20.4 + 18)(1/2) = 317.3 \text{ kNm}$

$M_{e(B)} = N^*(L/2) + 0.9(F_c + F_s)(L/2) = 600(1.5/2) + 0.9(20.4 + 18)(1.5/2) = 475.9 \text{ kNm}$

$e_p = t/3 \times (2a + b)/(a + b) = 0.4/3 \times (2 \times 40.2 + 60.3)/(40.2 + 60.3) = 0.187 \text{ m}$

$a = K_p(0.9 \times \gamma)(D - t) = 3.1(0.9 \times 18)(1.2 - 0.4) = 40.2$

$b = K_p(0.9 \times \gamma)D = 3.1(0.9 \times 18)1.2 = 60.3$

$$M_{p(L)} = F_{p(L)}e_p = 39.9 \times 0.187 = 7.4 \text{ kNm}$$

$$M_{p(B)} = F_{p(B)}e_p = 26.6 \times 0.187 = 5.0 \text{ kNm}$$

Stabilising overturning moment about L,

$$M_{d,stb(L)} = M_{e(L)} + \phi_g M_{p(L)} = 317.3 + 0.5 \times 7.4 = 321 \text{ kNm}$$

Stabilising overturning moment about B,

$$M_{d,stb(B)} = M_{e(B)} + \phi_g M_{p(B)} = 475.9 + 0.5 \times 5 = 478.4 \text{ kNm}$$

Pass (321 > 42)
Pass (478.4 > 70)

ULTIMATE BEARING CAPACITY

Resolving forces about the base,

$$N_b^* = 600 \text{kN} + 1.2 \times (20.4 \text{kN} + 18 \text{kN}) = 646.1 \text{kN}$$

(assuming design case is $1.2G + 1.5Q$, therefore adding footing mass with 1.2 factor)

Equivalent pad size,

$$e_L = M_B^*/N_b^* = 70/646.1 = 0.11 \text{ m}$$

$$e_B = M_L^*/N_b^* = 42/646.1 = 0.07 \text{ m}$$

$$L' = L - 2e_L = 1.5 - 2 \times 0.11 = 1.28 \text{ m}$$

$$B' = B - 2e_B = 1 - 2 \times 0.07 = 0.87 \text{ m}$$

$$A' = L' \times B' = 1.28 \times 0.87 = 1.12 \text{ m}^2$$

Ultimate bearing pressure,

$$P^* = [N^* + 1.2(F_c + F_s)] \div A' = N_b^*/A' = 646.1/1.12 = 577 \text{ kPa}$$

Bearing capacity factors,

$$N_q = e^{\pi \tan\phi} \tan^2(45 + \phi/2) = e^{\pi \tan 35} \tan^2(45 + 35/2) = 33$$

$$N_\gamma = 1.5(N_q - 1) \tan\phi = 1.5(20 - 1) \tan 33 = 34$$

$$i_{qB} = [1 - 0.5F_B^*/(V + A'c \cot\phi)]^5 = [1 - 0.5 \times 30/(627.6 + 1.11 \times 5 \cot 35)]^5 = 0.89$$

$$i_{qL} = [1 - 0.5 F_L^* / (V + A'c \cot \phi)]^5 = [1 - 0.5 \times 50/(627.6 + 1.11 \times 5 \cot 35)]^5 = 0.82$$

$$i_{\gamma B} = [1 - 0.7 F_B^* / (V + A'c \cot \phi)]^5 = [1 - 0.7 \times 30/(627.6 + 1.11 \times 5 \cot 35)]^5 = 0.85$$

$$i_{\gamma L} = [1 - 0.7 F_L^* / (V + A'c \cot \phi)]^5 = [1 - 0.7 \times 50/(627.6 + 1.11 \times 5 \cot 35)]^5 = 0.76$$

$$s_{qB} = 1 + \sin(\phi) B' i_{qB} / L' = 1 + \sin(35)\, 0.87 \times 0.89/1.28 = 1.35$$

$$s_{qL} = 1 + \sin(\phi) L' i_{qL} / B' = 1 + \sin(35)\, 1.28 \times 0.82/0.87 = 1.7$$

$$s_{\gamma B} = 1 - 0.4 B' i_{\gamma B} / (L' i_{\gamma L}) = 1 - 0.4 \times 0.87 \times 0.85/(1.28 \times 0.76) = 0.7$$

$$s_{\gamma L} = 1 - 0.4 L' i_{\gamma L} / (B' i_{\gamma B}) = 1 - 0.4 \times 1.28 \times 0.76/(0.87 \times 0.85) = 0.6$$

$$d_{qB} = 1 + 2 \tan(\phi(1 - \sin \phi)^2) \tan^{-1}(D/B') = 1 + 2 \tan(35(1 - \sin 35)^2) \tan^{-1}(1.2/0.87)$$
$$= 1.21 \quad \text{(for } D > B')$$

$$d_{qL} = 1 + 2 \tan(\phi(1 - \sin \phi)^2) D/L' = 1 + 2 \tan(35(1 - \sin 35)^2) 1.2/1.28 = 1.21 \quad \text{(for } D \leq L')$$

$$b_q = 1.0 \qquad b_\gamma = 1.0 \qquad \text{(for flat foundation and soil surface)}$$

$$\bar{q} = \gamma D = 18 \times 1.2 = 21.6 \text{ kPa}$$

Bearing capacity,

$$Q_{(L)}/A' = 0.5 \gamma N_\gamma B' s_{\gamma B} i_{\gamma B} b_\gamma + (\bar{q} + c \cot \phi) N_q d_{qB} s_{qB} i_{qB} b_q - c \cot \phi$$
$$= 0.5 \times 18 \times 34 \times 0.87 \times 0.7 \times 0.85 \times 1 + (21.6 + 5 \cot 35)33 \times 1.21 \times 1.35 \times 0.89 \times 1$$
$$-5 \cot 35 = 1539 \text{ kPa}$$

$$Q_{(B)}/A' = 0.5 \gamma N_\gamma L' s_{\gamma L} i_{\gamma L} b_\gamma + (\bar{q} + c \cot \phi) N_q d_{qL} s_{qL} i_{qL} b_q - c \cot \phi$$
$$= 0.5 \times 18 \times 34 \times 1.28 \times 0.6 \times 0.76 \times 1 + (21.6 + 5 \cot 35)33 \times 1.21 \times 1.7 \times 0.82 \times 1$$
$$-5 \cot 35 = 1786 \text{ kPa}$$

Ultimate (geotechnical) bearing capacity, $\quad P_{ug} = Q/A' = MIN[Q_{(L)}/A', Q_{(B)}/A'] = 1539 \text{ kPa}$

Factored ultimate bearing capacity, $\phi_g P_{ug} = 0.5 \times 1539 = 769 \text{ kPa}$

$$P^* \leq \phi_g P_{ug}$$

Pass $(577 \leq 769)$

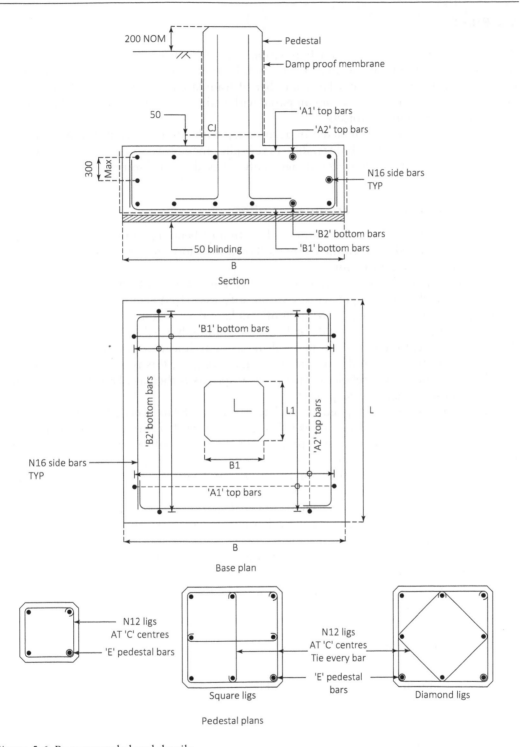

Figure 5.6 Recommended pad details.

5.2 PILES

Piles are commonly selected as a foundation system because of high lateral loads, strict deflection criteria or poor soil quality. The mobilisation cost of a piling rig is high; however, this cost can be offset when a large number of foundations are required. A much higher degree of prefabrication can occur compared with pad footings; therefore, piled footings are often used for greenfield projects.

The design of piles is completed in accordance with AS 2159. Concrete piles are commonly used due to their high durability. Screw piles may prove to be more economical for lightly loaded structures; however, design is often based on load testing.

5.2.1 Structural requirements for piles

The structural design of a pile requires calculation of bending moments and shear forces. This can be achieved through the methods shown for short and long piles or by modelling the pile in a structural program.

The modulus of subgrade reaction (refer to Section 5.2.4) can be multiplied by the width of a pile to calculate a very approximate spring stiffness [31]. A geotechnical engineer should be engaged for the design of piles with significant lateral loading or piles of high importance. Specialised software such as L-Pile or Plaxis is commonly used for geotechnical analysis. These geotechnical programs can be used to calculate linear or non-linear springs which can be used in the structural program. For a large project, this is generally done for each pile diameter and each borehole, so that a set of springs is then available for the structural engineer to use throughout the project.

Linear springs can be applied in packages such as Space Gass; however, non-linear springs require finite element analysis (FEA) programs such as Strand7. Linear springs are an approximation of a non-linear entity, and are therefore only accurate when the applied loads are similar to the typical loads used by the geotechnical engineer. Some piles may warrant individual consideration based on engineering judgement.

Piles should be modelled using beam elements, with the beam divided into a number of segments (ideally equal to n if also completing calculations to the Brinch–Hansen method). Each of the beam elements is assigned a horizontal spring depending on the stiffness of the soil, pile diameter and node spacing. Generally, a spacing of 0.5 or 1.0 m is sufficient to achieve a realistic bending moment diagram. The bending moment and shear force diagrams should appear smooth; if the diagram is jagged at the node locations, then further subdivision of beam elements is required. A useful check is to compare the spring reactions (for an applied load of H) with those calculated using the Brinch–Hansen method.

5.2.1.1 Concrete piles

AS 2159 specifies requirements for the structural design of concrete piles. Key aspects of the code are summarised in this section. Piles should generally be designed as columns; however, the compressive strength used in the design of cast-in-place concrete piles should be reduced by the concrete placement factor, k. A value should be selected ranging from 0.75 to 1.0 (refer to AS 2159 for further details). For details on the structural design of concrete piles, refer to Sections 4.4.5.8 and 4.2.4. Structural capacity tables are provided in Section 7.3.2 for $k = 1$.

$$f_c^{*\prime} = k f_c^\prime$$

$k = 0.75$ (no previous successful use of construction method, limited testing or monitoring)

  $= 1.0$ (successful use of construction method, 5% testing, monitoring)

Minimum concrete strength and cover requirements for concrete piles are dependent on the soil exposure classification and construction type. The required cover is shown in Table 5.2; refer to AS 2159 for further details. Details and recommendations for exposure classification are typically provided in the geotechnical report for a site.

The minimum longitudinal reinforcement in a precast pile is 1.4% of the cross-sectional area. Other piles require a minimum of 1% where above ground (and for the first three diameters below ground) and 0.5% for fully submerged piles (and depths beyond the top three diameters). The maximum cross-sectional reinforcement is 4% unless it can be shown that proper compaction can still be achieved.

Reinforcement may be terminated one development length below the depth at which tension, bending and shear loads become insignificant, provided that the compressive load is less than $0.5 \, k \, f'_c \, \phi A_g$. Piles may be entirely unreinforced for similar cases where the compressive loading is less than $0.45 \, k \, f'_c \, \phi A_g$. The reduction factor for concrete in compression without bending shall be taken as $\phi = 0.6$.

The ties or helices for piles that extend by two pile diameters or more above ground level are required to comply with the AS 3600 requirements for columns (down to a depth of three pile diameters below ground). Piles with a compressive strength greater than 65 MPa are also required to comply with AS 3600 confinement requirements (confining actions of soil or rock may be allowed for). All other reinforced piles (and the lower section of the aforementioned piles) shall comply with Table 5.3. It is also common to have two full turns

Table 5.2 Concrete strength and reinforcement cover in piles

| Exposure classification | Minimum concrete strength (MPa) | | Minimum cover to reinforcement (mm) | | | |
| | | | 50 year design life | | 100 year design life | |
	Precast and prestressed	Cast-in-place	Precast and prestressed	Cast-in-place	Precast and prestressed	Cast-in-place
Non-aggressive	50	25[a]	20	45	25	65
Mild	50	32	20	60	30	75
Moderate	50	40	25	65	40	85
Severe	50	50	40	70	50	100
Very severe	>50 (preferably >60)		40	75	50	120

Source: AS 2159, table 6.4.3. Copied by Mr L. Pack with the permission of Standards Australia under Licence 1607-c010.

Note: Refer to AS 2159 for notes and further details in relation to exposure classification.

[a] Use minimum f'_c = 32 MPa for reinforced piles.

Table 5.3 Minimum bar size for ties and helices for concrete piles ($f'_c < 65$ MPa)

Pile size (mm)	Longitudinal bar diameter (mm)	Minimum diameter of tie or helix (mm)
Up to 500	Less than 32	5
Up to 500	32–36	6
501–700	All	6
701 and above	All	10

Source: AS 2159, table 5.3.7.Copied by Mr L. Pack with the permission of Standards Australia under Licence 1607-c010.

of the helix at the top and bottom of the cage. Refer to Figure 5.7 for a graphical representation of piles with compressive strength less than or equal to 65 MPa.

5.2.1.2 Steel piles

Steel piles should be designed in accordance with AS 4100 (refer to Section 3.4.2); however, the cross-section of the pile used for calculations should be reduced by a corrosion allowance. The corrosion allowance is calculated by multiplying the uniform corrosion allowance (Table 5.4) by the design life of the pile. The section is reduced by this thickness around the perimeter.

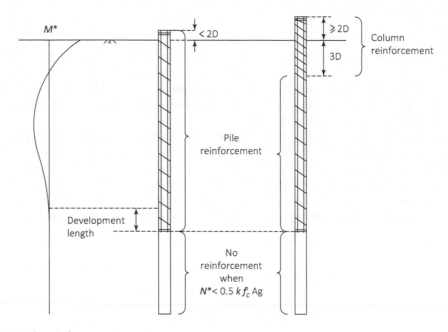

Figure 5.7 Pile reinforcement requirements.

Table 5.4 Corrosion allowance for steel piles

Exposure classification	Uniform corrosion allowance (mm/year)
Non-aggressive	<0.01
Mild	0.01–0.02
Moderate	0.02–0.04
Severe	0.04–0.1
Very severe[a]	>0.1

Source: AS 2159, table 6.5.3. Copied by Mr L. Pack with the permission of Standards Australia under Licence 1607-c010.

Notes:
1. Refer to AS 2159 for details on reducing the corrosion allowance by introducing coating systems or cathodic protection.
2. Refer to AS 2159 for additional details, including descriptions of exposure classifications.

[a] For very severe conditions, a site-specific assessment should be sought.

5.2.2 Vertically loaded piles

Vertical loads on piles are carried by skin friction and the end bearing of the pile. The axial capacity of a pile in compression is equal to the side contact surface area multiplied by the skin friction capacity, plus the end cross-sectional area multiplied by the end bearing capacity. Piles subject to tensile loads rely purely on the skin friction (and pile mass). The values for skin friction and end bearing capacities are typically provided in geotechnical reports, along with recommendations for reduction factors for serviceability loading. The following equations reduce skin friction by a factor of two and end bearing by a factor of three. The ULS capacity is calculated using a reduction factor, as required by AS 2159.

ULS requirement,	$N^* \leq R_{d,g(ULS)}$
SLS requirement,	$N \leq R_{d,g(SLS)}$
ULS vertical pile compression capacity,	$R_{d,g(ULS,+V)} = \phi_g(f_{m,s}A_s + f_b A_b)$
ULS vertical pile tension capacity,	$R_{d,g(ULS,-V)} = \phi_g(f_{m,s}A_s + W)$
SLS vertical pile compression capacity,	$R_{d,g(SLS,+V)} = f_{m,s}A_s/2 + f_b A_b/3$
SLS vertical pile tension capacity,	$R_{d,g(SLS,-V)} = f_{m,st}A_s/2 + W$

where:

A_b = Base area of pile = $\pi D^2/4$
A_s = Reduced skin friction area = $MIN[\pi D(L-L_s), \pi D(L-1.5D)]$
D = Pile diameter
f_b = End bearing capacity
$f_{m,s}$ = Skin friction capacity for compression
$f_{m,st}$ = Skin friction capacity for tension
L = Length of pile
L_s = Length of pile sleeve (if applicable)
W = Weight of pile

Reactive soils may cause uplift loading on piles. A plastic polythene sleeve liner (wrapped around the pile with a minimum of two layers) can be used to reduce swell-induced pressures. The sleeve section of the pile should be removed from the axial skin friction component of the calculation.

The geotechnical strength reduction factor is a risk-based calculation which uses a combination of site, design and installation factors to create a single factor for ultimate design. The resultant value ranges between 0.40 and 0.90; however, it is more commonly towards the lower end.

Geotechnical strength reduction factor, $\phi_g = \phi_{gb}$ (no pile testing)

$$= \phi_{gb} + (\phi_{tf} - \phi_{gb})K \geq \phi_{gb} \quad \text{(pile testing)}$$

Testing benefit factor, $K = 1.33p/(p+3.3) \leq 1$ (static of rapid load testing)

$$= 1.13p/(p+3.3) \leq 1 \quad \text{(dynamic testing)}$$

Test factor,

$\phi_{tf} = \phi_{gb}$ (for no testing),   0.9 (for static load testing),   0.85 (for bi-directional testing)

$= 0.8$ (for dynamic load testing of preformed piles)

$= 0.75$ (dynamic testing other than preformed piles, and for rapid load testing)

Table 5.5 Basic geotechnical strength reduction factor

Range of average risk rating (ARR)	Overall risk category	ϕ_{gb} for low redundancy systems	ϕ_{gb} for high redundancy systems
ARR ≤ 1.5	Very low	0.67	0.76
1.5 < ARR ≤ 2.0	Very low to low	0.61	0.70
2.0 < ARR ≤ 2.5	Low	0.56	0.64
2.5 < ARR ≤ 3.0	Low to moderate	0.52	0.60
3.0 < ARR ≤ 3.5	Moderate	0.48	0.56
3.5 < ARR ≤ 4.0	Moderate to high	0.45	0.53
4.0 < ARR ≤ 4.5	High	0.42	0.50
ARR > 4.5	Very high	0.40	0.47

Source: AS 2159, table 4.3.2(C). Copied by Mr L. Pack with the permission of Standards Australia under Licence 1607-c010.

Note: Low redundancy systems include isolated heavily loaded piles and piles set out at large centres. High redundancy systems include large pile groups under large caps, piled rafts and pile groups with more than four piles.

ϕ_{gb} = Basic geotechnical strength reduction factor   (refer to Table 5.5)

The average risk rating (ARR) is the average resultant of the weighting factor (w_i) multiplied by the individual risk rating (IRR) for each of the relevant risk factors. Intermediate individual risk ratings may be adopted. Refer to Table 5.6 for a simplified IRR table from AS 2159.

Average risk rating, $ARR = \sum (w_i \times IRR_i) \big/ \sum w_i$ (refer to Table 5.6)

5.2.2.1 Pile groups and spacing

The capacity of a pile group is taken as the minimum of: the sum of all individual pile capacities; or the capacity of a large equivalent pile (including all the piles and the soil between them). Generally, piles which rely on skin friction should not be spaced at centres which are less than 2.5 times the pile diameter. Piles which rely mainly on end bearing may be spaced at centres which are as close as 2.0 times the base diameter.

5.2.2.2 Induced bending moment

AS 2159 specifies a construction tolerance of 75 mm for the horizontal positioning of piles. Depending on the type of structure (such as column loads), this may induce a moment in the pile equal to the axial load multiplied by this eccentricity. This moment is added to any existing design moments in the pile. Pile groups (three or greater) with a rigid pile cap are typically not subject to this effect. A minimum design moment is also specified, equal to the axial force multiplied by 5% of the pile diameter.

$$M^* = MAX[M^*_{applied} + N^* \times 0.075\ m, N^* \times 0.05D]$$

5.2.3 Settlement

Piles carry load first in skin friction, then in end bearing. Only small deflections are required to engage full skin friction capacity compared with end bearing capacity. Therefore, piles which can carry working or serviceability loads with only the skin

Table 5.6 Weighting factors and individual risk ratings for risk factors

Risk factor	Weighting factor (w_i)	Typical description of risk circumstances for individual risk rating (IRR)		
		1 (Very low risk)	3 (Moderate)	5 (Very high risk)
Site				
Geological complexity	2	Well defined	Some variability	Highly variable
Extent of investigation	2	Extensive drilling	Moderate	Very limited investigation
Amount and quality of geotechnical data	2	Detailed information	Cone penetration test (CPT) probes over full depth of proposed piles or boreholes	Limited simple testing (e.g. SPT) or index tests only
Design				
Experience with similar foundations	1	Extensive	Limited	None
Method of assessment of geotechnical parameters	2	Laboratory or in situ tests or relevant load test data	Site-specific correlations or conventional testing	Based on non-site-specific correlations
Design method adopted	1	Well established	Simplified methods	Simple empirical methods
Method of using results of data	2	Design values based on minimum measured values to failure	Design methods based on average values	Design values based on maximum values up to working load, or indirect measurements
Installation				
Construction control	2	Professional geotechnical supervision, well established methods	Limited professional geotechnical involvement, conventional procedures	Very limited or no geotechnical supervision, not well established or complex construction processes
Performance monitoring	0.5	Detailed	Correlation with tests	No monitoring

Source: Modified from AS 2159, table 4.3.2(A). Copied by Mr L. Pack with the permission of Standards Australia under Licence 1607-c010.

Note: Only relevant rows of the table should be included in the calculation of IRR.

friction capacity will have small amounts of settlement. The calculation of pile settlement is beyond the scope of this text; refer to Tomlinson [15] for calculation methods for various soil and pile types.

5.2.4 Laterally loaded piles

Various methods are available for the design of piles under lateral loading. The Brinch–Hansen [22] method is used for the design of short piles and can also be partially used for the design of long piles. The design of pile groups is beyond the scope of this text; however, AS 2159 states that in the absence of an alternative method, the capacity of a pile group is taken as the minimum of: the sum of all individual pile capacities; or the capacity of a large equivalent pile (including all the piles and the soil between them). Strength limit states criteria are satisfied using the reduction factor (ϕ_g) calculated in Section 5.2.2. Serviceability limit states should be assessed by calculating deflections and comparing with criteria for the supported structure (refer to Section 2.5). If a pile sleeve is used, or if the pile is founded in reactive soils (typically clays), then the top portion of the soil resistance may need to be ignored. The depth of cracking for reactive soils is shown in Section 5.5.2.

ULS requirement, $V^* \le R_{d,g(ULS,H)}$

ULS lateral pile capacity, $R_{d,g(ULS,H)} = \phi_g H$

The first step in the design of individual piles is to determine whether the pile behaves as short or long. Short piles are those which suffer soil failure prior to structural failure. Long piles are those which suffer structural failure prior to soil failure.

A pile is classified as short if it is embedded to a depth less than the critical length, L_c, and long if it is deeper. The following approximations are for individual free-head piles; the method is based on the various research summarised by Tomlinson [16].

$$L_c = 2R \quad \text{(for a rigid pile and a constant soil modulus)}$$

$$= 2T \quad \text{(for a rigid pile and soil modulus linearly increasing with depth)}$$

$$= 3.5R \quad \text{(for an elastic pile with a constant soil modulus)}$$

$$= 4T \quad \text{(for an elastic pile with a linearly increasing soil modulus)}$$

where:

Stiffness factor, $T = \sqrt[5]{EI/n_h}$

Stiffness factor, $R = \sqrt[4]{EI/k_s b}$

Stiff over-consolidated clays are typically assumed to have a constant soil modulus. Normally consolidated clays and granular soils are typically assumed to have a modulus linearly increasing with depth [16]. Resultant values for the stiffness factors should be in units of length.

b = Pile width or diameter
E = Modulus for pile material (refer to Section 3.1 for steel and Table 4.1 for concrete)
I = Second moment of area (refer to Table 3.30)
k_s = Modulus of subgrade reaction = $n_h \times z/b$ (when modulus increases with depth)
    (refer to Table 5.7 when modulus is a constant and does not vary with depth)
n_h = Coefficient of modulus variation
    = 150 kN/m³ (for soft organic silts)

Table 5.7 Modulus of subgrade reaction

Soil description	Firm to stiff	Stiff to very stiff	Hard
Undrained shear strength, s_u (Kn/m^2)	50–100	100–200	>200
Modulus of subgrade reaction, k_s (MN/m^3)	15–30	30–60	>60

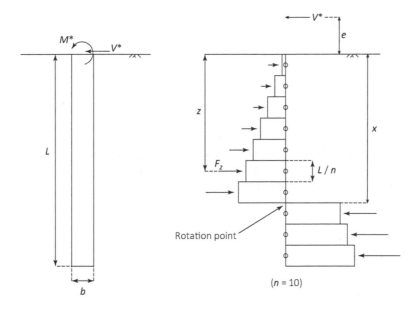

Figure 5.8 Laterally loaded pile.

$\quad$ = 350 to 700 kN/m³ (for normally consolidated clays)
$\quad$ = 4,000 to 9,000 kN/m³ (for loose sand)
$\quad$ = 9,000 to 26,000 kN/m³ (for medium dense sand)
$\quad$ = 26,000 to 41,000 kN/m³ (for dense sand)
$z\quad$ = Depth to the centre of segment being considered (refer to Figure 5.8)

5.2.4.1 Short piles

The geotechnical design of short piles can be completed using the Brinch–Hansen method [16,22]. The pile is divided into *n* segments (use *n* = 5 for a rough calculation, or *n* = 10 for an accurate calculation). The overburden pressure is calculated at the depth of each segment, and the lateral capacity of that segment is then calculated. Trial and error is then used to calculate the depth of pile rotation by taking moments about the point where the horizontal load is applied and changing the depth until the moment equals zero (segments below the rotation point have a negative sign). The capacity is calculated by taking moments about the point of rotation (all segments have positive signs about this point) and dividing by the distance from that point to the point where the horizontal load is applied.

$\quad$ Lateral loads and moments should be resolved into lateral loads at an equivalent eccentricity. The induced lateral moment from Section 5.2.2.2 should also be included in the design of piles for lateral loading.

Equivalent eccentricity of horizontal load, $\qquad e = M^*/V^*$

Passive pressure resisted by each segment (at depth of z), $\qquad p_z = p_{o_z}K_{q_z} + cK_{c_z}$

Passive force resisted by each segment (at depth of z), $\qquad F_z = p_z(L/n)b$

where:

c $\quad$ = Cohesion
L $\quad$ = Length of pile
n $\quad$ = Number of segments
p_{o_z} = Overburden pressure (at depth of z) = γz
γ $\quad$ = Soil density

The factors for K_q and K_c can be quickly read from Figure 5.9; however, if using a spreadsheet, the following formulas may be more appropriate. The arrows on the right side of each graph show the value that each line approaches as z/b reaches infinity. Where a stiff soil layer is beneath a soft layer, the factors for the stiff layer should be calculated with z being measured from the top of the stiff layer. Undrained soil parameters are typically used for short-term loads and drained parameters for sustained loads [16].

Earth pressure coefficient for overburden pressure,

$$K_q = \frac{K_q^o + K_q^\infty \alpha_q \dfrac{z}{b}}{1 + \alpha_q \dfrac{z}{b}}$$

Earth pressure coefficient for cohesion,

$$K_c = \frac{K_c^o + K_c^\infty \alpha_c \dfrac{z}{b}}{1 + \alpha_c \dfrac{z}{b}}$$

where:

$K_q^o = e^{(\phi+\pi/2)\tan\phi}\cos\phi\tan(\phi/2 + \pi/4) - e^{(\phi-\pi/2)\tan\phi}\cos\phi\tan(\pi/4 - \phi/2)$

$K_c^o = [e^{(\phi+\pi/2)\tan\phi}\cos\phi\tan(\phi/2 + \pi/4) - 1]/\tan\phi$

$N_c = [e^{\pi\tan\phi}\tan^2(\phi/2+\pi/4)-1]/\tan\phi$

$d_c^\infty = 1.58 + 4.09\tan^4\phi$

$K_o = 1 - \sin\phi$

$K_q^\infty = N_c d_c^\infty K_o \tan\phi$

$K_c^\infty = N_c d_c^\infty$

$\alpha_q = \dfrac{K_q^o}{K_q^\infty - K_q^o} \times \dfrac{K_o\sin\phi}{\sin(\phi/2 + \pi/4)}$

$\alpha_c = \dfrac{K_c^o}{K_c^\infty - K_c^o} \times 2\sin(\phi/2 + \pi/4)$

Taking moments about the point of load application (V^*) and using trial and error to find the rotation depth, x, at which the pile is in equilibrium (refer to Figure 5.8),

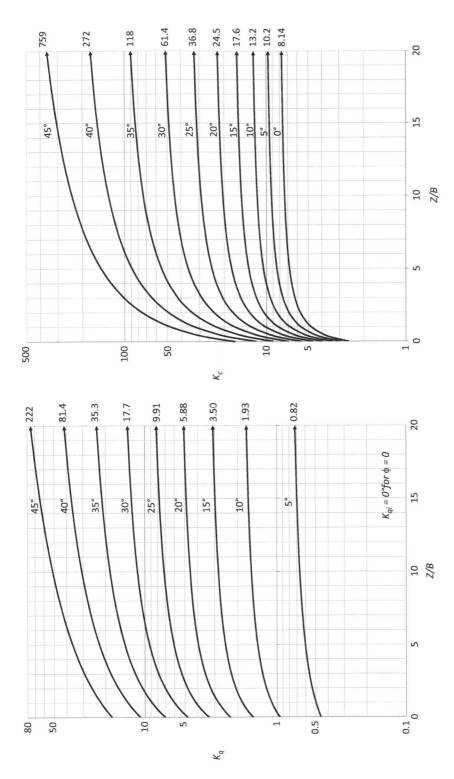

Figure 5.9 Earth pressure coefficients. (From Hansen, B.J., Geo (Former Danish Geotechnical Institute), 12, 5–9, 1961. Reproduced with permission from Geo (Former Danish Geotechnical Institute).)

$$\sum M = \sum_{z=0}^{z=x} F_z(e+z) - \sum_{z=x}^{z=L} F_z(e+z) \approx 0$$

Note: Segments below the rotation depth use a negative sign in the formula above. The depth should be changed to find the solution which is closest to zero.

A more accurate value can be calculated by either increasing the number of nodes or modifying the node spacing at the rotation point so that the formula above is closer to zero.

The horizontal capacity is calculated by taking moments about the point of rotation,

$$H = \left[\sum_{z=0}^{z=x} F_z(x-z) + \sum_{z=x}^{z=L} F_z(z-x) \right] \Big/ (e+x)$$

which can also be written as

$$H = \left[\sum_{z=0}^{z=L} F_z |x-z| \right] \Big/ (e+x)$$

Fixed-head piles can be designed using the dimensions for an equivalent free-pile [36], with the horizontal load applied at an eccentricity of e_1 above the point of virtual fixity, z_f (refer to Figure 5.10). This method requires a minimum pile embedment of $4R$ or $4T$ (as appropriate) and a minimum unsupported length greater than $2R$ or T above ground.

$$e_1 = 0.5(e + z_f)$$

$$z_f \approx 1.4R \quad \text{(for stiff over-consolidated clays)}$$

$$\approx 1.8T \quad \text{(for normally consolidated clays, granulated soils, silt, and peat)}$$

Shear force and bending moment diagrams for the pile length can be constructed by considering each of the element reactions (refer to Figure 5.11). The maximum bending moment occurs where the shear force equals zero.

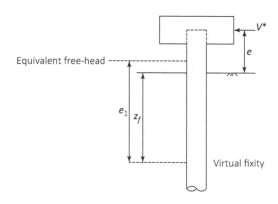

Figure 5.10 Fixed-head pile diagram.

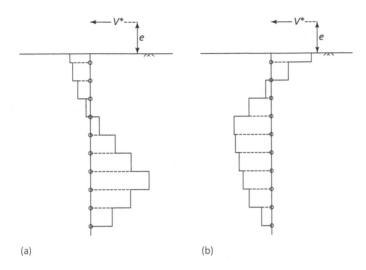

(a)   (b)

Figure 5.11 Pile reaction diagrams. (a) Shear force, (b) bending moment.

5.2.4.2 Long piles

Long piles fail structurally rather than geotechnically. The design of long piles can be completed by treating the pile as a cantilever, extending from the point of virtual fixity to the point where load is applied ($e + z_f$, refer to Figure 5.10) [16]. The capacity is then based on the bending capacity of the column (refer to Section 5.2.1).

Horizontal capacity, $\quad R_{d,g(ULS,H)} = \phi M_u / (e + z_f) \quad$ (for free-head piles)

$$R_{d,g(ULS,H)} = 2\phi M_u / (e + z_f) \quad \text{(for fixed-head piles)}$$

Values for e and z_f are as defined for short piles (refer to Section 5.2.4.1).

5.2.4.3 Pile deflections

Pile deflections can be estimated by solving them as cantilevers spanning from the depth of virtual fixity ($e + z_f$, refer to Figure 5.10) [16].

Deflection of free-head pile, $\quad \delta \approx V^*(e + z_f)^3 / (3EI)$

Deflection of fixed-head pile, $\quad \delta \approx V^*(e + z_f)^3 / (12EI)$

Alternatively, springs can be used in a linear or non-linear analysis program (refer to Section 5.2.1). More advanced geotechnical theory is available and can be adopted by hand calculations [16] or in the use of specialised geotechnical software if more accurate deflections are required. It is recommended that geotechnical engineers be engaged to check highly loaded or sensitive piles for greenfield projects. For brownfield works, it is recommended that piles of similar stiffness to those of adjacent structures are provided (with similar loading) to avoid differential deflection or stiffness issues.

5.2.5 Pile detailing

The following drawings are provided as recommended details (Figure 5.12). Piles are typically called up using a diameter with standard dimensions and reinforcement requirements.

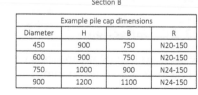

Figure 5.12 Recommended pile details.

Additional types (A and B) can be used to allow a lightly reinforced option. Anchor bolts (refer to Table 3.25 and Figure 3.30) and shear keys (Figure 3.58) should also be shown as required. The selection of shear ligature geometry should consider the geometry of cast-in items to prevent potential clashes. Required cover should consider product availability of circular bar chair clips: typically, 20, 25, 30, 45, 50, 60, 65 or 75 mm. The bar chairs are essentially plastic wheels which clip to the helix and spin against the soil as the cage is inserted into the hole to ensure cover.

Example 5.2: Check the vertical and horizontal geotechnical capacities of a rigid free-head pile, detailed below, founding in a stiff over-consolidated clay:

Pile, $L = 3500$ mm $b = 600$ mm $f'_c = 40$ MPa $k = 1$ $\phi M_u = 350$ kNm

$E_c = 32800$ MPa

Soil, $\qquad f_{m,s} = 60 \text{ kPa} \quad f_b = 1500 \text{ kPa} \quad k_s = 30000 \text{ kN/m}^3$

$\qquad \phi_g = 0.45 \qquad \gamma = 16 \text{ kN/m}^3 \qquad \phi = 35° \quad c = 2 \text{ kPa}$

Loads at top of pile,

$\qquad N = 250 \text{ kN} \quad N^* = 350 \text{ kN} \quad V^* = 50 \text{ kN} \quad M^*_{applied} = 10 \text{ kNm}$

Axial capacity,

$\qquad$ Shaft area, $\qquad A_s = \pi b L = \pi \times 600 \text{ mm} \times 3500 \text{ mm} = 6.6 \times 10^6 \text{ mm}^2$

$\qquad$ Base area, $\qquad A_b = \pi b^2 / 4 = \pi (600 \text{ mm})^2 / 4 = 2.8 \times 10^5 \text{ mm}^2$

$\qquad$ SLS compression, $R_{d,g(SLS,+V)} = f_{m,s} A_s / 2 + f_b A_b / 3 = 339 \text{ kN} \quad$ Pass(339 > 250)

$\qquad$ ULS compression, $R_{d,g(ULS,+V)} = \phi_g (f_{m,s} A_s + f_b A_b) = 369 \text{ kN} \quad$ Pass(369 > 350)

Lateral capacity,

$\qquad$ Design bending moment, $\quad M^* = MAX[M^*_{applied} + N^* \times 0.075 \text{ m}, N^* \times 0.05D]$

$\qquad\qquad\qquad = MAX[10 \text{ kNm} + 350 \text{ kN} \times 0.075 \text{ m}, 300 \text{ kN} \times 0.05 \times 600 \text{ mm}]$

$\qquad\qquad\qquad = MAX[36.3 \text{ kNm}, 9 \text{ kNm}] = 36.3 \text{ kNm}$

Pile classification,

$$R = \sqrt[4]{EI/kb} = \sqrt[4]{32800 \text{ MPa}/(30000 \text{ kN/m}^3 \times 600 \text{ mm})} = 1845 \text{ mm}$$

$$L_c = 2R = 3690 \text{ mm}$$

Pile = Short $\quad$ (3500 < 3690)

Brinch–Hansen analysis (using five segments of 700 mm),

$\quad e = M^*/V^* = (36.3 \text{ kNm})/(50 \text{ kN}) = 72.5 \text{ mm}$

Segment number	1	2	3	4	5
Depth to centre, z (mm)	350	1050	1750	2450	3150
z/b	0.6	1.8	2.9	4.1	5.3
$p_{o_z} = z\gamma$ (kPa)	5.6	16.8	28	39.2	50.4
K_{q_z}	8.1	10.1	11.7	13.2	14.5
K_{c_z}	17.4	31	41.4	49.5	56.1
$p_z = p_{o_z} K_{q_z} + c K_{c_z}$	80.4	231.1	411.6	617.2	843.9
$F_z = p_z (L/n) b$	33.7	97.1	172.9	259.2	354.4
Assuming a rotation point half way between Nodes 4 and 5 ($x = 2800$ mm)					
Moment about V^*, $F_z(e+z)$ (kNm)	36.3	172.3	427.9	823.1	(–)1373.5

$$\sum M = \sum_{z=0}^{z=x} F_z(e+z) - \sum_{z=x}^{z=L} F_z(e+z) = 36.3 + 172.3 + 427.9 + 823.1 - 1373.5 = 86 \text{ kN} \approx 0$$

Moment about rotation point, $F_z|x-z|$ (kNm)	82.7	169.9	181.5	90.7	124.1

$$H = \left[\sum_{z=0}^{z=L} F_z |x-z| \right] \bigg/ (e+x) = (82.7 + 169.9 + 181.5 + 90.7 + 124.1)/(725 + 2800) = 184 \text{ kN}$$

Horizontal capacity,

$$R_{d,g(ULS,H)} = \phi_g H = 0.45 \times 184 = 82.8 \text{ kN} \quad \text{Pass} (82.8 > 50)$$

Note: The variance is calculated as 86 kN, which is considered low in relation to the forces at each node. If a high variance is calculated, the rotation node should be changed to the node with the lowest variance. If this value remains high, it can be reduced by either solving the problem with more segments or changing the segment lengths. For the case above, Segment 4 would be decreased in length, and Segment 5 would be increased in length; the (L/n) component in the force equation would then be replaced by the length for each segment. Structural loads can be calculated by modelling the pile with springs.

Reactions can be calculated for the full (unfactored) capacity of the pile,

Segment number (n)	0	1	2	3	4	5
Shear force, $V = H - \sum_0^n F_z$	184	150	53	−120	−379	−24
Moment, $M = H(e+z) + \sum_1^n F_n(z_n - z_{n-1})$	133.5	197.9	303.1	340.4	257	−9

The tabulated values above are for the full capacity of the pile, and can be factored down by a ratio of V^*/H for the actual reactions (assuming a linear relationship between the applied load and the full capacity).

Example 5.3: Check the lateral geotechnical capacity and deflection if the pile length is increased to 4000 mm:

Pile = Long   (4000 > 3690)

$$z_f \approx 1.4R = 1.4 \times 1845 = 2583 \text{ mm}$$

Horizontal capacity,

$$R_{d,g(ULS,H)} = \phi M_u/(e + z_f) = 350/(725 + 2583) = 105.8 \text{ kN} \quad \text{Pass} (105.8 > 50)$$

Deflection,

$$I = \frac{\pi}{4}r^4 = \frac{\pi}{4}(600/2)^4 = 6.4 \times 10^9 \text{ mm}^4$$

$$\delta \approx V^*(e + z_f)^3/(3EI) = 50,000(725 + 2583)^3/(3 \times 32,800 \times 6.4 \times 10^9) = 2.9 \text{ mm}$$

Note: Increasing the pile length beyond the critical pile length will not affect pile deflection or capacity. If a higher capacity is required, a larger pile diameter should be adopted.

5.3 RETAINING WALLS

Retaining walls are typically designed using either the Rankine pressure method or the Coulomb wedge method. Design should be in accordance with AS 4678-2002, which nominates load and material factors. Geotechnical engineers should be engaged for the design of walls with high significance. A computer program called Wallap is recommended for analysis;—however, it does not include code provisions.

5.3.1 Code requirements

AS 4678 groups retaining walls into one of three possible classifications (refer to Table 5.8). The adopted classification is used to specify site investigation details, live loading, material design factors and performance monitoring. This chapter focuses only on design-related elements.

5.3.1.1 Loads and surcharges

A minimum wall surcharge (Q) is specified in relation to the structure classification (refer to Table 5.9). Walls which are subject to vehicular loading require a traffic surcharge of 10 kPa for temporary roads and 20 kPa for functional roads (this is similar to the bridge code (AS 5100.2, Cl 13.2) requirement of an equivalent allowance for the increase in pressure which would be caused by one additional metre of fill).

5.3.1.2 Material design factors

A significant, yet frequently overlooked, component of retaining wall analysis is the AS 4678 requirement for material design factors. In addition to the use of standard load factors, a material factor is used. This is similar to those for steel and concrete design (ϕ); however, different values are required for strength and serviceability analyses. The geotechnical material design factors affect the cohesion and friction angle for soil. The design values

Table 5.8 Retaining wall structure classification

Classification	Examples of structures
A	Where failure would result in minimal damage or loss of access (e.g. garden fence)
B	Where failure would result in moderate damage and loss of services (e.g. wall in public space or minor road)
C	Where failure would result in significant damage or risk to life (e.g. walls supporting structures identified for post-disaster recovery)

Source: Modified from AS 4678, table 1.1. Copied by Mr L. Pack with the permission of Standards Australia under Licence 1607-c010.

Notes:
1. Classification B includes structures not covered by Classification A or C.
2. Structures where failure would result in minimal damage and loss of access where the wall height is greater than 1.5 m are deemed to be Classification B structures.

Table 5.9 Wall surcharge

	Backfill slope (horizontal / vertical)	
Classification	Steeper than 4:1	4:1 or flatter
A	1.5 kPa	2.5 kPa
B, C	2.5 kPa	5 kPa

Source: AS 4678, table 4.1. Copied by Mr L. Pack with the permission of Standards Australia under Licence 1607-c010.

Table 5.10 Material design factors

Soil or material uncertainty factor		Soil or fill conditions			
		Controlled fill		Uncontrolled fill	In situ material
		Class I	Class II		
Effective (drained) parameters (values used for calculating c' and ϕ')					
Strength (ULS)	$\Phi_{u\phi}$	0.95	0.9	0.75	0.85
	Φ_{uc}	0.9	0.75	0.5	0.7
Serviceability (SLS)	$\Phi_{u\phi}$	1.0	0.95	0.9	1.0
	Φ_{uc}	1.0	0.85	0.65	0.85
Undrained parameters (values used for calculating c_u and ϕ_u)					
Strength (ULS)	$\Phi_{u\phi}$	0.0	0.0	0.0	0.0
	Φ_{uc}	0.6	0.5	0.3	0.5
Serviceability (SLS)	$\Phi_{u\phi}$	0.0	0.0	0.0	0.0
	Φ_{uc}	0.9	0.8	0.5	0.75

Source: Modified from AS 4678, tables 5.1(A) and 5.1(B). Copied by Mr L. Pack with the permission of Standards Australia under Licence 1607-c010.

Notes:
1. Class I fill is defined as an inert material, placed under appropriate supervision to ensure material is consistent in character and with an average density equivalent to 98% of the maximum dry density (and no test results falling below 95%). Cohesionless soils should be compacted to 75% Density Index. Refer to AS 4678, Cl 1.4.3.
2. Class II fill is defined as an inert material, placed in specified layers to achieve a material which is consistent in character and with an average density equivalent to 95% of the maximum dry density (and no test results falling below 92%). Cohesionless soils should be compacted to 65% Density Index with layers no thicker than 300 mm. Refer to AS 4678, Cl 1.4.3.

calculated below should be used in place of the original material values for both the Rankine pressure method and the Coulomb wedge method (Table 5.10).

Design cohesion, $\quad\quad c^* = \Phi_{uc}c$

Design internal friction, $\quad \phi^* = \tan^{-1}(\Phi_{u\phi}\tan\phi)$

5.3.1.3 Load combinations

Load factors for retaining walls vary depending on whether the action causes or resists stability. Combination cases are in accordance with AS/NZS 1170.0 for strength, stability and serviceability; however, the combination factors vary, as shown in Table 5.11. Notably, factors for dead loads range from 0.8 to 1.25, and hydrostatic loads are factored at 1.0. Wind and seismic loads should also be included. Refer to Section 1.7.1 for general combination cases.

5.3.2 Rankine pressure method

The Rankine method uses active and passive earth pressures to estimate the total force on a wall. This method provides a quick and easy calculation for retained soils; however, it is far less versatile than the Coulomb method. It is intended for the design of vertical walls with a flat upper surface, uniform surcharges and non-cohesive soil types. The Rankine method assumes a frictionless wall, a planar failure surface and a resultant force which is horizontal. The two methods will give identical results for cases which fit these criteria and have no surcharge.

Table 5.11 Retaining wall combination factors

| | Load case | | |
Description	Strength	Stability	Serviceability
Dead load of structure	1.25	0.8	1.0
Dead load of fill behind structure	1.25	1.25	1.0
Dead load of fill on structure	1.25	0.8	1.0
Dead load of fill in front of structure	0.8	0.8	1.0
Weight of water behind or in front of structure	1.0	1.0	1.0
Live load of Traffic load on structure	1.5	0	0.7[a] or 0.4[b] or 1.0[c]
Live load of Traffic load behind structure	1.5	1.5	0.7[a] or 0.4[b] or 1.0[c]
Live load of Traffic load in front of structure	0	0	0

Source: Modified from AS 4678, table J1. Copied by Mr L. Pack with the permission of Standards Australia under Licence 1607-c010.

[a] Refers to short-term case (AS/NZS 1170.0, ψ_s).
[b] Refers to long-term case (AS/NZS 1170.0, ψ_l).
[c] Refers to storage-term case (AS/NZS 1170.0, ψ_s).

Active pressures are those which are caused by a soil wanting to collapse away from a face. The active pressure is the force which is retained. Passive pressures are those which are caused by the compression of a soil face. It is essentially the resistance which a wall of soil can sustain. Passive pressures are used for keys under retaining walls and other items such as footings and thrust blocks. For cases where the wall is rigid or restrained from deflecting, the active pressure coefficient should be replaced with an at-rest pressure coefficient.

The Rankine method calculates the lateral pressure by multiplying the vertical pressure of each soil element by a coefficient for either of the active or passive pressures (Figure 5.13).

Active pressure coefficient,	$K_a = (1 - \sin\phi)/(1 + \sin\phi)$
At-rest pressure coefficient,	$K_o = (1 - \sin\phi)$
Passive pressure coefficient,	$K_p = (1 + \sin\phi)/(1 - \sin\phi)$
Lateral pressure,	$P = K(Q + \gamma z)$ (at a depth of z)
Lateral force,	$F = K_a Q h + K_a \gamma h^2 / 2$ (per metre width)
Moment at base of wall,	$M = (K_a Q h)(h/2) + (K_a \gamma h^2 / 2)(h/3)$ (per metre width)

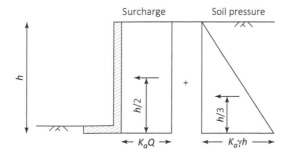

Figure 5.13 Rankine soil pressures (without water table).

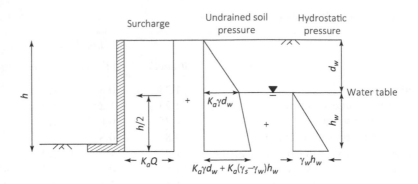

Figure 5.14 Rankine soil pressures (with water table).

where:

h = Height of wall

K = Active or passive pressure coefficient

Q = Surcharge (pressure at top of wall)

γ = Drained density of soil

ϕ = Internal friction angle of soil

In the presence of a water table, the calculation should be completed using the saturated density of the soil minus the density of water (for mass of soil below the water table level); refer to Figure 5.14. The hydrostatic water pressure should then be added.

Hydrostatic pressure, $\qquad P_w = \gamma_w h_w$ (at base of wall)

Hydrostatic lateral force, $\qquad F_w = \gamma_w h_w^2 / 2$ (per metre width)

Hydrostatic Moment at Base, $\quad M_w = (\gamma_w h_w^2 / 2)(h_w / 3)$ (per metre width)

where:

h_w = Height of water table from lower level

γ_w = Density of water = 10 kN/m³

The final solution for a retaining wall with a water table present can be written as,

Lateral force, $F = K_a Q h + K_a \gamma h^2 / 2 - K_a[\gamma - (\gamma_s - \gamma_w)]h_w^2 / 2 + \gamma_w h_w^2 / 2$ (per metre width)

Moment at base of wall, $M = K_a Q h^2 / 2 + K_a \gamma h^3 / 6 - K_a[\gamma - (\gamma_s - \gamma_w)]h_w^3 / 6 + \gamma_w h_w^3 / 6$

(per metre width)

where γ_s = saturated density of soil.

5.3.3 Coulomb wedge method

The Coulomb wedge method (trial wedge method) uses a more flexible approach, which can be applied to configurations with non-uniform ground and non-uniform surcharges. The method works by assuming a linear failure plane at an angle, θ, from horizontal (refer to

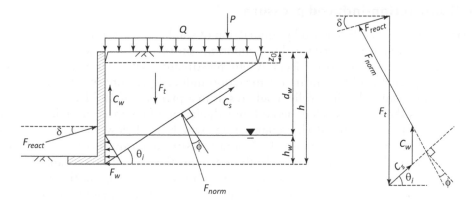

Figure 5.15 Coulomb wedge vectors.

Figure 5.15). A vector analysis is completed, with various failure plane angles being trialled to find the maximum wall reaction. This occurs at the angle at which the soil is most likely to fail.

The following vector solution is for the geometry shown in Figure 5.15. It can also be solved by drawing the vectors to scale (e.g. using 1 kN = 1 mm) and measuring the resultant F_{react} value.

Wedge reaction force,

$$F_{react} = [(F_t - C_w - C_s \cos\phi/\sin(\theta_i - \phi))\sin(\theta_i - \phi)]/\cos(\delta + \phi - \theta_i) \quad \text{(per metre width)}$$

Slip cohesion,	$C_s = (h - z_0)c/\sin\theta_i$
Wall cohesion,	$C_w = (h - z_0)$
Total wedge load,	$F_t = F_{soil} + F_{surc}$

Wedge soil force, $\quad F_{soil} = \gamma(h^2 - z_0^2)/(2\tan\theta_i) + (\gamma_s - \gamma - \gamma_w)(h_w^2/2\tan\theta_i)$

Wedge surcharge force, $\quad F_{surc} = QL + P \quad$ (for all loads acting above wedge)

Length of wedge, $\qquad L = (h - z_0)/\tan\theta_i$

Height of tension crack, $\qquad z_0 = 2c/\left(\gamma\sqrt{K_a}\right)$

Active pressure coefficient, $\qquad K_a = (1 - \sin\phi)/(1 + \sin\phi)$

where:
- c = Cohesion
- δ = Angle of wall friction
- θ_i = Failure angle for ith wedge
- ϕ = Internal friction angle of soil

Hydrostatic lateral force, $\qquad F_w = \gamma_w h_w^2/2$

Total lateral force, $\quad F = F_{react}\cos\delta + F_w \quad$ (per metre width)

Moment at base of wall, $\quad M = F_{react}\cos\delta \times [(h - z_0)/3] + F_w \times (h_w/3) \quad$ (per metre width)

5.3.4 Compaction-induced pressure

High lateral stresses can be caused by compaction of soil behind retaining walls. The effects are more significant with shorter walls. Limitations on the use of various compactors can be placed to reduce the negative effects. AS 4678 estimates that the use of a 125 kg vibrating plate compactor could result in a compaction-induced pressure of 12 kPa, and that a 10.2 tonne static smooth wheel roller could result in a pressure of 20 kPa.

Ingold [24] is the recommended method for calculation of compaction-induced pressure. Any pressures (prior to considering compaction) which are below the Ingold values are increased, as shown in Figure 5.16. The Ingold pressure diagram increases linearly from zero at the top of the wall to P'_{hm} at a depth of z_c, and then remains constant. The Ingold diagram is based on the soil being compacted in layers, with the concept that the soil yields laterally and does not fully recover when loading is removed. The diagram therefore represents a minimum pressure on the wall.

Critical depth, $$z_c = K_a\sqrt{2Q_1/(\pi\gamma)}$$

Lateral compaction-induced pressure, $\quad P'_{hm} = \sqrt{2Q_1\gamma/\pi}$

Effective line load imposed by compaction,

Q_1 = Roller weight/Roller width $\quad$ (for dead weight rollers)

$\quad$ = (Roller weight + Centrifugal force)/Roller width $\quad$ (for vibrating weight rollers)

$\quad \approx (2 \times$ Roller weight)/Roller width

Compaction-induced pressure should be factored as a live load; therefore, the roller load (Q_1) should be factored at 1.5 to calculate an ultimate pressure.

5.3.5 Stability

The stability of the wall can be checked after calculating the net forces acting on a retaining wall. Sliding and overturning should be checked about the front edge of the wall. A vertical key can be used to increase the sliding capacity. Increased heel and toe lengths can be used to increase sliding and overturning capacities. The heel of a retaining wall extends underneath the retained soil to engage the mass of the soil above. The toe of a wall extends the lever arm for resisting forces (refer to Figure 5.17).

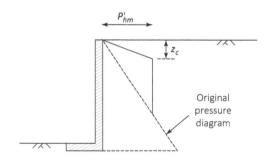

Figure 5.16 Compaction-induced pressure diagram. (From Ingold, T.S., *Gèotechnique*, 29, 265–283, 1979.)

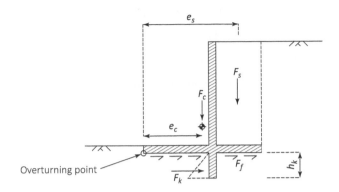

Figure 5.17 Retaining wall stability.

Sliding requirement, $F_{d,stb} \geq F_{d,dst}$

Overturning requirement, $M_{d,stb} \geq M_{d,dst}$

Destabilising forces should include retained earth pressure, surcharges, hydrostatic pressure and compaction-induced pressure. Calculations should include the material design factors for strength limit states and combination factors shown in Table 5.11. Any material which causes a stabilising effect should be factored at 0.8; those which cause destabilising effects should be factored at 1.25.

AS 5100.3 (Bridge Design–Foundations and soil-supporting structures), Cl 13.3.1 requires that where a retaining wall relies on the passive resistance from soil in front of the structure, the depth shall be decreased by the maximum of 10% of the wall height and 0.5 m. Consideration should always be given to whether unplanned excavations could occur near the wall.

Destabilising sliding force, $F_{d,dst} = F$

> (retained force, using Rankine or Coulomb method and factoring soil density at 1.25)

Stabilising sliding force, $F_{d,stb} = F_f + F_k$

Destabilising overturning moment, $M_{d,dst} = M$

> (retained moment, using Rankine or Coulomb method and factoring soil density at 1.25)

Frictional resistance, $F_f = 0.8\mu(F_c + F_s)$

Passive resistance of key, $F_k = K_p(0.8 \times \gamma)h_k^2/2$ (if applicable)

> (using Rankine method and factoring soil density at 0.8)

Coefficient of friction, $\mu = \tan(\delta)$

External friction, $\delta \approx 0.667\phi^*$ (for smooth concrete)

$\delta \approx 1.0\phi^*$ (for no-fines concrete)

Weight of concrete, $F_c = \text{Cross-sectional area of concrete} \times 24 \text{ kN/m}^3$

Weight of soil, F_s = Cross-sectional area of soil above heel $\times \gamma$

Stabilising overturning moment, $M_{d,stb} = 0.8(F_c e_c + F_s e_s)$ (self-weight $\times$ eccentricity)

Eccentricity of concrete, e_c = Dimension from centroid of concrete to overturning point

Eccentricity of soil, e_s = Dimension from centroid of soil to overturning point

5.3.6 Bearing pressure

The bearing pressure at the base of the wall should be checked using the design method shown in Section 5.1.2.

5.3.7 Typical soil properties

Typical soil parameters are provided in AS 4678, Appendix D and repeated in Tables 5.12 through 5.14. The values are intended for use in the absence of more reliable information.

5.3.8 Retaining wall detailing

Correct detailing is necessary to prevent damage induced by poor drainage. AS 4678, Appendix G provides guidance on the correct detailing for various types of retaining walls. Drainage is provided to the back of the wall by means of a coarse aggregate material in conjunction with an agricultural drain. Weepholes may also be used, especially for long walls which cannot provide drainage to the agricultural drain. Geotextile layers prevent the migration of fine-grained particulates into the more granular segments. Geocomposite drains can be used in addition to (or as a replacement for) the granular fill. A 300–500 mm thick compacted clay layer at the top surface seals the lower drainage system. A surface drain can be used at the top of the wall to prevent pooling of water at the clay capping layer.

Table 5.12 Typical rock internal friction

Stratum	ϕ' (degrees)
Chalk	35
Weathered granite	33
Fresh basalt	37
Weak sandstone	42
Weak siltstone	35
Weak mudstone	28

Source: AS 4678, table D3. Copied by Mr L. Pack with the permission of Standards Australia under Licence 1607-c010.

Notes:
1. The presence of a preferred orientation of joints, bedding or cleavage in a direction near that of a possible failure plane may require a reduction in the above values, especially if the discontinuities are filled with weaker materials.
2. Chalk is defined here as unweathered medium to hard, rubbly to blocky chalk.
3. Weathered basalt may have very low values of ϕ'.

Table 5.13 Typical soil cohesion and internal friction

Soil group	Typical soils in group	Soil parameters	
		c' (kPa)	φ' (degrees)
Poor	Soft and firm clay of medium to high plasticity, silty clays, loose variable clayey fill, loose sandy silts	0–5	17–25
Average	Stiff sandy clays, gravelly clays, compact clayey sands and sandy silts, compacted clay fill (Class II)	0–10	26–32
Good	Gravelly sands, compacted sands, controlled crushed sandstone and gravel fills (Class I), dense well-graded sands	0–5	32–37
Very good	Weak weathered rock, controlled fills (Class I) of roadbase, gravel and recycled concrete	0–25	36–43

Source: AS 4678, table D4. Copied by Mr L. Pack with the permission of Standards Australia under Licence 1607-c010.

Table 5.14 Unit weights of soils (and similar materials)

Material	Moist bulk weight, γ_m (kN/m³)		Saturated bulk weight, γ_s (kN/m³)	
	Loose	Dense	Loose	Dense
A: Granular				
Gravel	16.0	18.0	20.0	21.0
Well-graded sand and gravel	19.0	21.0	21.5	23.0
Coarse or medium sand	16.5	18.5	20.0	21.5
Well-graded sand	18.0	21.0	20.5	22.5
Fine or silty sand	17.0	19.0	20.0	21.5
Rock fill	15.0	17.5	19.5	21.0
Brick hardcore	13.0	17.5	16.5	19.0
Slag fill	12.0	15.0	18.0	20.0
Ash fill	6.5	10.0	13.0	15.0
B: Cohesive				
Peat (very variable)	12.0		12.0	
Organic clay	15.0		15.0	
Soft clay	17.0		17.0	
Firm clay	18.0		18.0	
Stiff clay	19.0		19.0	
Hard clay	20.0		20.0	
Stiff or hard glacial clay	21.0		21.0	

Source: AS 4678, table D1. Copied by Mr L. Pack with the permission of Standards Australia under Licence 1607-c010.

Damp proof membranes are typically used for all concrete which is cast directly against soil; they reduce cover requirements and prevent incorrect curing of concrete. Correct drainage can negate the requirement for the design of hydrostatic pressure on the back of the wall and justifies the use of drained design parameters (ϕ' and c') rather than undrained parameters (ϕ_u and c_u) (Figure 5.18).

Note: For details on damp proof membranes, geotextiles and agricultural drains, refer to Chapter 8.

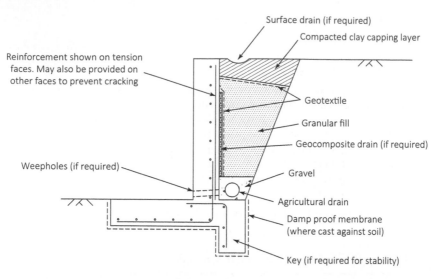

Figure 5.18 Typical retaining wall detail. (Modified from AS 4678, Figure G3. Copied by Mr L. Pack with the permission of Standards Australia under Licence 1607-c010.)

Example 5.4: Retaining Wall

Use the Rankine method to calculate the horizontal force and overturning moment on a 2 m high 'Class B' retaining wall with a 750 mm toe. The retained material is 'Class II' controlled fill, compacted using a 1000 kg static roller, with a width of 1200 mm. Design the wall for overturning and sliding.

SOIL PROPERTIES

$$\gamma = 18 \, \text{kN/m}^3 \quad \phi = 30°$$

ULS design requirements and combination factors,

$$\Phi_{u\phi} = 0.9 \quad \text{(refer to Table 5.10)}$$

$$\phi^* = \tan^{-1}(\Phi_{u\phi} \tan \phi) = \tan^{-1}(0.9 \tan 30) = 27.46°$$

$$K_a = (1 - \sin \phi^*)/(1 + \sin \phi^*) = (1 - \sin 27.46°)/(1 + \sin 27.46°) = 0.369$$

$$Q = 5.0 \, \text{kPa} \quad \text{(refer to Table 5.9)}$$

$$Q^* = 1.5 \times 5.0 \, \text{kPa} = 7.5 \, \text{kPa} \quad \text{(refer to Table 5.11)}$$

$$\gamma^* = 1.25\gamma = 1.25 \times 18 \, \text{kN/m}^3 = 22.5 \, \text{kN/m}^3 \quad \text{(refer to Table 5.11)}$$

Due to the fact that compaction-induced pressure is being considered, the pressure diagram will change. The pressure diagram is calculated for the soil load and the surcharge, and then it is modified to include compaction pressures.

PRESSURE FROM SOIL AND SURCHARGE

Surcharge pressure, $= K_a Q^* = 0.369 \times 7.5 \text{ kPa} = 1.85 \text{ kPa}$

Soil pressure at base, $= K_a \gamma^* z = 0.369 \times 22.5 \text{ kN/m}^3 \times 2\text{m} = 16.61 \text{ kPa}$

Net pressure at base, $P = 1.85 + 16.61 = 18.46 \text{ kPa}$

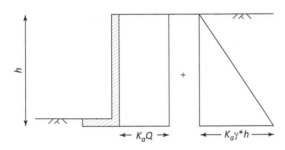
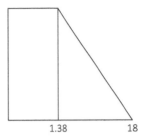

$\leftarrow K_a Q \rightarrow$ $\leftarrow K_a \gamma^* h \longrightarrow$ 1.38   18

COMPACTION-INDUCED PRESSURE

Roller load, $Q_1^* = 1.5(1000 \text{ kg} \times g/1200 \text{ mm}) = 12.26 \text{ kN/m}$

Critical depth, $z_c = K_a \sqrt{2Q_1^*/(\pi \gamma^*)} = 0.369\sqrt{2 \times 12.26 \text{ kN/m}/(\pi \times 22.5 \text{ kN/m}^3)} = 0.217 \text{ m}$

Pressure, $P'_{hm} = \sqrt{2Q_1^* \gamma^*/\pi} = \sqrt{2 \times 12.26 \text{ kN/m} \times 22.5 \text{ kN/m}^3/\pi} = 13.25 \text{ kPa}$

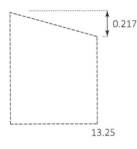
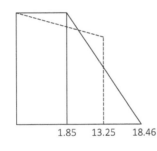
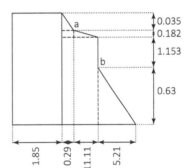

13.25   1.85   13.25   18.46

The lateral force is the area of the pressure diagram.

The points at which compaction pressure governs (a) and no longer governs (b) need to be calculated.

Resolving the pressures at point (a),

$$\left(\frac{13.25}{0.217}\right) \times h_a = 1.85 + \frac{(18.46 - 1.85)}{2} \times h_a$$

Height at point a, $h_a = 0.035 \text{ m}$

Pressure at point a, $P_a = \left(\frac{13.25}{0.217}\right) \times h_a = 2.14 \text{ kPa}$

Resolving the pressures at point (b),

$$13.25 = 1.85 + \frac{(18.46 - 1.85)}{2} \times h_b$$

Height at point b, $h_b = 1.37 \text{ m}$

Pressure at point b, $P_b = 13.25 \text{ kPa}$

To calculate the area, the shape can be dissected into rectangles and triangles.

$F = 1.85 \times 2 + 11.4 \times 1.783 + 0.29 \times 0.182 + 0.29 \times 0.035/2 + 11.11 \times 0.182/2$
$+ 5.21 \times 0.63/2 = 26.7$ kN   (per metre width)

The overturning moment is the area multiplied by the lever arms to the base,

$M = 1.85 \times 2^2/2 + 11.4 \times 1.783^2/2 + 0.29 \times 0.182 \times (0.63 + 1.153 + 1.182/2) + 0.29$

$\times 0.035/2 \times (0.63 + 1.153 + 0.182 + 0.035/3) + 11.11$

$\times 0.182/2 \times (0.63 + 1.153 + 0.182/3) + 5.21 \times 0.63/2 \times (0.63/3)$

$= 24.2$ kNm   (per metre width)

Note: For comparison purposes, without the effects of compaction the force would have been 20.3 kN and the moment would be 14.8 kNm.

STABILITY

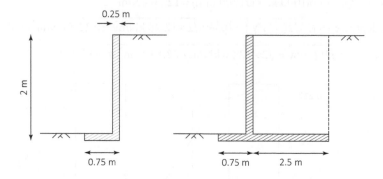

Sliding,

Destabilising reactions, $F_{d,dst} = 26.7$ kN $M_{d,dst} = 24.2$ kNm

Friction coefficient, $\mu = \tan(\delta) = \tan(18.3) = 0.33$

$\delta \approx 0.667 \phi^* = 0.667 \times 27.46 = 18.3$ (for smooth concrete)

Concrete mass, $F_c = 24(2 \times 0.25 + 0.5 \times 0.25) = 15$ kN

Sliding resistance, $F_{d,stb} = 0.8(F_c)\mu = 0.8 \times 15 \times 0.33 = 4$ kN   Fail $(4 < 26.7)$

The current arrangement fails in sliding. Sliding capacity is best increased by adding either a key or a heel. Adding a 2.5 m heel,

Revised mass, $(F_c + F_s) = 24(2 \times 0.25 + 3 \times 0.25) + 18(1.75 \; 2.5) = 109$ kN

Sliding resistance, $F_{d,stb} = 0.8(F_c + F_s)\mu = 0.8 \times 109 \times 0.33 = 29$ kN   Pass$(29 > 26.7)$

Overturning,

Overturning resistance, $M_{d,stb} = 0.8(F_c e_c + F_s e_s)$

$= 0.8(24(1.75 \times 0.25 \times 0.625 + 3.25 \times 0.25 \times 1.265) + 18(2.5 \times 1.75 \times 2))$

$= 151$ kNm   Pass$(151 > 24.2)$

Note: The slab and wall should be designed for a maximum moment of 24.2 kNm/m and a shear of 26.7 kN/m. Bearing should also be checked; refer to Section 5.1.2. For the values in this calculation and if the wall was 10 m in length: L = 10 m, B = 3.25 m, D = 0.25 m,

ϕ_g = 0.5, V = 1090 kN, H_B = 267 kN, M_L = 137 kNm (calculating moment about centre of base, and using 0.8 for soil mass and 1.25 for concrete mass), factored ultimate bearing capacity, $\phi_g\,p_{ug}$ = 99 kPa, maximum vertical load = 2981 kN Pass (2981 > 1090).

5.4 SLABS ON GRADE

The details provided in this section are aimed at providing a method which can be used with a structural analysis package (such as Space Gass or Strand7) to design structurally reinforced slabs on grade. This method can be used for complex arrangements or loading patterns.

Designs for typical slabs on grade are commonly completed in accordance with the *Guide to Industrial Floors and Pavements* [3]. The guide provides a method for determining concrete thickness in relation to subgrade and loading details. It should be used where a purely analytical approach is desired. The slab design relies on the flexural stiffness of concrete and provides reinforcement for crack control purposes only. It is useful for standard wheel configurations and post loads. The following details draw from information provided in the guide and other references to allow slab design using structural analysis programs.

The tensile strength of concrete is relied on for ground slabs with light loads or stiff subgrades. The following calculations are provided based on the flexural strength of the concrete only. A top layer of reinforcement is used to reduce cracking. For thin slabs, the cracking reinforcement can be placed centrally and used for strength calculations. High loads or poor sub-bases may result in a thicker slab which requires a top layer of reinforcement for crack control and a bottom layer for bending strength. For the structural design of reinforced slabs, refer to Section 4.3.

$$M^* \le \phi M_{uo}$$

Flexural tensile strength of concrete, $\qquad f'_{cf} = 0.7\sqrt{f'_c}$

Design tensile strength of concrete, $\qquad f'_{all} = k_1 k_2 f'_{cf}$

Design elastic bending capacity of unreinforced slab, $\qquad \phi M_{uo} = f'_{all}[Bt^2/6]$

Table 5.15 Load repetition factor, k_2

Expected load repetitions	Load repetition factor, k_2
Unlimited	0.50
400,000	0.51
300,000	0.52
200,000	0.54
100,000	0.56
50,000	0.59
30,000	0.61
10,000	0.65
2,000	0.70
1,000	0.73

Source: Cement Concrete & Aggregates Australia, *Guide to Industrial Floors and Pavements: Design, Construction and Specification*, 3rd ed., Illustration TechMedia Publishing, Australia, 2009. Reproduced with permission from Cement Concrete & Aggregates Australia.

where:
   B = 1m (to calculate result per metre width)
   f'_c = Compressive strength of concrete
   k_1 = 0.85 to 0.95   (vehicle loads)
      = 0.75 to 0.85   (sustained loads)
   k_2 = Load repetition factor (Table 5.15)
   t = Thickness of slab
   ϕ = 0.6

5.4.1 Preliminary sizing

A preliminary slab thickness can be selected from Table 5.16. This can be used for concept designs or for the first iteration of detailed design.

5.4.2 Soil parameters

The soil properties for a slab on grade are by far the most essential aspect of the design. A poor subgrade will inevitably lead to poor performance of the slab. A high-quality sub-base layer can be installed between the subgrade and the slab to increase the soil stiffness. A compaction to 95% of the maximum modified dry density (95% MMDD) is recommended under typical slabs and footings to achieve a high performance of the soil.

The California bearing ratio (CBR) for the slab location is typically required; however, an estimate can be used for slabs of low importance. The modulus of subgrade reaction is required when adopting a linear or non-linear analysis using plate elements. The Young's modulus and Poisson's ratio are required when completing a finite element analysis using brick elements. Slabs which are founded on reactive soils should be checked for shrink-swell in accordance with Section 5.5.

5.4.2.1 California bearing ratio (CBR)

The primary geotechnical variable is the CBR, which is obtained by an insitu CBR test or by cone penetrometer testing. For the design of lightly loaded slabs, a typical CBR value can be selected from Table 5.17. Many of the other parameters can be correlated from this value; however, numerous correlations are available for each parameter, and results vary in accuracy. The *Guide to Industrial Floors and Pavements* [3] provides a method of calculating an equivalent uniform layer for cases with varied layers of soil properties.

Where a 100 mm bound sub-base is specified (lean mix concrete blinding or cement stabilised gravel), the *CBR* value may be increased to an equivalent higher value.

Equivalent *CBR* (100 mm bound), $CBR_{bound} = 6.14 \times CBR^{0.7}$

Table 5.16 Preliminary slab thickness

Loading	Soil condition	Slab thickness
Light industrial, 5 kPa, cars	Poor	150
	Medium–good	130
Industrial, 5–20 kPa, commercial vehicles	Poor	200
	Medium–good	180

Source: Cement Concrete & Aggregates Australia, *Guide to Industrial Floors and Pavements: Design, Construction and Specification*, 3rd ed., Illustration TechMedia Publishing, Australia, 2009.

Table 5.17 Typical *CBR* values for soil

Soil type (and classification)	CBR (%)	
	Poor to fair drainage	Good to excellent drainage
Silt (ML)	2	4
Highly plastic clay (CH)	2–3	5
Silty/sandy-clay (CL/SC)	3–4	5–6
Sand (SW, SP)	10–18	10–18
Gravels, rocks and roadbase	10–100	15–100

Source: Jameson, G., *Guide to Pavement Technology Part 2: Pavement Structural Design,* 2nd ed., Austroads Incorporated, Australia, 2012. Reproduced with permission from Austroads Incorporated.

Note: Fill layers of roadbase may be specified by *CBR* value. Tabulated values for gravels, rocks and roadbase are provided based on experience.

5.4.2.2 Modulus of subgrade reaction

The modulus of subgrade reaction provides the simplest method of analysis; however, it may not provide the required accuracy for detailed design. A modulus of subgrade reaction (spring stiffness) can be estimated using the ultimate bearing pressure [2] with the equation,

Modulus of subgrade reaction, $\qquad k_s = P_{ult}/\delta$

where:

P_{ult} = Ultimate bearing capacity
δ = Allowable deflection

The equation is recommended for an allowable deflection of 25 mm; however, the design engineer can select a value which is appropriate for the purpose of the structure. Alternatively, a value can be estimated from Figure 5.19 (prescribed for U.S. airfield pavement design) or Table 5.18 (range of typical values).

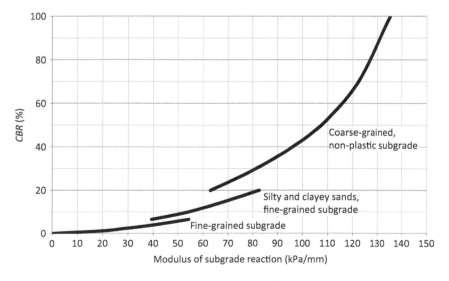

Figure 5.19 Correlation between CBR and modulus of subgrade reaction. (From UFC 3-260-03, 'Unified Facilities Criteria – Airfield Pavement Evaluation'.)

Table 5.18 Typical values for modulus of subgrade reaction

Soil description	Modulus of subgrade reaction, k_s (kPa/mm)
Loose sand	4.8–16
Medium dense sand	9.6–80
Dense sand	64–128
Clayey medium dense sand	32–80
Silty medium dense sand	24–48
Clayey soil $q_a \leq 200$ kPa	12–24
Clayey soil $200 < q_a \leq 800$ kPa	24–48
Clayey soil $q_a > 800$ kPa	>48

Source: Bowles, J.E., *Foundation Analysis and Design*, McGraw-Hill, Singapore, 1997. Reproduced with permission from McGraw-Hill Education.

Note: q_a = Allowable bearing capacity.

5.4.2.3 Young's modulus and Poisson's ratio

A more accurate solution can be reached using brick elements in a finite element analysis package. The model requires a Young's modulus and a Poisson's ratio. A long-term Young's modulus should be used for soils with sustained loads, and a short-term Young's modulus should be used for vehicle loads. Values can be estimated if specific data is not provided in the geotechnical report. The long-term Young's modulus can be selected from Table 5.19 or from Figure 5.20 using specified *CBR* layers or an estimated *CBR* from Table 5.17. The long-term modulus can be converted to a short-term value using the correlation factor from Table 5.20. Typical Poisson's ratios are provided in Table 5.21.

Short-term Young's modulus, $\qquad E_{ss} = E_{sl}/\beta$

5.4.3 Loads

The selection of design loads should be completed in conjunction with the client and is dependent on the intended use of the slab. Various loading configurations are available in accordance with different references. For typical slabs on Oil & Gas sites, it is recommended to use the W80 (wheel) and A160 (axle) loads (refer to Section 2.3.3 and Table 2.21) from the bridge code (AS 5100). Mine sites are highly variable, and wheel loading should be

Table 5.19 Typical Young's modulus, E (rock and cemented materials)

Material	Young's modulus, E (MPa)
Sub-base quality materials (road base)	150–400
Base quality gravel	150–400
Normal standard crushed rock	200–500
High standard crushed rock	300–700
Sub-base quality natural gravel (4%–5% cement)	1,500–3,000
Sub-base quality crushed rock (2%–4% cement)	2,000–5,000
Base (4%–5% cement)	3,000–8,000
Lean mix concrete	5,000–15,000

Source: Jameson, G., *Guide to Pavement Technology Part 2: Pavement Structural Design*, 2nd ed., Austroads Incorporated, Australia, 2012.

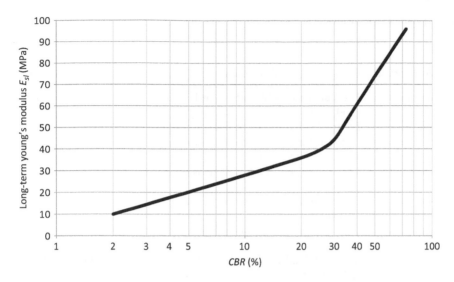

Figure 5.20 Long-term Young's modulus. (From Cement Concrete & Aggregates Australia, *Guide to Industrial Floors and Pavements: Design, Construction and Specification*, 3rd ed., Illustration TechMedia Publishing, Australia, 2009. Reproduced with permission from Cement Concrete & Aggregates Australia.)

Table 5.20 Correlation factor

Soil type	Correlation factor, β
Gravels	0.9
Sands	0.8
Silts, silty clays	0.7
Stiff clays	0.6
Soft clays	0.4

Source: Cement Concrete & Aggregates Australia, *Guide to Industrial Floors and Pavements: Design, Construction and Specification*, 3rd ed., Illustration TechMedia Publishing, Australia, 2009.

Table 5.21 Poisson's ratio

Soil type	Short-term Poisson's ratio	Long-term Poisson's ratio
Lean mix concrete and cement stabilised crushed rock	0.2	0.2
Crushed rock	0.35	0.35
Gravels	0.30	0.30
Sands	0.35	0.30
Silts, silty clays	0.45	0.35
Stiff clays	0.45	0.25
Soft clays	0.50	0.40
Compacted clay	0.45	0.30

Source: Cement Concrete & Aggregates Australia, *Guide to Industrial Floors and Pavements: Design, Construction and Specification*, 3rd ed., Illustration TechMedia Publishing, Australia, 2009; Jameson, G., *Guide to Pavement Technology Part 2: Pavement Structural Design*, 2nd ed., Austroads Incorporated, Australia, 2012.

Table 5.22 Load classifications

Load class	Description	Typical use	Nominal wheel load (kg)	SLS load (kN)	ULS load (kN)
A	Extra light duty	Footpaths accessible only to pedestrians and cyclists	330	6.7	10
B	Light duty	Footpaths that can be mounted by vehicles	2,670	53	80
C	Medium duty	Pedestrian areas open to slow-moving commercial vehicles	5,000	100	150
D	Heavy duty	Carriageways of roads and areas open to commercial vehicles	8,000	140	210
E	Extra heavy duty	General docks and aircraft pavements (trucks)	13,700	267	400
F	Extra heavy duty	High load docks and aircraft pavements (forklifts)	20,000	400	600
G	Extra heavy duty	Very high load docks and aircraft pavements	30,000	600	900

Source: AS 3996, table 3.1. Copied by Mr L. Pack with the permission of Standards Australia under Licence 1607-c010.

provided by the client. A more specific set of design loads is provided in AS 3996 (refer to Table 5.22), which can be adopted if specific loading details are known.

Typical dynamic factors are 1.4 for uncontrolled wheel or axle loads and 1.1 for controlled and slow-moving loads. Uncertainty factors are 1.8 for typical uncontrolled loads and 1.5 for controlled loads.

Design force for typical W80 load, $\quad F^* = 80\,\text{kN} \times 1.4 \times 1.8 = 201.6\,\text{kN}$

5.4.4 Analysis

Structural analysis should consider wheel or axle loads in a range of locations to maximise shear and bending forces. Separate cases should be solved with wheels at the edge, corner and centre of the slab, as well as at joint locations.

Joints can be modelled in Strand7 by applying edge releases to connecting plates, which provides a hinge connection. The same thing can be done in Space Gass by connecting the plates using master–slave restraints with directional (x,y,z) fixity.

5.4.4.1 Linear and non-linear analysis using modulus of subgrade reaction

A slab can be modelled using plate elements in any linear analysis package (such as Space Gass or Strand7). Plate elements should typically be a plan dimension of 200×200 mm.

Strand7 allows a spring to be applied directly to plate elements (face support); however, Space Gass requires springs to be applied at the nodes. For a node spring stiffness, the modulus should be multiplied by the tributary area of the plate elements (spring restraints are halved at edges and quartered at corner nodes). Lateral restraint can be provided along one or more edges to provide stability to the model.

Refer to Section 6.11.1 for a step-by-step linear analysis of a centrally loaded slab using Strand7 with the modulus of subgrade reaction method.

5.4.4.2 Finite element analysis using Young's modulus and Poisson's ratio

A more advanced analysis can be completed using a three-dimensional model of the slab. Concrete is modelled using plate elements, and soil is modelled using brick elements. The brick element matrix should extend several metres beyond the slab edges in all directions. The slab should be modelled above the soil and attached to the soil using beam point contact elements with an optional frictional restraint. Density of the soil elements should be set to zero to ignore existing overburden pressures.

5.4.5 Crack control

The design for crack control reinforcement (top layer) should consider AS 3600 reinforcement requirements for minimum steel (refer to Section 4.3.2) and the subgrade drag forces. The area of steel may be increased to as much as 0.15% where crack control is essential.

The subgrade drag theory assumes that the shrinkage of the slab is restrained entirely by reinforcement. The shrinkage force at the centre is equal to half of the slab mass multiplied by the coefficient of friction. A sufficient area of reinforcement should be provided [3] to restrain the force without exceeding $0.67 f_{sy}$.

Shrinkage force, $\quad\quad F_{sh} = \gamma_c (L/2) Bt\mu$

Minimum area of steel, $\quad A_s = \dfrac{\gamma_c (L/2) Bt\mu}{0.67 f_{sy}}$

where:
- f_{sy} = Yield strength of reinforcement = 500 MPa
- L = Length of slab between joints (in direction of bars being considered)
- γ_c = Density of slab = 24 kN/m^3
- μ = Friction and shear coefficient of friction = 1.5 (1.0 for smooth surfaces, 2.0 for very rough surfaces)

5.4.6 Joints

Joints are used in slabs to reduce the amount of unwanted or unplanned cracking and to facilitate construction. A joint layout should be completed using the continuous method or the strip method.

The continuous method (no joints) is preferable; however, it is not always practical, and it increases the amount of crack control reinforcement required. Reinforcement percentages [3] typically vary between 0.6% and 0.9%. This method is recommended for lengths up to 25 m, although it can be used for significantly longer dimensions. Longer lengths should be designed in consultation with the construction contractors. A detail for construction joints should be provided for unplanned stoppages.

The strip method is required for large pavement areas and can achieve more accurate floor levels. The geometry of strips should be carefully considered to minimise the number of joints and to align them with any required penetrations. Strips are 10–15 m in length with a width to suit construction requirements (pour volumes, float/screed length). Construction joints are used longitudinally, and control joints are used transverse to the long dimension of the slab (refer to Figure 5.21). Isolation joints should be used around adjacent structures and footings. A wide spacing of joints results in increased crack control reinforcement and increased movement at joint locations. Four types of joints are commonly used, as explained in the following sub-sections.

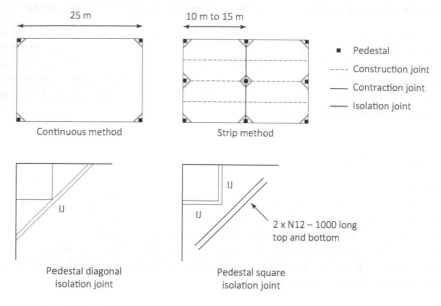

Figure 5.21 Joint layout diagram.

In cases where isolation joints cannot be placed diagonally, re-entrant bars should be detailed. Nominally, 1000 mm long N12 bars should be used (top and bottom layers) as a minimum.

5.4.6.1 Control joints

Control joints are used to induce cracking along a controlled line of the slab. The joint essentially creates a weak point that will crack before other sections in the slab. A groove in the top of the joint can be formed or sawn (after concrete is hard and prior to uncontrolled cracking) to 1/4 or 1/3 of the slab thickness. Reinforcement may be continuous through the joint; however, the top (compression) face is interrupted. Dowels should be used to transfer shear loads through control joints if reinforcement is not continuous. Control joints are also referred to as *saw joints* or *contraction joints* (Figure 5.22).

5.4.6.2 Isolation joints

Isolation joints are used to create an effective barrier between the edge of a slab and another object. They allow horizontal and vertical movement at the joint. Isolation joints are most commonly used around footing pedestals to prevent footing loads from being shared with the slab. They are also used to isolate slabs from footings with dynamic loads (Figure 5.23).

5.4.6.3 Construction joints

Construction joints are used between adjacent concrete pours in cases where expansion, contraction and isolation are not required. They are typically used for construction stoppages, such as bad weather or limited volumes of concrete supply. Lightly loaded slabs

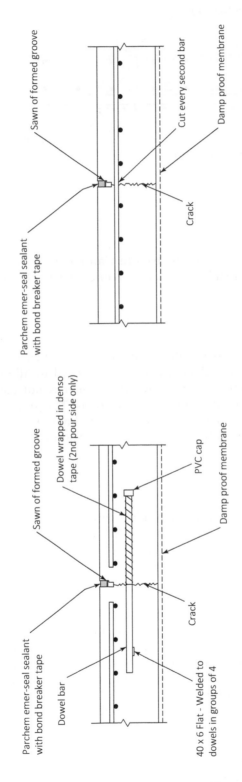

Figure 5.22 Control joint detail.

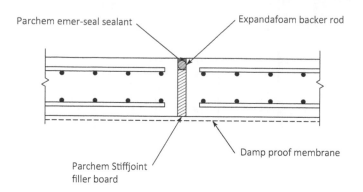

Figure 5.23 Isolation joint detail.

with high-quality subgrades can use tie bars to transfer shear loads. Highly loaded slabs or poor-quality subgrades require continuous reinforcement through construction joints (Figure 5.24).

5.4.6.4 Expansion joints

Expansion joints are designed to allow slabs to move when experiencing thermal or moisture variations. The joint thickness should be sized to allow the expected movement via a compressible joint material. Reinforcement is not continuous through the joint; however, dowels are used to transfer shear loads. The joint is therefore pinned and does not transfer moment. Expansion joints are not commonly used, because the initial contraction of a slab is typically less than the expansion caused by thermal variations [3] (Figure 5.25).

5.4.6.5 Joint armouring

Joints which are subject to hard-wheeled forklifts should be armoured to prevent damage to the opening. Joint armouring can be done by using a steel angle or by casting a recess and filling the joint with epoxy grout (for joints which do not require movement). Numerous vendor products are also available (refer to Chapter 8) (Figure 5.26).

5.4.6.6 Joint movement

Joint movement should be considered in the detailing of expansion and control joints to ensure that the calculated movement can be tolerated. Contraction at a joint can be estimated by multiplying the shrinkage strain by the effective length of slab on each side of the joint. The effective length is half the total length for slabs which are free to move in both directions and the full length for restrained slabs (Figure 5.27). This assumes that unrestrained slabs shrink towards their centre. Shrinkage strain is calculated in accordance with AS 3600, Clause 3.1.7. Time should be set to 30 years (11,000 days) for final shrinkage values.

Joint movement, $\delta = \varepsilon_{cs}(L_1 + L_2)$

Shrinkage strain, $\varepsilon_{cs} = \varepsilon_{cse} + \varepsilon_{csd}$

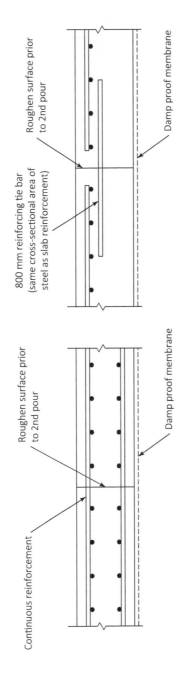

Figure 5.24 Construction joint detail.

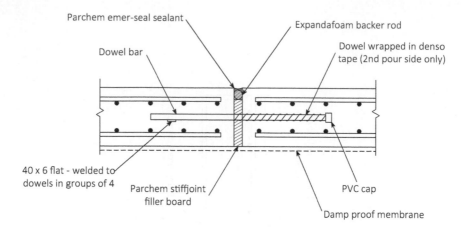

Figure 5.25 Expansion joint detail.

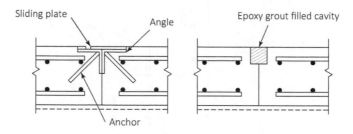

Figure 5.26 Additional joint armouring details.

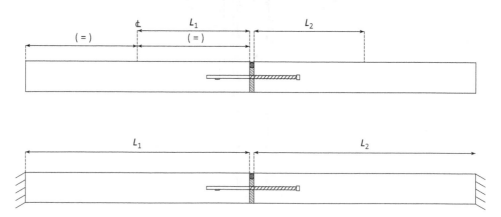

Figure 5.27 Joint movement.

Chemical shrinkage, $\qquad$ $\varepsilon_{cse} = \varepsilon_{cse}^*(1 - e^{-0.1t})$

Final chemical shrinkage, $\qquad$ $\varepsilon_{cse}^* = (0.06f_c' - 1) \times 50 \times 10^{-6}$

Drying shrinkage, $\qquad$ $\varepsilon_{csd} = k_1 k_4 \varepsilon_{csd.b}$

Basic drying shrinkage, $\qquad$ $\varepsilon_{csd.b} = \varepsilon_{csd.b}^*(1 - 0.008f_c')$

Final basic drying shrinkage, $\qquad$ $\varepsilon_{csd.b}^* = 800 \times 10^{-6}$ (Sydney, Brisbane)

$$= 900 \times 10^{-6} \text{ (Melbourne)}$$

$$= 1000 \times 10^{-6} \text{ (elsewhere)}$$

where:

k_1 $\quad = \alpha_1 t^{0.8}/(t^{0.8} + 0.15t_h)$
k_4 $\quad = 0.7$ (arid) $\qquad = 0.65$ (interior)
$\qquad = 0.6$ (temperate) $= 0.5$ (tropical or near-coastal)
$L_1, L_2 =$ Effective length of adjacent slabs
t $\qquad =$ Time since commencement of drying (days)
t_h $\qquad =$ Slab thickness
α_1 $\qquad = 0.8 + 1.2e^{-0.005t_h}$

5.4.7 Dowels

Dowels are used to transfer shear between joints without transferring moment. They hold adjacent slabs together vertically to prevent joints from becoming uneven. Typical dowels are constructed from round bars with a yield strength of 250 MPa (Table 5.23). One end of the dowel is wrapped in Denso tape to allow expansion and contraction in the longitudinal direction. Square dowels can be provided by vendors to also allow transverse movement. Diamond-shaped dowels are also available and can be used closer to corners, so that the slab is free to move in both horizontal directions (refer to Chapter 8).

The spacing of joints should be limited to approximately 15 m to ensure that the joint sealing is not damaged by slab movement, and a minimum slab thickness of 125 mm should be adopted for slabs with dowel joints to ensure adequate load transfer [3].

The design of dowels generally assumes that any applied point load is distributed between a minimum of two adjacent dowels, and therefore each dowel is subject to 50% of the total load. Alternatively, the load can be distributed over a distance of l to each side of the load position ($2l$ total width if not positioned near corners) [42], refer Figure 5.28.

The structural design of dowels needs to consider the shear loading as well as bending induced by the joint gap and opening (refer to Section 5.4.6) and concrete bearing

Table 5.23 Typical dowel dimensions

Slab thickness	Round dowels			Square dowels			Diamond dowels		
	Diameter	Length	Spacing	Length	Thickness	Spacing	Length	Thickness	Spacing
125–150	20	400	350	350	20	350	115	6	400
150–200	24	450	350	400	25	350	115	10	450
200–275	33	500	300	450	32	300	115	20	500

Notes:
1. All values are in millimetres.
2. Round dowel grade is 250R with bars spaced at 300 mm centres.
3. Square and diamond dowels are based on a maximum joint opening of 5 mm.

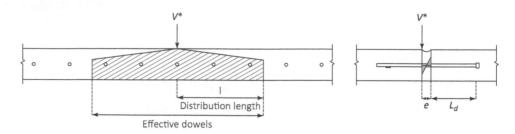

Figure 5.28 Dowel distribution.

failure. The dowel is fixed at both ends and therefore undergoes double-bending (refer to Figure 5.28). Design equations are shown for circular dowels.

Design bending moment, $M^* = V^*(e/2)$

Bending capacity, $M^* \leq \phi M = \phi f_y(d^3/6)$

Shear capacity, $V^* \leq \phi V_f = \phi 0.62 f_y A$ (refer to AS 4100, Cl 9.5)

Bearing capacity, $V^* \leq \phi V_b = \phi\, 0.85 f'_c(d \times L_d)$

Distribution length, $l = 0.9[(E_c t^3)/(12(1-\nu^2)k_s)]^{0.25}$

where:
  A = Cross-sectional area of dowel
  d = Dowel diameter
  e = Joint gap (specified gap plus joint movement)
  E_c = Young's modulus of concrete (refer to Table 4.1)
  f_y = Yield strength
  k_s = Modulus of subgrade reaction (Section 5.4.2)
  L_d = Embedment length of dowel into each side of slab
  t = Slab thickness
  V^* = Design shear load
  ϕ = 0.8 (for bending and shear) = 0.6 (for bearing)
  ν = Poisson's ratio of concrete = 0.2

5.5 SHRINK-SWELL MOVEMENT

Shrink-swell movement is the variation in elevation of a soil between wet and dry periods. 'Reactive' soils shrink during dry seasons and swell during wet seasons. This phenomenon can induce movements and stresses on structures. A common result is cracking in large slabs, as the soil in the centre is not affected by the seasonal change. It also causes differential movement where one structure is supported on deeper foundations and is less susceptible to movement compared with a shallow slab or pad. This regularly results in trip hazards and misalignment of equipment such as stairs.

Unfortunately, there is no standard focusing on industrial calculations for shrink-swell values. The residential standard, AS 2870, is commonly used in lieu. It therefore needs to be understood that the concept behind the standard is to minimise construction cost and accept a higher degree of risk than is usually adopted for industrial purposes. Hence, clients should be involved in decisions regarding the adoption of the code. It may be decided to use a factored approach for items of higher importance, such as sensitive equipment or bunds. Values calculated in the standard are based on a 5% chance of being exceeded in a design life of 50 years.

HB 28-1997 references a text by Cameron, 1989, which found that measured results range from 43% to 167% of the calculated values with a standard deviation of 25%.

5.5.1 Investigation

AS 2870 requests a minimum of one pit or borehole per building site, and a minimum of three where the depth of design soil suction change (H_s) is greater than or equal to 3.0 m. The shrinkage index (I_{ps}) is the variable required for calculations.

The depth of investigations shall be to the greater of 0.75 H_s and 1.5 m, or until rock is reached. Values are calculated in accordance with AS 1289.7.1.1, AS 1289.7.1.2 and AS 1289.7.1.3.

For industrial purposes, it is recommended to complete a greater number of tests to achieve a cross-section of the site. Measured values often differ by a reasonable amount. If a smaller number of tests are completed, the structures may be over or under-designed, resulting in additional cost. For a square site, 100 or 200 m in length, it is recommended to complete at least five tests. The exact locations may be specified prior to commencement if a plot plan exists at the investigation stage.

If other structures exist in the area, it is recommended that they are inspected for damage or movement. This can be used as a basis for calculations.

5.5.2 Calculation of characteristic surface movement

The following calculation method is based on AS 2870, Section 2.3. The characteristic surface movement (y_s) is the movement of the surface of a reactive site caused by moisture change.

5.5.2.1 Swelling profile

The swelling profile needs to be examined prior to beginning the calculation. This is the depth and shape of the soil, which will affect the surface movement. It is based on location and is modified if water or rock are encountered (refer to Figure 5.29). This value is generally provided in the geotechnical investigation.

If not provided in the geotechnical investigation, values for H_s and Δu can be selected from Table 5.24 or from maps of the area, such as are published in the *Australian Geomechanics*

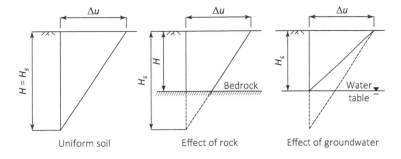

Figure 5.29 Shape and depth of suction. (From AS 2870, figure 2.1. Copied by Mr L. Pack with the permission of Standards Australia under Licence 1607-c010.)

Table 5.24   Soil suction change profiles for certain locations

Location	Change in suction at the soil surface, Δu (pF)	Depth of design soil suction change, H_s(m)
Adelaide	1.2	4.0
Albury/Wodonga	1.2	3.0
Brisbane/Ipswich	1.2	1.5–2.3 (see Note)
Gosford	1.2	1.5–1.8 (see Note)
Hobart	1.2	2.3–3.0 (see Note)
Hunter valley	1.2	1.8–3.0 (see Note)
Launceston	1.2	2.3–3.0 (see Note)
Melbourne	1.2	1.8–2.3 (see Note)
Newcastle	1.2	1.5–1.8 (see Note)
Perth	1.2	1.8
Sydney	1.2	1.5–1.8 (see Note)
Toowoomba	1.2	1.8–2.3 (see Note)

Source: AS 2870, table 2.4. Copied by Mr L. Pack with the permission of Standards Australia under Licence 1607-c010.

Note: The variation of H_s depends largely on climatic variation.

Table 5.25 Soil suction change profiles for climatic zones

Climatic zone	Description	Depth of design soil suction change, H_s(m)
1	Alpine/wet coastal	1.5
2	Wet temperate	1.8
3	Temperate	2.3
4	Dry temperate	3.0
5	Semi-arid	4.0
6	Arid	>4.0

Source: Fox, E., A climate-based design depth of moisture change map of Queensland and the use of such maps to classify sites under AS 2870-1996. *Australian Geomechanics Journal*, 53–60, 2000. Reproduced with permission from Australian Geomechanics.

Table 5.26 Available resources for maps of regions

Region	Author	Date
Victoria	Smith	1993
QLD	Fox	2000
Hunter Valley	Fityus et al.	1998
NSW	Barnett & Kingsland	1999
South-east QLD South-west WA	Walsh et al.	1998

Journal [21] (refer to Table 5.25). Other maps are available in AS 2870, Appendix D, and as listed in Table 5.26. The Australian Geomechanics Society is a useful source of information on the topic. The value can also be affected by local geography, such as dams or lakes.

5.5.2.2 Depth of cracking

The depth of cracking is taken as

$$H_c = 0.75 H_s \quad \text{for Adelaide and Melbourne}$$

$$H_c = 0.5 H_s \quad \text{for other areas} \quad \text{(AS 2870, Clause 2.3.2)}$$

5.5.2.3 Existence of cut or fill

Sites that have been cut or filled exhibit increased surface movement (in the order of 60%). Where fill is placed less than 5 years prior to construction, the depth of cracking (H_c) should be taken as zero. Where cut has occurred less than 2 years prior to construction, the depth of cracking (H_c) should be reduced by the depth of the cut. For this case, fill is defined as more than 800 mm of sand or more than 400 mm of other materials.

Provision is also made to allow the movement to be estimated by reference to established knowledge of similar fills in similar areas. However, the alternative classification should not be less than the natural site classification unless non-reactive, controlled fill is used for the greater of 1 m and 0.5 H_s.

5.5.2.4 Characteristic surface movement

The characteristic surface movement (y_s) is calculated by multiplying the area of each shape within the triangle by a factor and adding them together. It is defined in general terms as (AS 2870, Clause 2.3.1),

$$y_s = \frac{1}{100} \sum_{n=1}^{N} (I_{pt} \overline{\Delta u h})_n$$

where:
I_{pt} = Instability Index = αI_{ps}
I_{ps} = Shrinkage Index
α = 1.0 (for the cracked zone)
α = 2.0–z/5 (for the uncracked zone)
h = Thickness of shape being considered (mm)
N = Number of soil layers (shapes)
y_s = Characteristic surface movement (mm)
z = Depth to centre of shape being considered (m)
$\overline{\Delta u}$ = Soil suction for centre of shape being considered (pF)

5.5.2.5 Site classification

Classification of the site is based on the calculated value of the characteristic surface movement, as selected from Table 5.27. As explained in the example, the characteristic surface movement is rounded to the nearest 5 mm.

Table 5.27 Classification table

Characteristic surface movement, y_s (mm)	Class	Foundation
0	A	Most sand and rock sites with little or no ground movement from moisture changes.
$0 < y_s \leq 20$	S	Slightly reactive clay sites, which may experience only slight ground movement from moisture changes.
$20 < y_s \leq 40$	M	Moderately reactive clay or silt sites, which may experience moderate ground movement from moisture changes.
$40 < y_s \leq 60$	H1	Highly reactive clay sites, which may experience high ground movement from moisture changes.
$60 < y_s \leq 75$	H2	Highly reactive clay sites, which may experience very high ground movement from moisture changes.
$y_s > 75$	E	Extremely reactive sites, which may experience extreme ground movement from moisture changes.
–	P	Problem sites are given a 'P' classification, usually due to severe stability or bearing capacities (refer to AS 2870, Clause 2.1.3 for details).

Source: Based on AS 2870, tables 2.1 and 2.3. Copied by Mr L. Pack with the permission of Standards Australia under Licence 1607-c010.

Note: Deep-seated profiles ($H_s \geq 3.0$ m) are given an additional '-D' after the class. That is, a classification of E on a deep-seated site is classified as E-D.

Example 5.5: Shrink-swell

The example calculation is for two layers of soil in Queensland with a 3.0 m design soil suction change depth. The top layer of soil (A) is 1.0 m thick, and the shrinkage index (Ips) is 2%; soil layer B continues beyond the depth of suction and has a shrinkage index of 3%.

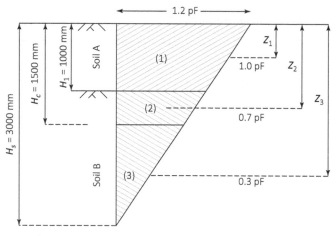

Depth of design soil suction change, $H_s = 3.0$ m

Depth of cracking, $H_c = 0.5H_s = 1.5$ m

Depth to centre of shape (1), $Z_1 = \dfrac{1}{2} = 0.5$ m

Note: Values for u and z are taken at the mid-height of the shapes being considered [21]. AS 2870 states that it should be the centroid; however, as the profile is divided into more sub-sections, it approaches the value for the mid-height, as recommended by Fox.

Soil suction for shape (1),
$$\Delta u_1 = 1.2\frac{H_s - Z_1}{H_s} = 1.2\frac{3-0.5}{3} = 1.0pF$$

Lateral restraint factor for cracked zone, $\alpha_c = 1.0$

Movement from cracked zone (1),

$$y_{(1)} = \frac{I_{pt}\overline{\Delta u h}}{100} = \alpha_c\frac{I_{ps(A)}}{100}\Delta u_1 H_1 = 1\times0.02\times1\times1000 = 20\text{ mm}$$

Depth to centre of shape (2),
$$Z_2 = 1 + \frac{0.5}{2} = 1.25\text{ m}$$

Soil suction for shape (2),
$$\Delta u_2 = 1.2\frac{H_s - Z_2}{H_s} = 1.2\frac{3-1.25}{3} = 0.7pF$$

Movement from cracked zone (2),

$$y_{(2)} = \frac{I_{pt}\overline{\Delta u h}}{100} = \alpha_c\frac{I_{ps(B)}}{100}\Delta u_2(H_c - H_1) = 1\times0.03\times0.7\times500 = 10.5\text{ mm}$$

Depth to centre of shape (3),
$$Z_3 = 1.5 + \frac{1.5}{2} = 2.25$$

Soil suction for shape (3),
$$\Delta u_3 = 1.2\frac{H_s - Z_3}{H_s} = 1.2\frac{3-2.25}{3} = 0.3pF$$

Lateral restraint factor for uncracked zone, $\alpha_u = 2.0 - \frac{z}{5} = 2 - \frac{2.25}{5} = 1.55$

Movement from uncracked zone (3),

$$y_{(3)} = \frac{I_{pt}\overline{\Delta u h}}{100} = \alpha_u\frac{I_{ps(B)}}{100}\Delta u_3(H_s - H_c) = 1.55\times0.03\times0.3\times1500 = 20.9\text{ mm}$$

Characteristic surface movement, $y_s = y_{(1)} + y_{(2)} + y_{(3)} = 20 + 10.5 + 20.9 = 51.4\text{ mm}$

In accordance with the supplement to AS 2870, the calculated value is rounded to the nearest 5 mm and classified in accordance with Table 5.27.

Reported characteristic surface movement, $y_s = 50\text{ mm}$ $\left(\text{Classification H1}-\text{D}\right)$

If a footing founds some depth below the surface level, the area of the triangle above the base of the footing can be removed from the calculation. It is often helpful for a project to calculate a graph showing the calculated movement versus the founding depth.

5.5.2.6 *Soil structure interaction: Heave*

Centre heave and edge heave are support conditions in which the edges of a foundation have shrunk or swelled due to the movement of reactive soils (refer to Figure 5.30). Structures should be either stiff enough to accommodate the induced stress caused by the support conditions or flexible enough to follow the contours of the ground. AS 2870, Appendix F provides two methods for analysis, the Walsh method [33] and the Mitchell method [26]. Details for the Walsh method are provided below.

The idealised mound shapes need to be detailed after the characteristic surface movement is calculated. The shapes are based on the depth of design soil suction change (H_s) and

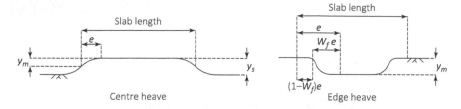

Figure 5.30 Idealised mound shapes to represent design ground movement. (From AS 2870, figure F1.Copied by Mr L. Pack with the permission of Standards Australia under Licence 1607-c010.)

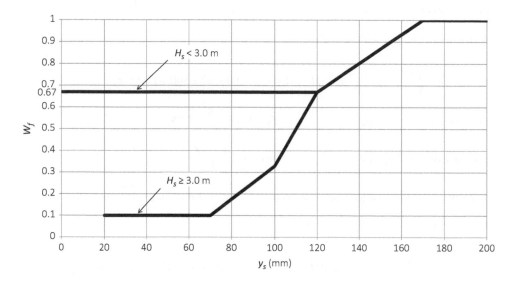

Figure 5.31 Mound shape factor.(From AS 2870, figure F2.Copied by Mr L. Pack with the permission of Standards Australia under Licence 1607-c010.)

mound movement (y_m), which is related to characteristic surface movement. The mound shape factor (W_f) is derived from Figure 5.31.

Edge heave is usually a transitory phase, which may only occur before the edge heave condition becomes established (AS 2870, Appendix CF). This is why the edge heave distance does not rely on H_s and may be reduced by up to 40% if construction is during wet periods (refer to AS 2870, Appendix F).

Mound movement, $y_m = 0.7y_s$ (for centre heave)

$y_m = 0.5y_s$ (for edge heave)

Edge distance, $e = \dfrac{H_s}{8} + \dfrac{y_m}{36}$ (for centre heave)

$e = 0.2L \leq 0.6 + \dfrac{y_m}{25}$ (for edge heave)

Note: if the slab length is less than 2e, then y_m may be reduced linearly with span/2e (e.g. if the slab length is equal to e, then y_m should be reduced to half of the originally calculated value).

The mound stiffness needs to be known if an advanced analysis is going to be completed on the structure. AS 2870, Appendix F provides information on stiffness values and explains that computed forces and displacements are generally not sensitive to the adopted value. Swelling soils range from $k = 0.4$ kPa/mm to $k = 1.5$ kPa/mm. Shrinking soils should use a minimum of 5 kPa/mm.

5.5.2.7 Load combinations

AS 2870, Clause 1.4.2 states that the strength and serviceability designs should be considered using the permanent action plus 0.5 times the imposed action. This is because the code is for residential use, and a relatively low cost of failure is considered.

The factor of 0.5 is similar to the long-term factor (ψ_l) and the combination factor (ψ_c) from AS/NZS 1170.0; therefore, for industrial applications, where additional loads need to be considered, it is suggested that the standard load combinations are used. It is also stated that the design bearing capacity should be not more than 0.33 times the ultimate bearing pressure, which is fairly typical for a serviceability analysis.

5.5.2.8 Modelling

After the mound shape is detailed, modelling of the different support scenarios can be completed. This may be done by either simplified hand calculations (treating the spans as cantilevers or slabs supported on four sides), analysis in a package such as Space Gass, or FEA (refer to Section 6.11) in a package such as Strand7. Experience and knowledge of the area should be considered above all else.

A detailed non-linear analysis can be completed in an FEA package (refer to Figure 5.32 for details) by creating the slab mesh, then extruding the nodes downwards (50 mm) into point contact beam elements. In Strand7, the zero-gap type should be selected in the 'Beam Element Property' section. This creates many small beams which provide support only when the gap is closed.

A translational stiffness (spring) should be applied to the bottom layer of nodes in the vertical axis (the spring stiffness should equal the mound stiffness, multiplied by the plate

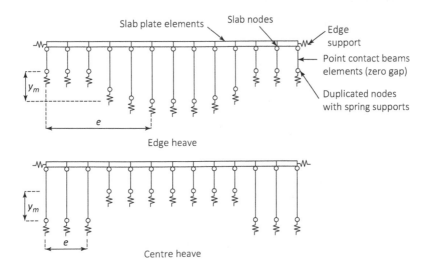

Figure 5.32 FEA modelling of slab-on-ground.

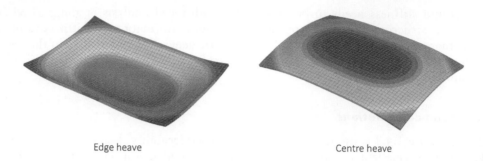

Edge heave   Centre heave

Figure 5.33 Example displaced shape for edge heave and centre heave.

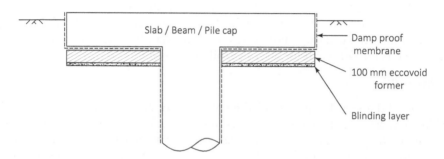

Slab / Beam / Pile cap

Damp proof
membrane

100 mm eccovoid
former

Blinding layer

Figure 5.34 Typical detail for swelling soils.

area). The plates should also be provided with an 'edge support' in the lateral direction to provide stability to the model.

The bottom layer of nodes should then be moved to represent the shape of the ground. This means that two models are required: one for edge heave and one for centre heave. When the model is solved, it will essentially lower the slab onto a soil profile with the calculated shape for edge heave or centre heave. The slab will then either span the required distance or deflect to the shape of the ground (refer to Figure 5.33 for an example solution). The model can then be used to extract displacements, moments, shear forces and bearing pressures.

Note: The actual displacement will be 50 mm less than the reported value due to the length of the point contact beams.

5.5.3 Shrink-swell detailing

The preference is often to detail a solution which separates the foundations from swelling soils. This may be done by founding the slab on piles. However, a sufficient gap is required to prevent upward pressure on the slab. This problem is often solved by using a void former, such as Eccovoid (refer to Chapter 8). The concept is to create a free-spanning slab, raised 100–200 mm above the soil. This is achieved by the use of a temporary cardboard former, which degrades after the concrete has cured.

Detailing and construction should be in strict accordance with the technical data sheet. Void former depths are provided in relation to various mound movements.

A blinding layer should be provided on the base for a flat, stable work platform. The void former rests on the blinding layer and is protected from the concrete by a damp proof membrane (refer to Figure 5.34).

Chapter 6

Design items

This chapter describes the design of numerous items which commonly occur in both the Mining and the Oil & Gas industries. There are often many ways to complete these designs. Those which rely heavily on Australian Standards are reasonably straightforward, and results should be similar, despite assumptions made throughout the project. Designs which are not covered by Australian Standards have a less definitive path, and results may vary depending on the design code which is selected. The information contained within this chapter is based on commonly used methods within the industry. A combination of national and international codes is generally used throughout typical projects.

6.1 PIPE RACKS (PIPE STRESS)

One of the most common functions of a structural engineer in the Oil & Gas industry is the design of pipe support structures. This involves working closely with the pipe stress engineers to understand the specific requirements of the pipe system. The designs are generally completed in structural packages such as Space Gass, because the member design inputs can be completed, and the program can check code compliance. Space Gass has some capacity to check connections; however, specialised programs, such as Limcon, are recommended.

The most common support configurations are t-posts, goal posts and braced frames (refer to Figure 6.1). The t-post is a cantilever support and therefore requires a moment base plate and typically heavy section sizes. The goal post uses a portal frame configuration; however, it still requires a moment base plate to resist load in the longitudinal direction. The braced frame is typically a set of two or more goal post supports which are supported longitudinally by struts and consists of at least one bay of diagonal bracing. The braced frame uses pinned base plates and typically smaller section sizes. Plan bracing may also be used in the braced frame if required.

6.1.1 Pipe stress

The principal goal for pipe stress engineers is to create a system with sufficient flexibility to prevent excessive stresses with respect to code (e.g. ASME B31.3 and ASME B31.8)

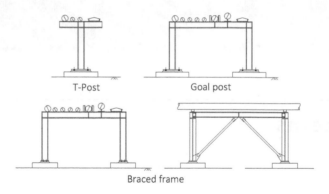

Figure 6.1 Typical pipe support configurations.

limitations. Due to stress concentrations and particular code requirements, the points of interest for pipe stress engineers include pipe support locations and nozzle interfaces with equipment. The structural engineer needs to work with the piping team to produce a support system that meets their requirements.

Concept discussions should occur prior to commencement of the design, including

1. Support spacing
2. Technical requirements
3. Load cases
4. Load combinations
5. Support types
6. Global coordinate system

6.1.1.1 Support spacing

The initial item which should be agreed on is support spacing. For small pipes (<100 mm dia.), this can be as little as 3 m, and for larger pipes, it is commonly 6 or even 8 m. The frame spacing is governed by the span of the smallest, or weakest, pipe on the rack. In some cases, it may be more feasible to create an alternate support type for the smaller pipes. For example, if two pipes on the rack are only capable of spanning 3 m, but all others are capable of spanning 6 m, the optimum solution may be to space the frames at 6 m centres and to provide intermediate supports for the two smaller pipes. This can be done by adding trimmer steelwork between the frames; running the smaller pipes in cable trays; or simply installing isolated supports between the bays. It is also possible to support small pipes from larger pipes using a piggyback arrangement. In the absence of more specific information, assume a support spacing of 6 m.

6.1.1.2 Technical requirements

Technical inputs for each of the design programs need to be agreed between the piping and structural teams. Pipe stress analysis programs (e.g. Caesar II) rely on the user to input stiffness values for supports. These values are very frequently created without any input from the structural engineers and are often excessively high (leading to unrealistically high stresses and low deflections). The piping programs are not hugely reliant on accurate values; however, it is a good idea to calculate some typical values. This is done by creating a rough model of a pipe rack and applying a unit load where the pipe will rest. The stiffness is the load divided by the calculated deflection. The stiffness should be calculated for each direction. If there are significantly different pipe racks or support types, you may want to create several different values; however, it is much more efficient to use a standard stiffness where possible. Alternatively, the pipe stress engineer can model steelwork in some piping programs; however, this complicates the design, and restraint types (pinned/rigid) are often modelled incorrectly.

The piping team should also commit to an allowable deflection limitation for the pipe supports. A value of 10 mm is commonly used; however, it is not uncommon for a smaller value (such as 1 or 3 mm) to be required at areas of significance, such as vessel or compressor nozzle tie-in locations. The structural calculations should then include a deflection check at pipe support locations to ensure that the deflection is within the allowable limit.

Using the same coordinate system and units will simplify the transfer of data between disciplines. This should ideally match the coordinate system for the 3D design model.

6.1.1.3 Load cases

The piping team should provide the wind and seismic factors which were used, so that the structural team is aware of whether they are ultimate or serviceability loads. Piping seismic loads are often calculated using AS 1170.4, Section 8; however, the structural supports should be designed to Section 2; therefore, it may be possible to reduce the values provided by the pipe stress engineers. Any upset loads, such as slug loads, hammer loads or pigging loads, should be supplied with a frequency or probability, so that the structural engineer can decide whether or not the load should be factored to be considered an ultimate load case. If the frequency or probability is similar to an ultimate wind or seismic case then no additional factors are necessary.

Caesar II is a commonly used pipe stress analysis program. A list of typical loads and their abbreviations for Caesar II is provided in Table 6.1.

Some pipes may not be considered 'stress critical', and therefore no input data will be provided by the pipe stress engineers. This is generally the case for small service piping, which does not exceed predefined diameter, class or pressure criteria. It is common to adopt a pressure (such as 2 kPa) for these pipes. However, this should be confirmed by looking at the spacing and weight of the pipes (including contents). An allowance should also be

Table 6.1 Specific piping load designations

Abbreviation	Description
CS	Cold spring
D1	Displacement (i.e. soil movement or other forced movement)
F1	Force (any specific applied force)
H	Hanger load
HP	Hydrostatic test pressure
P1	Pressure
T1	Thermal
U1	Uniform loads (such as seismic)
W1	Weight
WAV	Wave
WIN1	Wind
WNC	Weight, no contents
WW	Weight, water filled

included for wind, seismic and thermal loading. A horizontal value of 30% of the vertical load is generally conservative.

6.1.1.4 Load combinations

Typical load combinations are described in Section 1.7.1, based on AS/NZS 1170, the 'Process Industry Practices – Structural Design Criteria' (PIP STC01015) [47] and common practice.

Ideally, the piping team will commit to a standard set of load combinations for the project. It can be difficult if you have more than one piping system on a rack, and they have been provided with different load cases. For this purpose, piping programs work in the opposite way to structural programs. Structural engineers would usually solve each load case individually and then combine the results linearly. Pipe stress engineers solve each combination case non-linearly and then subtract cases from each other to find the individual values (refer to Table 6.2). This is because the piping program relies heavily on non-linear data (specifically friction). If a horizontal load is applied without a vertical load, the pipe stress will travel until it reaches a 'line stop'; however, with friction, some of the load is resisted at each 'rest' support and the remainder is taken at the line stop. Depending on the load type, the pipe stress engineer may apply it with or without friction.

The pipe stress engineer will want to do an allowable stress design; however, AS 4100 requires ultimate limit states (ULS) for the structural design. Therefore, the structural engineer will need to factor the individual loads provided by the piping team and combine them back in the original order. It is very important to always place the loads back in the original order to avoid errors, as shown in the following example.

For example, a support is only capable of resisting 10 kN (via friction), and the remaining load will travel along the pipe to the line stop. Two unknown loads, each equal to 8 kN, are applied in unison, and the reaction at the support is 10 kN (with the remaining 6 kN travelling along the pipe to the line stop). If you then look at another combination case with only one of the 8 kN loads, you see a reaction of 8 kN, and you may incorrectly think that the other load was 10 kN – 8 kN = 2 kN. You may then, incorrectly, apply that 2 kN on its own in a separate load case. However, if you separate the 10 kN reaction into components of 8 kN and 2 kN, then apply ultimate factors and combine them again, the resultant will, correctly, be a factored version of the 10 kN load. Ideally, permanent loads should be taken first, and then transient loads (refer to Table 6.2 for another example).

Caesar II operates with three load combination categories: operational (OPE), sustained (SUS) and expansion (EXP). Operational cases are those which represent the pipe during

Table 6.2 Pipe stress load combination case example

Combination case number	Details	Description
L1 (OPE)	W + T1 + P1	Weight, temperature and pressure
L2 (SUS)	W + P1	Weight and pressure
L3 (EXP)	L1 – L2 = T1	Increase from temperature, in the presence of weight and pressure

Note: Refer to Caesar II User Guide.

hot operation, typically including both primary (weight, pressure and force) and secondary (displacement and thermal) load cases. Sustained cases are those that represent the pipe in a cold (as-installed) position, with force-driven loads (weight and pressure), usually to calculate installed self-weight and meet code requirements. Expansion cases are created from subtracting one case from another to calculate an isolated load. However, care should be taken, as this only provides the increase in load caused by one case, in the presence of the other case (as explained in the previous example).

6.1.1.5 Support types

Pipe stressing programs, such as Caesar II, have numerous non-linear support configurations. A list is provided in Table 6.3 along with the restraint directions. The piping team would generally detail the specific aspects of how these supports are achieved; however, the structural team should be aware of the requirements (as they often require welding to structural steelwork).

Rests are detailed with specific materials to achieve an appropriate coefficient of friction. Generally, a value is used for steel to steel, and this can be reduced by adopting polytetrafluoroethylene (PTFE) or high-density polyethylene (HDPE) (refer to Table 2.22). The designers need to be aware of exact support details so that the correct top of steel level is achieved. Examples of pipe support configurations are shown in Figure 6.2. Rests, guides and line stops can all be used either directly, with sliding pads or with shoes. If high loads are expected, stiffeners should be located directly under the pipe. Spring supports are often

Table 6.3 Types of pipe support

Support type	Description
Guide	Transverse restraint (horizontal)
Hangers	Vertical restraint (from above)
Line stop	Longitudinal restraint (horizontal)
Rest	Vertical restraint, plus lateral friction
Spring	Vertical restraint (with or without specified stiffness)

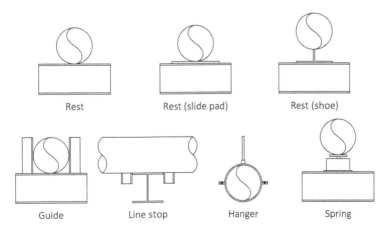

Figure 6.2 Pipe supports.

used under pipes which are subject to high temperature variations to allow flexibility in the support system and minimise forces at equipment nozzles.

Care should be taken in applying the pipe loads to the structural steelwork, as loads are generally applied to the top flange; therefore, torsion can be induced. In programs such as Space Gass, the simplest way to model the offset is to move the node where the force is applied to the top of steel and then offset the beam downwards to the original location. Web stiffeners may be required in the steel beam, especially at line stop locations and for pipes with large vertical loads (refer to Section 3.5.5). Steel material selection needs to be considered for pipes which operate at low or high temperatures (refer to Section 6.6.2).

6.1.2 Other pipe rack loads

In addition to pipe stressing values, pipe rack loads may also include the following values, depending on client preference:

1. Cable tray and ladders (often 2.0–3.0 kPa; refer to Chapter 8)
2. Live platform loads (often 2.5 kPa for walkways and 5.0 kPa for platforms; refer to Chapter 8)
3. Wind and seismic (structural components as well as piping components)
4. Transportation (horizontal loads range from 10% up to 80% depending on specific methods, if pre-assembled)
5. Thermal variation (for large welded structures, refer to Section 6.6)

6.1.3 Pre-assembled units (PAUs)

PAUs are used for either delivering packaged vendor items or minimising site work. The increased upfront design, material and fabrication costs are offset by savings in site work. PAUs are therefore ideally used in remote locations where site costs are higher.

A typical single-level PAU is shown in Figure 6.3, and a double-level PAU is shown in Figure 6.4. The double-level PAU is shown installed at ground level and in an elevated position in Figure 6.5. For elevated PAUs, the support structure should be stable prior to installation of the PAU to ensure safe positioning. Stairs and/or ladders may be required depending on client access requirements. Platforms are used on the PAUs to provide operators with access to valves and other process equipment.

The main (longitudinal) beams are generally governed by the lifting case. Lugs are placed symmetrically about the centre of gravity, ideally at locations with transverse beams. Refer to Section 6.3 for details on lifting design. More lifting points will result in a decreased amount of steelwork; however, this increases the complexity of the lift, and may require a

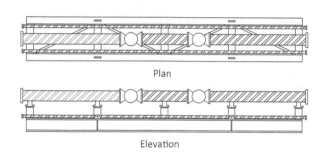

Plan

Elevation

Figure 6.3 Single-level PAU (skid).

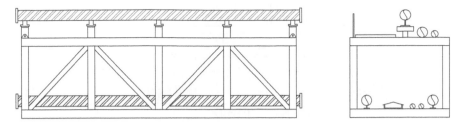

Figure 6.4 Double-level PAU example.

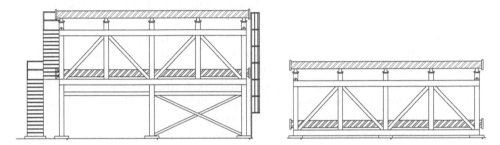

Figure 6.5 PAU installed configurations.

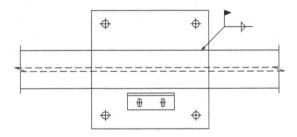

Figure 6.6 Landing plate detail.

larger crane and lifting (spreader) beams. Site installation typically requires site welding of the structure to base plates, which are bolted to the foundations. Angles with slotted holes can be used to aid in the alignment of the PAU (Figure 6.6).

Transverse beams are governed by either trucking cases or lifting cases. The units should be designed to be supported on the transverse beams for trucking cases. A transportation drawing should be completed to ensure that the correct location of supports is adopted. Temperature variation should also be considered in accordance with Section 6.6.

6.1.3.1 Transportation and load restraint

Logistics need to be considered, as the transportation envelope and accelerations are key design inputs. A route needs to be planned from the fabrication yard to the site; refer to Section 3.2 for dimensional requirements. Early involvement with fabrication and transportation contractors is important, as it minimises change in the design. Site installation, trucking and sea fastening methods should be considered.

Logistics and transportation companies should be involved in choosing appropriate acceleration factors for design. The *Load Restraint Guide* [10] requires that transported loads

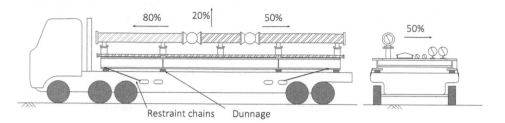

Figure 6.7 Load restraint.

are able to resist an 80% braking load, a 50% accelerating load, a 50% transverse load and a 20% uplift load. Shipping loads generally vary from 50% to 80% laterally. The structure should be checked for these ultimate load cases to ensure that structural failure does not occur. The load also needs to be fastened to the truck/ship to ensure that the structure does not slide or overturn. Chains with hooks are used to fasten the structure to the truck bed (refer to Figure 6.7). The chains both act as a direct restraint and also increase the vertical reaction and therefore the frictional capacity (clamping force). Dunnage should be used between the underside of the structure and the truck bed to provide a significant increase in the frictional resistance of the load (refer to Table 6.4). Vendor supplied rubber mats are the most effective and generally increase the friction coefficient to 0.6 (refer to Chapter 8).

Sacrificial sea fastening brackets can be welded to the structure and to the deck of ships to secure the PAU during shipping (refer to Figure 6.8). Tall PAUs may also require ties from the top of the structure to the deck. Temporary bracing can also be installed within the structure to provide transportation restraint and removed once the structure arrives at site. These braces are often painted pink to show that they are temporary and should be removed once the structure is installed.

Table 6.4 Recommended load restraint friction coefficients

Description	Friction coefficient, μ
Smooth steel: smooth steel	0.1–0.2
Smooth steel: timber dunnage	0.3–0.4
Smooth steel: rubber mats	0.6–0.7

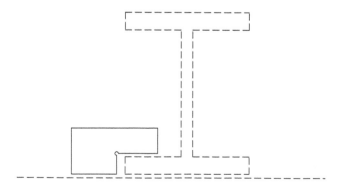

Figure 6.8 Sea fastening bracket.

6.2 VESSELS AND TANKS

Supports for vessels and tanks should be designed in conjunction with the requirements from other disciplines, such as piping, mechanical and environmental. The role of the structural engineer is to provide support to the vessel or tank and to ensure that deflection is limited to an acceptable level. Depending on the structure and contents of the equipment, a bund may be required (refer to Section 6.8). Access may also need to be provided by means of ladders or platforms (refer to Section 6.5).

A vessel is shown in Figure 6.9 with typical components. It is shown with a manhole on the left for maintenance. Interconnecting piping is shown on the right and above the vessel. A valve is provided at the base to drain liquid contents from the vessel. A bund is provided underneath the vessel, with walls located wider than any flanges, to collect any spills or drips. The vessel is supported by two saddles. One is bolted down to create a pinned connection, and the other uses a sliding plate and slotted holes to form a sliding support.

Loads on vessels and tanks are generally categorised as

1. Self-weight
2. Live platform loads (if applicable)
3. Wind
4. Seismic
5. Snow
6. Liquid
7. Piping loads
8. Hydrotest loads
9. Thermal loads

For combination load cases and load factors, refer to Section 1.7.1.

Loads are generally provided by the vendor, although most are easily calculated in accordance with AS/NZS 1170. The vendor should be informed of important design inputs and provided with a copy of the Structural Design Criteria (refer to Section 1.7) prior to commencing the design of the vessel or tank.

Piping loads should be included in the foundation design for vessels and tanks which have stress-critical piping attached. Depending on the preference of the client, these may

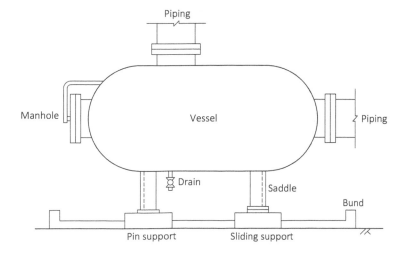

Figure 6.9 Vessel support arrangement.

be adopted as either the allowable nozzle loads or the calculated pipe stress loads. Using the allowable loads is the most conservative approach and allows for future modifications; however, it may lead to an overly conservative design.

Hydrotesting should be discussed with the client, as it is not always a design requirement for the foundation. It is the process of filling the tank with a liquid to check the tightness. Many vessels are hydrotested in fabrication shops and will never be tested in the installed location. This is often the governing case for vertical loads if the vessel is designed for contents which are less dense than water (such as gas). If a foundation is not designed for hydrotesting, this should be clearly documented and noted on design drawings, as future modifications to the site could potentially result in a scenario where the vessel requires further testing. The 'Process Industry Practices – Structural Design Criteria' (PIP STC01015) [47] recommends that hydrotest loads are factored at 1.4 (although this is the same factor used for maximum dead loads and is considered equivalent to the 1.35 factor adopted in AS/NZS 1170.0). A reduced factor of 1.2 is recommended when combined with wind loads. Wind is recommended as an operating speed (20 m/s) to be managed procedurally, or a serviceability speed assuming the installation period is less than 6 weeks.

Thermal loads result from expansion and contraction of the vessel, and are an important consideration for horizontal vessels. For these scenarios, the vessel is often supported with a pinned connection at one end and a sliding connection at the other end, as shown in Figure 6.9. The pinned connection is a base plate, bolted directly to the foundation. The sliding connection is usually a slotted hole with a sliding plate. Slotted holes should be in the direction of expansion, and the holes should be dimensioned for the calculated growth. The slots are often covered by an additional plate to prevent the holes from filling with water or other materials. The sliding plate is often fabricated from mild steel; however, other materials such as a Teflon (PTFE) pad can be used to reduce friction (refer to Table 2.22). For these cases, the saddle is usually detailed shorter on the sliding end to allow for the additional plate thickness.

Loads for thermal expansion are calculated in the same way as pipe stressing and similarly to the method presented in Section 6.6.4. The temperature range for the vessel should be available from the mechanical or process engineers. The total force is calculated; however, for sliding configurations, the supports can only restrain the vessel by the vertical reaction at the sliding end of the support multiplied by the coefficient of friction for the adopted support (refer to Table 2.22). As soon as the force reaches the frictional capacity, the vessel moves, and the load is relieved (unless restrained by a pipe). The reaction at the pinned end, due to thermal expansion, can only equal a maximum of the reaction calculated at the friction end, because the system is in equilibrium. Therefore, the reaction at each end is the minimum of the expansive force divided by two and the self-weight on the friction end multiplied by the coefficient of friction. Friction is almost always the governing case. The friction force acts outwards at each saddle support. Friction loads should be factored the same as the vertical load which causes the friction (typically 1.35).

Total vertical reaction, $\qquad F_y = 1.35 \times w$

Vertical reaction at each side (A and B), $\quad F_{yA} = F_{yB} = F_y/2 \qquad$ (for a symmetrical tank)

Friction capacity (and maximum thermal force), $\quad R_{\text{friction}} = \mu \times F_{yB}$

where:
$\quad w \quad$ = Full weight of tank
$\quad \mu \quad$ = Friction coefficient

Where the vessel has saddles supported on separated foundations, the thermal load can also be limited by modelling the foundations and allowing a deflection.

It is good practice to include the design of hold down bolts in the vendor's scope of work to avoid unnecessary interfaces in design. Often, international bolt grades are specified, such as B7, and therefore supply of the bolts is ideally provided by the vendor, as long as procurement timing allows the supply prior to foundation construction.

Example 6.1: Vessel

A vessel weighs 1000 kg, the total capacity is 4 m³ of water, and it operates at 80% capacity. It is supported in a pin and sliding configuration with support saddles spaced at 3 m. The vessel is subject to a 10% seismic load, a 20 kN piping live load and a 40° thermal load. The cross-sectional area of the vessel is 0.03 m². The centre of mass and piping load are both 4 m above ground. Calculate the foundation reactions for the following load combinations:

a. Hydrotest = 1.35 G
b. Live piping load = 1.2 G + 1.5 F_{pipe}
c. Ultimate seismic = $G + Eu_{(+x)}$
d. Max self-weight + thermal = 1.35 G + 1.35 F_f
e. Self-weight + live load + thermal = 1.2 G + 1.5 F_{pipe} + 1.2 F_f

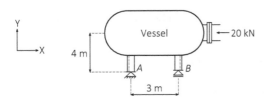

a) Hydrotest = 1.35 G:
The supports are spaced evenly about the centre of the vessel, and therefore the vertical reaction at each leg is assumed to be equal.

Total vertical reaction, $\quad F_y = 1.35 \times \left(1000 \text{ kg} + 4 \text{ m}^3 \times 1000 \, \dfrac{\text{kg}}{\text{m}^3}\right) \times 9.81 \, \dfrac{\text{m}}{\text{s}^2} = 66.2 \text{ kN}$

Reaction at each side, $\quad F_{yA} = F_{yB} = \dfrac{66.2 \text{ kN}}{2} = 33.1 \text{ kN}$

b) Live piping load = 1.2 G + 1.5 F_{pipe}:
For cases which rely on frictional restraints, the vertical reactions should be calculated first. This is done by calculating the component from self-weight (G) and the component from the pipe load (P).

Vertical reaction (G), $\quad F_{y(G)} = 1.2 \times \left(1000 \text{ kg} + 0.8 \times 4 \text{ m}^3 \times 1000 \, \dfrac{\text{kg}}{\text{m}^3}\right)$
$$\times 9.81 \, \dfrac{\text{m}}{\text{s}^2} = 49.4 \text{ kN}$$

Vertical reaction (G) at each side, $\quad F_{yA(G)} = F_{yB(G)} = \dfrac{49.4 \text{ kN}}{2} = 24.7 \text{ kN}$

The vertical reaction for piping is calculated by taking moments about support A and resolving it as zero.

$$\sum M @ A = 0 = 1.5 \times 20 \text{ kN} \times 4 \text{ m} + F_{yB(P)} \times 3 \text{ m}$$

Vertical reaction (P) at side B, $F_{yB(P)} = -40 \text{ kN}$ (uplift)

Vertical reaction (P) at side A, $F_{yA(P)} = 40 \text{ kN}$ $\left(\text{using } \sum F_y = 0 \right)$

Now, the friction capacity at support B can be calculated; however, it is in uplift and therefore does not resist friction, and all shear load is taken at support A. Net results are summarised as

Vertical reaction at side B, $F_{yB} = F_{yB(G)} + F_{yB(P)} = -15.3 \text{ kN}$ (uplift)

Vertical reaction at side A, $F_{yA} = F_{yA(G)} + F_{yA(P)} = 64.7 \text{ kN}$

Horizontal reaction at side A, $F_{xA} = 1.5 \times 20 \text{ kN} = 30 \text{ kN}$

c) Ultimate seismic = $G + Eu_{(+x)}$:

Vertical reaction (G), $F_{y(G)} = \left(1000 \text{ kg} + 0.8 \times 4 \text{ m}^3 \times 1000 \dfrac{\text{kg}}{\text{m}^3} \right) \times 9.81 \dfrac{\text{m}}{\text{s}^2} = 41.2 \text{ kN}$

Vertical reaction (G) at each side, $F_{yA(G)} = F_{yB(G)} = \dfrac{41.2 \text{ kN}}{2} = 20.6 \text{ kN}$

The vertical reaction for seismic (E) is calculated by taking moments about support A, and resolving it as zero.

$$\sum M @ A = 0 = -10\% \times G \times 4 \text{ m} + F_{yB(E)} \times 3 \text{ m}$$

Vertical reaction (E) at side B, $F_{yB(E)} = 5.5 \text{ kN}$

Vertical reaction (E) at side A, $F_{yA(E)} = -5.5 \text{ kN}$ (uplift) $\left(\text{using } \sum F_y = 0 \right)$

Net vertical results are summarised as

Vertical reaction at side B, $F_{yB} = F_{yB(G)} + F_{yB(E)} = 26.1 \text{ kN}$

Vertical reaction at side A, $F_{yA} = F_{yA(G)} + F_{yA(E)} = 15.1 \text{ kN}$

Now, the friction capacity at support B can be calculated using a coefficient of 0.3 for steel–steel contact.

Friction capacity, $R_{\text{friction}} = 0.3 \times 26.1 \text{ kN} = 7.8 \text{ kN}$

The net horizontal load (4.1 kN) can therefore be distributed evenly between each support. If the applied load were greater than double the friction capacity, the additional load would be attracted to restraint A.

Horizontal reactions, $F_{xA} = F_{xB} = \dfrac{-10\% \times G}{2} = -2.1 \text{ kN}$

d) Max self-weight + thermal = $1.35 \ G + 1.35 \ F_f$:
This case is considered at operating volume, as it is unlikely that the vessel would see a full temperature load during hydrotest. First, the vertical reactions are calculated.

Total vertical reaction, $F_y = 1.35 \times \left(1000 \text{ kg} + 0.8 \times 4 \text{ m}^3 \times 1000 \dfrac{\text{kg}}{\text{m}^3} \right) \times 9.81 \dfrac{\text{m}}{\text{s}^2} = 55.6 \text{ kN}$

Reaction at each side, $F_{yA} = F_{yB} = \dfrac{55.6\text{ kN}}{2} = 27.8\text{ kN}$

Now, the friction capacity at support B can be calculated using a coefficient of 0.3 for steel–steel contact.

Friction capacity, $R_{\text{friction}} = 0.3 \times 27.8\text{ kN} = 8.3\text{ kN}$

This is the maximum load which can be caused by thermal growth of the tank, as exceeding the load will cause the support to slide.
The thermal load can now be considered by calculating the unrestrained thermal growth of the vessel (refer to Section 6.6).

Unrestrained growth, $\Delta L = \propto_T L\Delta°C = \left(11.7\times10^{-6}/°C\right)\times 3000\text{ mm}$

$$\times\left(1.35\times40°C\right) = 1.9\text{ mm}$$

This value can be used to calculate the load which would be required to restrain the vessel and prevent expansion.

Strain, $\varepsilon = \Delta L/L = 1.9\text{ mm}/3000\text{ mm} = 6.3\times10^{-4}$

Stress, $\sigma = E\times\varepsilon = 200{,}000\text{ MPa}\times6.3\times10^{-4} = 126\text{ MPa}$

Force, $F = \sigma\times A = 94\text{ MPa}\times0.03\text{ m}^2 = 3780\text{ kN}$

It can be seen that the vessel will slide 1.9 mm on the friction support (B) and cause a reaction of 8.3 kN inwards on the supports.

Horizontal reactions, $F_{xA} = 8.3\text{ kN}$ $F_{xB} = -8.3\text{ kN}$

e) Self-weight + live load + thermal = 1.2 G + 1.5 F_{pipe} + 1.2 F_f:
   This case can be solved rationally by examining the solutions to cases b) and d). The vertical reactions will be the same as case b), therefore placing restraint B in uplift and preventing any friction. This means that no load will be attracted by the thermal movement of the vessel, which will move freely. The reactions are therefore identical to case b).

6.3 LIFTING LUGS

A lifting lug is a structural element which is attached to the element being lifted. It usually consists of a steel plate with a hole for the shackle. The plate is welded to the lifted element, and in the case of heavy lifts, the hole is often stiffened with a cheek plate (refer to Figure 6.10).

Lifting should generally be conducted with a minimum of four points for stability purposes. A six-point lift requires complicated rigging, and therefore, preferred lifting arrangements are four- or eight-point configurations (refer to Figures 6.11 and 6.12).

The vertical lifting arrangements impart only vertical reactions to the structure and are often preferred by engineers due to simplicity. Inclined arrangements are more commonly used on site; they are easier to rig and can be more stable. Induced compressive forces in the support members should be considered when using inclined slings.

The calculation of design loads is usually based on either an '$n - 1$' analysis of the loads (i.e. if using four lugs, the weight is divided by three) or if lifting beams are used, an unbalance between each pair of lugs (i.e. 60% and 40% weight under each lifting beam).

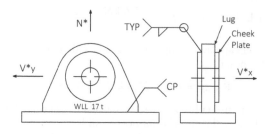

Figure 6.10 Typical lifting lug detail.

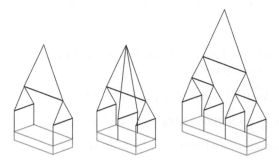

Figure 6.11 Typical vertical lifting arrangements.

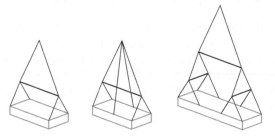

Figure 6.12 Typical inclined lifting arrangements.

6.3.1 Design factors

The design parameters that follow have been based on AS 4100 and AS 2741 (Shackles). The suggested philosophy is to calculate the design loads and then adopt a shackle with a capacity higher than the required loads (refer to Chapter 8). The lug should then be designed for the capacity and dimensions of the adopted shackle.

Shackles are specified using a working load limit and are required to be capable of supporting six times the specified working load limit (up to a total value of 1000 t), in accordance with AS 2741. Lifting lugs should be designed in accordance with strength limit states, and therefore the working load limit of the shackle should be factored for the design load of the lug.

The design factors for lifting typically include a contingency factor, a dynamic factor and an uncertainty factor. The working load should be calculated using the contingency factor and the dynamic factor, as this is the maximum expected load. The design of lifting lugs and

Table 6.5 Suggested design load factors

Design item	Contingency factor	Dynamic factor	Uncertainty factor	Total factor
Shackles	1.10	(1.5)[a]	N/A	1.65 (WLL)
Lugs	1.10	(1.5)[a]	1.5	2.48 (ULS)
Members	1.10	(1.25)[a]	1.5	2.06 (ULS)

[a] Values range from 1.10 to 2.0.

members should include the additional uncertainty factor for strength limit states design. A summary of suggested factors is provided in Table 6.5.

The contingency factor should be adopted in conjunction with input from the client; however, a minimum of 1.10 is recommended.

Dynamic factors depend on the configuration and location of lifts (onshore/offshore). This should be reviewed in consultation with the transportation contractor. A value of 2.0 is recommended for vertical actions in AS/NZS 1170.1; however, smaller dynamic factors are often used. For the design of members, a value of 1.25 is suggested for masses under 75 t (adopting the factor recommended for girders in AS/NZS 1170.1, Table 3.4). Heavier structures should require smaller dynamic factors, as they are moved more slowly, and therefore it is recommended that the factor should be reduced linearly to a value of 1.10 for 100 t lifts. However, lifts onto and off ships are recommended to use a minimum dynamic factor of 1.30. A standard uncertainty factor of 1.5 is recommended for strength design of the lug.

6.3.2 Placement of lugs

Lugs should ideally be placed symmetrically about the centre of gravity in each plane to maintain similar loads between slings. For cases where lugs are not symmetrical, the hook should be placed over the centre of gravity and steeper slings will attract higher loads.

The centre of gravity for each axis is calculated by summing the mass multiplied by the distance for each item and then dividing by the total mass, as shown in Figure 6.13. The example shows three masses with a length being measured from an arbitrary location.

$$X_{COG} = \frac{L_1 m_1 + L_2 m_2 + L_3 m_3 + \ldots}{m_1 + m_2 + m_3 + \ldots}$$

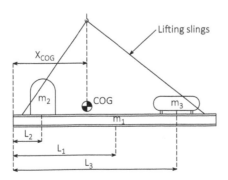

Figure 6.13 Centre of gravity.

6.3.3 Marking

The 'working load limit' (WLL) or 'rated capacity' (RC) of the structure should be clearly marked; refer to AS 4991 and AS(/NZS) 1418 for details. WLL is the load limit of an individual item used for lifting, such as a lug or a shackle. 'Rated capacity' is defined as the maximum gross load that may be applied to the crane, hoist or lifting attachment while in a particular working configuration and under a particular condition of use. 'Rated capacity' is more relevant for lifting beams or frames, for which the frame should be marked rather than the lugs.

6.3.4 Dimensional requirements

The following dimensional guidelines should be considered when sizing lifting lugs (although it may not be possible to conform to all guidelines for small lugs) (Figures 6.14 and 6.15).

1. The lifting lug thickness (including cheek plates) should be thicker than 75% of the inside width of the shackle:

$$T + 2T_p \geq 0.75\,W \qquad\qquad \text{(for stability of shackle)}$$

2. The lifting lug thickness (including cheek plates) should be thinner than the inside width of the shackle minus 5 mm:

$$T + 2T_p \leq W - 5\,\text{mm} \qquad\qquad \text{(to prevent binding)}$$

3. The hole diameter should equal the pin diameter plus the maximum of 3 mm or 4% to the nearest millimetre:

$$d_h = \text{MAX}\big[(\phi D + 3\text{mm}),(1.04\phi D)\big] \qquad\qquad \text{(to minimise bearing)}$$

4. The radius of the lug should be less than the dimension from the centre of the hole to the inside of the shackle body, minus the pin diameter, minus 5 mm (clearance):

$$r \leq L + \frac{d_h}{2} - \phi D - 5$$

(to allow a cable or connecting shackle to attach to the primary shackle)
5. After complying with the first four requirements, aim to ensure that a smaller shackle would not physically be able to attach to the lug, therefore preventing incorrect use.

6.3.5 Calculations

The following calculations are recommended for lifting lug design (Table 6.6). Bolted and welded connections should also be checked (refer to Sections 3.10.2 and 3.10.5).

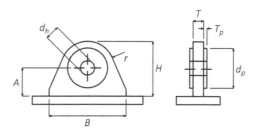

Figure 6.14 Lifting lug dimensions.

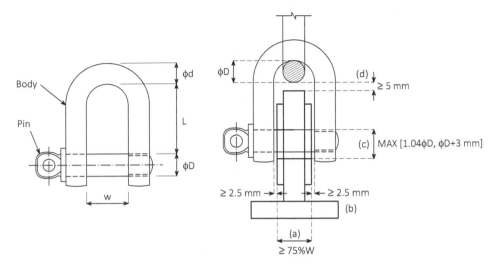

Figure 6.15 Shackle clearance details.

Table 6.6 Design calculations for lifting lugs

Design check	Calculation
Check 1: Shear capacity of a rectangular component (for lateral loads)	Design shear capacity,
	$$\phi V_v = 0.9 \times 0.5\, f_y BT$$

	where:
	  B = Width of lug
	  f_y = Yield strength
	  T = Thickness of lug
	(Refer to AS 4100, Cl 5.11.3)

Check 2: Major axis bending for a rectangular component (lateral loads or eccentricity)	Design major axis bending capacity,
	$$\phi M_{sx} = \frac{0.9 f_y TB^2}{4}$$

	where:
	  B = Width of lug
	  f_y = Yield strength
	  T = Thickness of lug
	(Assuming compact section)
	(Refer to AS 4100, Cl 5.2.1)

Check 3: Minor axis bending for a rectangular component (transverse loads)	Design minor axis bending capacity,
	$$\phi M_y = \frac{0.9 f_y BT^2}{4}$$

	where:
	  B = Width of lug
	  f_y = Yield strength
	  T = Thickness of lug
	(Compact section)
	(Refer to AS 4100, Cl 5.2.1)

(Continued)

Table 6.6 (Continued) Design calculations for lifting lugs

Design check	Calculation
Check 4: Design capacity in axial tension for a rectangular component (magnitude of vertical and lateral loads) 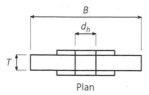 Plan	Design tensile capacity, $$\phi N_t \le 0.9 A_g f_y \le 0.9 \times 0.85 A_n f_u$$ where: Gross area, $A_g = BT$ Net area, $A_n = (B - d_h)T$ d_h = Diameter of hole B = Width of lug f_u = Tensile strength f_y = Yield strength T = Thickness of lug (Assuming $k_t = 1$; refer to AS 4100, Cl 7.2)
Check 5: Design against rupture due to block shear failure for rectangular component (magnitude of vertical and lateral loads) 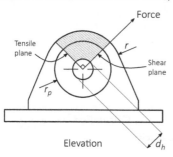 Elevation	Design rupture capacity of lug, $$\phi N_t \le 0.75 \times \left[0.6 f_y A_{gv} + f_u A_{nt} k_{bs} \right]$$ $$\le 0.75 \times \left[0.6 f_u A_{nv} + f_u A_{nt} k_{bs} \right]$$ where: Gross shear area, $A_{gv} = rT$ Net shear area, $A_{nv} = \left(r - \dfrac{d_h}{2} \right) T$ Net tension area, $A_{nt} = \left(r - \dfrac{d_h}{2} \right) T$ f_u = Tensile strength f_y = Yield strength K_{bs} = 0.5 (non-uniform stress), 1 (uniform stress) Additional rupture capacity from cheek plates can be included in areas. (Refer to AS 4100, Cl 9.1.9)
Check 6: Ply in bearing (tear-out and bearing) 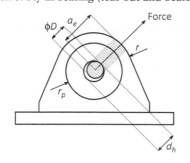	Design capacity of ply in bearing, $$\phi V_b \le 0.9 \times 3.2 (\phi D) T f_u \quad \text{(local bearing)}$$ $$\le 0.9 a_e T f_u \quad \text{(tear-out)}$$ where: (ϕD) = Shackle pin diameter $$a_e = r - \frac{d_h}{2} + \frac{\phi D}{2}$$ (centre of pin to edge of lug) d_h = Diameter of hole T = Thickness of lug (Refer to AS 4100, Cl 9.3.2.4)

(Continued)

Table 6.6 (Continued) Design calculations for lifting lugs

Design check	Calculation
Check 7: Pin connection requirements 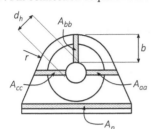	AS 4100 includes three requirements for the design of pin connections, which are relevant to lugs, $T \geq 0.25b$ (for unstiffened elements) $A_{bb} \geq A_n$ $A_{aa} + A_{cc} \geq 1.33\,A_n$ where: $$A_n = \frac{N^*}{0.9\,f_y} \geq \frac{N^*}{0.9 \times 0.85\,f_y}$$ [Minimum area required for tension] $$A_{bb} = \left(r - \frac{d_h}{2} \right) T$$ [top cross-sectional area] $$A_{aa} = A_{cc} \approx \left(r - \frac{d_h}{2} \right) T$$ [side areas] T = Thickness of lug
Von Mises stress criterion: combined stress can be calculated by combining the stress caused by each force at specific locations. Corner edge: 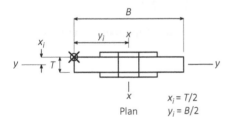 Centre edge: 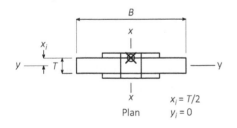	Example: Combined stress at any location on base of lug, $\phi f_y \geq \sigma_{\text{MAX}}$ Vertical stress caused by axial load, $$\sigma_{z1} = \frac{N^*}{BT}$$ Vertical stress caused by major axis bending, $$\sigma_{z2} = \frac{M^*_x\, y_i}{I_{xx}}$$ where: $$I_{xx} = \frac{TB^3}{12}$$ Vertical stress caused by minor axis bending, $$\sigma_{z3} = \frac{M^*_y\, x_i}{I_{yy}}$$ where: $$I_{yy} = \frac{BT^3}{12}$$

(*Continued*)

Table 6.6 (Continued) Design calculations for lifting lugs

Design check	Calculation
	Total vertical stress,
	$\sigma_{zz} = \sigma_{z1} + \sigma_{z2} + \sigma_{z3}$
	Shear stress caused by major shear,
	$$\tau_{yz} = \frac{V^*_y Q_y}{I_{xx} T}$$
	where:
	$$I_{xx} = \frac{TB^3}{12}$$
	$$Q_y = T\left(\frac{B}{2} - y_i\right)\left(\frac{B}{4} - \frac{y_i}{2}\right)$$
	(First moment area for shape above y_i and about x-axis)
	Shear stress caused by minor shear,
	$$\tau_{xz} = \frac{V^*_x Q_x}{I_{yy} T}$$
	where:
	$$I_{yy} = \frac{BT^3}{12}$$
	$$Q_y = B\left(\frac{T}{2} - x_i\right)\left(\frac{T}{4} - \frac{x_i}{2}\right)$$
	(First moment area for shape above x_i and about y-axis)
	Combined stress:
	$$\sigma_{MAX} = \sqrt{\frac{1}{2}\left[2(\sigma_{zz})^2 + 6\left(\tau_{yz}^2 + \tau_{xz}^2\right)\right]}$$

Source: From AS 4100, Section 9.

6.3.6 Lifting lug detailing

The following details are provided as recommended dimensions with calculations based on grade 250 steel and SP welds. As with all standard details, calculations should be produced to confirm the validity of the design for each specific application (Figure 6.16; Table 6.7).

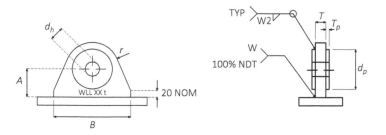

Figure 6.16 Example lifting lug detail.

Table 6.7   Example lifting lug details

WLL	A	B	T	r	d_h	W	T_p	d_p	W2
1	40	70	12	20	12	6	–	–	–
1.5	40	90	12	25	14	6	–	–	–
2	45	100	16	28	17	8	–	–	–
3.2	45	115	20	35	20	10	–	–	–
4.7	55	135	25	40	23	12	–	–	–
6.5	55	150	32	45	26	CP	–	–	–
8.5	60	160	25	50	31	CP	6	80	6
9.5	65	185	25	55	34	CP	8	90	6
12	70	200	25	60	37	CP	10	100	6
13.5	75	220	25	65	40	CP	10	110	6
17	85	230	32	70	43	CP	10	120	6
25	105	280	40	85	54	CP	12	150	8
35	120	295	50	105	60	CP	12	170	8
42.5	130	340	50	115	67	CP	12	190	8
55	150	360	50	130	74	CP	16	220	10

6.4 MACHINE FOUNDATIONS

The function of a machine foundation is to transmit the load of the equipment to the ground without enabling unacceptable vibrations or settlement. A dynamic design may be required to achieve this criterion. There is no Australian Standard for the design of machine foundations. The details presented in this chapter are largely based on the British (CP 2012) [43] and American (ACI 351) [35] Standards, although there are many additional international standards, as detailed in Table 6.8.

6.4.1 Rule of thumb sizing

The American Standard (ACI 351) explains that the rule of thumb methods are often considered sufficient for the final design of block type foundations for small machinery (up to 22 kN in weight and having small unbalanced forces). For reciprocating machinery and movement sensitive machinery, a dynamic design is often necessary, and the rules of thumb are only useful for preliminary sizing. The most commonly used rule of thumb for machine foundation design is to size the foundation based on the mass of the machine. Suggested mass ratios are shown in Table 6.9.

Other commonly used rules of thumb are

1. The resultant vertical and horizontal loads fall within the middle third of the foundation and do not cause uplift.
2. The distribution of live and dead loads on supporting soil or piles is uniform.
3. The minimum foundation width should be 1.5 times the vertical distance from the machine centre line to the bottom of the foundation block.
4. The width and length should be adjusted so that the centre of gravity for the machine coincides with the centre of gravity for the foundation block in plan (typically within 5%).

Table 6.8 Standards of practice related to machine foundation design

Standard	Country	Title
ACI 351.3R-04 2004	American	Foundations for Dynamic Equipment
CP-2012-1 1974	British	Foundations for Machinery: Part 1 – Foundations for reciprocating machines
DIN 4024 Part 1 1988	German	Machine Foundations: Flexible structures that support machines with rotating elements
DIN 4024 Part 2 1991	German	Machine Foundations: Rigid foundations for machinery subject to periodic vibration
IS: 2974 Part I 1998	Indian	Code of Practice for Design and Construction of Machine Foundations: Part I – Foundations for Reciprocating Type Machines
IS: 2974 Part II 1998	Indian	Code of Practice for Design and Construction of Machine Foundations: Part II – Foundations for Impact Type Machines (Hammer Foundations)
ISO 10816/1 1995	International	Mechanical Vibration: Evaluation of Machine Vibration by Measurements on Non-rotating Parts – Part 1: General guidelines
ISO 10816/3 1998	International	Mechanical Vibration: Evaluation of Machine Vibration by Measurements on Non-rotating parts – Part 3: Industrial machines with nominal power above 15 kW and nominal speeds between 120 r/min and 15 000 r/min when measured in situ
ISO 1940/1 2003	International	Balance Quality Requirements of Rigid Rotors

Table 6.9 Rule-of-thumb mass ratios

Foundation type	Machine type	Ratio (foundation mass/machine mass)
Block	Rotating	3
	Reciprocating	5
Piled	Rotating	2.5
	Reciprocating	4

Note: Machine mass is inclusive of moving and stationary parts.

5. The thickness of the foundation block should be the maximum of 20% of the width and 10% of the length, and a minimum of 600 mm. Alternatively, the thickness can be sized as length/30 + 600 mm.
6. A maximum of 50% of bearing pressure under static loading should be used.
7. The foundation should be isolated from surrounding structures.

6.4.2 Natural frequency analysis

The natural frequency of a structure is the frequency at which it oscillates when not subject to external forces. Structures have an infinite number of natural frequencies in different directions and different modes. A nominal beam supported at both ends has a first mode natural frequency in the direction of movement shown by a single arch. The next mode in this direction would be a double arch and is at a higher frequency. The same principles apply to movement in different directions and also to rotations. The first six modes to consider for block type foundations are the three translational modes and three rotational modes. For thick foundations which are short in length and with relatively low operating frequencies, the assumption is often made that the block behaves in a dynamically rigid manner. This assumption means that only these six modes are considered in the design. The assumption can be checked by modelling the foundation in a finite element analysis (FEA) package such as Strand7.

Natural frequency is proportional to stiffness and inversely proportional to mass, as shown in the following formula:

$$f_n = \frac{1}{2\pi}\sqrt{\frac{k}{m}}$$

Therefore, the frequency can be increased by adding stiffness or decreased by adding mass. This process is called *tuning*.

Some vendors only require a natural frequency analysis to be completed for the dynamic design of the foundation. This is often the case for equipment of a low importance or with low dynamic forces. During completion of the natural frequency analysis, a separation ratio is specified to prevent resonance. This ratio is the operating frequency of the machine (f_m) divided by the natural frequency being considered (f_n). The natural frequencies under consideration should only include those which are excited by the direction of the applied force. The American standard recommends that this ratio is provided by the vendor; however, typical values are provided. The British code also provides ratios; however, the values are much more conservative (possibly due to the age of the code) (Table 6.10).

Table 6.10 Natural frequency separation ratios

Standard	Separation ratio
CP 2012 (For general foundations)	$\frac{f_m}{f_n} < 0.5$ or $\frac{f_m}{f_n} > 2.0$
CP 2012 (For low importance foundations)	$\frac{f_m}{f_n} < 0.6$ or $\frac{f_m}{f_n} > 1.5$
ACI 351 (Based on vendor data with typical values as shown)	$\frac{f_m}{f_n} < 0.5$ or 0.7 or 0.8 or 0.9
	$\frac{f_m}{f_n} > 1.5$ or 1.3 or 1.2 or 1.1

Mass participation can also be extracted from the *Natural Frequency Analysis*. This is a representation of the percentage of the mass involved in each specific analysis. For example, a natural frequency with 97% mass participation in the z-direction will be excited by loading in the z-direction. However, if the load is applied in the x-direction, this frequency is not of significant importance. This factor is useful in determining whether an adequate number of frequencies have been analysed. If the first 10 natural frequencies have been analysed, and the sum of the mass participations in the y-direction is 5%, the primary mode in the y-direction has not yet been found. This output can be found in Strand7 by using the text report function when the option is selected prior to solving the model.

6.4.3 Harmonic response analysis

A harmonic response analysis uses a load applied at an operating frequency to excite the foundation once the natural frequency analysis is completed. If the applied load is in a similar direction and frequency to the natural frequency, resonance will occur. The result of the analysis is generally the node velocities or displacements, usually at anchor points and the centre of the compressor. Velocities and displacements are related using the following formula:

$$V = 2\pi f \times A$$

where:

A = Amplitude (mm) (peak amplitude, measured from the axis to the peak)
f = Frequency (H_Z)
V = Velocity (mm/s)

If the foundation is assumed to act as a rigid block, only the six primary modes of excitation (DX, DY, DZ, RX, RY and RZ) are relevant. Specialist programs such as Dyna6 are able to calculate the natural frequencies of these modes and also perform a harmonic response analysis. The limitation of programs such as Dyna6 is that they do not include the higher-order modes. Experience has shown that when complying with rules of thumb, these modes can occur at frequencies above 30 Hz; however, this is highly dependent on the geometry of the foundation.

Where higher-order frequencies are to be considered, the analysis of the foundation should be completed in an FEA program such as Strand7. The spring stiffness values for the foundation should be extracted from the preliminary analysis in Dyna6. These spring stiffness values are dependent on excitation frequency. Therefore, different freedom cases are to be completed for analysis at each frequency considered.

6.4.3.1 Damping

Damping is the energy dissipation of an oscillating system. It effectively reduces the magnitude of resonance in a system which is caused by the external force.

Damping can be considered when completing the harmonic response analysis; however, it should be considered with extreme care, as it may mask the results of the analysis. It is only relevant where resonance is occurring, and reduces the amplitude and velocity, as shown in Figure 6.17. Dyna6 outputs damping factors for each mode of excitation; however, they

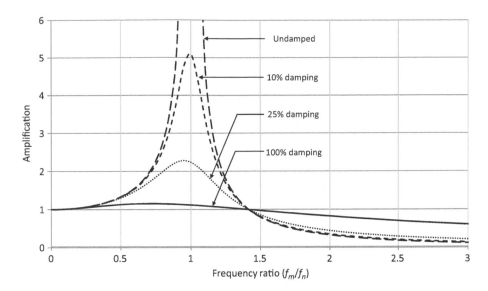

Figure 6.17 Force transmissibility. (Authorised reprint from ACI 351.3R:2004, Foundations for Dynamic Equipment. Reproduced with permission from the American Concrete Institute.)

would need to be applied separately for each natural frequency mode. For a conservative value, the German standard (DIN 4024-1) recommends a damping factor of 2% for the entire system, although it is possible to use values as high as 10% or 25%, depending on the damping mode.

6.4.4 Dynamic load

Dynamic loads should be provided by the vendor for movement sensitive equipment. They are usually provided as loads at anchor points. The loads should be applied in each axis as separate load cases and should be solved with a phase shift provided by the vendor or calculated to give the worst result. If the load is caused by a single rotating mass, the maximum vertical load is concurrent with a zero horizontal load; therefore, the phase shift is 90°. This also means that the maximum horizontal load is concurrent with a zero vertical load.

If the loads are not provided by the vendor, and an estimate is required, the following procedure can be used for rotating equipment, using the machine classification:

1. Obtain machine classification from ISO-1940; refer to Table 6.11.
2. Calculate maximum allowable eccentricity of mass:

    Eccentricity, $e = \dfrac{G}{\omega}$

    G = Balance quality grade (rad.mm/s)
    ω = Angular frequency (rad/s)
3. Calculate out-of-balance force:
    Force, $F = m \times e \times \omega^2$
    m = total mass of rotating parts (kg)

Table 6.11 Machine classifications

Balance quality grade	Product of the relationship $(e_{per}x \omega)^{(1)(2)}$ mm/s	Rotor types: General examples
G 4000	4000	Crankshaft drives for large slow marine diesel engines (piston speed below 9 m/s), inherently unbalanced
G 1600	1600	Crankshaft drives for large slow marine diesel engines (piston speed below 9 m/s), inherently balanced
G 630	630	Crankshaft drives, inherently unbalanced, elastically mounted
G 250	250	Crankshaft drives, inherently unbalanced, rigidly mounted
G 100	100	Complete reciprocating engines for cars, trucks and locomotives
G 40	40	Cars: wheels, wheel rims, wheel sets, drive shafts Crankshaft drives, inherently balanced, elastically mounted
G 16	16	Agricultural machinery Crankshaft drives, inherently balanced, rigidly mounted Crushing machines Drive shafts (cardan shafts, propeller shafts)
G 6.3	6.3	Aircraft gas turbines Centrifuges (separators, decanters) Electric motors and generators (of at least 80 mm shaft height), of maximum rated speeds up to 950 r/min Electric motors of shaft heights smaller than 80 mm Fans Gears Machinery, general Machine-tools Paper machines Process plant machines Pumps Turbo-chargers Water turbines
G 2.5	2.5	Compressors Computer drives Electric motors and generators (of at least 80 mm shaft height), of maximum rated speeds above 950 r/min Gas turbines and steam turbines Machine-tool drives Textile machines
G 1	1	Audio and video drives Grinding machine drives
G 0.4	0.4	Gyroscopes Spindles and drives of high-precision systems

Source: ISO-1940, table 1. Copied by Mr L. Pack with the permission of Standards Australia under Licence 1607-c010.

Notes:

1. Typically completely assembled rotors are classified here. Depending on the particular application, the next higher or lower grade may be used instead. For components, see Clause 9 (ISO 1940/1).
2. All items are rotating if not otherwise mentioned (reciprocating) or self-evident (e.g. crankshaft drives).
3. For limitations due to set-up conditions (balancing machine, tooling), see Notes 4 and 5 in 5.2 (ISO 1940/1).
4. For some additional information on the chosen balance quality grade, see ISO 1940/1, figure 2. It contains generally used areas (service speed and balance quality grade G) based on common experience.
5. Crankshaft drives may include crankshaft, flywheel, clutch, vibration damper and rotating portion of connecting rod. Inherently unbalanced crankshaft drives theoretically cannot be balanced; inherently balanced crankshaft drives theoretically can be balanced.
6. For some machines, specific International Standards stating balance tolerances may exist (see ISO 1940/1 Bibliography).
7. $\omega = 2\pi n/60 \approx n/10$, for n (revolutions per minute) and ω (radians per second).

6.4.5 Acceptance criteria

Acceptance of the harmonic response analysis is achieved by extracting node velocities (or displacements) from the model and comparing them with the vendor specified limits. The equations of simple harmonic motion can also be used to convert from node amplitude to velocity and acceleration. In the absence of vendor limitations, ACI-351 provides guidelines on the severity of calculated values (refer to Figure 6.18). The vertical axis represents the peak-to-peak displacement amplitude of the foundation. The horizontal axis is the vibration frequency in revolutions per minute. The acceptance criterion is typically a velocity and therefore runs diagonally through the graph. Past experience has shown that larger machines and machines with higher operating frequencies have more stringent limits. Centrifugal compressors generally require categories between 'extremely smooth' and 'very good'; screw compressors are often 'good' or 'fair'; and reciprocating compressors may be even higher.

6.4.6 General design requirements

Additional requirements include those to aid construction and those which help achieve the assumptions made during the dynamic design. Compressors and the associated piping are generally movement sensitive, and therefore, settlement and seasonal movement of the soil need to be considered in the design of the foundation. This often leads to the choice of piled foundations, unless the structure is founding on rock. Where reactive soils are encountered,

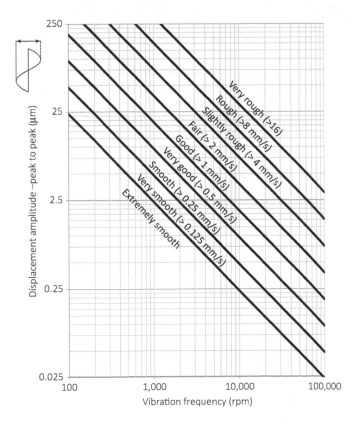

Figure 6.18 General machine vibration severity chart. (From ACI-351, figure 3.10. Authorised reprint from ACI 351.3R:2004. Reproduced with permission from the American Concrete Institute.)

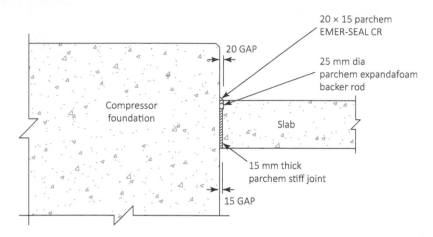

Figure 6.19 Detail of isolation joint.

the structure should be designed in accordance with Section 5.5. Noise attenuation may also be required depending on client requirements for noise limitations.

The foundation should be separated from other structures as much as possible. An isolation joint is recommended around the perimeter of the slab (refer to Figure 6.19). This joint prevents additional stiffness from being provided to the slab and changing the frequency, and also prevents vibrations from travelling to the adjacent foundations. Platforms should ideally found on the adjacent slabs and butt against the skid platforms with a nominal 20 mm gap.

Grading of the slab is generally required to facilitate drainage and bunding (refer to Section 6.8 for bunding requirements). The slope should generally be away from the skid and is typically at 1% fall. Grout block-outs may be detailed on the grouting plan if oil can be spilled within the skid area and needs to drain to another location. Grout block-outs are generally a 50 mm gap in the grout, which is located at low points, depending on the grade of the slab (formed by using polystyrene strips under the skid beam prior to grouting).

Epoxy grout should be specified under skids for significant structures, as cementitious grout can become damaged after long periods of cyclical loading. Conbextra EP120 (refer to Chapter 8) is recommended, as it is flowable and can be used for thicknesses ranging from 10 to 120 mm.

A low-heat concrete mix should be specified where the foundation is thicker than 1000 mm to prevent cracking. For thicker foundations, two pours should be specified, and a methodology should be written to prevent shrinkage cracking or other damage from the high temperatures reached during curing.

The first pipe supports adjacent to the compressor should have a higher stiffness than general pipe supports to provide a gradient from the extremely stiff compressor foundation to the more flexible pipe racks. Allowable displacements or stiffness should be provided by the pipe stress engineers. Highly sensitive machines may require deflections as low as 1 mm, and less sensitive equipment may allow as much as 10 mm.

6.4.6.1 Construction requirements

Skids are often supplied with many hold-down bolts and specified jacking locations, which may require cast-in plates. The complexity of these locations usually results in construction tolerances being higher than allowed for the skid. Therefore, the use of polystyrene

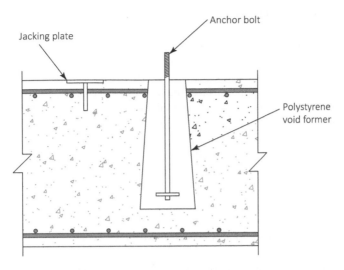

Jacking plate

Anchor bolt

Polystyrene
void former

Figure 6.20 Detail of polystyrene pocket and jacking plate.

pockets is recommended, so that bolts can be installed after the foundation is cast (refer to Figure 6.20). The pockets should be a trapezoidal shape to help prevent slippage at the bonding surface. The polystyrene should be removed mechanically after the concrete is cured, and a non-shrink grout should be used to fill the pockets once the bolts are installed in the correct locations. This can be done either by using the skid or by using a template for the base of the skid. Reinforcement bar spacing should be at least as wide as the top of the pocket to allow the pockets to fit and be fastened to the bars. Cover at the edge of the foundation should also be checked to ensure the trapezoid does not encroach on the location of the side bars, as hold-down bolts are usually at the edges of the skid.

6.4.7 Design methodology

The general design methodology for a machine foundation should use the following steps:

1. Calculate a preliminary size of the foundation using rule-of-thumb methods.
2. Calculate dynamic loads.
   For Dyna6, the loads will have to be resolved about a single point; therefore, a load which is offset from the centre of gravity will become a load and a moment. The FEA package will allow loads to be applied in exact locations.
3. Create and solve a Dyna6 model of the foundation, soil and masses.
   The required input parameters for soil layers are shear wave velocity, density, Poisson's ratio and damping (assume 2%). If the foundation behaves as a block, the Dyna6 model can be used to examine velocities.
4. Extract support stiffness values from the Dyna6 solution file for operating frequency (±50%).
   These stiffness values are frequency dependent. For piled foundations, it will provide springs at each pile location for each direction.
5. Create a model in Strand7 (or other FEA package). Ensure that non-structural masses are modelled and connected to the foundation using multi-point links. The spring stiffness values should be used from the operating frequency analysis.

6. Run the natural frequency analysis, and determine which frequencies have high mass participation.
7. If there are frequencies which are close to the operating range that have high mass participation in a direction which may be excited, a harmonic response analysis is required. Apply the dynamic loads, with loads in each direction in separate load cases, and then run the analysis with the appropriate phase shift. Extract maximum node velocities.
8. Repeat Steps 5 through to 7 with a range of frequencies to understand how the foundation will react at, say, 90% of operating speed, or 80%.
9. Compare the extracted node velocities with the American Concrete Institute (ACI) criteria and the allowable limits provided by the vendor. If limits are exceeded, the structure needs to be changed (vary thickness, plan dimensions, pile size or location, etc.).
10. Check ULS and serviceability limit state (SLS) criteria for foundation.

6.5 ACCESS (STAIRS, LADDERS AND HANDRAILS)

Access requirements are based on legislation. Where access is open to the public, an act or regulation often requires compliance with the Building Code of Australia (or National Construction Code). However, in the absence of this requirement, the Australian Standard for 'Fixed platforms, walkways, stairways and ladders – Design, construction and installation' (AS 1657) is usually specified. The standard provides very detailed requirements; therefore, this section only provides an outline of the important details which need to be considered during structural design.

The limits for each means for access are summarised in Figure 6.21, and minimum general requirements are summarised in Table 6.12. Guardrails should be provided where access is categorised as a walkway or stairway (refer to Chapter 8) and a fall of 300 mm is possible. Kick plates are required for walkways and not for stairways.

6.5.1 Walkways

Walking surfaces should be slip resistant, and therefore Webforge C325MPG grating is recommended (refer to Chapter 8). Refer to Table 6.12 for additional requirements. The fundamental natural frequency of pedestrian crossovers (bridges) can be checked to ensure resonance does not occur from pedestrian loading. AS 5100.2, Cl 12.4 recommends that vertical frequencies between 1.5 and 3.5 Hz and horizontal frequencies of less than 1.5 Hz should be avoided unless further investigated.

6.5.2 Stairs

The requirements for stairways can be avoided for changes in elevation where the access between levels is provided by a single drop of up to 300 mm, or a drop of 300–450 mm with only one intermediate step.

Stairway flights are limited to a maximum of 18 risers, followed by a 900 mm landing. Where more than 36 risers occur, another means of fall prevention is required, such as a barrier, a 2 m landing or a change in direction of at least 90°. Refer to Figure 6.23 for typical details.

Risers and goings should be of uniform dimensions and within a tolerance of ±5 mm. The angle should be between 20° and 45°, complying with the formula

$$540 \leq (2R + G) \leq 700$$

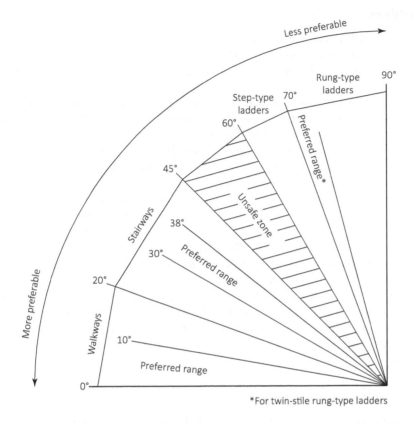

Figure 6.21 Selection of access: Limits of scope. (From AS 1657, figure 2.1. Copied by Mr L. Pack with the permission of Standards Australia under Licence 1607-c010.)

where:

R   = Riser (130 mm to 225 mm) = Vertical spacing of stair treads
G   = (215 mm to 355 mm) = Horizontal spacing of stair treads

Treads should be slip resistant, and the nosing should be clearly visible; therefore, Webforge Type T5 (welded) and Type T6 (bolted) are recommended (refer to Chapter 8).

Care should be taken to ensure that differential settlement is limited. For example, where a stairway is attached to a large structure and lands on a shallow pad, settlement and reactive soil movement (refer to Section 5.5) should be considered. Refer to Table 6.12 for other requirements.

Table 6.12 General access requirements

Height	2000 mm
Width	600 mm minimum (900 mm typical)
Landing length	900 mm
Handrail height	900–1100 mm
Load	2.5 kPa pressure, 1.1 kN point load
Deflection	Minimum of L/100 or 40 mm

6.5.3 Ladders

Ladders are preferably inclined between 70° and 75°; however, vertical ladders are sometimes used for vessels and equipment. Either step-through or side-access ladders are acceptable. Twin-stile rung-type ladders are the most commonly used ladder system; refer to AS 1657, Clause 7.4.

Landings are required to be provided at a maximum vertical spacing of 6 m, and each ladder should be staggered or in a different direction to prevent long falls. The landing should be a minimum of 900 mm long and 600 mm wide.

The clear width between stiles is required to be between 375 and 525 mm, widening to between 525 and 675 mm (for step-through ladders) from the landing and extending for a height of 1000 mm. The individual stiles can be any shape that is larger than a 40 mm diameter and smaller than an 80 mm diameter. Rungs shall be slip resistant, between 20 and 50 mm diameter, and spaced between 250 and 300 mm. Rungs shall be uniformly spaced within a tolerance of ±5 mm. The bottom space between the bottom rung and the landing can be between 90% and 100% of the general rung spacing. Ladders should be designed for a concentrated load of 1.5 kN for each 3 m of vertical height on the same flight. Typical clearances are shown in Figure 6.22.

A ladder cage is required where a fall of more than 6 m is possible, irrespective of landings. Cages should start between 2000 and 2200 mm above the base, and extend to the elevation of the handrail or 1000 mm above the platform. The cage banding shall be reinforced with cross bars spaced at maximum centres of 2000 mm. Refer to Figure 6.24 for typical details.

6.5.3.1 Stair and ladder detailing

The details in Figures 6.23 and 6.24 are provided as recommended dimensions. As with all standard details, calculations should be produced to confirm the validity of the design for each specific application.

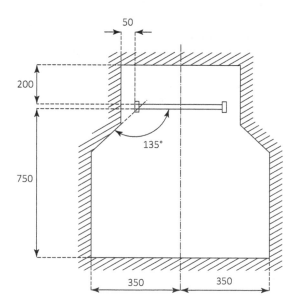

Figure 6.22 Typical minimum clearances for rung-type ladders. (From AS 1657, figure 7.6. Copied by Mr L. Pack with the permission of Standards Australia under Licence 1607-c010.)

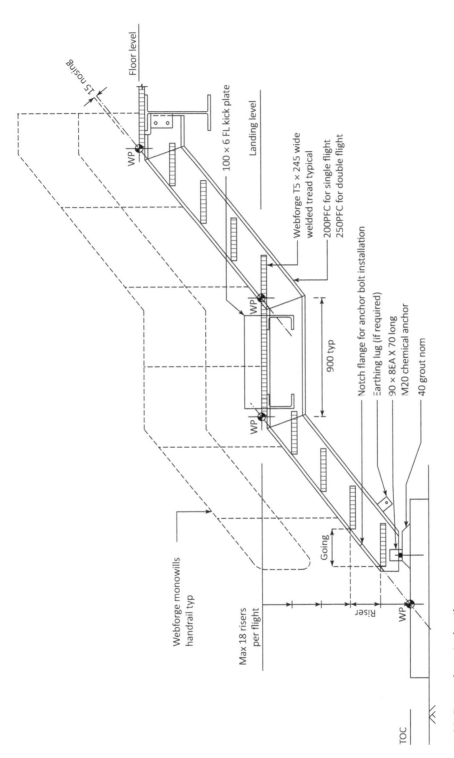

15 nosing

Floor level

WP

100 × 6 FL kick plate

Landing level

Webforge TS × 245 wide
welded tread typical

200PFC for single flight
250PFC for double flight

WP

900 typ

Notch flange for anchor bolt installation

Earthing lug (if required)

90 × 8EA X 70 long

M20 chemical anchor

40 grout nom

WP

Going

Riser

WP

Webforge monowills
handrail typ

Max 18 risers
per flight

TOC

Figure 6.23 Example stair detail.

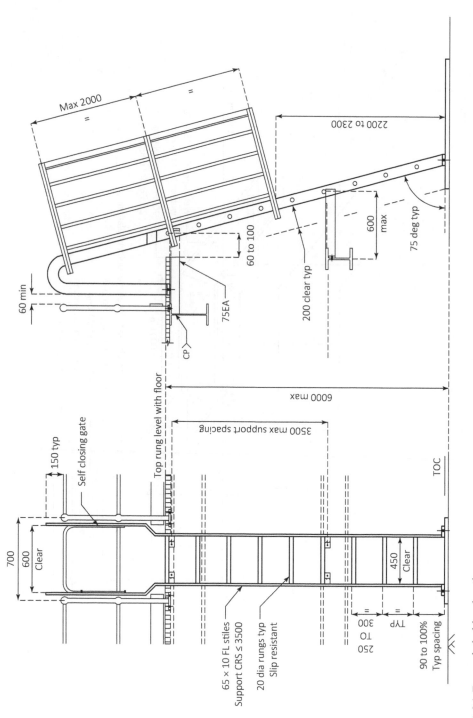

Figure 6.24 Example ladder detail.

6.6 TEMPERATURE VARIATION

Steel grades should be selected based on the minimum service temperature. Large structures should also be assessed for a temperature variation in conjunction with design loads. Of course, in the event of external temperature sources, the design range should be modified appropriately. For temperatures over 215°C, yield strength decreases with temperature. Young's modulus is also affected by temperature, as detailed in this chapter. The following details are a suggested method of calculating reasonable temperatures on a steel structure.

6.6.1 Minimum temperature (AS 4100)

AS 4100 defines the 'lowest one day mean ambient temperature' (LODMAT) as the minimum design temperature for brittle fracture of structural steel. This is a location-based temperature (Figure 6.25), and the supplement explains that the contour map outlines the extremities of the zones, such that a location between the 0° and 5° contours results in a 5° classification. This should be lowered by 5° where the structure may be subjected to low local ambient temperatures. Temperature values should be obtained from the Bureau of Meteorology for critical structures.

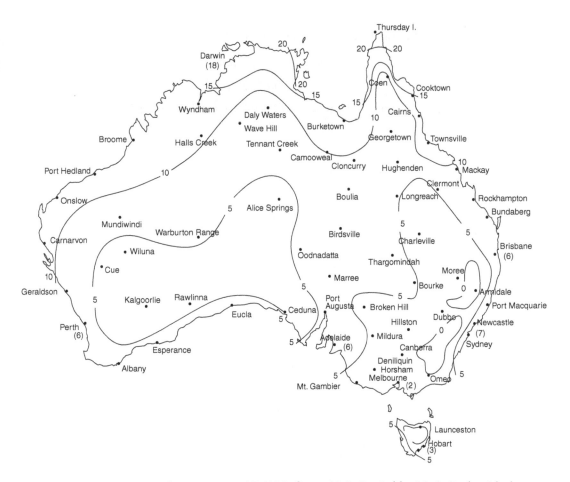

Figure 6.25 LODMAT isotherms. (From AS 4100, figure 10.3. Copied by Mr L. Pack with the permission of Standards Australia under Licence 1607-c010.)

6.6.2 Steel grade selection

Steel grade should be selected on the basis of the minimum service temperature to prevent brittle fracture. AS 4100, Section 10, details the permissible service temperature for each grade available (Tables 6.13 and 6.14). The permissible service temperature is generally based on a correlation from a Charpy test of the material and is dependent on plate thickness. However, some materials are accepted for certain temperatures without any testing based on statistical data from notch toughness characteristics of steel from mills in Australia and New Zealand. The standard prohibits the use of steels for which the permissible service temperature is unknown (unless significant testing is completed), and therefore, care should be taken when using imported steel without Charpy testing.

6.6.3 Temperature range (bridges)

A reasonable reference for maximum temperatures of structures can be found in AS 5100.2 (Bridge Design: Part 2 – Loads). The standard details extreme shade air temperatures and the associated average bridge temperatures (Tables 6.15 and 6.16).

For structures such as steel girder bridges with steel decks, the minimum should be decreased by 10°C, and the maximum should be increased by 20°C.

Table 6.13 Permissible service temperatures according to steel type and thickness

Steel type	Permissible service temperature (°C) (see Note 1)					
	Thickness (mm)					
	≤6	>6 ≤12	>12 ≤20	>20 ≤32	>32 ≤70	>70
1	−20	−10	0	0	0	5
2	−30	−20	−10	−10	0	0
2S	0	0	0	0	0	0
3	−40	−30	−20	−15	−15	−10
4	−10	0	0	0	0	5
5	−30	−20	−10	0	0	0
5S	0	0	0	0	0	0
6	−40	−30	−20	−15	−15	−10
7A	−10	0	0	0	0	—
7B	−30	−20	−10	0	0	—
7C	−40	−30	−20	−15	−15	—
8C	−40	−30	—	—	—	—
8Q	−20	−20	−20	−20	−20	−20
9Q	−20	−20	−20	−20	−20	−20
10Q	−20	−20	−20	−20	−20	−20

Source: AS 4100, table 10.4.1. Copied by Mr L. Pack with the permission of Standards Australia under Licence 1607-c010.

Notes:
1. The permissible service temperature for steels with a L20, L40, L50, Y20 or Y40 designation shall be the colder of the temperature shown in table 10.4.1 and the specified impact test temperature.
2. This table is based on available statistical data on notch toughness characteristics of steels currently made in Australia or New Zealand. Care should be taken in applying this table to imported steels, as verification tests may be required. For a further explanation, see WTIA Technical Note 11.
3. (—) indicates that material is not available in these thicknesses.

Table 6.14 Steel type relationship to steel grade

Steel type (see note)	Specification and grade of parent steel				
	AS/NZS 1163	AS/NZS 1594	AS/NZS 3678 AS/NZS 3679.2	AS/NZS 3679.1	AS 3597
1	C250	HA1 HA3 HA4N HA200 HA250 HA250/1 HU250 HA300 HA300/1 HU300 HU300/1	200 250 300	300	–
2	C250L0	–	–	300L0	–
2S	–	–	250S0 300S0	300S0	–
3	–	XF300	250L15 250L20 250Y20 250L40 250Y40 300L15 300L20 300Y20 300L40 300Y40	300L15	–
4	C350	HA350 HA400 HW350	350 WR350 400	350	–
5	C350L0	–	WR250L0	350L0	–
5S	–	–	350S0	350S0	–
6	–	XF400	350L15 350L20 350Y20 350L40 350Y40 400L15 400L20 400Y20 400L40 400Y40	–	–
7A	C450	–	450	–	–
7B	C450L0	–	–	–	–
7C	–	–	450L15 450L20 450Y20 450L40 450Y40	–	–
8C	–	XF500	–	–	–
8Q	–	–	–	–	500
9Q	–	–	–	–	600
10Q	–	–	–	–	700

Source: AS 4100, table 10.4.4. Copied by Mr L. Pack with the permission of Standards Australia under Licence 1607-c010.

Note: Steel types 8Q, 9Q and 10Q are quenched and tempered steels currently designated as steel types 8, 9 and 10, respectively, in AS/NZS 1554.4.

Table 6.15 Shade air temperatures

Location	Height above sea level (m)	Shade air temperature (°C)					
		Region I North of 22.5 °S		Region II South of 22.5 °S		Region III Tasmania	
		Max.	Min.	Max.	Min.	Max.	Min.
Inland	≤1000	46	0	45	−5	37	−5
	>1000	36	−5	36	−10	32	−10
Coastal	≤1000	44	4	44	−1	35	−1
	>1000	34	−1	34	−6	30	−6

Source: AS 5100.2, table 17.2(1). Copied by Mr L. Pack with the permission of Standards Australia under Licence 1607-c010.

Note: Coastal locations are locations that are less than 20 km from the coast.

Table 6.16 Average bridge temperatures

Min.	
Shade air temp (°C)	Average bridge temp (°C)
−8	2
−2	4
4	8
10	12
Max.	
50	54
46	50
42	46
38	43
34	40
30	37

Source: AS 5100.2, table 17.2(2). Copied by Mr L. Pack with the permission of Standards Australia under Licence 1607-c010.

Note: Linear interpolation of intermediate values is permitted.

6.6.4 Installation temperature and design range

For structures where the temperature range is considered, it is important to specify the installation temperature. This is often the case for structures such as PAUs, where large welded structures are site welded to base plates.

For example,

1. A minimum temperature of 5°C is adopted.
2. A maximum temperature of 50°C is adopted.
3. An installation range of 20°C–30°C is specified.

Therefore, the cold case would be the structure being installed at 30°C and decreasing to 5°C, resulting in a differential of −25°C. The hot case would be for the structure being installed at 20°C and increasing to 50°C, resulting in a differential of +30°C.

Reactions can then be calculated using programs such as Strand7 or Space Gass. Alternatively, a relatively simple calculation can be adopted as follows (refer to Example 6.1):

1. Calculate expansive/contractive length:

$$\Delta L = \alpha_T \, L \, \Delta T$$

2. Calculate stress required to restrain movement (i.e. stress required to compress a length of steel by the same amount as the expansion/contraction):

$$E_s = \sigma / \varepsilon$$

The strain is equal to the expansive length divided by the total length,

$$\varepsilon = \Delta L / L$$

Stress can therefore be calculated,

$$\sigma = E_s \Delta L / L = E_s \alpha_T \Delta T$$

3. Calculate force required to restrain movement:

$$F_f = \sigma A = E_s \, \alpha_T \, \Delta T \, A$$

where:
  A = Cross-sectional area of member
  E_s = 20000 MPa (at 20°C)
  L = Initial length (at 20°C)
  α_T = Coefficient of thermal expansion for steel = 11.7×10^{-6} per °C (at 20°C)
  ΔT = Change in temperature

This method of calculating results computes the force required to entirely prevent movement. It is generally a conservative value. Stiffness of the foundations may be modelled to represent the expansion of the restraints and therefore reduce the expansive force.

AS/NZS 1170.0, Appendix A, recommends a special study for structures subjected to temperature changes and gradients, including appropriate load factors. The 'Process Industry Practices – Structural Design Criteria' (PIP STC01015) [47] recommends that the same factor be used for thermal loading as adopted for self-weight and that it be included in combinations with operating loads. It is therefore reasonable to adopt a combination of 1.35 G + 1.35 F_f for max vertical and 1.2 G + 1.5 Q + 1.2 F_f for operating cases.

6.6.5 Change in properties with high temperatures

AS 4100, Section 12 details the change in mechanical properties of steel for the purposes of fire design. Yield strength and Young's modulus both decrease as temperature increases.

The yield strength of steel can be calculated at different temperatures:

$$f_y(T) = 1.0 \times f_y(20) \quad \text{when } 0°C < T \leq 215°C; \text{ and}$$

$$f_y(T) = f_y(20)\left[\frac{905 - T}{690}\right] \quad \text{when } 215°C < T \leq 905°C$$

where:

$f_y(T)$ = Yield stress of steel at $T°C$
$f_y(20)$ = Yield stress of steel at $20°C$
T = Temperature of the steel in $°C$

The Young's modulus changes due to temperature:

$$E(T) = E(20)\left[1.0 + \left[\frac{T}{2000\left[ln\left(\frac{T}{1100}\right)\right]}\right]\right] \quad \text{when } 0°C < T \leq 600°C$$

$$E(T) = E(20)\frac{690\left(1 - \frac{T}{1000}\right)}{T - 53.5} \quad \text{when } 600°C < T \leq 1000°C$$

where:

$E(T)$ = Modules of elasticity of steel at $T°C$
$E(20)$ = Modules of elasticity of steel at $20°C$

The limiting steel temperature for beams and columns subject to fire loading is

$$T_l = 905 - 690\, r_f$$

where:

T_l = Limiting steel temperature
r_f = Radio of design action to design capacity of member

For specific details on the application of this formula, refer to AS 4100, Section 12.

6.7 COMPOSITE BEAMS AND SLABS

Composite structures use steel beams to add strength and stiffness to reinforced concrete slabs. The design of composite beams and slabs should be completed in accordance with AS 2327.1 or AS 5100.6. The information presented in this chapter represents a combination of the two standards, as each one has particular strengths and pitfalls. AS 2327.1 is intended for simply supported beams, and AS 5100.6 is intended for bridge design. Each of the standards contains significantly more detail than that presented in this section, especially in relation to practical and detailing requirements as well as various construction stages. The following summary focuses only on the technical aspects required to complete basic design calculations for the final installed capacity of simply supported composite beams (Figure 6.26).

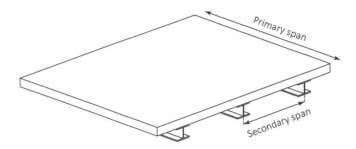

Figure 6.26 Composite structure.

Composite design is based on a concrete slab having sufficient capacity to span in the secondary direction (between steel beams) and the combined steel/concrete section spanning in the primary direction. Shear studs need to be checked to ensure composite action can be transferred in shear flow between the steel and concrete sections. Primary shear is checked for the web of the steel sections (refer to Section 3.5). Bending and shear (especially punching shear) for the slab should also be checked (refer to Section 4.3). Steel beam connections should be designed for the entire shear force. The concrete strength is limited to 40 MPa, because higher strengths do not have the required ductility to enable effective transfer of longitudinal shear forces through the studs.

6.7.1 Bending design

The bending capacity of the composite section is calculated using a plastic analysis of the effective cross-section (Figure 6.27). Only a portion of the concrete slab is assumed to act together with the steel beam in bending. Steel sections which are 'compact' or 'non-compact' should be used to support the slab. Steel sections with a 'slender' classification are not permitted for composite design.

$$M^* \le \phi M_b$$

$\phi = 0.9$ (after concrete reaches design strength, refer to AS 2327.1 for information on prior stages)

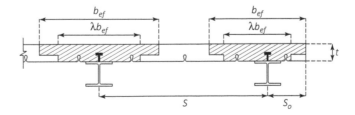

Figure 6.27 Effective composite cross-section.

Effective width of concrete slab, $\quad b_{ef} = MIN\big[(L/5),(S),(S/2+S_o),(12t)\big]$

(including both AS 2327.1 and AS 5100.6 requirements)

where:

L	= Length of primary span
S	= Steel beam spacing
S_o	= Edge overhang dimension
t	= Slab thickness

The strength limit state analysis is completed using a plastic analysis of the cross-section. Member effects are not considered for sagging moments, because the concrete slab restrains the compressive flange from lateral torsional buckling. The plastic neutral axis is calculated first by calculating the tensile capacity of the steel section and comparing it with the compressive capacity of the slab. Calculations are presented with the option for permanent steel sheeting formwork, which is used if the wet concrete cannot be supported by falsework below. To facilitate the transparency of the original theory, no simplifications have been included.

Note that the section calculations provided in the appendix for AS 2327.1 contain errors, and the calculations provided in AS 5100.6 do not allow for steel sheeting. AS 2327.1 also requires the bending contribution from the web to be reduced in areas where the shear load is greater than 50% of the design shear capacity of the web. This reduction should be linear from full bending capacity, where 50% of the design shear capacity is used, down to a bending capacity which completely excludes the web for locations where the design shear capacity is fully used.

Max Tensile capacity of steel, $\qquad F_{st} = Af_y = 2b_f t_f f_{yf} + d_1 t_w f_{yw}$

Max Compression capacity of concrete slab, $\qquad F_{c1} = b_{ef}(t-h_r)0.85f_c'$ (above sheeting)

Max Compression capacity of concrete slab, $\qquad F_{c2} = \lambda b_{ef}\, h_r\, 0.85f_c'$ (inside sheeting)

Max Compression capacity of top flange, $\qquad F_{scf} = b_f t_f f_{yf}$

where:

A	= Area of steel
b_f	= Width of steel flange
d_1	= Depth of steel web
f_y	= Yield strength of steel
f_{yf}	= Yield strength of steel flange
f_{yw}	= Yield of steel web
f_c'	= Compressive strength of concrete slab
h_r	= Height of sheeting ribs
t_f	= Thickness of steel flange
t_w	= Thickness of steel web

The concrete between sheeting ribs is factored to account for a loss of strength due to the size and angle of the sheeting ribs. The concrete is completely excluded for ribs which are

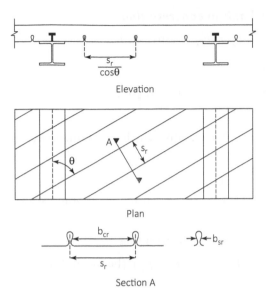

Figure 6.28 Sheeting factor. (From AS 2327.1, figure 5.2.2.2. Copied by Mr L. Pack with the permission of Standards Australia under Licence 1607-c010.)

perpendicular to the supporting beams. Refer to AS 2327.1 for detailed sheeting dimension requirements (Figure 6.28).

$$\text{Sheeting Factor,} \quad \lambda = 1 \qquad (\text{for} \quad 0 < \theta \leq 15°)$$

$$= \left[b_{cr} \left(\cos\theta \right)^2 \right] / s_r \quad (\text{for} \quad 15° < \theta \leq 60°)$$

$$= 0 \qquad (\text{for} \quad \theta > 60°)$$

Width of concrete between mid-height of adjacent steel ribs:

$b_{cr} = s_r - b_{sr}$
b_{sr} = Width of sheeting ribs at mid height
S_r = Spacing between sheeting ribs
θ = Angle between sheeting ribs and beam longitudinal axis
   = 90° (for sheeting spanning directly transverse to beams)

Four cases are checked for possible locations of the neutral axis. Sections with thick slabs and comparatively shallow sections will have the neutral axis in the slab, whereas sections with thin slabs and deep sections will have the neutral axis in the web. Slenderness effects should be considered for any components of the non-compact steel cross-sections which are in compression (refer to Section 3.4.1.3). Compressive flanges are reduced to an effective width which would be considered compact. The compressive section of a web may be reduced in effective thickness to achieve a compact classification; alternatively, a segment with a depth of $30t_w\sqrt{250/f_y}$ and a thickness of t_w may be excluded from the centre of the web compression zone in the cross-section (however, this further complicates the analysis).

6.7.1.1 Case 1: Neutral axis in concrete slab

Case 1 is applicable when $F_{st} \leq F_{c1}$ (Figure 6.29)

Equilibrium, $\quad F_{cc} = F_{st}$

Depth of compression zone, $\quad d_c = (t-h_r)(F_{st}/F_{c1})$

Moment capacity, $\quad M_b = F_{st} e$

$M_b = F_{st}(D/2 + t - d_c/2)$ (for sections which are symmetrical about the major axis)

where:
- D = Depth of steel section
- e = distance between centroid of tension and compression blocks

6.7.1.2 Case 2: Neutral axis in steel sheeting

Case 2 is applicable when $F_{c1} < F_{st} < (F_{c1} + F_{c2})$ (Figure 6.30)

Equilibrium, $\quad F_{cc} = F_{st}$

Depth of compression zone, $\quad d_c = (t - h_r) + h_r \left((F_{st} - F_{c1})/F_{c2} \right)$

Moment capacity, $\quad M_b = F_{st} e$

$$M_b = F_{c1}(D/2 + t - (t - h_r)/2) + (F_{cc} - F_{c1})(D/2 + t - (t - h_r) - (d_c - t + h_r)/2)$$

(for sections which are symmetrical about the major axis)
e = distance between centroid of tension and compression blocks

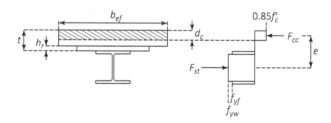

Figure 6.29 Case 1 force diagram.

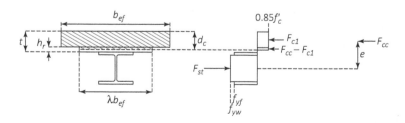

Figure 6.30 Case 2 force diagram.

6.7.1.3 Case 3: Neutral axis in top flange

Case 3 is applicable when $\left(F_{c1}+F_{c2}\right)<F_{st}<\left(F_{c1}+F_{c2}+2F_{scf}\right)$ (Figure 6.31)

Equilibrium, $\quad F_{cc}=F_{st}-F_{sc}$

also, $\quad F_{cc}=F_{c1}+F_{c2}+F_{sc}$

therefore, $\quad F_{sc}=\left(F_{st}-F_{c1}-F_{c2}\right)/2$

Depth of compression zone, $\quad d_{c}=t+F_{sc}/\left(b_{f}f_{yf}\right)$

Moment capacity, $\quad M_{b}=\left(F_{st}-F_{sc}\right)e$

$$M_{b}=F_{st}\left(D/2+t\right)-2F_{sc}\left(t+d_{c}\right)/2-F_{c2}\left(t-h_{r}/2\right)-F_{c1}\left(t-h_{r}\right)/2$$

(taking moments about the top of the slab for sections which are symmetrical about the major axis)

e = distance between centriod of tension and compression blocks)

6.7.1.4 Case 4: Neutral axis in web

Case 4 is applicable when $\quad F_{st}>\left(F_{c1}+F_{c2}+2F_{scf}\right)$ (Figure 6.32)

Equilibrium, $\quad F_{cc}=F_{st}-F_{scf}-F_{scw}$

also, $\quad F_{cc}=F_{c1}+F_{c2}+F_{scf}+F_{scw}$

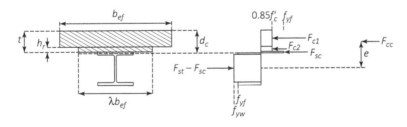

Figure 6.31 Case 3 force diagram.

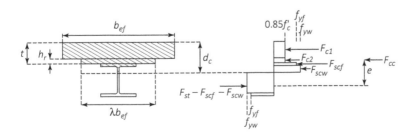

Figure 6.32 Case 4 force diagram.

therefore, $F_{scw} = (F_{st} - 2F_{scf} - F_{c1} - F_{c2})/2$

Depth of compression zone, $d_c = t + t_f + F_{scw}/(t_w f_{yw})$

Moment capacity, $M_b = (F_{st} - F_{scf} - F_{scw})e$

$$M_b = F_{st}(D/2 + t) - 2F_{scw}(t + t_f + d_c)/2 - 2F_{scf}(t + t_f/2) - F_{c2}(t - h_r/2) - F_{c1}(t - h_r)/2$$

(taking moments about the top of the slab for sections which are symmetrical about the major axis)

e = distance between centriod of tension and compression blocks)

6.7.2 Shear stud design

The two Australian Standards present the capacity for a single shear stud using the same calculation; however, the application of the capacity is entirely different. The calculations shown below are for headed shear studs, which are most commonly used.

Longitudinal shear capacity of a single shear stud, $f_{vs} = \text{MIN}\left[0.63d_{bs}^2 f_{uc}, 0.31d_{bs}^2 \sqrt{f_{cj}' E_{cj}}\right]$

where:
$\quad d_{bs}$ = Diameter of shear stud
$\quad E_{cj}$ = Young's Module of concrete at time considered = E_c for final state
$\quad f_{uc}$ = Tensile strength of stud, limited to 500MPa
$\quad f'_{cj}$ = compressive strength of concrete at time considered = f_c' for the final state

6.7.2.1 AS 2327.1 Shear stud design

AS 2327.1 uses a strength limit state calculation to check the number of studs required to achieve full moment capacity, also enabling the user to provide as little as half the shear capacity when complying with specific code requirements. The simplified calculation effectively chooses a point where bending is zero (at the support) and another point where bending is at a maximum. The slab increases from zero to its maximum longitudinal compression along this distance, and therefore, the number of shear studs required between those points is equal to maximum compression in the slab divided by the factored shear capacity of the studs. The studs are then distributed evenly along each segment of the beam. The simplified analysis is applicable to uniformly loaded prismatic beams where the design moment is less than 2.5 times the bending capacity of the steel section.

Factored capacity of shear studs (AS 2327.1), $f_{ds} = \phi k_n f_{vs}$

where:

$$k_n = 1.18 - (0.18/\sqrt{n_{s1}})$$

$\quad n_{s1}$ = Minimum of studs provide between support and mid-span
$\quad \phi = 0.85$

Number of shear studs required between support and mid-span, $n_{ic} = F_{c\Delta}/f_{ds}$

where, $F_{c\Delta} = \text{MIN}\left[(F_{c1} + F_{c2}), F_{cc}\right]$

The longitudinal shear capacity of concrete should also be checked for a failure around the studs. Refer to AS 2327.1, Section 9 for details on failure planes and the calculation procedure. The design of longitudinal concrete shear planes becomes increasingly significant when sheeting is used, as it decreases the shear perimeter.

6.7.2.2 AS 5100.6 Shear stud design

AS 5100.6 uses an SLS analysis of the shear flow to design the shear studs. The calculation is based on the number of studs required for a specific cross-section rather than along a specified length. AS 5100.6 has an error in the calculation for f_{ks} (referred to as f_{vs} in AS 2327.1), although the tabulated values are correct. The SLS shear force applied to the studs must be less than the design capacity for the group of studs for all locations along the beam.

$$V_L \le \phi v_{Ls}$$

$$\phi = 1.0 \qquad \text{(for shear connectors under SLS loading)}$$

Factored capacity of shear studs (AS 5100.6), $v_{Ls} = 0.55 n_{s2} f_{vs}$

where, n_{s2} = Number of shear studs per unit length

Shear studs are often spaced at closer centres near the ends of a beam due to the increased shear force. The closer spacing should be continued for a minimum of 10% of the beam length if this approach is adopted. The beam is therefore checked for locations 1 and 2 as shown in Figure 6.33. Fatigue should also be checked for composite bridges in accordance with AS 5100.6 (Figure 6.33).

Serviceability analysis requires an analysis using a transformed section, similar to that for concrete design (refer to Section 4.5), although in this scenario, the effective composite cross-section is reduced to a transformed effective elastic cross-section, which is an equivalent (entirely steel) section with the same stiffness as the effective section of both materials (Figure 6.34).

$\qquad b_t$ = Effective width of transformed slab = $b_{ef} n$
$\qquad n$ = Transformation ration = E_c / E_s (refer Section 4.5)
$\qquad n_{32}$ = 30100/200000 = 0.151 (for 32 MPa concrete)
$\qquad n_{40}$ = 32800/200000 = 0.164 (for 40 MPa concrete)

The shear force applied to the studs is calculated using shear flow theory (the same theory is covered for shear stresses caused by major axis beam shear; refer to Section 3.11).

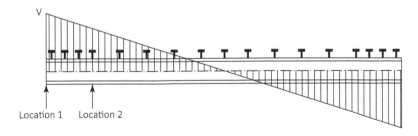

Figure 6.33 Shear stud design locations.

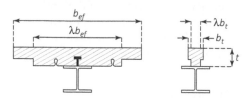

Figure 6.34 Transformed effective elastic composite cross-section.

Longitudinal shear force, $\quad V_L = VQ/I_x$

where:

$I_x \quad$ = Second moment of area for entire transformed cross-section (Section 3.4.1.1)
$Q \quad$ = Statical moment of area = $A_t e$
$A_t \quad$ = Transformed area = $(t-h_r)\, b_t + h_r \lambda b_t$
$e \quad$ = Distance between neutral axis of cross-section and neutral axis of transformed cross-section = $D + Y_{ct} - Y_c$
$Y_c \quad$ = Neutral axis from base
$Y_{ct} \quad$ = Neutral axis of transformed section from top steel beam
$V \quad$ = SLS Shear force in composite beam

Longitudinal concrete shear planes can be checked readily using this method, because shear force is expressed as force per unit length. The shear capacity along any failure plane, drawn from the steel beam and surrounding the shear studs (such as the full depth of the slab on each side), should be greater than the shear force under strength limit states rather than serviceability limit states (translated to be perpendicular to the failure plane for cases where the plane is interrupted by inclined steel sheeting). Refer to AS 5100.6, Section 6.6 for numerous additional requirements.

ULS Longitudinal shear force, $\quad V_{Lp}^* \leq \phi \times \mathrm{MIN}\left[0.9\,\mathrm{MPa} \times u + 0.7 A_{ts} f_{ry}, 0.15 u f_c'\right]$

$\phi = 1.0$ (for longitudinal shear reinforcement under ULS loading)

where:

$A_{ts} \quad$ = Area of effective transverse reinforcement per unit length
$f_{ry} \quad$ = Yield stress of reinforcement, limited to 450MPa
$u \quad$ = Length of shear plane transverse to span

6.7.3 Elastic transformed stress analysis

Elastic tensile stress at the base of the steel section and compression at the top of the concrete may be calculated using Figure 6.35, although this is not a code requirement. AS 2327.1 allows yielding of the steel under serviceability loading.

Longitudinal stress, $\quad \sigma = My/I_x$

where:

$M \quad$ = SLS moment

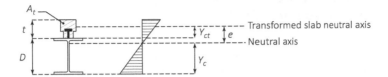

Figure 6.35 Transformed elastic stresses.

y = Distance from neutral axis to point being considered

σ = Stress

Maximum concrete stress, $\sigma_{cc} = n\,M\left(Y_{ct} - t - e\right)/I_x$

Tension steel stress, $\sigma_{st} = M\,Y_c/I_x$

6.8 BUNDS

Bunds are generally provided around tanks or vessels to prevent leakage or spillage from contaminating the surrounding soil or watercourse. Bunding is a legislative requirement. Most large projects are required to produce an environmental impact statement (EIS). The EIS details bunding requirements for different scenarios. Legislation may also set out requirements depending on the applicable code. In the absence of more specific information, Australian Standards should be adopted along with good practice. Common references include

1. NOHSC 1015:2001 National Standard – Storage and Handling of Workplace Dangerous Goods
2. AS 1940 The storage and handling of flammable and combustible liquids
3. AS 2067 Substations and high-voltage installations exceeding 1 kV a.c.
4. AS 3780 The storage and handling of corrosive substances
5. AS/NZS 4452 The storage and handling of toxic substances
6. The Australian Dangerous Goods Code (for transport of dangerous goods)

For the purpose of this section, the bunds being discussed are concrete slabs surrounded by concrete walls to contain the liquid. Three common options for containment are

1. Standard bund:
   a.   110% of stored volume
   b.   100% of largest volume, plus 25% of remaining storage up to 10,000 L, plus 10% of storage between 10,000 and 100,000 L, plus 5% above 100,000 L
2. 20% containment in immediate bund followed by 100% containment in secondary bund
3. Liquid separation (e.g. Humeceptor)

Rainfall should also be considered in the bund sizing depending on the adopted drainage philosophy.

6.8.1 The storage and handling of flammable and combustible liquids, AS 1940

The requirements of AS 1940 are specific for above-ground tanks which are not classified as 'minor' storage (refer AS 1940, Section 2). This standard provides useful guidelines for the design of storage compounds and is considered good practice, regardless of whether there is a legislative requirement to comply.

AS 1940, Section 5 details numerous additional design requirements (refer to the standard for the full detailed list of requirements).

Important aspects of concrete bund walls are:

1. To prevent spillage from the tank, the angle from the bund wall to the edge of the tank should be greater than 26.5°. This minimum clearance should be greater than 1 m (refer to Figure 6.36).
2. Storage facilities should be impervious to retain liquids (refer to Section 6.9).
3. The ground should slope away from the tank to a sump, which is drained from its low point.
4. Different bunds should not drain into each other.
5. Drain valves should be located outside the bund.
6. Structural integrity (fire, hydrostatic loads) should be included in the design.
7. Safe entry and exit locations need to be provided.
8. Penetrations for pipes and other joints should be sealed/waterproofed (refer to Chapter 8).
9. There should be adequate separation distances to safe locations.

6.8.2 Substations and high-voltage installations exceeding 1 kV a.c., AS 2067

Details provided in AS 2067 are specific to high-voltage installations (such as transformer compounds). Some aspects of the code are easily applied to other bunding compounds, such as the way in which the code details and allows drainage of immediate bunds to a common point.

AS 2067 requires installations with over 500 L of oil to have provisions which can contain the full volume. It also allows underflow drainage, as referenced in AS 1940, Appendix H. Underflow drainage is a method of allowing spilled oil to be retained while letting additional water (more dense liquids) drain from the bottom of the containment area (refer to Figure 6.37).

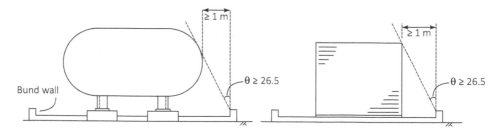

Figure 6.36 Bund location limits.

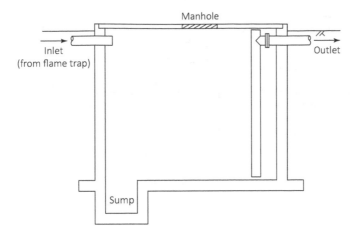

Figure 6.37 Oil interception tank. (From AS 2067. Copied by Mr L. Pack with the permission of Standards Australia under Licence 1607-c010.)

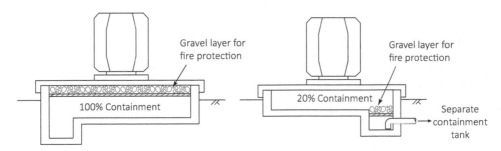

Figure 6.38 Sump with integrated containment tank. (From AS 2067, figure 6.3. Copied by Mr L. Pack with the permission of Standards Australia under Licence 1607-c010.)

The code shows that the bund can be provided locally with full containment or in a separate containment tank (refer to Figure 6.38). For the case of separate containment, a minimum containment of 20% is required in the immediate bund, with the separate container capable of holding the largest volume which drains to it.

Flame traps are required for combustible liquids to prevent flames being spread from one bund to another (refer to Figure 6.39). This is often provided by a grate with gravel placed over the top.

6.9 CONCRETE STRUCTURES FOR RETAINING LIQUIDS

The design of concrete structures for retaining liquids in Australia should be in accordance with AS 3735; however more detail is provided in the New Zealand code, NZS 3106, which is therefore sometimes adopted. The New Zealand code provides the user with the ability to design to different levels of 'tightness'; however, the Australian code provides a simpler solution with no sub-categories. This section provides a summary of the Australian code requirements with references to other codes where necessary (refer to AS 3735 for further detail).

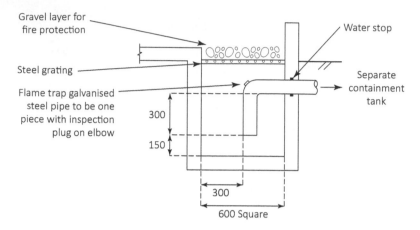

Gravel layer for fire protection

Water stop

Steel grating

Flame trap galvanised steel pipe to be one piece with inspection plug on elbow

Separate containment tank

300

150

300

600 Square

Figure 6.39 Flame trap. (From AS 2067, figure 6.7. Copied by Mr L. Pack with the permission of Standards Australia under Licence 1607-c010.)

6.9.1 Loads

6.9.1.1 Hydrostatic pressure

The horizontal force exerted by the water on a wall is linearly dependent on the depth under consideration (refer to Figure 6.40):

Hydostatic Pressure, $\quad P = \gamma H$

where, $\quad \gamma = \text{Density of liquid}\left(10 \, \frac{\text{kN}}{\text{m}^3} \text{for water}\right)$

Note: The resultant force for the triangular load is $\gamma H^2/2$, located at H/3 from the base. Tanks are usually designed for a water level at the top of the wall, even if overflow devices exist, as failure of the device should not lead to a structural failure of the tank.

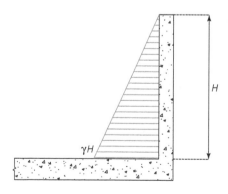

H

γH

Figure 6.40 Hydrostatic pressure.

6.9.1.2 Temperature

Temperature loads are calculated from a positive and negative thermal gradient. Walls and roofs which are subject to direct solar radiation, and which contain liquids at ambient temperature, are subject to the following thermal gradients:

1. Roofs
   a. Variation of $\pm20°C$ from the mean temperature
   b. 5°C per 100 m thickness (hot outside)
   c. 10°C per 100 m thickness (snow, cold outside)
2. Walls (refer to Figure 6.41)
   a. +30°C and −20°C when tank is full
   b. +20°C and −12°C when tank is empty

Special consideration should be given to shielded or buried tanks, which may have varied thermal gradients (refer AS 3735, Supplement 1).

6.9.1.3 Moisture variation

Minimum effects due to moisture variation are provided in Table 6.17. The adopted values can be converted into equivalent temperature variations for ease of analysis. Using this method, they can be combined with the temperature loads. Shrinkage corresponds to a temperature decrease, and swelling corresponds to a temperature increase.

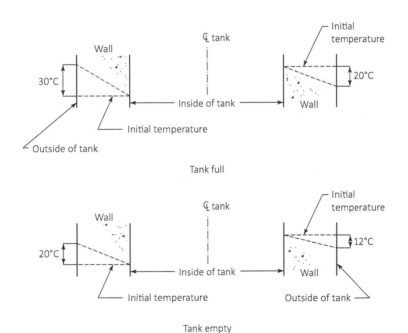

Figure 6.41 Temperature distribution in tank walls. (From AS 3735, figure 2.1. Copied by Mr L. Pack with the permission of Standards Australia under Licence 1607-c010.)

Table 6.17 Moisture variation: shrinkage and swelling strains

| Wall thickness (mm) | Mean shrinkage and swelling strain (Creep adjusted) × 10⁻⁶ | | | |
| | Shrinkage (ε_{sh}) | | Swelling (ε_{sw}) | |
	Precast	Cast in situ	Precast	Cast in situ
100	70	120	300	250
150	50	85	205	170
200	45	70	160	135
250	35	60	135	110

Source: AS 3735, table 2.2. Copied by Mr L. Pack with the permission of Standards Australia under Licence 1607-c010.

Table 6.18 Typical coefficients of thermal expansion for water-cured concrete for different aggregate types

Aggregate	Coefficient of thermal expansion (α_c) × 10⁻⁶/°C	Aggregate	Coefficient of thermal expansion (α_c) × 10⁻⁶/°C
Andesite	6.5	Greywacke	11
Basalt	9.5	Limestone	6
Dolerite	8.5	Pumice	7
Foamed slag	9	Quartzite	13
Granite	9	Sandstone	10

Source: AS 3735 Supplement 1, table A4. Copied by Mr L. Pack with the permission of Standards Australia under Licence 1607-c010.

Note: In the absence of information on the aggregate type, assume α_c = 11 × 10⁻⁶/°C.

Equivalent temperature variation, $T_{strain} = \varepsilon_s / \alpha_c$

where:

α_c = Coefficient of thermal expansion for concrete (refer to Table 6.18 for values of α_c)

ε_s = Shrinkage or swelling strain

6.9.1.4 Seismic

Seismic calculations may need to consider earthquake-induced liquid and earth loads. AS 3735 refers the user to NZS 3106, Appendix A for details on loads due to earthquakes. API Standard 650 also provides useful information for the seismic design of tanks. The seismic design of liquid-retaining structures is beyond the scope of this book. Refer to NZS 3106 for a detailed explanation.

6.9.1.5 Earth pressures

If the structure is embedded in the ground, horizontal earth pressures need to be considered. Refer to Section 5.3 for calculation details. The seismic mass of soil may also need to be considered, as detailed in the previous sub-section.

6.9.1.6 Wind

Wind loading will rarely affect the design of walls or slabs; however, it may have an effect on the roof of the tank (refer to Section 2.3.1).

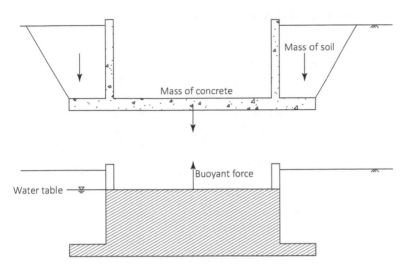

Figure 6.42 Buoyancy analysis.

6.9.1.7 Buoyancy

If the tank is embedded in the ground and there is a possibility of a high water table, buoyancy may be a design consideration. For stability of the structure, the mass of the tank should be enough to prevent it from lifting out of the ground when empty. Archimedes' principle defines the upward buoyant force as equal to the weight of the fluid displaced (i.e. volume $\times$ 10 kN/m^3; refer to Figure 6.42).

A collar can be used to engage the soil around the tank. Buoyancy also causes upward pressure on the base slab and therefore needs to be considered in the flexural design.

6.9.2 Load combinations

Loads and combinations should generally be in accordance with AS 3600, which refers to AS/NZS 1170; however, additional serviceability load combinations are provided.

6.9.2.1 Serviceability combination cases

Experience shows that these combinations frequently govern the design, and therefore, the initial sizing should be based on serviceability. The combinations shown in Table 6.19 are a mixture of the Australian and New Zealand combinations.

6.9.2.2 Strength combination cases

No specific guidance is provided in the Australian Standard for strength combinations. It is therefore recommended that the New Zealand combinations are used. The strain induced loads (thermal and moisture) are not significant for strength design and are therefore excluded from the combinations (NZS 3106, Clause 4.6).

AS/NZS 1170.0 specifies that where the liquid type and density are well defined, and the volume cannot be exceeded, the static liquid pressure should be factored by 1.2, and the self-weight should be factored the same as for permanent actions. Where this is not the case, the static liquid pressure should be factored by 1.5, and the self-weight should be factored

Table 6.19 Suggested serviceability load combinations

Item	*Description*	*Combination*
Long-term service load cases (Group A)[1]		
Roof	Shrinkage	$G + F_P + F_{sh}$
Walls: tank full	Swelling, not backfilled	$G + F_{lp} + F_P + 0.5\, F_{sw}$
	Swelling, backfilled	$G + F_{lp} + F_{ep} + F_{gw} + F_P + 0.5\, F_{sw}$
Walls: tank empty	Shrinkage, backfilled	$G + F_{ep} + F_{gw} + F_P + F_{sh}$
	Swelling, backfilled	$G + F_{ep} + F_{gw} + F_P + 0.5\, F_{sw}$
Short-term service load cases (Group B)[1]		
Roof	Live, temperature	$G + Q + F_P + T$
	Wind	$0.8\, G + W_s + F_P$
	Shrinkage, temperature	$G + T + F_P + 0.7\, F_{sh}$
	Swelling, temperature	$G + T + F_P + 0.7\, F_{sw}$
	Seismic[2]	$0.8\, G + 0.8\, F_{eq}$
Walls: tank full	Seismic[2], swelling, backfilled	$G + F_{lp} + F_{ep} + F_{gw} + F_P + 0.8\, F_{eq} + 0.5\, F_{sw}$
	Temperature, swelling, backfilled	$G + F_{lp} + F_{ep} + F_{gw} + F_P + 0.7\, F_{sw} + T$
Walls: tank empty	Temperature, shrinkage, backfilled	$G + F_{ep} + F_{gw} + F_P + T + 0.7\, F_{sh}$
	Temperature, swelling, backfilled	$G + F_{ep} + F_{gw} + F_P + T + 0.35\, F_{sw}$

Notes:

1. Combinations should be considered without transient load cases (temperature and shrinkage/swelling) where worse effects are calculated.
2. $0.8\, F_{eq}$ from the Australian Standard is replaced by (E_{s1} or E_{s2}) in the New Zealand standard. The seismic load should consider soil and liquid masses. Either 80% of the ultimate load should be applied or a service load should be calculated.

where:

G = Dead load
F_{ep} = Earth pressure
E_{eq} = Earthquake
F_{gw} = Ground water
F_{lp} = Liquid pressure
F_P = Prestress
F_{sh} = Shrinkage force
F_{sw} = Swelling force
Q = Live load
T = Temperature
W_s = Wind load

the same as for imposed (live) actions. The combinations shown in Table 6.20 are based on NZS 3106.

6.9.3 Durability

Durability should satisfy the more stringent requirements obtained using AS 3600 and AS 3735. The requirements from AS 3735 are presented in this section. An exposure classification is chosen and used to determine concrete and reinforcement requirements.

6.9.3.1 Exposure classification

The adopted exposure classification should be the larger of those obtained using AS 3600 (refer Table 4.5) and AS 3735 (refer Table 6.21).

Table 6.20 Suggested strength load combinations

Item	Description	Combination
Long-term service load cases (Group A)		
Roof	Self-weight	$1.35\ G$
	Live	$1.2\ G + 1.5\ Q$
	Live	$1.2\ G + 1.5\ \psi_l\ Q$
	Seismic	$G + F_{eq} + \psi_c\ Q$
	Snow/liquid pressure/ rainwater ponding	$1.2\ G + S_u + \psi_c\ Q$
	Wind (heavy)	$1.2\ G + \psi_c\ Q + W_u$
	Wind (light)	$0.9\ G + W_u$
Walls: tank full	Self-weight	$1.2\ G + 1.2\ F_{lp}$
	Self-weight, backfilled	$1.2\ G + 1.2\ F_{lp} + 1.5\ F_{ep} + 1.5\ F_{gw}$
	Seismic	$G + F_{eq} + F_{lp}$
	Seismic, backfilled	$G + F_{eq} + F_{lp} + F_{ep} + F_{gw}$
Walls: tank empty	Backfilled	$1.2\ G + 1.5\ F_{ep} + 1.5\ F_{gw}$

Notes:
1. Wind cases are only necessary for lightweight roofs.
2. Seismic loads should consider soil and liquid masses.

where:
G = Dead load
F_{ep} = Earth pressure
E_{eq} = Earthquake (ultimate)
F_{gw} = Ground water
F_{lp} = Liquid pressure
Q = Live load
S_u = Snow/liquid pressure/rainwater ponding
W_u = Wind (ultimate)
$\psi_c = \psi_l$ = 0.4 for roofs used as a floor, otherwise taken as zero

6.9.3.2 *Concrete requirements*

Minimum cover shall be selected from Table 6.22 and increased where necessary to account for being cast against ground, chemical or mechanical surface treatment, formed upper surface, embedded items or abrasion. Cover should consider concrete placement and compaction (i.e. larger than 1.5 times the aggregate size is recommended and at least one bar diameter).

Required cover modifications:

Cast against ground,
+10 mm where protected by a damp proof membrane.
+20 mm where cast in direct contact.
Chemical or mechanical surface treatment,
Increase by the depth that is degraded.
Formed upper surface,
+10 mm where using impermeable shutters to form sloping surfaces or other upper surfaces.
Embedded items in cover zone,
No items that can be corroded should be embedded into the cover zone. Non-corrodible items may be embedded where cover is still achieved (refer to AS 3735, Clause 4.4.4.4).

Table 6.21 Exposure classifications

		Exposure classification		
		Predominantly submerged		Alternate wet and dry (condensation, splashing or washing)
Item	Characteristic of liquid in contact with concrete surface	Generally quiescent	Agitated or flowing	
1	Freshwater: (Notes 1, 2, 3)			
	a LI positive or pH >7.5	B1	B1	B1
	b LI negative and pH 6.5 to 7.5	B1	B2	B1
	c LI negative and pH 5.5 to 6.5	B2	C	B2
2	Sewage and waste water: (Note 4)			
	a Fresh – low risk of H_2S corrosion	B1	B1	B2
	b Stale – high risk of (Note 8) H_2S corrosion	B2	B2	D
	c Anaerobic sludge	B1	B1	B1
3	Sea water: (Notes 5, 6)			
	a General immersion and pH $\geq$7.5	B1(7)	B2(7)	C
	b Retaining or excluding situations or pH <7.5	C	C	C
4	Corrosive liquids, vapours (Note 8) or gases Severity:			
	a Slight/mild	B1	B2	B2
	b Moderate (Note 9)	B2	C	C
	c Severe/extreme (Note 9)	D	D	D
5	Other liquids: (Note 10)			
	a Water containing chloride, sulfate, magnesium or ammonium	B1-D	B1-D	B1-D
	b Wine, non-corrosive vegetable oils, mineral oils and coal tar products	B1-D	B1-D	B1-D
6	Ground water (in ground) (Notes 10, 11)	B1-D	—	—

Source: Based on AS 3735, table 4.1. Copied by Mr L. Pack with the permission of Standards Australia under Licence 1607-c010.

Notes
1. An approximate value of Langelier Saturation Index (LI) may be obtained from the equation
   LI = pH of water – pH when in equilibrium with calcium carbonate
   LI = pH $-12.0 + \log_{10}(2.5 \times Ca_2+$ [mg/L] $\times$ total alkalinity [as $CaCO_3$ mg/L]).
   (A negative value for LI means the water has a demand for $CaCO_3$).
2. For lower pH values, see Item 4.
3. For water containing significant quantities of aggressive dissolved materials, see Item 5(b).
4. Industrial sewage and waste water may contain aggressive chemicals. The designer shall refer to other liquids as given in AS 3735, table 4.1 (see also AS 3735 Supp1).
5. The use of galvanised or epoxy-coated reinforcement or a waterproofing agent should be considered. Details are given in AS 3735 Supp1.
6. The use of sulfate-resisting cement is discouraged.
7. Only applicable for submergence greater than 1 m below low water ordinary spring tide.
8. Typical examples of severities are given in AS 3735 Supp1.
9. The use of calcareous aggregate should be considered. Details are specified in AS 3735 Supp1.
10. Guidance on the selection of an appropriate exposure classification from within the range indicated is specified in AS 3735 Supp1.
11. For members in contact with extracted ground water, see Item 1 or 5.

Table 6.22 Minimum cover

Exposure classification	Required cover (mm)[a]		
	Characteristic strength, f'_c (MPa)		
	32	40	50
Standard formwork and compaction			
B1	45	40	40
B2	70	50	40
C	N/A (Note 1)	75	55
D (Note 1)	N/A (Note 1)	75	55
Rigid formwork and intense compaction (precast)			
B1	35	30	25
B2	55	40	30
C	N/A (Note 1)	60	45
D (Note 1)	N/A (Note 1)	60	45

Source: AS 3735, tables 4.2 and 4.3. Copied by Mr L. Pack with the permission of Standards Australia under Licence 1607-c010.

Notes:
1. The concrete surface shall be isolated from the attacking environment.
2. Protective coatings (see AS 3735 Clause 5.3) do not permit a reduction in cover requirement.

[a] Cover shall be increased where subject to cast against ground, chemical or mechanical surface treatment, formed upper surface, embedded items or abrasion. Refer to text for details.

Allowance for abrasion,
  Additional cover shall be allowed for, equal to the expected abrasion. Protection shall be considered where concrete is in contact with liquids moving at speeds greater than 4 m/s.

6.9.4 Crack control

The reinforcement ratio (p) is calculated using an effective concrete area (AS 3735 Clause 3.2.1):

$$p = A_s / A_{c,eff}$$

where:

A_s = Area of steel on face being considered
$A_{c,eff}$ = Effective concrete area (refer to Figure 6.43)
α_1 = Bar spacing
α_2 = $\dfrac{D}{2} \leq 250$ mm (or ≤ 100 mm when in contact with ground)

The minimum reinforcement ratio (p_{min}) is required to ensure even shrinkage cracking of acceptable width (AS 3735 Clause 3.2.2):
  For unrestrained concrete,

$$p_{min} = f_{ct.3} / f_{sy}$$

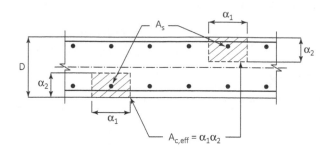

Figure 6.43 Effective concrete area. (From AS 3735, figure 3.1. Copied by Mr L. Pack with the permission of Standards Australia under Licence 1607-c010.)

where:
$f_{ct.3}$ = Principal tensile strength of concrete at 3 days
f_{sy} = Yield strength of bars being considered

For restrained concrete, p_{min} is shown in Table 6.23 with the principal tensile strength estimated using the correlated regression below. Calculated values may be reduced in proportion to the degree of restraint (up to 25%) where movement joints are provided at maximum 15 m spacing, or where partial movement joints are provided at maximum 7.5 m spacing. Refer to NZS 3106 for further details on principal tensile strength.

$$f_{ct.3} \approx 0.1768 f_c'^{0.6707} \quad \left(20\ \text{MPa} \le f_c' \le 50\ \text{MPa}\right)$$

For thick sections, or extremely hot climates, with high cement content, higher values may be required (refer AS 3735, Supplement 1, Clause C3.2.2).

6.9.5 Analysis

Thermal and shrinkage loads are important for serviceability design. The analysis is generally completed using an uncracked (elastic) analysis. AS 3735 allows a cracked section to be used for partially prestressed designs and for the hoop direction of a circular tank. Refer to AS 3735, Clauses 3.3.5 and 3.2.4 for details.

NZS 3106 allows a cracked section to be used for doubly reinforced circular walls. This is done by factoring down the forces by a value obtained from NZS 3106, figure 3. Alternatively, an FEA may be completed.

The calculation can be solved in an FEA package such as Strand7 by using plate elements. The walls should be connected to the floor and roof by using link elements (master–slave) which model the connectivity provided in the construction details (sliding, hinged or fixed).

Table 6.23 Deformed bars: Percentage for fully restrained concrete

Bar diameter, d_b (mm)	8–12	16	20	24	28	32
Minimum reinforcement ratio, p_{min} (%)	0.48	0.64	0.80	0.96	1.12	1.28

Source: AS 3735, table 3.1. Copied by Mr L. Pack with the permission of Standards Australia under Licence 1607-c010.

Note: The most efficient way of modelling the tank in an FEA package is to create a cylindrical coordinate system located at the centre of the base of the tank (global -> coordinate systems). Then, create a node at the base of the wall and extrude it (tools -> extrude -> by increments) in the direction of θ (perhaps 10°, repeated 36 times), creating a ring of beams. Select the beams and extrude them in the vertical direction, creating plates for walls. Select the beams again, and extrude them towards the centre of the base to create the floor slab. Some meshing may have to be completed to prevent the plates near the centre from having small angles (refer to Section 6.11). Ensure that plates are 'Quad 8' elements for circular plates to ensure that hoop behaviour is adequately achieved.

A temperature gradient can be applied to plate elements (attributes -> plates -> temperature gradient), making one side of the plate hotter than the other. The gradient is applied as 'temperature per thickness'; therefore, the total temperature variation should be divided by the wall thickness. Figure 6.44 shows the expected deflection and moments for temperature changes.

The hydrostatic pressure should be applied as 'normal face pressure' (attributes -> plate -> face pressure -> normal) with a pressure calculated for the centre of the plate. This ensures that the pressure is always directly outwards. The plates should be checked to ensure that they are all facing in the correct direction (attributes -> entity display -> plates -> property colour). If some plates are upside-down, the force will be applied in the opposing direction. When this display configuration is selected, the inside of the tank should be one colour, and the outside should be a different colour. Upside-down plates should be flipped to correct the orientation (tools -> align -> flip elements).

The best way to check that the model is working correctly is to analyse it for each load case, and look at the deflected shape of the structure. Once the model is behaving as expected, the tensile and flexural forces should be inspected. Figure 6.45 shows the expected deflected shape and moments for a rectangular tank. Positive moments are in the centres and negative moments on the corners.

6.9.6 Serviceability

Serviceability stress in the reinforcing bars ($f'_{s\,max}$) is limited to control cracking:

$$f'_{s\,max} = Y_1 Y_2 Y_3 f_{so} \qquad \text{(refer to Tables 6.24 through Tables 6.27)}$$

This formula should be used on the inside face of tanks where the inside bars are in tension. Both sides should be checked for members that are less than 225 mm thick. If ground water is present for a tank which is embedded in the ground, both sides should be checked,

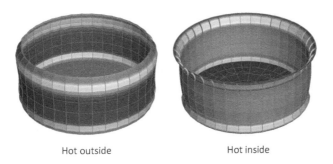

Hot outside   Hot inside

Figure 6.44 Example FEA solution.

Figure 6.45 Bending and deflection for a rectangular tank.

Table 6.24 Nominal limiting stresses in steel reinforcement

Bar diameter, d_b (mm)	8–12	16	20	24	28–32
Limited yield strength, f_{so} (MPa)	150	140	130	120	110

Source: AS 3735, table 3.2. Copied by Mr L. Pack with the permission of Standards Australia under Licence 1607-c010.

Table 6.25 Coefficient for bar type, Y_1

Exposure coefficient	Type of reinforcement	
	Plain bar	Deformed bar and welded wire fabric
Y_1	0.85	1.00

Source: AS 3735, table 3.3. Copied by Mr L. Pack with the permission of Standards Australia under Licence 1607-c010.

Table 6.26 Load combination coefficient, Y_2

Load combination	Y_2
Long-term effects (Group A)	1.0
Short-term effects (Group B)	1.25

Source: AS 3735, table 3.4. Copied by Mr L. Pack with the permission of Standards Australia under Licence 1607-c010.

Table 6.27 Coefficient for stress state and type of exposure, Y_3

Exposure coefficient	Predominant stress state	Type of exposure	
		Continuously submerged	Intermittent wetting and drying
Y_3	Tension	1.00	1.00
	Flexure	1.25	

Source: AS 3735, table 3.5. Copied by Mr L. Pack with the permission of Standards Australia under Licence 1607-c010.

Table 6.28 Limiting mean crack widths, b_m

Predominant stress state	Type of exposure	
	Continuously submerged	Intermittent wetting and drying
Tension	0.10 mm	0.10 mm
Flexure	0.15 mm	

Source: AS 3735, Supplement 1, table C3.1. Copied by Mr L. Pack with the permission of Standards Australia under Licence 1607-c010.

as they both retain water. For welded fabric, welded intersections should have a spacing of 200 mm or less.

The intention of the stress limitation is to limit the mean crack width. For Group A loads, complying with the stress limit is intended to limit cracks to the values provided in Table 6.28.

6.9.7 Design

Rectangular tanks will be found to have large vertical and horizontal moments. Circular tanks will have considerably smaller bending moments; however, significant tensile loads will exist (hoop stress). For circular tanks, the concrete is assumed to be fully cracked in the hoop direction, and the reinforcement is designed to take the full hoop force. Shear forces should be checked for rectangular and circular tanks.

After the analysis is completed, the extracted serviceability loads should be used to check the bar stresses against $f'_{s\,max}$. An elastic analysis is appropriate to analyse bending stresses; refer to Section 4.5. For the hoop tension, the bar stress is simply calculated as the force divided by the area of steel.

Standard methods of strength calculations should be used for strength limit states (refer to Section 4.2). For the hoop stress case, a capacity reduction factor of 0.8 should be used for bars in tension.

A hand method is presented in *Circular Concrete Tanks without Prestressing*, Portland Cement Association, 1993 [46]. Tables are used to solve hand calculations for pinned, hinged and fixed tanks. The actual solution is generally between hinged and fixed; therefore, they can be adopted as upper and lower bounds. This reference provides a quick method of sizing where the geometry is simple.

6.9.8 Concrete structures for retaining liquids detailing

Detailing of tanks is important, as it justifies the design assumptions. Specific consideration should be given to reinforcing, connections, joints and waterproofing.

The example provided in Figure 6.46 shows a tank foundation with an integral wall, separate slab and isolated roof. The foundation for the wall is cast first with a concrete upstand. This aids in waterproofing and allows a central waterstop to be used (other methods are also acceptable). Depending on the height of the wall, the starter bars may have to be lapped with the wall reinforcement. The central slab for the tank is isolated from the wall foundation and is shown with a rearguard waterstop (required where ground water is present). A blinding layer is provided under the foundations to ensure a flat and stable working platform for the waterstops to be installed correctly. The slab may be keyed into the foundation or supported by dowel bars if shear transfer is required. For smaller tanks, the slab may be integral to the wall foundation. The roof is supported on a neoprene bearing strip to allow for thermal movements. Additional details are provided in Chapter 8 and AS 3735, Supplement 1.

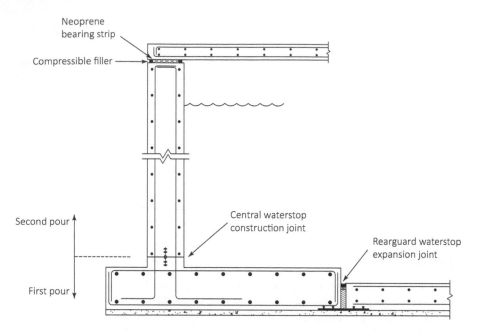

Figure 6.46 Typical tank details.

For an improved surface finish and durability, controlled permeability formwork products such as Zemdrain (refer to Chapter 8) may be used. If additional resistance is required (especially for oils or acids), a coating system such as Nitocote EP405, Vendex Cemelast or Nitocote EP410 may be adopted (refer to Chapter 8), although AS 3735 does not permit a reduction in cover based on protective coatings. These coating systems are common for potable water systems and for chemical bunds.

Note: For details on waterstops, refer to Parchem catalogues: Supercast PVC Waterstops, Superswell 47B, Emer-Seal 200 and Emer-Seal CR (refer to Chapter 8).

6.9.9 Construction and testing

Concrete shall be cured initially for 7 days under ambient conditions, or by accelerated methods which achieve 75% of f'_c on completion of curing. Additional information on concrete and reinforcement requirements are available in AS 3735, Section 5.

Inspection and testing should be conducted on completion of construction and at maximum intervals of 5 years. Testing requirements are detailed in AS 3735, Section 7. Essentially, the tank is kept full for 7 days and then the levels are measured for another 7 days. After accounting for evaporation and rainfall, the drop in level should be less than depth/500 and 10 mm.

6.10 LINEAR AND NON-LINEAR ANALYSIS (SPACE GASS)

This section gives a brief introduction to the use of Space Gass as a design tool. Space Gass is probably the most commonly used steel design package in Australia. The package can be used for the design of concrete and steel connections; however, it is traditionally a tool for the design of steel members in accordance with AS 4100.

6.10.1 T-post design model

The example is for a 1.0 m high 150UC23 T-post supporting three pipes. Each of the pipes has a dead load of 3 kN and a live load of 5 kN. The design is to be checked for a wind speed of 43 m/s and a seismic equivalent static force of 12.5%.

6.10.1.1 Create geometry of model

Open a new Space Gass file.

Press the 'Draw Members' button to begin drawing the column.

Draw the column for the t-post by creating a node at the base and extending it vertically to create a column, then sideways to create a beam:

Type '0,0,0' > Press Ok
Type '@0,1,0' > Press Ok
Type '@0.3,0,0' > Press Ok > Press Escape

Right click on the beam and copy it to the left to create the second beam at the top of the t-post.

Copy > Along Line > Click the node on the right side of the beam then on the left side
  > Press Ok (Figure 6.47)

Choose a section for the t-post and add material properties:

Right click on any member > 'View/Edit Member Properties (Form)' > Press the 'Library
  Editor' button in the 'Section' box > Choose a 150UC23 section in the 'Aust300'
  universal columns folder
Click the 'Library Editor' button in the 'Material' box and select steel

Change the 'Display Mode' to 'Rendered' on the left hand side:

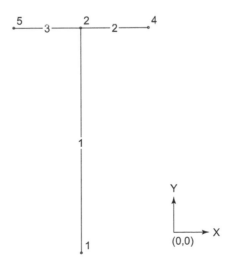

Figure 6.47 t-post geometry.

Press the 'Display Mode' icon and change the selection to 'Rendered'

Offset the top beams of the t-post (Figure 6.48) so that the nodes are at the top of steel (so that loads are applied at the top flange):

Select the two beams at the top of the t-post > Right click > 'View/Edit Member Properties (Form)' > input a global offset of '–0.075' in the DY direction at ends A and B
Select the column > Right click > 'View/Edit Member Properties (Form)' > input a global offset of '–0.15' in the DY direction at ends B only

Input load titles to the model (Table 6.29):

Select the 'Load' tab > 'Load Case Titles' > 'Datasheet' > Input the four load case titles shown

Add self-weight to the model, including a 12.5% seismic load in the x-direction to load case 4 (Table 6.30):

Select the 'Load' tab > 'Self Weight Loads' > add the self-weight loading shown

Add node loads and member loads to the model ($G = 3$ kN/pipe, $Q = 5$ kN/pipe, $W_u = 8$ kN windward pipe, $E_u = 1$ kN/pipe) (Table 6.31):

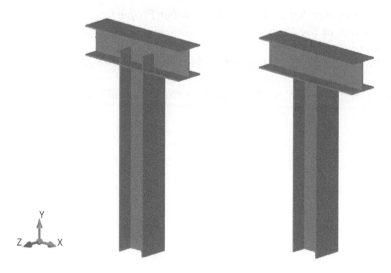

Figure 6.48 Three-dimensional offset.

Table 6.29 Load case titles

Case	Title
1	G: Dead load
2	Q: Live load
3	Wu: ULS wind load
4	Eu: ULS seismic load

Table 6.30 Self-weight

Case	X Acceleration(g)	Y Acceleration(g)	Z Acceleration (g)
1	0	−1	0
4	0.125	0	0

Table 6.31 Node loads

Case	Node	X Force (kN)	Y Force (kN)	Z Force (kN)	X Moment (kNm)	Y Moment (kNm)	Z Moment (kNm)
1	4	0	−3	0	0	0	0
1	5	0	−3	0	0	0	0
1	6	0	−3	0	0	0	0
2	4	0	−5	0	0	0	0
2	5	0	−5	0	0	0	0
2	6	0	−5	0	0	0	0
3	6	8	0	0	0	0	0
4	4	1	0	0	0	0	0
4	5	1	0	0	0	0	0
4	6	1	0	0	0	0	0

Select the top three nodes > Right click > 'Node Loads' > Enter '1,2,3,4' in the load case list > Click Ok > Enter the node loads shown

Add the wind load to the column using a member distributed force (Table 6.32):

Right click on the column and select 'Distributed Forces' > Enter '3' in the load case list > Click Ok > Enter the distributed force shown

Create a set of ULS and SLS load combinations (100 series cases are ULS, 400 series cases are SLS) (Table 6.33):

Select the 'Load' tab > 'Combination Load Cases' > Enter the table shown

Apply restraint to the model by fully fixing the base of the column (Direction X,Y,Z fixed, Rotation X,Y,Z fixed):

Select the node at the base of the column > Right click > 'View/Edit Node Properties (Form)' > add a restraint of 'FFFFFF'

Solve the model using a non-linear analysis (includes p-delta effects):

Select the 'Analysis' tab > 'Non-linear analysis' > Ok

Table 6.32 Wind loads

Case	Member	Sub-load	Axes	Units	Start position (m or %)	Finish position (m or %)	Start X force (kN/m)	Finish X force (kN/m)
3	1	1	G-Incl	Percent	0	100	0.186	0.186

Table 6.33 Combination cases

Combination case	Title	1 (G)	2 (Q)	3 (Wu)	4 (Eu)
100	1.35G	1.35			
101	1.2G + 1.5Q	1.2	1.5		
102	1.2G + 0.6Q + Wu	1.2	0.6	1	
103	0.9G + Wu	0.9		1	
104	G + Eu	1			1
400	G	1			
401	G + 0.6Q	1	0.6		
402	G + Ws	1		0.74	

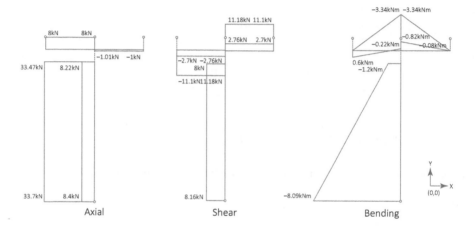

Figure 6.49 Analysis results.

Axial forces, shear forces, bending moments and deflections can now be viewed using the icons on the left (Figure 6.49).

Check the model in accordance with AS 4100. Input member design parameters:

Right click on the column > 'Input/Edit Steel Member' > Enter the Steel Member Design Data

The load height position is at the 'Centre'. The column is a cantilever and therefore has an effective length factor of 2.2 (ratio) in the major and minor directions. The top and bottom flanges are each only restrained at the base plate; therefore, the restraint type is 'FU' for both.

Turn on 'Show Steel Member Top Flanges' and 'Show Steel Member Flange Restraints' using the icons on the left, and ensure that the base of the column shows 'FF' and the top shows 'UU'.

Input steel member properties for the top beams (flange restraints for the beam on the left will be UF rather than FU, because the beam is restrained on the right hand side). Note that a slender column would not be able to provide lateral flange restraint; however, short columns can provide sufficient stiffness (Figure 6.50).

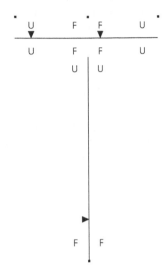

Figure 6.50 Steel member design data.

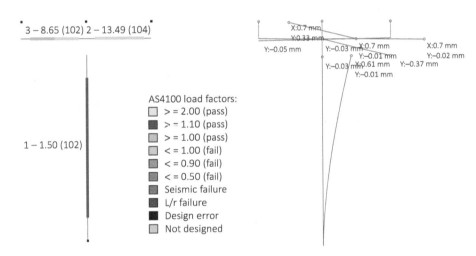

Figure 6.51 AS 4100 load factors and deflections.

Perform the AS 4100 member design check:

Select the 'Design' tab > 'Design/Check Steel Members' > Input '100–500' in the 'Load cases' box

Turn on 'Show Steel Member Design Results' using the icons on the left

The colour contour plot shows that all members pass the design check. Any value greater than 1.0 is a pass (Figure 6.51).

Deflections should also be checked for SLS combination load cases:

Use the drop down list at the top to filter the load cases for 'Selected Load Cases' > Input '400–500' > Turn on deflections using the icon on the left > Use the envelope icon on the left to filter maximum values

The maximum deflection is 0.7 mm, which is equal to Height/1428, which is much lower than any typical deflection limits (such as Height/125 for cantilevers and Height/250 for beams; refer to Table 2.25).

6.11 FINITE ELEMENT ANALYSIS (STRAND7)

This section gives a brief introduction to the use of FEA in structural engineering. It presents basic concrete and steel designs with a linear and non-linear example. The examples are shown in Strand7, a commonly used package for Australia. A demo of the software is available at www.strand7.com. FEA packages use the finite element method (FEM) to integrate and closely approximate solutions to complex structural arrangements. Think of it as drawing a circle using nodes and straight lines, and then calculating the area and adding more and more nodes to refine the answer.

The most relevant reference to FEA in Australian Standards is in AS 1210, Pressure Vessels. The standard allows either linear or non-linear analysis and recommends a Tresca or von Mises stress plot.

6.11.1 Linear analysis

A linear analysis is the fastest means of reaching a solution with FEA. Material and geometry are considered linearly. The solver does not consider buckling or plastic deformations. A linear analysis is appropriate for lightly loaded elements which are expected to remain elastic.

AS 1210 recommendations include a maximum meshing aspect ratio of 3 (long edge to short edge of elements) and a maximum transition size of 2 (adjacent elements should not vary in size by large factors), and four-sided elements are recommended over three-sided elements. The mesh should be sufficiently fine to ensure that linear stress varies by a maximum of 30% across an element (second-order elements). High-stress singularities (such as sharp internal angles) may be ignored, provided that adjacent elements are not required for integrity.

6.11.1.1 Concrete slab model (linear)

A concrete slab is used as an example for a simple linear FEA problem. An A160 axle load (refer to Table 2.21) is parked in the middle of a 4000 mm wide, 3750 mm long concrete slab, 200 mm thick. The soil provides an elastic restraint equal to 10 kPa/mm. This example will show how to calculate bending moments in the slab.

6.11.1.1.1 Create geometry of model

Open a new Strand7 file and set the default units to newtons, millimetres and megapascals (using these settings simplifies calculations).

Create the first node at the origin:

Create > Node > Apply (0, 0, 0)

Select the new node by toggling 'node select' and then using the 'select' cursor tool to capture it.

Extrude the node into a group of beams (15 beams, 250 mm long); this creates the 3750 mm edge, with a convenient length equal to the 250 mm tyre length:

Tools > Extrude > by Increments > Set the y direction to 250 (mm) and repeat to 15 times

Select all the beams by 'toggling beam select' and then using 'Ctrl + A' to select all. Extrude the beams into a grid of plates to represent the entire slab:

Tools > Extrude > by Increments > Set the x direction to 400 (mm) and repeat to 10 times, with the source selected to erase (this deletes the beams and replaces them with plates)

Press F3 to refresh the view and fit the entire model on the screen.

Hold shift and hover the cursor over an item to view the attributes (change the toggle from nodes, beams or plates to view different entities).

6.11.1.1.2 Configure material properties

Set the material properties to 32 MPa concrete and the thickness to 200 mm:

Properties > Plate > Materials > Concrete > 'Compressive Strength f_c = 32'
Properties > Plate > Geometry > Set membrane thickness to 200 mm

6.11.1.1.3 Apply restraints to model

Select all plates and then apply the vertical spring support to all plates and apply a horizontal restraint:

Attributes > Plate > Face Support > Use a value of 0.01 (MPa/mm)

We are completing a linear analysis; therefore, the solver will not consider 'compression only' values. If there were uplift loads on the slab, a compression only spring could be used in conjunction with solving the model in the non-linear solver.

Select all plates again, and apply a horizontal spring support to the edges of the slab:

Attributes > Plate > Edge Support > Use a value of 0.001 (MPa/mm) and select 'free edges'

This support is nominal, as there are no lateral loads; however, the model will not solve without the restraint.

6.11.1.1.4 Apply load to the model

Select the two plates which are located in the wheel positions by 'toggling plate select' and using the 'select' cursor again (refer to Figure 6.52).

Apply the wheel pressure of, $\dfrac{\text{Force}}{\text{Area}} = \dfrac{80000\,\text{N}}{400\,\text{mm} \times 250\,\text{mm}} = 0.8\,\text{MPa}$ at each wheel location:

Attributes > Plate > Face Pressure > Global > set Z to –0.8 (MPa)

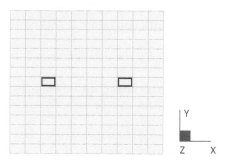

Figure 6.52 Grid of plates with wheel load positions selected.

6.11.1.1.5 Check the model

Clean the mesh (Ctrl + Alt + C) and then click 'Apply'.

Turn on 'Show Plate Free Edges' to ensure that the only free edges are the perimeter of the slab. If any lines are shown inside the slab, it means that the nodes of one plate are not connecting to the adjacent plate. This is fixed by cleaning the mesh with a sufficient tolerance or redrawing the plate.

6.11.1.1.6 Solve the model

Save the model in a directory. The model file will use a '*.ST7' file extension, and any solution files will be saved in the same directory with different file extensions. The linear static analysis file used in this example will use '*.LSA'.

A linear static analysis is used, because non-linear behaviour is not expected:

> *Solver > Linear Static > Solve*

Always view the log after solving the model, as there may be errors. The only error expected in this example is a warning about 'rigid body motion'. This is because only spring supports were used to restrain the model, and Strand7 does not check stability. The results are therefore acceptable.

6.11.1.1.7 View the results

Open the results file:

> *Results > Open Results File > Open*

Press F4 to reposition the model. Left click and drag the mouse to rotate the model. Click both mouse buttons and drag the mouse to move the model. Right click and drag the mouse to zoom the model. Hold 'Ctrl' and then left click and drag the mouse to show the deflected shape of the slab. Press 'Enter' when the model is positioned correctly. The deflected shape should show the slab bending along the x-axis.

Additional buttons are now available to view details from the results file. Select the 'Results Settings' box and choose 'Contour', 'Moment', 'Global' and 'YY'. This will contour the bending moment values for those which cause stress in the global y-direction (i.e. about the x-axis). The maximum moment (refer to Figure 6.53) is shown as 17,907 Nmm/mm

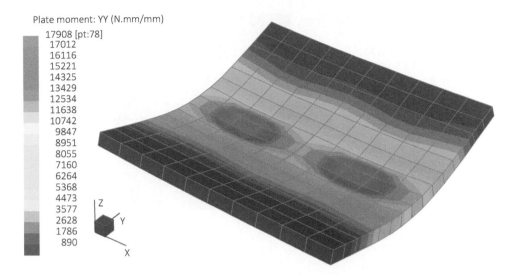

Plate moment: YY (N.mm/mm)

Figure 6.53 Iteration 1: Slab bending about *x*-axis.

(= 17.9 kNm/m). Bending about the *y*-axis can be shown by changing the setting to 'XX', resulting in 9.2 kNm/m.

This is only the first iteration of the model. It is good practice to then refine the model to check that the estimated stresses are accurate. A good indication that the model needs to be refined is that there are relatively high stress changes across individual plates.

6.11.1.1.8 Refine the model

Close the results file (*Results > Close Results File*) and then save the Strand7 file under a new name. Press F12 and set each view angle to zero to reset the viewing angle and look at the slab in plan. The plates which had high stress changes were those around the wheel locations. Select the plates where the wheel load is applied along with the plates above and below, and then subdivide the plates:

Tools > Subdivide > A = 2, B = 2, C = 1

The 'Plate Free Edges' tool used earlier will now show that the subdivided plates are not properly connected to the larger plates. This is because the nodes do not align, and therefore, the adjacent plates need to be graded.

Use various shapes to cut up the larger plates so that the nodes align with smaller plates. This is done by splitting up the larger shapes. The dotted lines represent which side(s) of the plate is selected (refer to Figure 6.54). Sharp angles and slender elements should be avoided. A good general rule is to keep the aspect ratios below 3.

Tools > Grade Plates and Bricks > 1 × 2 Grade
Tools > Grade Plates and Bricks > Quarter Quad Grade

Repeat the 'Check the Model' and 'Solve the Model' steps from earlier, and then open the solution file.

The bending moments have now increased to 23.16 kNm/m.

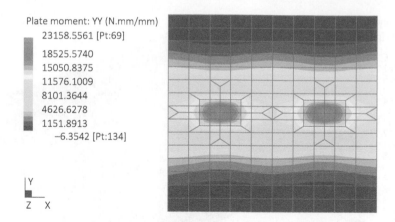

Plate moment: YY (N.mm/mm)
23158.5561 [Pt:69]
18525.5740
15050.8375
11576.1009
8101.3644
4626.6278
1151.8913
−6.3542 [Pt:134]

Figure 6.54 Iteration 2: Slab bending about *x*-axis.

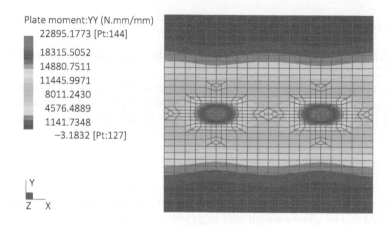

Plate moment:YY (N.mm/mm)
22895.1773 [Pt:144]
18315.5052
14880.7511
11445.9971
8011.2430
4576.4889
1141.7348
−3.1832 [Pt:127]

Figure 6.55 Iteration 3: Slab bending about *x*-axis.

A third iteration can be completed to ensure that the model is converging on an accurate value.

For the third iteration, a different method can be shown. Select all elements, and then subdivide the plates again (refer to Figure 6.55):

Tools > Subdivide > A = 2, B = 2, C = 1

This method of refinement is quicker for the user; however, it slows down the time required for the computer to solve the model. Remember to save the model under a new name, and check the model again prior to solving. The third iteration calculates a bending moment of 22.895 kNm/m. Repeating the process will give a fourth iteration with a value of 22.844 kNm/m. Results are summarised in Table 6.34. The bending moment in the slab is accepted as 23 kNm/m.

Other values, such as shear forces or displacements, can also be contoured in the same way. The shear values in the same direction would be shown by selecting 'Contour', 'Force', 'Global' and 'YZ'.

Table 6.34 Bending moment for each
model iteration

Iteration	Bending moment (kNm/m)
1	17.907
2	23.159
3	22.895
4	22.844

6.11.2 Non-linear analysis

A non-linear analysis can be completed with plastic deformations and buckling. The load is increased incrementally until the yield stress is reached; then plastic deformation will occur until the strain limit is exceeded or the model buckles (fails to converge). If the model fails to converge, it typically means that the structure is buckling; if the model solves, yield and strain values should be checked. SLS stresses and strains should remain in the elastic region, and ULS stresses and strains can go as high as the strain limit (refer to Figure 6.56). Both non-linear geometry and non-linear material properties need to be selected.

A stress-strain curve needs to be created when completing a non-linear analysis with non-linear material properties. A maximum strain equal to seven times the strain at yield is recommended. This represents a strain plateau equal to six times the maximum elastic strain (AS 4100, Cl 4.5.2). This is a conservative limit, as most steels require an elongation of 0.16–0.25 under tensile testing (AS/NZS 3678 & AS/NZS 3679.1). The solver will provide a warning and continue linearly if the specified maximum is exceeded. Research into the exact type of steel is recommended when exceeding the code limit.

Note: AS 1210 recommends a maximum of 0.01 strain when remote from discontinuities and as high as the smaller of 0.05 or one-third of the material's failure elongation when including discontinuities.

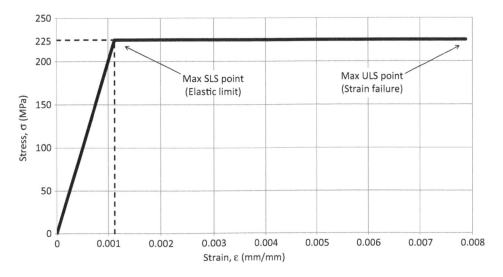

Figure 6.56 FEA stress-strain diagram (Grade 250).

6.11.2.1 Steel connection model (linear and non-linear)

A lifting lug is used for the steel example. The lug is 170 mm in diameter, with a 150 mm cheek plate and a 55 mm hole. The load is 460 kN at 45° from vertical. Proficiency in the skills gained from the concrete example is assumed in this section.

6.11.2.1.1 Create geometry of model

Create a node at (0, 0, 0) and extrude it by 280 mm in the x-direction. Create another node at (140, 100, 0).

A cylindrical coordinate system can be used to create the shape of the lug:

Global > Coordinate System > Node: 3, System: Cylindrical, Type: XY, Points: 1

Copy node 3 by increment using the new coordinate system three times:

Tools > Copy > by Increments > R = 85, R = 75 then R = 27.5

Select the three new nodes, and extrude them around the cylindrical coordinate system by increment (refer to Figure 6.57):

Tools > Extrude > by Increments > θ = 22.5 (Repeat = 16)

Create two new beams to join the base of the lug to the edge of the radius, and delete 10 beams at the base of the outermost circle:

Create > Elements > Beam

The beam geometry can now be converted to a face and then automeshed. Select all beams, convert to faces, and then graft the edges to faces and remove the unnecessary faces:
Select all beams

Tools > Geometry Tools > Face from Beam Polygon

(Three polygons should be converted)

Tools > Geometry Tools > Graft edges to faces

(There are now six faces, including copies of the cheek plate and hole. Care needs to be taken in the meshing of the model to ensure that the correct faces are selected.)

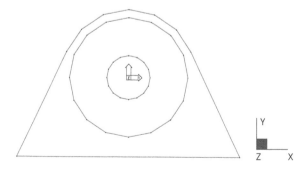

Figure 6.57 Preliminary lifting lug beam geometry.

6.11.2.1.2 Mesh the model

Select the cheek plate face (this should also include the hole in the one face):

Tools > Automeshing > Surface Mesh
Maximum edge length: 10 mm

Select the outside face (lug perimeter)
Tools > Automeshing > Surface Mesh
Maximum edge length: 10 mm
Target > Plate property: Specified: 2

Clean the mesh
Tools > Clean > Mesh

The circular edges of the lug can be corrected using the cylindrical coordinate system. Select nodes on the outer edge of the lug and move them to absolute radius of 85. Select nodes on the outer edge of the cheek plate and move them to absolute radius of 75 (refer to Figure 6.58).

6.11.2.1.3 Configure material properties

Now, the properties for steel can be applied to each of the plate types:
Properties > Plate > Materials > Steel (Structural) > 'Structural Steelwork (AS 4100-1998)'
Properties > Plate > Geometry > Set membrane thickness to 40 mm
Repeat the process for the lug plate property (2), using a thickness of 60 mm.
The model can now be viewed in 3D (refer to Figure 6.59):
View > Entity Display > Plate Display Options > Solid

6.11.2.1.4 Apply restraints to model

Select the nodes along the base of the lug, and then provide them with translational restraint:
Attributes > Node > Restraint > Choose translational X, Y & Z, then click 'Apply'

6.11.2.1.5 Apply load to the model

The load could be applied using several different methods (links, node forces or edge pressure). Select five nodes (refer to Figure 6.60) on the inside of the lug hole (around the top right corner):

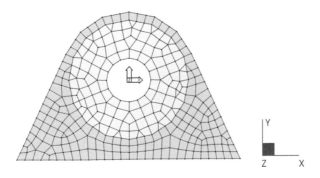

Figure 6.58 Lifting lug plate geometry.

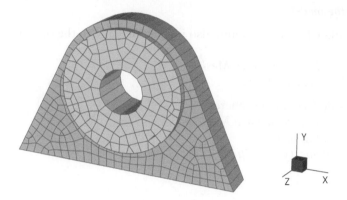

Figure 6.59 3D view of lifting lug.

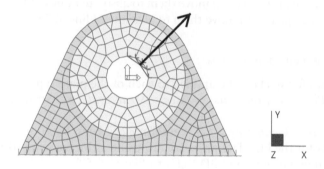

Figure 6.60 Lifting lug load.

> *Tools > Auto Assign > Restraints*
> Type: Multi-Point Link
> DoF: T
> Automatic Slave Node
> Select a new node on the inside of the multi-point link, and apply a node force:
> *Attributes > Node Force*
> $R = 460\,000$ (using UCS 1:[Cylindrical])

6.11.2.1.6 Check, save and solve the model

Check, save and solve the model as per the previous example.

6.11.2.1.7 View the results

The most important result for this model will be combined stress. The ideal way to view it is using the von Mises stress state. This is a combination of stresses into a single value, to be compared against yield stress.

Select the 'Results Settings' box and choose 'Contour', 'Stress', 'Combined' and 'Von Mises'. Deselect 'Extrapolate to Nodes' if it is selected. If the model is being viewed as a 'surface', you will have to check the mid-surface as well as the $+Z$ and $-Z$ faces of the plate. Alternatively, swap the model over to 3D (solid) view again to have the entire lug contoured. The maximum stress is 203 MPa, occurring at the shackle hole (refer to Figure 6.61).

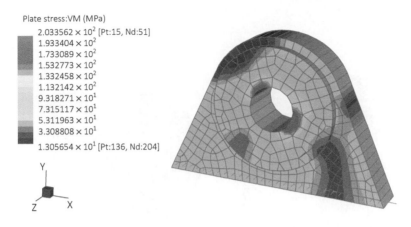

Plate stress:VM (MPa)

	2.033562×10^2 [Pt:15, Nd:51]
	1.933404×10^2
	1.733089×10^2
	1.532773×10^2
	1.332458×10^2
	1.132142×10^2
	9.318271×10^1
	7.315117×10^1
	5.311963×10^1
	3.308808×10^1
	1.305654×10^1 [Pt:136, Nd:204]

Figure 6.61 Lifting lug linear combined stress.

6.11.2.1.8 Non-linear Analysis

A stress-strain curve will need to be created to solve a non-linear material model. Close the results viewer and create the table (refer to Figure 6.56 and Table 6.35):

Tables > Stress vs. Strain

The table then needs to be applied to each of the plate types (Plate Property 1 and 2):

Property > Plate > Non-linear Stress vs. Strain Table: 'Stress vs. Strain Table 1'

A non-linear analysis can now be completed:

Solver > Non-linear Static Analysis

Select 'Load Increments', and then add six cases and input the data from Table 6.36.
The purpose of the load increments is to gradually load the model. This configuration will apply 10%, then 50%, then 100%. The additional cases are to check when the model will

Table 6.35 Stress vs. strain table (Grade 250)

Description	Strain	Stress (MPa)
Zero point	0	0
$0.9 \times$ Yield stress	$\dfrac{0.9\,f_y}{E_s} = \dfrac{0.9 \times 250}{200000} = 0.001125$	$0.9\,f_y = 0.9 \times 250 = 225$
Plastic range	$7 \times \dfrac{0.9\,f_y}{E_s} = 0.007875$	$0.9\,f_y = 0.9 \times 250 = 225$

Table 6.36 Suggested load increments

Title	Include	0.1	0.5	1.0	1.5	1.75	2.0
Load case 1	Y	0.1	0.5	1.0	1.5	1.75	2.0
Freedom case 1	Y	1.0	1.0	1.0	1.0	1.0	1.0

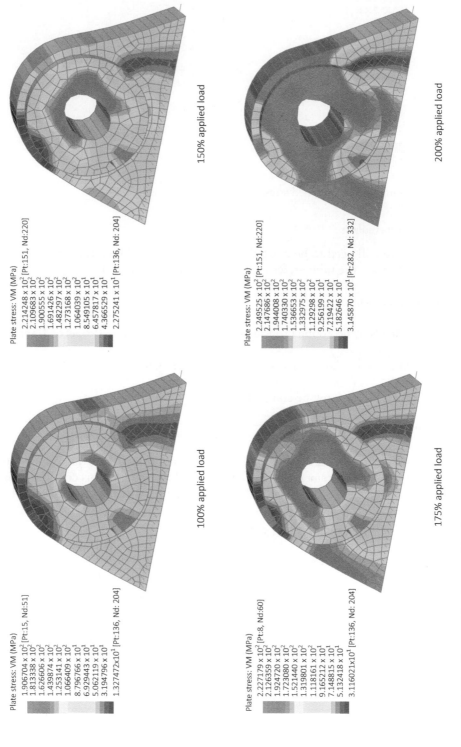

Figure 6.62 Non-linear lifting lug combined stress.

fail. If the gaps are too far apart, Strand7 can automatically add sub-increments; however, this slows down the solution time. The freedom case should always be 1.0 (unless springs are used), as this is the restraint for the model.

The load case needs to be selected in the top left corner after solving the model and opening the results. All six load cases should be available; however, the final case has not converged. The stress for 100% of the design load is 191 MPa, as shown in Figure 6.62. The stress zone is much wider in the 150% and 175% load cases, because the steel is behaving plastically. The 200% load case does not converge; therefore, the lug will fail at a point between 175% and 200%. More cases can be added to find the exact point of failure. The lug passes with a factor of 150% by following the calculation method in Section 6.3.

Chapter 7

Design aids

This chapter includes design aids in the form of section calculations, force diagrams and capacity tables. The equations and values provided are based on the theory provided in previous chapters.

7.1 SECTION CALCULATIONS

Section calculations are provided for both generic and specific cross-sections. The generic calculations provided in Table 7.1 can be used for any shape. The section is broken down into numerous rectangles for the purpose of calculations. Sections which are constructed of rectangular forms will have accurate values calculated, those which do not fit perfectly (such as the example images) should be considered as an approximation. Specific section calculations are provided in Table 7.2. These are accurate, however they do not include a root radius, therefore values will not exactly match tabulated standard section values.

7.2 FORCE DIAGRAMS

Force diagrams are provided in this section for the calculation of idealised arrangements. Table 7.3 shows bending, shear, reactions and deflections for single span beams. Table 7.4 provides the same details for multispan continuous arrangements, all of which are rounded for simplicity. All deflection calculations are also simplified values which provide a reasonably accurate solution. Multiple loading patterns should be checked to ensure that worst case values are calculated.

7.3 DESIGN CATALOGUES AND CAPACITY TABLES

Design capacity tables are provided for various steel and concrete items. Refer to the full design chapters for the detailed design of each specific item. Assumptions are made for typical arrangements as explained in each section. The catalogues which have been provided are based on sections and materials which are commonly used.

Table 7.1 Generic section calculations

	Generic calculations	
Item	*(about x-axis)*	*(about y-axis)*
Centroid	$x_c = \sum_{1}^{N}(A_i x_i) / \sum_{1}^{N}(A_i)$	$y_c = \sum_{1}^{N}(A_i y_i) / \sum_{1}^{N}(A_i)$
Distance to extreme fibre	$x_e = \text{MAX}[x_c, b - x_c]$	$y_e = \text{MAX}[y_c, d - y_c]$
Second moment of area	$I_x = \sum_{1}^{N}\left[\dfrac{b_i h_i^3}{12} + b_i h_i \overline{y_i}^2\right]$	$I_y = \sum_{1}^{N}\left[\dfrac{h_i b_i^3}{12} + h_i b_i \overline{x_i}^2\right]$
Elastic section modulus	$Z_x = I_x / y_e$	$Z_y = I_y / x_e$
Plastic section (see note 1)	$S_x = \sum_{1}^{N} A_{cx} e_{cy} + \sum_{1}^{N} A_{tx} e_{ty}$ Note: $\sum A_{cx} = \sum A_{tx}$	$S_y = \sum_{1}^{N} A_{cy} e_{cx} + \sum_{1}^{N} A_{ty} e_{tx}$ Note: $\sum A_{cy} = \sum A_{ty}$
Radius of gyration	$r_x = \sqrt{I_x / A}$	$r_y = \sqrt{I_y / A}$

Note: 1. Effective section modulus is calculated in accordance with Section 3.4.1.3, to include slenderness effects.

7.3.1 Steel catalogues and capacity tables

The section catalogues provided within this chapter are based on a limited selection of products currently available from OneSteel. Refer to the website (Chapter 8) to check availability and additional products.

7.3.1.1 Bolt capacity

For bolt capacity tables, refer to Tables 3.19 through 3.21.

7.3.1.2 Weld capacity

For weld capacity tables, refer to Table 3.26.

Table 7.2 Specific section calculations

Section	Calculation	
	(about x-axis)	*(about y-axis)*
Solid Circle	Area, $A = \pi r^2$	
	Distance to extreme fibre, $x_e = y_e = r$	
	Second moment of area, $I_x = I_y = \pi r^4/4$	
	Elastic section modulus, $Z_x = Z_y = \pi r^3/4$	
	Plastic section modulus, $S_x = S_y = 4r^3/3$	
	Radius of gyration, $r_x = r_y = r/2$	
Circular Hollow Section	$A = \pi(r_o^2 - r_i^2)$	
	$x_e = y_e = r_o$	
	$I_x = I_y = \pi(r_o^4 - r_i^4)/4$	
	$Z_x = Z_y = \pi(r_o^4 - r_i^4)/(4r_o)$	
	$S_x = S_y = 4(r_o^3 - r_i^3)/3$	
	$r_x = r_y = 0.5\sqrt{r_o^2 + r_i^2}$	
Solid Square	$A = a^2$	
	$x_e = y_e = a/2$	
	$I_x = I_y = a^4/12$	
	$Z_x = Z_y = a^3/6$	
	$S_x = S_y = a^3/4$	
	$r = r_x = r_y = a/\left(2\sqrt{3}\right)$	
Square Hollow Section	$A = a_o^2 - a_i^2$	
	$x_e = y_e = a_o/2$	
	$I = I_x = I_y = (a_o^4 - a_i^4)/12$	
	$Z_x = Z_y = (a_o^4 - a_i^4)/(6a_o)$	
	$S_x = S_y = (a_o^3 - a_i^3)/4$	
	$r = r_x = r_y = \sqrt{I/A}$	

(*Continued*)

Table 7.2 (Continued) Specific section calculations

Section	Calculation	
	(about x-axis)	*(about y-axis)*
Solid Rectangle	$A = bh$	
	$x_e = b/2$	$y_e = h/2$
	$I_x = bh^3/12$	$I_y = hb^3/12$
	$Z_x = bh^2/6$	$Z_y = hb^2/6$
	$S_x = bh^2/4$	$S_y = hb^2/4$
	$r_x = h/\left(2\sqrt{3}\right)$	$r_y = b/\left(2\sqrt{3}\right)$
Rectangle Hollow Section	$A = b_o h_o - b_i h_i$	
	$x_e = b_o/2$	$y_e = h_o/2$
	$I_x = (b_o h_o^3 - b_i h_i^3)/12$	$I_y = (h_o b_o^3 - h_i b_i^3)/12$
	$Z_x = (b_o h_o^3 - b_i h_i^3)/(6h_o)$	$Z_y = (h_o b_o^3 - h_i b_i^3)/(6b_o)$
	$S_x = (b_o h_o^2 - b_i h_i^2)/4$	$S_y = (h_o b_o^2 - h_i b_i^2)/4s$
	$r_x = \sqrt{I_x/A}$	$r_y = \sqrt{I_y/A}$
I Section	$A = 2t_f b_f + t_w d_1$	
	$x_e = b_f/2$	$y_e = D/2$
	$I_x = (b_f D^3 - b_f d_1^3 + t_w d_1^3)/12$	$I_y = (2t_f b_f^3 + d_1 t_w^3)/12$
	$Z_x = (b_f D^3 - b_f d_1^3 + t_w d_1^3)/(6D)$	$Z_y = (2t_f b_f^3 + d_1 t_w^3)/(6b_f)$
	$S_x = b_f t_f (D - t_f) + t_w d_1^2/4$	$S_y = t_f b_f^2/2 + d_1 t_w^2/4$
	$r_x = \sqrt{I_x/A}$	$r_y = \sqrt{I_y/A}$
PFC	$A = 2t_f b_f + t_w d_1$	
	$x_c = [t_f b_f^2 + d_1 t_w^2/2]/A$	$y_c = D/2$
	$x_e = b_f - x_c$	$y_e = D/2$
	if, $(b_f - A/4t_f) \geq t_w$, $x_{PNA} = b_f - A/4t_f$ otherwise, $x_{PNA} = A/(2D)$	$y_{PNA} = D/2$
	$I_x = (b_f D^3 - b_f d_1^3 + t_w d_1^3)/12$	$I_y = \dfrac{Db_f^3 - d_1(b_f - t_w)^3}{3} - A(b_f - x_c)^2$
	$Z_x = (b_f D^3 - b_f d_1^3 + t_w d_1^3)/(6D)$	$Z_y = \dfrac{Db_f^3 - d_1(b_f - t_w)^3}{3x_e} - \dfrac{A(b_f - x_c)^2}{x_e}$

(Continued)

Table 7.2 (Continued) Specific section calculations

Section	Calculation	
	(about x-axis)	*(about y-axis)*
	$S_x = b_f t_f (D - t_f) + t_w d_1^2 / 4$	if $(b_f - A/4t_f) \geq t_w$, $$S_y = \frac{(b_f - t_w)^2 t_f}{2} - \frac{D^2 t_w^2}{8t_f} + \frac{Dt_w b_f}{2}$$ otherwise, $$S_y = \frac{t_w^2 D}{4} + t_f (b_f - t_w)\left(b_f - \frac{t_f (b_f - t_w)}{D} \right)$$
	$r_x = \sqrt{I_x/A}$	$r_y = \sqrt{I_y/A}$
Angle	$A = t(B + D - t)$	
	$\theta = \frac{1}{2} \tan^{-1}\left(\frac{-2I_{np}}{I_n - I_p} \right)$ where, $I_{np} = \pm \dfrac{bBdDt}{4(B+d)}$	
	$n_c = \dfrac{B^2 + dt}{2(B+d)}$	$P_c = \dfrac{D^2 + bt}{2(D+b)}$
	$n_e = B - n_c$	$P_e = D - P_c$
	if $2Dt \geq A$ $n_{PNA} = A/2D$ otherwise, $n_{PNA} = B - A/2t$	if $2Bt \geq A$, $P_{PNA} = A/2B$ otherwise, $P_{PNA} = D - A/2t$
	$I_x = \dfrac{I_n + I_P}{2} + \dfrac{I_n - I_P}{2\cos 2\theta}$	$I_y = \dfrac{I_n + I_P}{2} - \dfrac{I_n - I_P}{2\cos 2\theta}$
	$I_n = \dfrac{1}{3}\left[tP_e^3 + BP_c^3 - b(P_c - t)^3 \right]$	$I_P = \dfrac{1}{3}\left[tn_e^3 + Dn_c^3 - d(n_c - t)^3 \right]$
	$Z_x = I_x/Y_e$	$Z_y = I_y/x_e$
	$Z_n = I_n/P_e$	$Z_P = I_P/n_e$
	if $2Dt \geq A$, $$S_n = \frac{1}{2}[BP_{PNA}^2 + b(t - P_{PNA})^2 + t(D - P_{PNA})^2]$$ otherwise, $$S_n = Bt\left(P_c - \frac{t}{2} \right) + \frac{t}{2}(P_c - t)^2 (D - P_c)^2$$	if $2Bt \geq A$, $$S_P = \frac{1}{2}\left[Dn_{PNA}^2 + d(t - n_{PNA})^2 + t(B - n_{PNA})^2 \right]$$ otherwise, $$S_P = Dt\left(n_c - \frac{t}{2} \right) + \frac{t}{2}(n_c - t)^2 (B - n_c)^2$$
	$r_n = \sqrt{I_n/A}$ $r_x = \sqrt{I_x/A}$	$r_P = \sqrt{I_P/A}$ $r_y = \sqrt{I_y/A}$

Note: For torsion calculations, refer to Section 3.9.

Table 7.3 Bending, shear, reaction and displacement diagrams

Diagram	Equations
Force Diagram 1	Moment, $M = PL/4$ Shear, $V = P/2$ Reactions, $R_A = P/2$ Reactions, $R_B = P/2$ Deflection, $\delta = \dfrac{PL^3}{48EI}$
Force Diagram 2	[For $L_B \geq L_A$] $M = PL_A L_B / L$ $V = R_A = PL_B / L$ $R_B = PL_A / L$ $\delta \approx \dfrac{PL_A}{3EIL}\left(\dfrac{L^2 - L_A^2}{3}\right)^{3/2}$ $x = L - \left(\dfrac{L^2 - L_A^2}{3}\right)^{1/2}$
Force Diagram 3	[For $L_B \geq L_A$] $M_A = PL_A(2L_B + L_C)/L$ $M_B = PL_B(2L_A + L_C)/L$ $V = R_A = P(2L_B + L_C)/L$ $R_B = P(2L_A + L_C)/L$ $\delta \approx \dfrac{PL^3}{48EI}\left(\dfrac{3L_A}{L} - \dfrac{4L_A^3}{L^3}\right) + \dfrac{PL^3}{48EI}\left(\dfrac{3L_B}{L} - \dfrac{4L_B^3}{L^3}\right)$
Force Diagram 4	$M = -PL_A$ $V = R_A = P$ $\delta_1 \approx \dfrac{PL_A^3}{3EI}$ $\delta \approx \dfrac{PL_A^3}{3EI}\left(1 + \dfrac{3L_B}{2L_A}\right)$
Force Diagram 5	$M = PL/8$ $M_A = M_B = -PL/8$ $V = R_A = R_B = P/2$ $\delta = \dfrac{PL^3}{192EI}$

(Continued)

Table 7.3 (Continued) Bending, shear, reaction and displacement diagrams

Diagram	Equations
Force Diagram 6	$$M = \frac{2PL_A^2}{L^3}(L - L_A)^2$$ $$M_A = -\frac{PL_A}{L^2}(L - L_A)^2$$ $$M_B = -\frac{PL_A^2}{L^2}(L - L_A)$$ $$V = \text{MAX}[R_A, R_B]$$ $$R_A = \frac{P}{L^3}(L - L_A)^2(L + 2L_A)$$ $$R_B = \frac{PL_A^2}{L^3}(3L - 2L_A)$$ $$\delta = \frac{2P(L - L_A)^2 L_A^3}{3EI(L + 2L_A)^2}$$ $$x = 2LL_A/(L + 2L_A)$$
Force Diagram 7	$$M = \frac{PL_A}{2L^3}(L - L_A)^2(2L + L_A)$$ $$M_A = 0$$ $$M_B = -\frac{PL_A}{2L^2}(L^2 - L_A^2)$$ $$V = \text{MAX}[R_A, R_B]$$ $$R_A = \frac{P}{2L^3}(L - L_A)^2(2L + L_A)$$ $$R_B = \frac{PL_A}{2L^3}(3L^2 - L_A^2)$$ $$\delta = \frac{PL_A(L^2 - L_A^2)^3}{3EI(3L^2 - L_A^2)^2}$$ $$x = L(L^2 + L_A^2)/(3L^2 - L_A^2)$$
Force Diagram 8	$$M = wL^2/8$$ $$V = R_A = R_B = wL/2$$ $$\delta = \frac{5wL^4}{384EI}$$
Force Diagram 9	$$M = R_B L_B$$ $$V = R_A = wL_A - R_B$$ $$R_B = \frac{w}{2L}(L - L_B)^2$$

(*Continued*)

Table 7.3 (Continued) Bending, shear, reaction and displacement diagrams

Diagram	Equations
Force Diagram 10	$M = -wL^2/2$ $V = R_A = wL$ $\delta = \dfrac{wL^4}{8EI}$
Force Diagram 11	$M = wL^2/24$ $M_A = M_B = -wL^2/12$ $V = R_A = R_B = wL/2$ $\delta = \dfrac{wL^4}{384EI}$
Force Diagram 12	$M = 9wL^2/128$ $x_M = 3L/8$ $M_B = -wL^2/8$ $V = R_B = 5wL/8$ $R_A = 3wL/8$ $\delta = 0.0054\dfrac{wL^4}{EI}$ $x = 0.4215L$

Source: Gorenc, BE et al., *Steel Designers' Handbook*, 7th ed., University of New South Wales Press Ltd., Australia, 2005; Young, WC et al., *Roark's Formulas for Stress and Strain*, 8th ed., McGraw-Hill, 2012 [5,18].

Note: E = Young's modulus of material (MPa), I = Second moment of area (mm⁴), L = Length (mm), M = Moment as shown (Nmm), P = Point load (N), R_A = Reaction at A (N), R_B = Reaction at B (N), V = Maximum shear (N), w = Uniform distributed load (N/mm), x = Maximum deflection position (mm), x_M = Maximum moment position (mm), δ = Displacement (mm).

7.3.1.3 Steel plates

Plates are available in large sheets and can therefore be used to create profiles of various sizes and shapes. They are commonly used to create base plates and any other connections. Plates are available in widths that vary from 2.4 to 3.1 m and lengths that vary from 5.5 to 9.6 m depending on the specified thickness. Table 7.5 provides a list of available plate thicknesses.

7.3.1.4 Steel flats

Flats are used to fabricate steelwork to standard widths and are typically cheaper than plates, especially when the available widths can be adopted. Flats are commonly used as stiffeners and cleats. Table 7.6 provides a list of available flat dimensions.

Table 7.4 Loading patterns

Diagram	Equations
Load Pattern Diagram 1	$M = 0.096wL^2$ $x_M = 0.438\,L$ $M_B = -0.063wL^2$ $R_A = 0.438wL$ $R_B = 0.625wL$ $R_C = -0.063wL$ $V = wL - R_A$ $\delta = 0.0092\dfrac{wL^4}{EI}$ $x = 0.472\,L$
Load Pattern Diagram 2 *Note* : $\lvert M_B \rvert > \lvert M \rvert$	$M = 0.07wL^2$ $x_M = 0.375L$ $M_B = -0.125wL^2$ $R_A = R_C = 0.375wL$ $R_B = 1.25wL$ $V = wL - R_A$ $\delta = 0.0054\dfrac{wL^4}{EI}$ $x = 0.421L$
Load Pattern Diagram 3	$M = 0.094wL^2$ $x_M = 0.433L$ $M_B = -0.067wL^2$ $M_C = 0.017wL^2$ $R_A = 0.433wL$ $R_B = 0.65wL$ $R_C = -0.1wL$ $R_D = 0.017wL$ $V = wL - R_A$ $\delta = 0.0089\dfrac{wL^4}{EI}$ $x = 0.471L$
Load Pattern Diagram 4	$M = 0.101wL^2$ $x_M = 0.45L$ $M_B = -0.05wL^2$ $R_A = R_D = 0.45wL$ $V = R_B = R_C = 0.55wL$ $\delta = 0.0099\dfrac{wL^4}{EI}$ $x = 0.479L$

(Continued)

Table 7.4 (Continued) Loading patterns

Diagram	Equations
Load Pattern Diagram 5	$M = 0.08wL^2$ $x_M = 0.4L$ $M_B = -0.1wL^2$ $R_A = R_D = 0.4wL$ $R_B = R_C = 1.1wL$ $V = wL - R_A$ $\delta = 0.0069\dfrac{wL^4}{EI}$ $x = 0.446L$
Load Pattern Diagram 6	$M = 0.073wL^2$ $x_M = 0.383L$ $M_2 = -0.054wL^2$ $M_B = -0.117wL^2$ $M_C = -0.003wL^2$ $R_A = 0.383wL$ $R_B = 1.2wL$ $R_C = 0.45wL$ $R_D = -0.033wL$ $V = wL - R_A$ $\delta = 0.0059\dfrac{wL^4}{EI}$ $x = 0.43L$
Load Pattern Diagram 7	$M = 0.075wL^2$ $M_B = -0.05wL^2$ $R_A = R_D = -0.05wL$ $R_B = R_C = 0.55wL$ $V = R_A + R_B$ $\delta = 0.0068\dfrac{wL^4}{EI}$

Source: Gorenc, BE et al., *Steel Designers' Handbook*, 7th ed., University of New South Wales Press Ltd., Australia, 2005; Young, WC et al., *Roark's Formulas for Stress and Strain*, 8th ed., McGraw-Hill, 2012.

Note: E = Young's modulus of material (MPa), I = Second moment of area (mm⁴), L = Length (mm), M = Moment as shown (Nmm), R_A = Reaction at A (N), R_B = Reaction at B (N), R_C = Reaction at C (N), R_D = Reaction at D (N), V = Maximum shear (N), w = Uniform distributed load (N/mm), x = Maximum deflection position (mm), x_M = Maximum moment position (mm), δ = Displacement (mm).

Table 7.5 Available plate sizes

Available thicknesses (mm)	5, 6, 8, 10, 12, 16, 20, 25, 32, 40, 50, 60, 70, 80

Source: OneSteel. Grades 250 & 350, AS/NZS 3678.

Table 7.6 Available flat sizes

Thickness (mm)	Available widths (mm)
5	25, 32, 40, 50, 65, 75, 100, 130, 150
6	25, 32, 40, 50, 65, 75, 90, 100, 110, 130, 150, 180, 200, 250, 300
8	25, 32, 40, 50, 65, 75, 90, 100, 110, 130, 150, 200, 250, 300
10	20, 25, 32, 40, 50, 65, 75, 90, 100, 110, 130, 150, 180, 200, 250, 300
12	25, 32, 40, 50, 65, 75, 90, 100, 110, 130, 150, 180, 200, 250, 300
16	40, 50, 65, 75, 100, 130, 150
20	40, 50, 65, 75, 100, 130, 150
25	50, 75, 100, 130, 150

Source: OneSteel. Grade 300, Standard length = 6.0 m.

7.3.1.5 Steel square sections

Available square section dimensions are provided in Table 7.7 for a standard length of six metres.

7.3.1.6 Steel round sections

Available round section dimensions are provided in Table 7.8 for a standard length of six metres.

7.3.1.7 Plate capacities

Plate capacity tables are provided in Table 7.9 based on the theory shown in Section 3.10.6.

7.3.1.8 Pin capacities

Pin capacity tables provided in Table 7.10 are based on the theory shown in Section 3.10.4. Ply in bearing and pin in bearing are not included in the capacities.

Table 7.7 Available square sizes

Available dimensions (mm)	10[a], 12, 16, 20, 25, 32, 40

Source: OneSteel. Grade 300, Standard length = 6.0 m.

[a] Check availability.

Table 7.8 Available round sizes

Available diameters (mm)	10, 12, 13, 14, 15, 16, 17, 18, 19, 20, 22, 24, 27, 30, 33, 36, 39, 42, 45, 48, 50, 56, 60, 65, 75, 90, 100, 110, 120, 130, 140, 150, 160, 170, 180, 190, 200

Source: OneSteel. Grade 300, Standard length = 6.0 m.

Table 7.9 Plate capacity table

Capacity	Grade	Plate thickness										
		5	6	8	10	12	16	20	25	30	32	40
ϕV_v (kN)	250	75.6	90.7	121	140	168	216	270	338	405	432	540
	300	86.4	104	138	173	194	259	302	378	454	484	605
	350	97.2	117	156	194	233	302	378	459	551	588	734
ϕM_y (kNm)	250	0.158	0.227	0.403	0.585	0.842	1.44	2.25	3.52	5.06	5.76	9.00
	300	0.180	0.259	0.461	0.720	0.972	1.73	2.52	3.94	5.67	6.45	10.1
	350	0.203	0.292	0.518	0.810	1.17	2.02	3.15	4.78	6.89	7.83	12.2
ϕN_t (kN)	250	126	151	202	234	281	360	450	563	675	720	900
	300	144	173	230	288	324	432	504	630	756	806	1008
	350	162	194	259	324	389	504	630	765	918	979	1224

Notes:

1. Capacities are for a 100 mm wide plate. Capacity of other plate widths can be calculated by multiplying by the ratio of widths. For example, 200 mm plate capacity is tabulated value multiplied by $200/100 = 2$.

2. Grade 250 (AS/NZS 3678), Grade 300 (AS/NZS 3679.1), Grade 350 (AS/NZS 3678).

3. Tension capacity is based on a symmetric connection with no holes.

4. Bending capacity is about the minor axis.

Table 7.10 Pin capacity table

Capacity	Grade	Yield strength (MPa)	Pin diameter (mm)										
			8	10	11.2	12.5	16	17	18	20	24	30	36
ϕV_f (kN)	300	290	7.23	11.3	14.2	17.7	28.9	32.6	36.6	45.2	65.1	102	146
	350	340	8.48	13.2	16.6	20.7	33.9	38.3	42.9	53.0	76.3	119	172
	4140(t)	665	16.6	25.9	32.5	40.5	66.3	74.9	83.9	104	149	233	336
ϕM_u (kNm)	300	290	0.020	0.04	0.05	0.08	0.16	0.19	0.23	0.31	0.53	1.04	1.80
	350	340	0.023	0.05	0.06	0.09	0.19	0.22	0.26	0.36	0.63	1.22	2.12
	4140(t)	665	0.05	0.09	0.12	0.17	0.36	0.44	0.52	0.71	1.23	2.39	4.14

Note: Grade 300 (AS/NZS 3679.1), Grade 350 (AS/NZS 3679.1), Grade 4140 Condition T (AS 1444).

7.3.1.9 Steel sections (welded, hot rolled and cold formed)

Open structural sections are available in grades 300Plus®, 300PlusL0® and 300PlusS0®. Welded beams (WB) and columns (WC) are also available in grade 400, and all other sections (UC, UB, PFC, EA, UA) are also available in grades 350 and 350L0. Tables for tapered flange beams (TFB) and universal bearing piles (UBP) are not provided, as they are less commonly used. The naming convention for sections is that the nominal dimensions appear before the section type, which is followed by the weight. Tabulated data and graphs are shown for the grade 300 due to its strong prevalence in design and construction. Standard supply lengths range from 7.5 to 20 m depending on the section designation. Refer to the OneSteel website for details of all available sections and grades. Open section capacity tables are provided in Tables 7.11 through 7.31 and Figures 7.1 through 7.19.

Hollow structural sections are cold formed and are available in grades C250L0, C350L0 and C450L0. Section sizes and grades were changed several years ago, and a full section table book is available online for all current sections through the OneSteel website. Standard supply lengths range from 6.5 to 12 m depending on the section size. Hollow section capacity tables are provided in Tables 7.32 through 7.61 and Figures 7.20 through 7.31.

Tabulated section capacities for bending (ϕM_s), shear (ϕV_v), compression (ϕN_s), tension (ϕN_t) and torsion (ϕT_u) are calculated based on the vendor provided section details and the theory provided in Chapter 3.

Web bearing capacities are provided as $\phi R_{by}/b_{bf}$ and $\phi R_{bb}/b_b$ and therefore need to be multiplied by the relevant width of stress distribution; refer to Section 3.5.4.

7.3.1.10 Members subject to bending

The graphs and tables provided are for members which are subject to bending about their major axis without full lateral restraint. Values of ϕM_b are based on the tabulated section properties and the theory provided in Section 3.4.

The effective length for bending (L_e) of the member should be calculated as shown in Section 3.4.2.4, based on the twist restraint, load height and lateral rotation restraint. Members with full lateral restraint can be checked against the section capacity (ϕM_s) or the member capacity shown in this section with an effective length of zero. The smallest tabulated effective length is also the section capacity.

The critical variable included in the data is that $\alpha_m = 1$; where this is not the case, the adopted bending capacity (ϕM_b) should be multiplied by the actual value of α_m; however, the resultant shall not exceed ϕM_s. Refer to Tables 3.7 and 3.8 for values of α_m (this is a critical step and should not be neglected).

7.3.1.11 Members subject to axial compression

Graphs and tables provided are for members which are subject to axial compression. Values of ϕN_{cy} are for members which are free to buckle about their minor axis, and ϕN_{cx} are for the major axis. Members which are unrestrained in both directions will always be governed by the minor axis capacity. Values of ϕN_{cx} and ϕN_{cy} are based on the tabulated section properties and the theory provided in Section 3.7.2.

The effective length (L_e) of the member for compression is calculated differently from bending and should be completed as shown in Section 3.7.2 based on the buckling shape of the member. The smallest tabulated effective length is also the section capacity.

Table 7.11 Welded Beams: 1200 WB

Geometry and properties

Section Name	1200 WB 455	1200 WB 423	1200 WB 392	1200 WB 342	1200 WB 317	1200 WB 278	1200 WB 249
Weight (kg/m)	455	423	392	342	317	278	249
d (mm)	1200	1192	1184	1184	1176	1170	1170
b_f (mm)	500	500	500	400	400	350	275
t_f (mm)	40	36	32	32	28	25	25
t_w (mm)	16	16	16	16	16	16	16
d_1 (mm)	1120	1120	1120	1120	1120	1120	1120
A_g (mm²)	57900	53900	49900	43500	40300	34400	31700
I_x (10⁶ mm⁴)	15300	13900	12500	10400	9250	7610	6380
Z_x (10³ mm³)	25600	23300	21100	17500	15700	13000	10900
S_x (10³ mm³)	28200	25800	23400	19800	17900	15000	12900
r_x (mm)	515	508	500	488	479	464	449
I_y (10⁶ mm⁴)	834	750	667	342	299	179	87
Z_y (10³ mm³)	3330	3000	2670	1710	1500	1020	633
S_y (10³ mm³)	5070	4570	4070	2630	2310	1600	1020
r_y (mm)	120	118	116	88.6	86.1	71.1	52.4
J (10³ mm⁴)	22000	16500	12100	9960	7230	5090	4310
I_w (10⁹ mm⁶)	280000	251000	221000	113000	98500	58700	28500
f_{yf} (MPa)	280	280	280	280	280	280	280
f_{yw} (MPa)	300	300	300	300	300	300	300
Form Factor k_f	0.837	0.825	0.811	0.783	0.766	0.733	0.701
X Compactness	C	C	C	C	C	C	C
Z_{ex} (10³ mm³)	28200	25800	23400	19800	17900	15000	12900
Y Compactness	C	C	N	C	C	C	C
Z_{ey} (10³ mm³)	5000	4500	4000	2560	2240	1530	949
s_{gf} (mm)	140 / 90 [280]						140/90
s_{gw} (mm)	140 / 90 / 70						
$\phi R_{bb}/b_b$ (kN/mm)	0.783						
$\phi R_{by}/b_{bf}$ (kN/mm)	5.4						

Capacity table

Section Name	Weight (kg/m)	ϕM_{sx} (kNm)	ϕM_{sy} (kNm)	ϕV_v (kN)	ϕN_s (kN)	ϕN_t (kN)
1200 WB 455	455	7106	1260	3110	12212	14591
1200 WB 423	423	6502	1134	3090	11206	13583
1200 WB 392	392	5897	1008	3069	10198	12575
1200 WB 342	342	4990	645	3069	8583	10962
1200 WB 317	317	4511	564	3048	7779	10156
1200 WB 278	278	3780	386	3033	6539	8921
1200 WB 249	249	3251	239	3033	5600	7988

Member Bending Capacity ϕM_{bx} ($\alpha_m = 1$) (kNm) — Effective length (m)

Section Name	1	2	4	6	8	10	12	14	16	18	20
1200 WB 455	7106	7106	6855	6279	5608	4925	4292	3737	3269	2882	2562
1200 WB 423	6502	6502	6261	5719	5085	4438	3838	3315	2877	2517	2223
1200 WB 392	5897	5897	5666	5159	4564	3957	3394	2907	2502	2172	1905
1200 WB 342	4990	4990	4562	3935	3274	2682	2202	1830	1548	1331	1163
1200 WB 317	4511	4511	4103	3514	2894	2344	1902	1565	1311	1119	972
1200 WB 278	3780	3746	3263	2644	2057	1592	1254	1014	842	715	620
1200 WB 249	3251	3107	2457	1772	1263	934	724	585	488	418	365

Member Axial Capacity (Minor Axis) ϕN_{cy} (kN) — Effective length (m)

Section Name	1	2	4	6	8	10	12	14	16	18	20
1200 WB 455	12212	12050	11057	9981	8750	7422	6158	5077	4203	3510	2962
1200 WB 423	11206	11048	10128	9128	7983	6752	5588	4599	3802	3172	2675
1200 WB 392	10198	10047	9202	8282	7227	6097	5035	4137	3416	2848	2400
1200 WB 342	8583	8254	7314	6214	4995	3892	3032	2395	1926	1576	1311
1200 WB 317	7779	7466	6596	5574	4450	3449	2677	2110	1695	1386	1151
1200 WB 278	6539	6146	5245	4161	3101	2294	1731	1341	1064	863	713
1200 WB 249	5600	5025	3897	2672	1797	1257	919	698	547	440	361

Member Axial Capacity (Major Axis) ϕN_{cx} (kN) — Effective length (m)

Section Name	1	2	4	6	8	10	12	14	16	18	20
1200 WB 455	12212	12212	12212	12212	12118	11887	11658	11427	11195	10958	10716
1200 WB 423	11206	11206	11206	11206	11113	10901	10689	10476	10261	10042	9818
1200 WB 392	10198	10198	10198	10198	10108	9913	9719	9524	9326	9126	8920
1200 WB 342	8583	8583	8583	8583	8503	8338	8173	8008	7840	7670	7496
1200 WB 317	7779	7779	7779	7779	7702	7551	7401	7250	7097	6941	6781
1200 WB 278	6539	6539	6539	6539	6469	6341	6213	6085	5955	5822	5687
1200 WB 249	5600	5600	5600	5600	5535	5425	5314	5203	5090	4975	4858

Note: Geometry and capacity table, Grade 300, $f_u = 440$ MPa

Gauge lines listed are for M24 and M20 bolts in order of preference (refer Figure 3.29). Square brackets are used for a second, wider bolt gauge, s_{gf2}.

Table 7.12 Welded Beams: 900 and 1200 WB

Geometry and section properties

Section Name / Weight (kg/m)	1000 WB 322	1000 WB 296	1000 WB 258	1000 WB 215	900 WB 282	900 WB 257	900 WB 218	900 WB 175
d (mm)	1024	1016	1010	1000	924	916	910	900
b_f (mm)	400	400	350	300	400	400	350	300
t_f (mm)	32	28	25	20	32	28	25	20
t_w (mm)	16	16	16	16	12	12	12	12
d_1 (mm)	960	960	960	960	860	860	860	860
A_g (mm²)	41000	37800	32900	27400	35900	32700	27800	22300
I_x (10^6 mm⁴)	7480	6650	5430	4060	5730	5050	4060	2960
Z_x (10^3 mm³)	14600	13100	10700	8120	12400	11000	8930	6580
S_x (10^3 mm³)	16400	14800	12300	9570	13600	12200	9960	7500
r_x (mm)	427	420	406	385	399	393	382	364
I_y (10^6 mm⁴)	342	299	179	90.3	341	299	179	90.1
Z_y (10^3 mm³)	1710	1490	1020	602	1710	1490	1020	601
S_y (10^3 mm³)	2620	2300	1590	961	2590	2270	1560	931
r_y (mm)	91.3	89	73.8	57.5	97.5	95.6	80.2	63.5
J (10^3 mm⁴)	9740	7010	4870	2890	8870	6150	4020	2060
I_w (10^9 mm⁶)	84100	73000	43400	21700	67900	58900	35000	17400
f_{yf} (MPa)	280	280	280	300	280	280	280	300
f_{yw} (MPa)	300	300	300	300	310	310	310	310
Form Factor k_f	0.832	0.817	0.79	0.738	0.845	0.83	0.8	0.744
X Compactness	C	C	C	C	C	C	C	C
Z_{ex} (10^3 mm³)	16400	14800	12300	9570	13600	12200	9960	7500
Y Compactness	C	C	C	C	C	C	C	C
Z_{ey} (10^3 mm³)	2560	2240	1530	903	2560	2240	1530	901
s_{gf} (mm)	140 / 90 [280]	140 / 90 [280]	140/90	140/90	140 / 90 [280]	140 / 90 [280]	140/90	140/90
s_{gw} (mm)	140 / 90 / 70	140 / 90 / 70	140 / 90 / 70	140 / 90 / 70	140 / 90 / 70	140 / 90 / 70	140 / 90 / 70	140 / 90 / 70
$\phi R_{bb}/b_b$ (kN/mm)	1.02	1.02	0.567	0.567	1.02	1.02	0.567	0.567
$\phi R_{by}/b_{bf}$ (kN/mm)	5.4	5.4	4.19	4.19	5.4	5.4	4.19	4.19

Capacity table

Section Name / Weight (kg/m)	ϕM_{sx} (kNm)	ϕM_{by} (kNm)	ϕV_v (kN)	ϕN_s (kN)	ϕN_t (kN)
1000 WB 322	4133	645	2654	8596	10332
1000 WB 296	3730	564	2633	7782	9526
1000 WB 258	3100	386	2618	6550	8291
1000 WB 215	2584	244	2592	5460	7398
900 WB 282	3427	645	1856	7645	9047
900 WB 257	3074	564	1840	6840	8240
900 WB 218	2510	386	1828	5604	7006
900 WB 175	2025	243	1808	4480	6021

Member Bending Capacity ϕM_{bx} ($\alpha_m = 1$) (kNm) — Effective length (m)

Section Name	1	2	4	6	8	10	12	14	16	18	20
1000 WB 322	4133	4133	3802	3309	2791	2324	1940	1638	1404	1221	1078
1000 WB 296	3730	3730	3414	2949	2461	2025	1669	1393	1182	1020	894
1000 WB 258	3100	3079	2702	2219	1757	1385	1110	910	764	656	573
1000 WB 215	2584	2491	2015	1488	1077	800	620	500	416	356	310
900 WB 282	3427	3427	3184	2804	2400	2031	1720	1470	1272	1116	991
900 WB 257	3074	3074	2844	2489	2109	1762	1472	1243	1064	924	814
900 WB 218	2510	2504	2226	1862	1504	1205	977	807	682	587	514
900 WB 175	2025	1970	1633	1242	916	688	535	431	358	306	266

Member Axial Capacity (Minor Axis) ϕN_{cy} (kN) — Effective length (m)

Section Name	1	2	4	6	8	10	12	14	16	18	20
1000 WB 322	8596	8265	7324	6222	5001	3897	3035	2397	1928	1578	1312
1000 WB 296	7782	7470	6600	5579	4456	3454	2682	2114	1698	1388	1154
1000 WB 258	6550	6156	5254	4167	3106	2297	1733	1342	1066	864	714
1000 WB 215	5381	4930	3879	2708	1840	1294	949	722	566	456	374
900 WB 282	7645	7397	6613	5711	4690	3721	2933	2335	1887	1550	1292
900 WB 257	6840	6611	5901	5081	4157	3287	2585	2054	1659	1361	1134
900 WB 218	5604	5321	4619	3781	2908	2194	1675	1307	1042	848	702
900 WB 175	4448	4114	3356	2469	1733	1239	917	702	553	446	367

Member Axial Capacity (Major Axis) ϕN_{cx} (kN) — Effective length (m)

Section Name	1	2	4	6	8	10	12	14	16	18	20
1000 WB 322	8596	8596	8596	8593	8398	8204	8008	7811	7608	7400	7183
1000 WB 296	7782	7782	7782	7776	7598	7420	7242	7062	6877	6686	6488
1000 WB 258	6550	6550	6550	6536	6384	6232	6080	5925	5766	5602	5431
1000 WB 215	5460	5460	5460	5428	5294	5160	5025	4888	4747	4600	4447
900 WB 282	7645	7645	7645	7601	7414	7228	7039	6848	6650	6445	6232
900 WB 257	6840	6840	6840	6797	6629	6461	6292	6119	5941	5756	5564
900 WB 218	5604	5604	5604	5566	5427	5288	5147	5004	4856	4703	4543
900 WB 175	4480	4480	4480	4433	4316	4200	4082	3961	3835	3705	3568

Note: Geometry and capacity table, Grade 300, $f_u = 440$ MPa.

Gauge lines listed are for M24 and M20 bolts in order of preference (refer Figure 3.29). Square brackets are used for a second, wider bolt gauge, s_{gf2}.

Table 7.13　Welded Beams: 700 and 800 WB

Section Name	Weight (kg/m)	d (mm)	b_f (mm)	t_f (mm)	t_w (mm)	d_1 (mm)	A_g (mm²)	I_x (10⁶ mm⁴)	Z_x (10³ mm³)	S_x (10³ mm³)	r_x (mm)	I_y (10⁶ mm⁴)	Z_y (10³ mm³)	S_y (10³ mm³)	r_y (mm)	J (10³ mm⁴)	I_w (10⁹ mm⁶)	f_yf (MPa)	f_yw (MPa)	Form Factor k_f	X Compactness	Z_ex (10³ mm³)	Y Compactness	Z_ey (10³ mm³)	s_gf (mm)	s_gw (mm)	φR_bb/b_b (kN/mm)	φR_by/b_bf (kN/mm)
800 WB 192	192	816	300	28	10	760	24400	2970	7290	8060	349	126	840	1280	71.9	4420	19600	280	310	0.824	C	8060	C	1260	140/90	140/90/70	0.425	3.49
800 WB 168	168	810	275	25	10	760	21400	2480	6140	6840	341	86.7	631	964	63.7	2990	13400	280	310	0.799	C	6840	C	946				
800 WB 146	146	800	275	20	10	760	18600	2040	5100	5730	331	69.4	505	775	61.1	1670	10600	300	310	0.763	N	5710	C	757				
800 WB 122	122	792	250	16	10	760	15600	1570	3970	4550	317	41.7	334	519	51.7	921	6280	300	310	0.718	N	4530	N	498			0.544	3.49
700 WB 173	173	716	275	28	10	660	22000	2060	5760	6390	306	97.1	706	1080	66.4	4020	11500	280	310	0.85	C	6390	C	1060				
700 WB 150	150	710	250	25	10	660	19100	1710	4810	5370	299	65.2	521	798	58.4	2690	7640	280	310	0.828	C	5370	C	782				
700 WB 130	130	700	250	20	10	660	16600	1400	3990	4490	290	52.1	417	642	56	1510	6030	300	310	0.795	C	4490	C	626				
700 WB 115	115	692	250	16	10	660	14600	1150	3330	3790	281	41.7	334	516	53.5	888	4770	300	310	0.767	C	3790	N	498				

Section Name	Weight (kg/m)	φM_sx (kNm)	φM_sy (kNm)	φV_v (kN)	φN_s (kN)	φN_t (kN)	Member Bending Capacity φM_bx [α_m=1] (kNm) Effective length (m)											Member Axial Capacity (Minor Axis) φN_cy (kN) Effective length (m)											Member Axial Capacity (Major Axis) φN_cx (kN) Effective length (m)										
							1	2	4	6	8	10	12	14	16	18	20	1	2	4	6	8	10	12	14	16	18	20	1	2	4	6	8	10	12	14	16	18	20
800 WB 192	192	2031	318	1282	5067	6149	2031	2004	1733	1410	1122	900	737	619	531	464	412	5067	4731	3989	3097	2265	1636	1242	958	758	614	506	5067	5067	5067	4990	4850	4710	4567	4420	4267	4107	3939
800 WB 168	168	1724	238	1273	4309	5393	1724	1679	1404	1093	837	652	525	436	371	323	285	4279	3958	3232	2380	1671	1195	885	678	534	430	354	4309	4309	4309	4241	4121	4001	3879	3752	3621	3483	3338
800 WB 146	146	1542	204	1257	3832	5022	1542	1490	1213	907	665	501	392	319	267	230	201	3790	3488	2794	1999	1378	976	719	548	431	347	285	3832	3832	3832	3758	3647	3535	3421	3303	3179	3049	2912
800 WB 122	122	1223	134	1245	3024	4212	1223	1154	880	610	423	308	236	190	158	135	118	2956	2681	2020	1338	884	613	446	338	264	212	174	3024	3024	3024	2962	2874	2784	2693	2598	2499	2394	2284
700 WB 173	173	1610	267	1199	4712	5544	1610	1575	1337	1073	853	690	571	485	421	371	331	4684	4336	3553	2631	1855	1330	986	755	595	480	395	4712	4712	4712	4579	4429	4276	4118	3953	3779	3594	3401
700 WB 150	150	1353	197	1189	3985	4813	1353	1303	1062	812	621	489	398	335	288	253	225	3925	3594	2818	1960	1328	933	684	520	408	328	269	3985	3985	3985	3869	3740	3610	3475	3333	3184	3025	2859
700 WB 130	130	1212	169	1172	3563	4482	1212	1155	911	664	485	368	292	241	204	178	157	3493	3178	2429	1636	1090	759	554	420	329	264	217	3563	3563	3563	3443	3323	3200	3073	2938	2795	2644	2486
700 WB 115	115	1023	134	1158	3024	3942	1023	969	750	531	376	278	217	176	148	127	112	2956	2681	2021	1340	886	615	447	339	265	213	174	3024	3021	3021	2918	2814	2708	2598	2482	2358	2228	2091

Note:　Geometry and capacity table, Grade 300, $f_u = 440$ MPa

Gauge lines listed are for M24 and M20 bolts in order of preference (refer Figure 3.29). Square brackets are used for a second, wider bolt gauge, s_{gf2}.

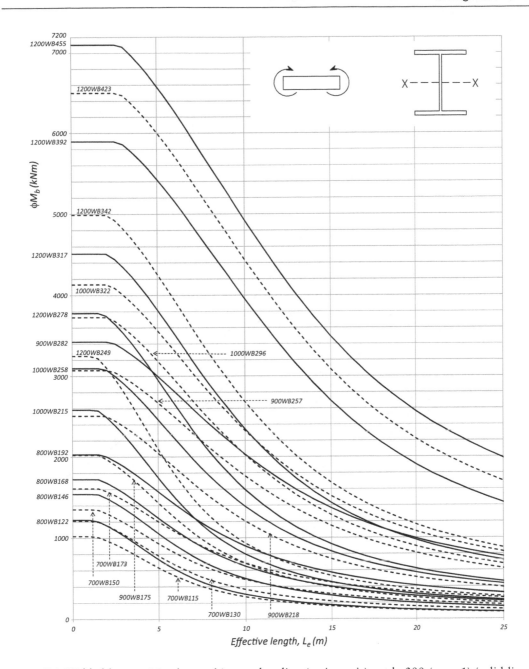

Figure 7.1 Welded beams: Members subject to bending (major axis) grade 300 ($\alpha_m = 1$) (solid lines labelled on the y-axis).

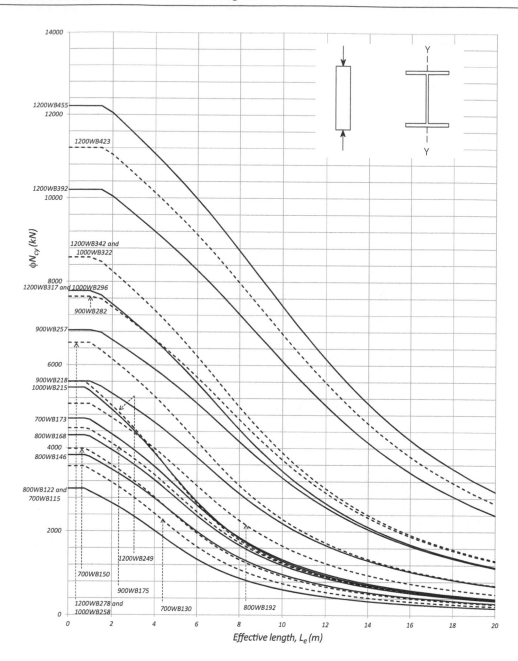

Figure 7.2 Welded beams: Members subject to axial compression (minor axis) grade 300 (solid lines labelled on the y-axis).

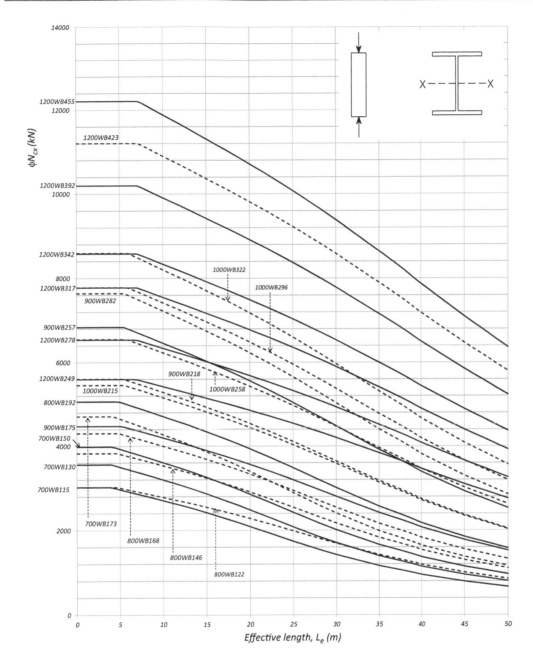

Figure 7.3 Welded beams: Members subject to axial compression (major axis) grade 300 (solid lines labelled on the *y*-axis).

Table 7.14 Welded Columns: 500 WC

Geometry and section properties

Section Name	Weight (kg/m)	d (mm)	b_f (mm)	t_f (mm)	t_w (mm)	d_1 (mm)	A_g (mm²)	I_x (10^6 mm⁴)	Z_x (10^3 mm³)	S_x (10^3 mm³)	r_x (mm)	I_y (10^6 mm⁴)	Z_y (10^3 mm³)	S_y (10^3 mm³)	r_y (mm)	J (10^3 mm⁴)	I_w (10^9 mm⁶)	f_{yf} (MPa)	f_{yw} (MPa)	Form Factor k_f	X Compactness	Z_{ex} (10^3 mm³)	Y Compactness	Z_{ey} (10^3 mm³)	s_{gl} (mm)	s_{gw} (mm)	$\phi R_{bb}/\phi b_b$ (kN/mm)	$\phi R_{by}/\phi b_{bf}$ (kN/mm)
500 WC 440		480	500	40	40	400	56000	2150	8980	10400	196	835	3340	5160	122	30100	40400	280	280	1	C	10400	C	5010	140 [280]	140 / 90 / 70	9.42	12.6
500 WC 414		480	500	40	32	400	52800	2110	8800	10100	200	834	3340	5100	126	25400	40400	280	280	1	C	10100	C	5010			7.27	10.1
500 WC 383		472	500	36	32	400	48800	1890	7990	9130	197	751	3000	4600	124	19900	35700	280	280	1	C	9130	C	4510			7.27	10.1
500 WC 340		514	500	32	25	450	43200	2050	7980	8980	218	667	2670	4070	124	13100	38800	280	280	1	C	8980	C	4000			5.18	7.88
500 WC 290		506	500	28	20	450	37000	1750	6930	7700	218	584	2330	3540	126	8420	33300	280	300	1	N	7570	N	3410			3.97	6.75
500 WC 267		500	500	25	20	450	34000	1560	6250	6950	214	521	2080	3170	124	6370	29400	280	300	1	N	6700	N	2970			3.97	6.75
500 WC 228		490	500	20	20	450	29000	1260	5130	5710	208	417	1670	2540	120	3880	23000	300	300	1	N	5210	N	2200			3.97	6.75

Section capacities

Section Name	Weight (kg/m)	ϕM_{sx} (kNm)	ϕM_{sy} (kNm)	ϕV_v (kN)	ϕN_s (kN)	ϕN_t (kN)
500 WC 440		2621	1263	2419	14112	14112
500 WC 414		2545	1263	1935	13306	13306
500 WC 383		2301	1137	1935	12298	12298
500 WC 340		2263	1008	1701	10886	10886
500 WC 290		1908	859	1458	9324	9324
500 WC 267		1688	748	1458	8568	8568
500 WC 228		1407	594	1458	7830	7830

Member Bending Capacity, ϕM_{bx} [$\alpha_m = 1$] (kNm) — Effective length (m)

Section Name	1	2	4	6	8	10	12	14	16	18	20
500 WC 440	2621	2621	2563	2427	2292	2166	2048	1940	1841	1749	1664
500 WC 414	2545	2545	2489	2353	2216	2087	1967	1858	1757	1664	1579
500 WC 383	2301	2301	2246	2117	1986	1862	1747	1643	1548	1461	1381
500 WC 340	2263	2263	2201	2056	1902	1755	1619	1497	1389	1292	1206
500 WC 290	1908	1908	1856	1730	1592	1458	1334	1223	1125	1038	962
500 WC 267	1688	1688	1642	1527	1401	1277	1162	1059	969	890	821
500 WC 228	1407	1407	1362	1258	1142	1028	922	829	748	679	620

Member Axial Capacity (Minor Axis), ϕN_{cy} (kN) — Effective length (m)

Section Name	1	2	4	6	8	10	12	14	16	18	20
500 WC 440	14112	13931	13065	12004	10614	8933	7254	5830	4717	3866	3214
500 WC 414	13306	13159	12374	11425	10192	8683	7127	5770	4688	3853	3208
500 WC 383	12298	12151	11412	10511	9336	7907	6455	5207	4222	3464	2882
500 WC 340	10886	10757	10102	9305	8265	7000	5714	4609	3737	3067	2552
500 WC 290	9324	9221	8671	8006	7142	6085	4994	4043	3285	2700	2248
500 WC 267	8568	8466	7951	7323	6505	5509	4497	3628	2941	2414	2008
500 WC 228	7830	7705	7194	6552	5703	4705	3758	2989	2404	1963	1628

Member Axial Capacity (Major Axis), ϕN_{cx} (kN) — Effective length (m)

Section Name	1	2	4	6	8	10	12	14	16	18	20
500 WC 440	14112	14112	13728	13188	12580	11866	11017	10035	8969	7901	6907
500 WC 414	13306	13306	12963	12466	11910	11260	10491	9598	8620	7627	6692
500 WC 383	12298	12298	11968	11500	10974	10357	9625	8777	7854	6927	6061
500 WC 340	10886	10886	10671	10304	9902	9443	8909	8287	7586	6842	6102
500 WC 290	9324	9324	9139	8825	8481	8088	7630	7097	6498	5860	5227
500 WC 267	8568	8568	8388	8093	7768	7396	6960	6453	5885	5286	4697
500 WC 228	7830	7830	7631	7340	7015	6636	6187	5667	5095	4513	3963

Note: Geometry and capacity table, Grade 300, f_u = 440 MPa

Gauge lines listed are for M24 and M20 bolts in order of preference (refer Figure 3.29). Square brackets are used for a second, wider bolt gauge, s_{g2}.

Table 7.15 Welded Columns: 400 WC

Geometry and section properties

Section Name / Weight (kg/m)	d (mm)	bf (mm)	tf (mm)	tw (mm)	d1 (mm)	Ag (mm²)	Ix (10⁶mm⁴)	Zx (10³mm³)	Sx (10³mm³)	rx (mm)	Iy (10⁶mm⁴)	Zy (10³mm³)	Sy (10³mm³)	ry (mm)	J (10³mm⁴)	Iw (10⁹mm⁶)	fyf (MPa)	fyw (MPa)	Form Factor kf	X Compactness	Zex (10³mm³)	Y Compactness	Zey (10³mm³)	sgw (mm)	sgf (mm)	φRbb/bb (kN/mm)	φRbt/bbf (kN/mm)
400 WC 361	430	400	40	40	350	46000	1360	6340	7460	172	429	2140	3340	96.5	24800	16300	280	280	1	C	7470	C	3210	140/90/70	140[280]	9.59	12.6
400 WC 328	430	400	40	28	350	41800	1320	6140	7100	178	427	2140	3270	101	19200	16200	280	280	1	C	7100	C	3200	140/90/70	140[280]	6.36	8.82
400 WC 303	422	400	36	28	350	38600	1180	5570	6420	175	385	1920	2950	99.8	14800	14300	280	280	1	C	6420	C	2880	140/90/70	140[280]	6.36	8.82
400 WC 270	414	400	32	25	350	34400	1030	4950	5660	173	342	1710	2610	99.8	10400	12500	280	280	1	C	5660	C	2560	140/90/70	140[280]	5.55	7.88
400 WC 212	400	400	25	20	350	27000	776	3880	4360	169	267	1330	2040	99.4	5060	9380	280	300	1	N	4360	N	2000	140/90/70	140[280]	4.43	6.75
400 WC 181	390	400	20	20	350	23000	620	3180	3570	164	214	1070	1640	96.4	3080	7310	300	300	1	N	3410	N	1510	140/90/70	140[280]	4.43	6.75
400 WC 144	382	400	16	16	350	18400	486	2550	2830	163	171	854	1300	96.3	1580	5720	300	300	1	N	2590	N	1120	140/90/70	140[280]	3.23	5.4

Section capacities

Section Name / Weight (kg/m)	φMsx (kNm)	φMsy (kNm)	φVv (kN)	φNs (kN)	φNt (kN)
400WC361	1882	809	2117	11592	11592
400WC328	1789	806	1482	10534	10534
400WC303	1618	726	1482	9727	9727
400WC270	1426	645	1323	8669	8669
400WC212	1099	504	1134	6804	6804
400WC181	921	408	1134	6210	6210
400WC144	699	302	907	4968	4968

Member Bending Capacity φMbx (αm = 1) (kNm)

Section Name	1	2	4	6	8	10	12	14	16	18	20
					Effective length (m)						
400WC361	1882	1882	1801	1690	1587	1493	1408	1329	1257	1190	1129
400WC328	1789	1789	1710	1598	1492	1396	1310	1231	1159	1093	1033
400WC303	1618	1618	1540	1432	1330	1238	1155	1080	1012	951	896
400WC270	1426	1426	1353	1250	1151	1063	983	913	850	794	744
400WC212	1099	1099	1035	943	854	773	703	642	589	543	503
400WC181	921	921	860	773	689	613	548	494	447	408	375
400WC144	699	699	653	584	513	449	395	350	313	282	256

Member Axial Capacity (Minor Axis) φNcy (kN)

Section Name	1	2	4	6	8	10	12	14	16	18	20
					Effective length (m)						
400WC361	11592	11263	10300	8979	7257	5555	4225	3269	2587	2092	1723
400WC328	10534	10270	9447	8344	6886	5368	4125	3210	2549	2065	1703
400WC303	9727	9475	8703	7662	6290	4879	3739	2905	2304	1866	1539
400WC270	8669	8444	7756	6828	5605	4348	3332	2589	2054	1663	1371
400WC212	6804	6626	6083	5349	4383	3395	2599	2018	1600	1296	1068
400WC181	6210	6017	5474	4718	3748	2830	2137	1647	1300	1050	864
400WC144	4968	4813	4378	3773	2995	2261	1707	1315	1038	838	690

Member Axial Capacity (Major Axis) φNcx (kN)

Section Name	1	2	4	6	8	10	12	14	16	18	20
					Effective length (m)						
400WC361	11592	11592	11157	10632	10018	9273	8381	7388	6391	5479	4694
400WC328	10534	10534	10168	9713	9185	8550	7789	6930	6047	5219	4492
400WC303	9727	9727	9376	8946	8445	7840	7115	6301	5475	4709	4044
400WC270	8669	8669	8347	7958	7504	6952	6292	5556	4813	4131	3541
400WC212	6804	6804	6538	6223	5852	5400	4859	4263	3671	3137	2681
400WC181	6210	6210	5931	5617	5239	4772	4220	3635	3083	2605	2209
400WC144	4968	4968	4742	4489	4183	3805	3359	2888	2446	2065	1750

Note: Geometry and capacity table, Grade 300, f_u = 440 MPa

Gauge lines listed are for M24 and M20 bolts in order of preference (refer Figure 3.29). Square brackets are used for a second, wider bolt gauge, s_{g2}.

Table 7.16 Welded Columns: 350 WC

Section Name Weight (kg/m)	d (mm)	b_f (mm)	t_f (mm)	t_w (mm)	d_1 (mm)	A_g (mm²)	I_x (10⁶mm⁴)	Z_x (10³mm³)	S_x (10³mm³)	r_x (mm)	I_y (10⁶mm⁴)	Z_y (10³mm³)	S_y (10³mm³)	r_y (mm)	J (10³mm⁴)	I_w (10⁹mm⁶)	f_{yf} (MPa)	f_{yw} (MPa)	Form Factor k_f	X Compactness	Z_{ex} (10³mm³)	Y Compactness	Z_{ey} (10³mm³)	s_{g1} (mm)	s_{gw} (mm)	$\phi R_{bb}/b_b$ (kN/mm)	$\phi R_{by}/b_{bf}$ (kN/mm)
350 WC 280	355	350	40	28	275	35700	747	4210	4940	145	286	1640	2500	89.6	16500	7100	280	280	1	C	4940	C	2450	140	140 / 90 / 70	6.61	8.82
350 WC 258	347	350	36	28	275	32900	661	3810	4450	142	258	1470	2260	88.5	12700	6230	280	280	1	C	4450	C	2210			6.61	8.82
350 WC 230	339	350	32	25	275	29300	573	3380	3910	140	229	1310	2000	88.4	8960	5400	280	280	1	C	3910	C	1960			5.81	7.88
350 WC 197	331	350	28	20	275	25100	486	2940	3350	139	200	1140	1740	89.3	5750	4600	280	300	1	C	3350	C	1720			4.74	6.75

Section Name Weight (kg/m)	ϕM_{sx} (kNm)	ϕM_{sy} (kNm)	ϕV_v (kN)	ϕN_s (kN)	ϕN_t (kN)	Member Bending Capacity ϕM_{bx} [$\alpha_m = 1$] (kNm) — Effective length (m)											Member Axial Capacity (Minor Axis) ϕN_{cy} (kN) — Effective length (m)											Member Axial Capacity (Major Axis) ϕN_{cx} (kN) — Effective length (m)										
						1	2	4	6	8	10	12	14	16	18	20	1	2	4	6	8	10	12	14	16	18	20	1	2	4	6	8	10	12	14	16	18	20
350 WC 280	1245	617	1164	8996	8996	1245	1245	1182	1109	1043	983	928	877	830	787	747	8996	8689	7858	6681	5204	3875	2904	2230	1757	1417	1165	8996	8963	8513	7996	7355	6555	5642	4736	3941	3288	2765
350 WC 258	1121	557	1164	8291	8291	1121	1121	1059	990	926	868	816	768	724	684	647	8291	7999	7219	6109	4729	3506	2622	2011	1583	1276	1049	8291	8252	7827	7336	6722	5956	5090	4249	3521	2930	2459
350 WC 230	985	494	1040	7384	7384	985	985	926	858	797	742	693	648	607	571	538	7384	7123	6427	5437	4206	3117	2331	1787	1407	1134	932	7384	7344	6959	6512	5951	5251	4467	3714	3069	2550	2138
350 WC 197	844	433	891	6325	6325	844	844	788	724	665	612	566	525	489	456	427	6325	6107	5520	4687	3645	2711	2031	1559	1228	990	814	6325	6289	5956	5569	5082	4475	3797	3151	2601	2159	1809

Note: Geometry and capacity table, Grade 300, f_u = 440 MPa

Gauge lines listed are for M24 and M20 bolts in order of preference (refer Figure 3.29). Square brackets are used for a second, wider bolt gauge, s_{g2}.

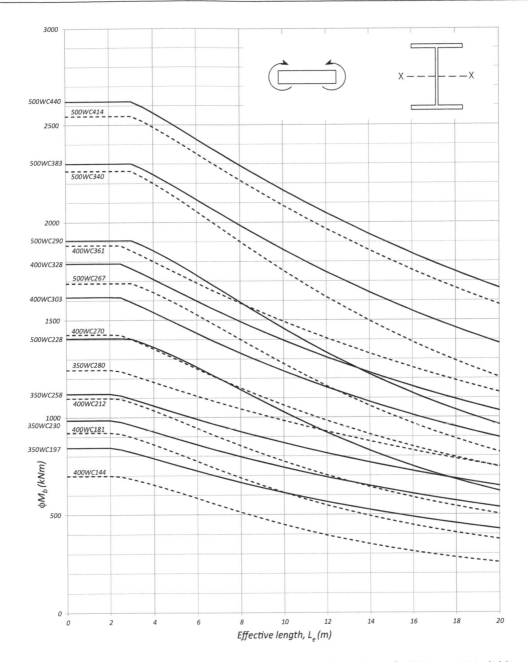

Figure 7.4 Welded columns: Members subject to bending (major axis) grade 300 (α_m = 1) (solid lines labelled on the *y*-axis).

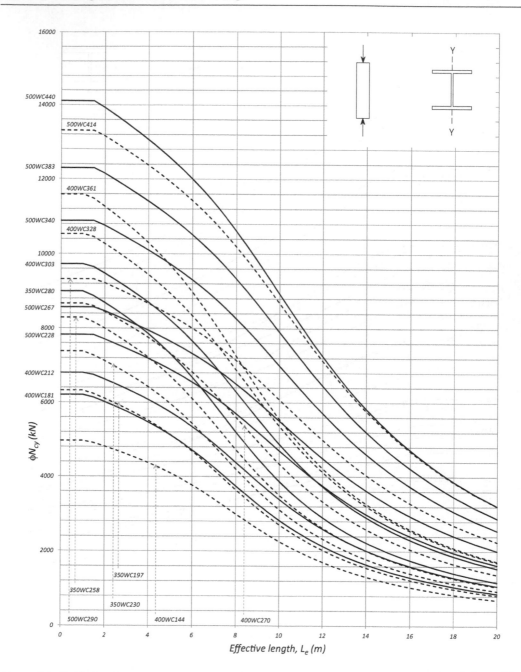

Figure 7.5 Welded columns: Members subject to axial compression (minor axis) grade 300 (solid lines labelled on the *y* axis).

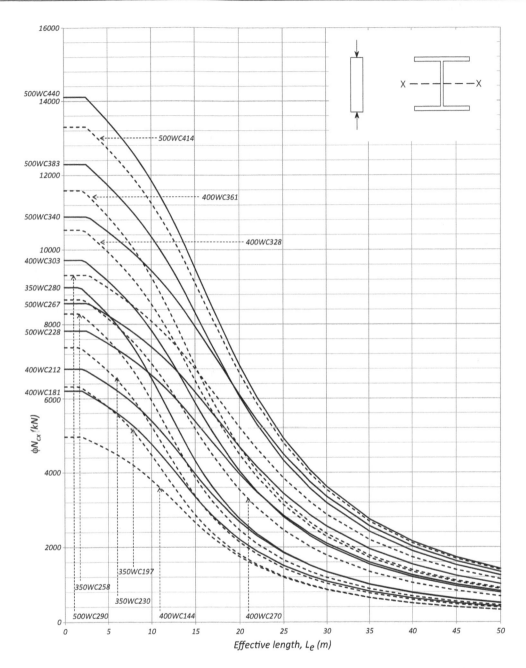

Figure 7.6 Welded columns: Members subject to axial compression (major axis) grade 300 (solid lines labelled on the *y* axis).

Table 7.17 Universal Beams: 460, 530 and 610 UB

Section Name Weight (kg/m)	d (mm)	b_f (mm)	t_f (mm)	t_w (mm)	r_1 (mm)	d_1 (mm)	A_g (mm²)	I_x (10^6 mm⁴)	Z_x (10^3 mm³)	S_x (10^3 mm³)	r_x (mm)	I_y (10^6 mm⁴)	Z_y (10^3 mm³)	S_y (10^3 mm³)	r_y (mm)	J (10^3 mm⁴)	I_w (10^9 mm⁶)	f_{yf} (MPa)	f_{yw} (MPa)	Form Factor k_f	X Compactness	Z_{ex} (10^3 mm³)	Y Compactness	Z_{ey} (10^3 mm³)	s_{g1} (mm)	s_{gw} (mm)	$\phi R_{bb}/b_b$ (kN/mm)	$\phi R_{by}/b_{bf}$ (kN/mm)
610 UB 125	612	229	19.6	11.9	14	572	16000	986	3230	3680	249	39.3	343	536	49.6	1560	3450	280	300	0.95	C	3680	C	515	140 / 90	140 / 90 / 70	1.07	4.02
610 UB 113	607	228	17.3	11.2	14	572	14500	875	2880	3290	246	34.3	300	469	48.7	1140	2980	280	300	0.926	C	3290	C	451	140 / 90	140 / 90 / 70	0.921	3.78
610 UB 101	602	228	14.8	10.6	14	572	13000	761	2530	2900	242	29.3	257	402	47.5	790	2530	300	320	0.888	C	2900	C	386	140 / 90	140 / 90 / 70	0.81	3.82
530 UB 92.4	533	209	15.6	10.2	14	502	11800	554	2080	2370	217	23.8	228	355	44.9	775	1590	300	320	0.928	C	2370	C	342	140 / 90	140 / 90 / 70	0.903	3.67
530 UB 82	528	209	13.2	9.6	14	502	10500	477	1810	2070	213	20.1	193	301	43.8	526	1330	300	320	0.902	C	2070	C	289	90 / 140 ^{M20}	90 / 70 / 140	0.773	3.46
460 UB 82.1	460	191	16	9.9	11.4	428	10500	372	1610	1840	188	18.6	195	303	42.2	701	919	300	320	0.979	C	1840	C	292	90 / 140 ^{M20}	90 / 70 / 140	1.06	3.56
460 UB 74.6	457	190	14.5	9.1	11.4	428	9520	335	1460	1660	188	16.6	175	271	41.8	530	815	300	320	0.948	C	1660	C	262	90 / 140 ^{M20}	90 / 70 / 140	0.861	3.28
460 UB 67.1	454	190	12.7	8.5	11.4	428	8580	296	1300	1480	186	14.5	153	238	41.2	378	708	300	320	0.922	C	1480	C	230	90 / 140 ^{M20}	90 / 70 / 140	0.724	3.06

Section Name Weight (kg/m)	ϕM_{sx} (kNm)	ϕM_{sy} (kNm)	ϕV_v (kN)	ϕN_s (kN)	ϕN_t (kN)	Member Bending Capacity, ϕM_{bx} [$\alpha_m = 1$] (kNm) Effective length (m)											Member Axial Capacity (Minor Axis), ϕN_{cy} (kN) Effective length (m)											Member Axial Capacity (Major Axis), ϕN_{cx} (kN) Effective length (m)										
						1	2	4	6	8	10	12	14	16	18	20	1	2	4	6	8	10	12	14	16	18	20	1	2	4	6	8	10	12	14	16	18	20
610 UB 125	927	130	1180	3830	4032	927	872	675	492	367	286	232	195	168	148	132	3737	3441	2521	1518	939	628	447	334	259	207	169	3830	3830	3791	3683	3569	3443	3302	3140	2956	2751	2531
610 UB 113	829	114	1101	3384	3654	829	777	593	424	310	238	192	160	137	120	107	3299	3036	2216	1330	822	549	391	292	226	181	147	3384	3384	3349	3254	3153	3042	2917	2774	2612	2431	2237
610 UB 101	783	104	1103	3117	3510	783	725	535	367	260	195	155	128	109	94.7	83.8	3031	2775	1969	1153	707	471	335	250	194	154	126	3117	3117	3080	2989	2892	2786	2664	2525	2367	2191	2005
530 UB 92.4	640	92	939	2957	3186	640	584	422	290	209	160	129	108	92.7	81.3	72.5	2857	2586	1720	963	583	386	274	204	158	126	103	2957	2957	2898	2798	2688	2564	2418	2248	2058	1855	1654
530 UB 82	559	78	876	2557	2835	559	507	359	239	168	127	101	84	71.8	62.8	55.7	2469	2231	1470	818	494	328	232	173	134	107	86.9	2557	2557	2505	2419	2323	2215	2088	1940	1774	1597	1423
460 UB 82.1	497	79	787	2775	2835	497	446	316	219	161	126	103	87.1	75.4	66.6	59.6	2661	2370	1451	778	465	307	217	162	125	99.5	81	2775	2775	2686	2571	2438	2280	2091	1875	1649	1432	1239
460 UB 74.6	448	71	719	2437	2570	448	401	280	190	137	106	86.4	72.7	62.8	55.3	49.4	2338	2085	1285	691	414	273	193	144	111	88.4	72	2437	2437	2361	2262	2149	2014	1853	1668	1473	1284	1113
460 UB 67.1	400	62	667	2136	2317	400	356	244	161	114	87.1	70.1	58.6	50.4	44.2	39.4	2049	1827	1125	605	362	239	169	126	97.3	77.4	63.1	2136	2136	2070	1984	1885	1767	1627	1466	1296	1130	980

Note: Geometry and capacity table, Grade 300, f_u = 440 MPa

Gauge lines listed are for M24 and M20 bolts in order of preference (refer Figure 3.29). Maximum bolt size is shown in superscript where limited for each gauge.

Table 7.18 Universal Beams: 310, 360 and 410 UB

Section Name	Weight (kg/m)	d (mm)	bf (mm)	tf (mm)	tw (mm)	r1 (mm)	d1 (mm)	Ag (mm²)	Ix (10⁶mm⁴)	Zx (10³mm³)	Sx (10³mm³)	rx (mm)	Iy (10⁶mm⁴)	Zy (10³mm³)	Sy (10³mm³)	ry (mm)	J (10³mm⁴)	Iw (10⁹mm⁶)	fyf (MPa)	fyw (MPa)	Form Factor kf	X Compactness	Zex (10³mm³)	Y Compactness	Zey (10³mm³)	sg (mm)	sgw (mm)	φRbb/bb (kN/mm)	φRby/bbf (kN/mm)
410 UB 59.7	59.7	406	178	12.8	7.8	11.4	381	7640	216	1060	1200	168	12.1	135	209	39.7	337	467	300	320	0.938	C	1200	C	203	90/70 M20	90/70/140	0.698	2.81
410 UB 53.7	53.7	403	178	10.9	7.6	11.4	381	6890	188	933	1060	165	10.3	115	179	38.6	234	394	320	320	0.913	C	1060	C	173	90/70 M20	90/70/140	0.654	2.74
360 UB 56.7	56.7	359	172	13	8	11.4	333	7240	161	899	1010	149	11	128	198	39	338	330	300	320	0.996	C	1010	C	193	90/70 M20	90/70/140	0.907	2.88
360 UB 50.7	50.7	356	171	11.5	7.3	11.4	333	6470	142	798	897	148	9.6	112	173	38.5	241	284	300	320	0.963	C	897	N	168	90/70 M20	90/70/140	0.725	2.63
360 UB 44.7	44.7	352	171	9.7	6.9	11.4	333	5720	121	689	777	146	8.1	94.7	146	37.6	161	237	320	320	0.93	N	770	C	140	90/70 M20	90/70/140	0.63	2.48
310 UB 46.2	46.2	307	166	11.8	6.7	11.4	284	5930	100	654	729	130	9.01	109	166	39	233	197	300	320	0.991	C	729	N	163	90/70 M20	90/70/140	0.74	2.41
310 UB 40.4	40.4	304	165	10.2	6.1	11.4	284	5210	86.4	569	633	129	7.65	92.7	142	38.3	157	165	320	320	0.952	N	633	C	139	90/70 M20	90/70/140	0.588	2.2
310 UB 32	32	298	149	8	5.5	13	282	4080	63.2	424	472	124	4.42	59.3	91.8	32.9	86.5	92.9	320	320	0.915	N	467	N	86.9	70 M20	90/70/140	0.456	1.98

Section Name	Weight (kg/m)	φMsx (kNm)	φMsy (kNm)	φVv (kN)	φNs (kN)	φNt (kN)
410 UB 59.7	59.7	324	55	547	1935	2063
410 UB 53.7	53.7	305	50	529	1812	1984
360 UB 56.7	56.7	273	52	496	1947	1955
360 UB 50.7	50.7	242	45	449	1682	1747
360 UB 44.7	44.7	222	40	420	1532	1647
310 UB 46.2	46.2	197	44	355	1587	1601
310 UB 40.4	40.4	182	40	320	1428	1500
310 UB 32	32	134	25	283	1075	1175

Member Bending Capacity, φMbx (αm = 1) (kNm) — Effective length (m)

Section Name	1	2	4	6	8	10	12	14	16	18	20
410 UB 59.7	322	285	193	129	93	71.9	58.4	49.2	42.6	37.5	33.5
410 UB 53.7	302	263	169	108	75.7	57.5	46.3	38.7	33.3	29.2	26.1
360 UB 56.7	271	239	164	113	83.2	65.5	53.9	45.8	39.8	35.2	31.6
360 UB 50.7	240	211	141	94.3	68.4	53.1	43.4	36.7	31.8	28	25.1
360 UB 44.7	219	189	121	77.1	54.3	41.5	33.6	28.2	24.3	21.4	19.1
310 UB 46.2	195	172	118	82	61.1	48.4	40	34.1	29.7	26.3	23.6
310 UB 40.4	180	156	102	67.4	48.8	38	31.1	26.3	22.8	20.1	18
310 UB 32	131	109	64.1	40.1	28.4	21.8	17.8	15	13	11.4	10.2

Member Axial Capacity (Minor Axis), φNcy (kN) — Effective length (m)

Section Name	1	2	4	6	8	10	12	14	16	18	20
410 UB 59.7	1848	1632	960	506	302	199	140	104	80.7	64.2	52.3
410 UB 53.7	1722	1504	843	437	259	170	120	89.3	69	54.9	44.7
360 UB 56.7	1850	1616	904	468	278	183	129	95.8	74	58.9	47.9
360 UB 50.7	1599	1398	786	408	242	159	112	83.4	64.4	51.2	41.7
360 UB 44.7	1451	1256	677	347	205	135	95	70.5	54.5	43.3	35.2
310 UB 46.2	1508	1318	740	383	227	150	106	78.5	60.6	48.2	39.2
310 UB 40.4	1354	1174	638	327	194	127	89.7	66.6	51.4	40.9	33.3
310 UB 32	1003	833	392	194	114	74.3	52.3	38.8	29.9	23.8	19.3

Member Axial Capacity (Major Axis), φNcx (kN) — Effective length (m)

Section Name	1	2	4	6	8	10	12	14	16	18	20
410 UB 59.7	1935	1935	1858	1767	1660	1530	1373	1202	1033	881	753
410 UB 53.7	1812	1812	1734	1644	1538	1406	1251	1083	923	782	665
360 UB 56.7	1947	1939	1841	1729	1589	1414	1215	1018	846	706	593
360 UB 50.7	1682	1677	1593	1497	1378	1230	1060	891	742	620	521
360 UB 44.7	1532	1525	1446	1355	1241	1098	938	782	648	539	452
310 UB 46.2	1587	1569	1476	1363	1216	1037	851	689	560	460	383
310 UB 40.4	1428	1411	1325	1220	1083	916	747	602	488	401	333
310 UB 32	1075	1061	995	913	806	677	548	440	356	291	242

Note: Geometry and capacity table, Grade 300, f_u = 440 MPa

Gauge lines listed are for M24 and M20 bolts in order of preference (refer Figure 3.29). Maximum bolt size is shown in superscript where limited for each gauge.

Table 7.19 Universal Beams: 200 and 250 UB

Geometry properties

Section Name, Weight (kg/m)	d (mm)	b_f (mm)	t_f (mm)	t_w (mm)	r_1 (mm)	d_1 (mm)	A_g (mm²)	I_x (10⁶mm⁴)	Z_x (10³mm³)	S_x (10³mm³)	r_x (mm)	I_y (10⁶mm⁴)	Z_y (10³mm³)	S_y (10³mm³)	r_y (mm)	J (10³mm⁴)	I_w (10⁹mm⁶)
250 UB 37.3	256	146	10.9	6.4	8.9	234	4750	55.7	435	486	108	5.66	77.5	119	34.5	158	85.2
250 UB 31.4	252	146	8.6	6.1	8.9	234	4010	44.5	354	397	105	4.47	61.2	94.2	33.4	89.3	65.9
250 UB 25.7	248	124	8	5	12	232	3270	35.4	285	319	104	2.55	41.1	63.6	27.9	67.4	36.7
200 UB 29.8	207	134	9.6	6.3	8.9	188	3820	29.1	281	316	87.3	3.86	57.5	88.4	31.8	105	37.6
200 UB 25.4	203	133	7.8	5.8	8.9	188	3230	23.6	232	260	85.4	3.06	46.1	70.9	30.8	62.7	29.2
200 UB 22.3	202	133	7	5	8.9	188	2870	21	208	231	85.5	2.75	41.3	63.4	31	45	26
200 UB 18.2	198	99	7	4.5	11	184	2320	15.8	160	180	82.6	1.14	23	35.7	22.1	38.6	10.4

Section Name, Weight (kg/m)	f_{yf} (MPa)	f_{yw} (MPa)	Form Factor k_f	X Compactness	Z_{ex} (10³mm³)	Y Compactness	Z_{ey} (10³mm³)	s_g (mm)	s_{pw} (mm)	$\phi R_{bb}/b_b$ (kN/mm)	$\phi R_{by}/b_{bf}$ (kN/mm)
250 UB 37.3	320	320	1	C	486	C	116	70^{M20} / 90^{M20} / 70^{M16}	70 / 90 / 140	0.857	2.3
250 UB 31.4	320	320	1	N	395	N	91.4			0.769	2.2
250 UB 25.7	320	320	0.949	C	319	C	61.7			0.484	1.8
200 UB 29.8	320	320	1	C	316	C	86.3	70^{M20}	70 / 90	1.06	2.27
200 UB 25.4	320	320	1	N	259	N	68.8			0.897	2.09
200 UB 22.3	320	320	1	N	227	N	60.3			0.65	1.8
200 UB 18.2	320	320	0.99	C	180	C	34.4	50^{M16}		0.522	1.62

Member Bending Capacity, ϕM_{bx} [$\alpha_m = 1$] (kNm)

Section Name, Weight (kg/m)	ϕM_{ox} (kNm)	ϕM_{oy} (kNm)	ϕV_v (kN)	ϕN_s (kN)	ϕN_t (kN)	Effective length (m) 1	2	4	6	8	10	12	14	16	18	20
250 UB 37.3	140	33	283	1368	1368	137	116	75.4	51.8	38.8	30.9	25.7	22	19.2	17	15.3
250 UB 31.4	114	26	266	1155	1155	111	92.6	57	37.4	27.3	21.4	17.6	14.9	13	11.5	10.3
250 UB 25.7	92	18	214	894	942	86.9	68.2	38.1	24.4	17.7	13.9	11.5	9.75	8.49	7.52	6.75
200 UB 29.8	91	25	225	1100	1100	87.9	73.3	47.6	33.4	25.4	20.4	17	14.6	12.8	11.4	10.2
200 UB 25.4	75	20	203	930	930	71.7	58.8	36.2	24.3	18.1	14.4	11.9	10.2	8.89	7.89	7.09
200 UB 22.3	65	17	175	827	827	62.9	51.5	30.9	20.3	14.9	11.7	9.69	8.25	7.19	6.37	5.72
200 UB 18.2	52	10	154	661	668	46.7	33.6	17.9	11.7	8.64	6.86	5.69	4.86	4.25	3.77	3.39

Member Axial Capacity (Minor Axis), ϕN_{cy} (kN)

Section Name, Weight (kg/m)	Effective length (m) 1	2	4	6	8	10	12	14	16	18	20
250 UB 37.3	1276	1061	501	248	145	95.1	67	49.7	38.3	30.4	24.8
250 UB 31.4	1073	881	402	198	116	75.5	53.1	39.4	30.4	24.1	19.6
250 UB 25.7	810	615	240	115	66.7	43.4	30.5	22.6	17.4	13.8	11.2
200 UB 29.8	1015	815	353	172	100	65.4	46	34.1	26.3	20.9	17
200 UB 25.4	854	675	283	137	79.8	52	36.6	27.1	20.9	16.6	13.5
200 UB 22.3	759	602	254	123	71.8	46.8	32.9	24.4	18.8	14.9	12.1
200 UB 18.2	566	350	113	52.6	30.3	19.6	13.7	10.2	7.81	6.2	5.03

Member Axial Capacity (Major Axis), ϕN_{cx} (kN)

Section Name, Weight (kg/m)	Effective length (m) 1	2	4	6	8	10	12	14	16	18	20
250 UB 37.3	1368	1334	1227	1084	895	697	536	417	331	268	221
250 UB 31.4	1155	1124	1030	903	736	567	433	336	266	215	177
250 UB 25.7	894	871	800	704	578	449	344	267	212	172	142
200 UB 29.8	1100	1054	937	766	568	411	303	231	181	146	120
200 UB 25.4	930	889	787	637	467	335	247	188	147	118	97.1
200 UB 22.3	827	790	699	566	416	299	220	167	131	105	86.5
200 UB 18.2	661	630	554	442	320	228	167	127	99.4	79.8	65.4

Note: Geometry and capacity table, Grade 300, $f_u = 440$ MPa

Gauge lines listed are for M24 and M20 bolts in order of preference (refer Figure 3.29). Maximum bolt size is shown in superscript where limited for each gauge.

Table 7.20 Universal Beams: 150 and 180 UB

Section Name / Weight (kg/m)	d (mm)	b_f (mm)	t_f (mm)	t_w (mm)	r_1 (mm)	d_1 (mm)	A_g (mm²)	I_x (10⁶ mm⁴)	Z_x (10³ mm³)	S_x (10³ mm³)	r_x (mm)	I_y (10⁶ mm⁴)	Z_y (10³ mm³)	S_y (10³ mm³)	r_y (mm)	J (10³ mm⁴)	I_w (10⁹ mm⁶)	f_{yf} (MPa)	f_{yw} (MPa)	Form Factor k_f	X Compactness	Z_{ex} (10³ mm³)	Y Compactness	Z_{ey} (10³ mm³)	s_g (mm)	s_{gw} (mm)	$\phi R_{bb}/b_b$ (kN/mm)	$\phi R_{by}/b_{bf}$ (kN/mm)
180 UB 22.2	179	90	10	6	8.9	159	2820	15.3	171	195	73.6	1.22	27.1	42.3	20.8	81.6	8.71	320	320	1	C	195	C	40.7	50 ᴹ¹²	70 / 90	1.12	2.16
180 UB 18.1	175	90	8	5	8.9	159	2300	12.1	139	157	72.6	0.975	21.7	33.7	20.6	44.8	6.8	320	320	1	C	157	C	32.5			0.788	1.8
180 UB 16.1	173	90	7	4.5	8.9	159	2040	10.6	123	138	72	0.853	19	29.4	20.4	31.5	5.88	320	320	1	C	138	C	28.4			0.63	1.62
150 UB 18	155	75	9.5	6	8	136	2300	9.05	117	135	62.8	0.672	17.9	28.2	17.1	60.5	3.56	320	320	1	C	135	C	26.9	N/A	70	1.24	2.16
150 UB 14	150	75	7	5	8	136	1780	6.66	88.8	102	61.1	0.495	13.2	20.8	16.6	28.1	2.53	320	320	1	C	102	C	19.8			0.912	1.8

Section Name / Weight (kg/m)	ϕM_{sx} (kNm)	ϕM_{sy} (kNm)	ϕV_v (kN)	ϕN_s (kN)	ϕN_t (kN)
180 UB 22.2	56	12	186	812	812
180 UB 18.1	45	9	151	662	662
180 UB 16.1	40	8	135	588	588
150 UB 18	39	8	161	662	662
150 UB 14	29	6	130	513	513

Member Bending Capacity, ϕM_{bx} $[\alpha_m = 1]$ (kNm) — Effective length (m)

Section Name	1	2	4	6	8	10	12	14	16	18	20
180 UB 22.2	50.2	37.9	23.1	16.2	12.4	9.99	8.36	7.19	6.31	5.61	5.06
180 UB 18.1	40.1	29	16.4	11.1	8.38	6.71	5.6	4.8	4.2	3.74	3.36
180 UB 16.1	35	24.8	13.4	8.91	6.65	5.31	4.42	3.78	3.31	2.94	2.64
150 UB 18	33.2	24.2	14.7	10.3	7.89	6.37	5.34	4.59	4.03	3.59	3.23
150 UB 14	24.4	16.6	9.21	6.25	4.71	3.78	3.15	2.71	2.37	2.11	1.9

Member Axial Capacity (Minor Axis), ϕN_{cy} (kN) — Effective length (m)

Section Name	1	2	4	6	8	10	12	14	16	18	20
180 UB 22.2	681	394	123	57	32.7	21.2	14.8	11	8.43	6.68	5.43
180 UB 18.1	554	317	98.2	45.6	26.2	17	11.9	8.78	6.75	5.35	4.34
180 UB 16.1	490	277	85.3	39.7	22.8	14.8	10.3	7.64	5.87	4.65	3.78
150 UB 18	512	239	69.3	31.9	18.2	11.8	8.24	6.08	4.67	3.7	3
150 UB 14	390	177	50.7	23.3	13.3	8.61	6.01	4.44	3.41	2.7	2.19

Member Axial Capacity (Major Axis), ϕN_{cx} (kN) — Effective length (m)

Section Name	1	2	4	6	8	10	12	14	16	18	20
180 UB 22.2	807	764	650	483	329	229	166	125	97.5	78.1	63.9
180 UB 18.1	658	622	527	388	263	182	132	99.4	77.5	62.1	50.8
180 UB 16.1	583	551	465	341	230	159	115	86.9	67.7	54.2	44.3
150 UB 18	652	610	487	324	208	141	101	75.9	58.9	47.1	38.4
150 UB 14	504	470	370	241	154	104	74.4	55.7	43.3	34.6	28.2

Note: Geometry and capacity table, Grade 300, $f_u = 440$ MPa

Gauge lines listed are for M24 and M20 bolts in order of preference (refer Figure 3.29). Maximum bolt size is shown in superscript where limited for each gauge.

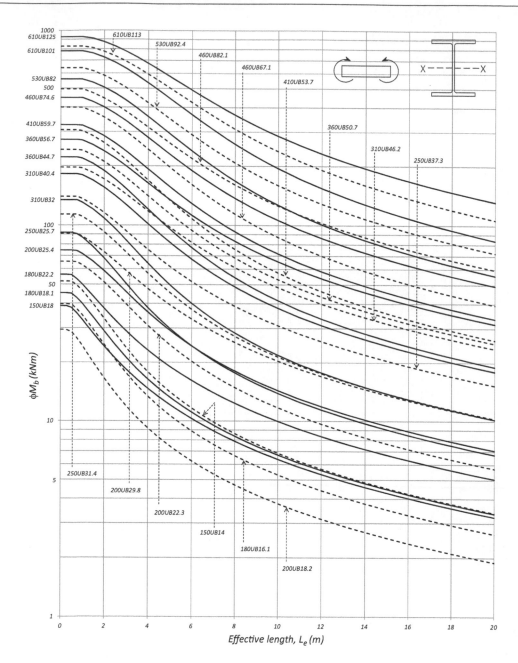

Figure 7.7 Universal beams: Members subject to bending (major axis) grade 300 ($\alpha_m = 1$) (solid lines labelled on the y axis).

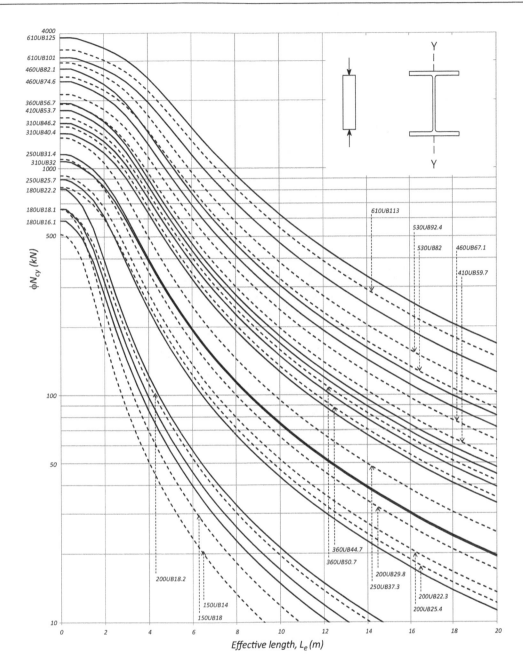

Figure 7.8 Universal beams: Members subject to axial compression (minor axis) grade 300 (solid lines labelled on the *y* axis).

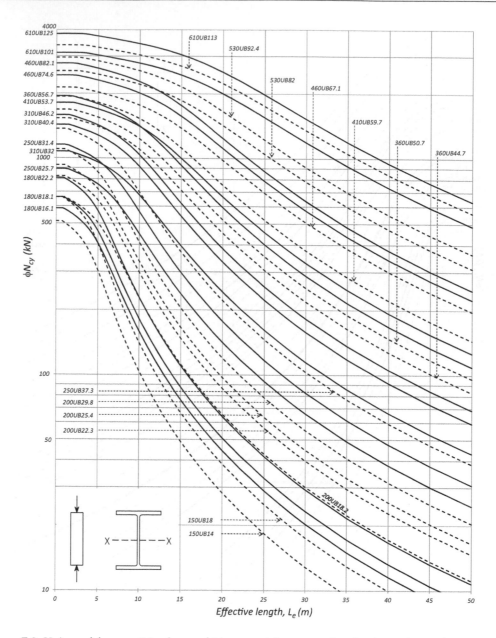

Figure 7.9 Universal beams: Members subject to axial compression (major axis) grade 300 (solid lines labelled on the *y* axis).

Table 7.21 Universal Columns: 250 and 310 UC

Geometry and section properties

Section Name	d (mm)	b_f (mm)	t_f (mm)	t_w (mm)	r_1 (mm)	d_1 (mm)	A_g (mm²)	I_x (10⁶ mm⁴)	Z_x (10³ mm³)	S_x (10³ mm³)	r_x (mm)	I_y (10⁶ mm⁴)	Z_y (10³ mm²)	S_y (10³ mm³)	r_y (mm)	J (10³ mm⁴)	I_w (10⁹ mm⁶)	f_yf (MPa)	f_yw (MPa)	k_f	X Comp	Z_ex (10³ mm³)	Y Comp	Z_ey (10³ mm³)	s_gf (mm)	s_gw (mm)	φR_bb/b_b (kN/mm)	φR_by/b_bf (kN/mm)
310 UC 158	327	311	25	15.7	16.5	277	20100	388	2370	2680	139	125	807	1230	78.9	3810	2860	280	300	1	C	2680	C	1210	140/90	90/70/140	3.47	5.3
310 UC 137	321	309	21.7	13.8	16.5	277	17500	329	2050	2300	137	107	691	1050	78.2	2520	2390	280	300	1	C	2300	C	1040	140/90	90/70/140	2.9	4.66
310 UC 118	315	307	18.7	11.9	16.5	277	15000	277	1760	1960	136	90.2	588	893	77.5	1630	1980	280	300	1	C	1960	C	882	140/90	90/70/140	2.32	4.02
310 UC 96.8	308	305	15.4	9.9	16.5	277	12400	223	1450	1600	134	72.9	478	725	76.7	928	1560	300	320	1	N	1560	N	694	140/90	90/70/140	1.76	3.56
250 UC 89.5	260	256	17.3	10.5	14	225	11400	143	1100	1230	112	48.4	378	575	65.2	1040	713	280	320	1	C	1230	C	567	140/90	90/70/140	2.24	3.78
250 UC 72.9	254	254	14.2	8.6	14	225	9320	114	897	992	111	38.8	306	463	64.5	586	557	300	320	1	N	986	N	454	140/90	90/70/140	1.62	3.1

Section capacities

Section Name	φM_sx (kNm)	φM_sy (kNm)	φV_v (kN)	φN_s (kN)	φN_t (kN)
310 UC 158	675	305	832	5065	5065
310 UC 137	580	262	718	4410	4410
310 UC 118	494	222	607	3780	3780
310 UC 96.8	421	187	527	3348	3348
250 UC 89.5	310	143	472	2873	2873
250 UC 72.9	266	123	377	2516	2516

Member Bending Capacity, φM_bx [α_m = 1] (kNm)

Section Name	Effective length (m)										
	1	2	4	6	8	10	12	14	16	18	20
310 UC 158	675	673	613	551	497	450	410	375	345	319	296
310 UC 137	580	577	521	463	412	368	331	300	274	251	232
310 UC 118	494	491	440	385	337	296	263	236	213	194	178
310 UC 96.8	421	417	368	314	267	229	199	175	156	140	127
250 UC 89.5	310	303	265	229	199	174	155	139	125	114	104
250 UC 72.9	266	258	220	184	154	132	114	101	89.7	80.8	73.4

Member Axial Capacity (Minor Axis), φN_cy (kN)

Section Name	Effective length (m)										
	1	2	4	6	8	10	12	14	16	18	20
310 UC 158	5065	4836	4268	3431	2501	1790	1316	1001	784	630	516
310 UC 137	4410	4206	3705	2965	2151	1536	1128	857	671	539	442
310 UC 118	3778	3602	3166	2522	1821	1298	952	723	566	454	373
310 UC 96.8	3339	3176	2760	2144	1514	1067	779	590	461	370	303
250 UC 89.5	2847	2684	2241	1605	1070	736	531	400	311	249	204
250 UC 72.9	2487	2337	1916	1329	871	595	428	322	250	200	164

Member Axial Capacity (Major Axis), φN_cx (kN)

Section Name	Effective length (m)										
	1	2	4	6	8	10	12	14	16	18	20
310 UC 158	5065	5036	4770	4460	4070	3583	3041	2523	2083	1729	1449
310 UC 137	4410	4381	4146	3870	3521	3086	2606	2154	1774	1470	1230
310 UC 118	3780	3754	3550	3311	3008	2630	2216	1828	1503	1245	1041
310 UC 96.8	3348	3316	3125	2896	2602	2238	1854	1510	1232	1015	847
250 UC 89.5	2873	2821	2625	2375	2043	1662	1313	1038	832	678	561
250 UC 72.9	2516	2464	2282	2046	1731	1383	1080	847	676	549	454

Note: Geometry and capacity table, Grade 300, $f_u = 440$ MPa

Gauge lines listed are for M24 and M20 bolts in order of preference (refer Figure 3.29).

Table 7.22 Universal Columns: 100, 150 and 200 UC

Section Name		200 UC 59.5	200 UC 52.2	200 UC 46.2	150 UC 37.2	150 UC 30	150 UC 23.4	100 UC 14.8
Weight	(kg/m)	59.5	52.2	46.2	37.2	30	23.4	14.8
d	(mm)	210	206	203	162	158	152	97
b_f	(mm)	205	204	203	154	153	152	99
t_f	(mm)	14.2	12.5	11	11.5	9.4	6.8	7
t_w	(mm)	9.3	8	7.3	8.1	6.6	6.1	5
r_1	(mm)	11.4	11.4	11.4	8.9	8.9	8.9	10
d_1	(mm)	181	181	181	139	139	139	83
A_g	(mm²)	7620	6660	5900	4730	3860	2980	1890
I_x	(10^6 mm⁴)	61.3	52.8	45.9	22.2	17.6	12.6	3.18
Z_x	(10^3 mm³)	584	512	451	274	223	166	65.6
S_x	(10^3 mm³)	656	570	500	310	250	184	74.4
r_x	(mm)	89.7	89.1	88.2	68.4	67.5	65.1	41.1
I_y	(10^6 mm⁴)	20.4	17.7	15.3	7.01	5.62	3.98	1.14
Z_y	(10^3 mm³)	199	174	151	91	73.4	52.4	22.9
S_y	(10^3 mm³)	303	264	230	139	112	80.2	35.2
r_y	(mm)	51.7	51.5	51	38.5	38.1	36.6	24.5
J	(10^3 mm⁴)	477	325	228	197	109	50.2	34.9
I_w	(10^9 mm⁶)	195	166	142	39.6	30.8	21.1	2.3
f_{yf}	(MPa)	300	300	300	300	320	320	320
f_{yw}	(MPa)	320	320	320	320	320	320	320
Form Factor k_f		1	1	1	1	1	1	1
X Compactness		C	C	N	C	C	N	C
Z_{ex}	(10^3 mm³)	656	570	494	310	250	176	74.4
Y Compactness		C	C	N	C	C	N	C
Z_{ey}	(10^3 mm³)	299	260	223	137	110	73.5	34.4
s_{g1}	(mm)	140 / 90	140 / 90	140 / 90	90 / 70 [M20]	90 / 70 [M20]	90 / 70 [M20]	60 [M20]
s_{g2}	(mm)	90 / 70	90 / 70	90 / 70	70	70	70	N/A
$\phi R_{bb}/b_b$	(kN/mm)	2.08	1.66	1.43	1.91	1.42	1.26	1.19
$\phi R_{by}/b_{bf}$	(kN/mm)	3.35	2.88	2.63	2.92	2.38	2.2	1.8

Section Name	Weight (kg/m)	ϕM_{sx} (kNm)	ϕM_{sy} (kNm)	ϕV_v (kN)	ϕN_s (kN)	ϕN_t (kN)
200 UC 59.5	59.5	177	80.7	337	2057	2057
200 UC 52.2	52.2	154	70.2	285	1798	1798
200 UC 46.2	46.2	133	60.2	256	1593	1593
150 UC 37.2	37.2	83.7	37	227	1277	1277
150 UC 30	30	72	31.7	180	1112	1112
150 UC 23.4	23.4	50.7	21.2	160	858	858
100 UC 14.8	14.8	21.4	9.91	83.8	544	544

Member Bending Capacity, ϕM_{bx} [$\alpha_m = 1$] (kNm) — Effective length (m)

Section Name	1	2	4	6	8	10	12	14	16	18	20
200 UC 59.5	177	167	139	115	97.3	83.6	73	64.5	57.7	52.1	47.5
200 UC 52.2	154	144	118	95.8	79.5	67.4	58.2	51.1	45.4	40.9	37.1
200 UC 46.2	133	125	100	79.9	65.2	54.6	46.7	40.7	36.1	32.3	29.3
150 UC 37.2	83	74.6	58.8	47.3	39.1	33.1	28.6	25.1	22.3	20	18.2
150 UC 30	71	62.6	46.4	35.5	28.4	23.5	20	17.3	15.3	13.7	12.4
150 UC 23.4	49.9	43.5	30.6	22.3	17.3	14	11.8	10.2	8.91	7.94	7.15
100 UC 14.8	20	16.7	12	9.11	7.26	6	5.1	4.42	3.9	3.49	3.15

Member Axial Capacity (Minor Axis), ϕN_{cy} (kN) — Effective length (m)

Section Name	1	2	4	6	8	10	12	14	16	18	20
200 UC 59.5	2004	1841	1331	792	488	326	232	173	134	107	87.3
200 UC 52.2	1751	1608	1160	688	424	283	201	150	116	92.8	75.7
200 UC 46.2	1550	1422	1018	600	369	246	175	131	101	80.7	65.8
150 UC 37.2	1212	1054	581	299	177	116	82.2	61.1	47.2	37.5	30.5
150 UC 30	1050	903	475	241	142	93.5	65.9	49	37.8	30	24.4
150 UC 23.4	807	685	345	173	102	66.8	47.1	35	27	21.4	17.4
100 UC 14.8	478	323	111	52.2	30.1	19.6	13.7	10.1	7.8	6.19	5.03

Member Axial Capacity (Major Axis), ϕN_{cx} (kN) — Effective length (m)

Section Name	1	2	4	6	8	10	12	14	16	18	20
200 UC 59.5	2057	1981	1782	1495	1145	843	628	481	378	305	250
200 UC 52.2	1798	1731	1554	1301	993	729	543	415	327	263	216
200 UC 46.2	1593	1532	1373	1144	868	636	473	361	284	229	188
150 UC 37.2	1266	1195	1003	725	486	335	242	182	142	114	92.9
150 UC 30	1100	1034	853	597	393	269	194	146	113	90.7	74.1
150 UC 23.4	847	794	645	440	286	195	140	105	81.7	65.3	53.4
100 UC 14.8	518	455	259	135	80.3	52.8	37.3	27.8	21.4	17	13.9

Note: Geometry and capacity table, Grade 300, $f_u = 440$ MPa

Gauge lines listed are for M24 and M20 bolts in order of preference (refer Figure 3.29). Maximum bolt size is shown in superscript where limited for each gauge.

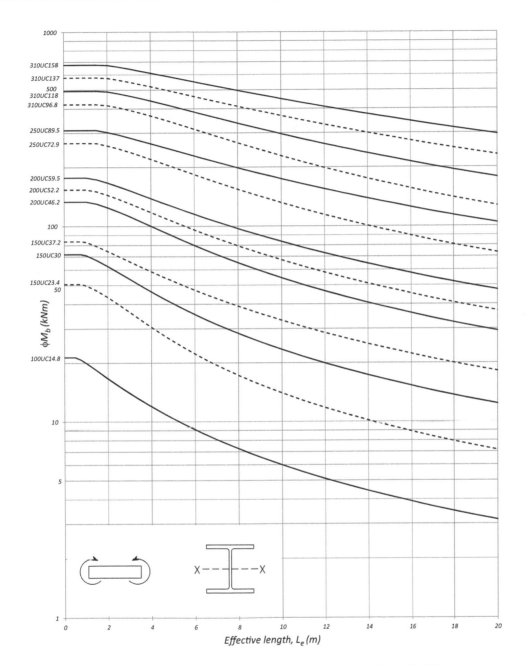

Figure 7.10 Universal columns: Members subject to bending (major axis) grade 300 ($\alpha_m = 1$).

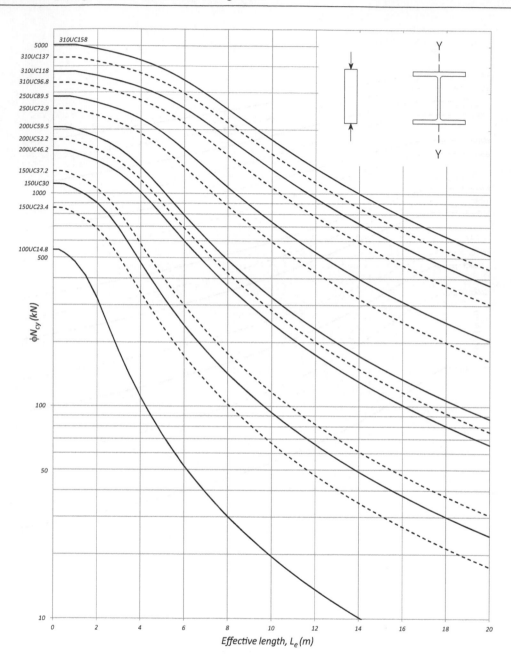

Figure 7.11 Universal columns: Members subject to axial compression (minor axis) grade 300.

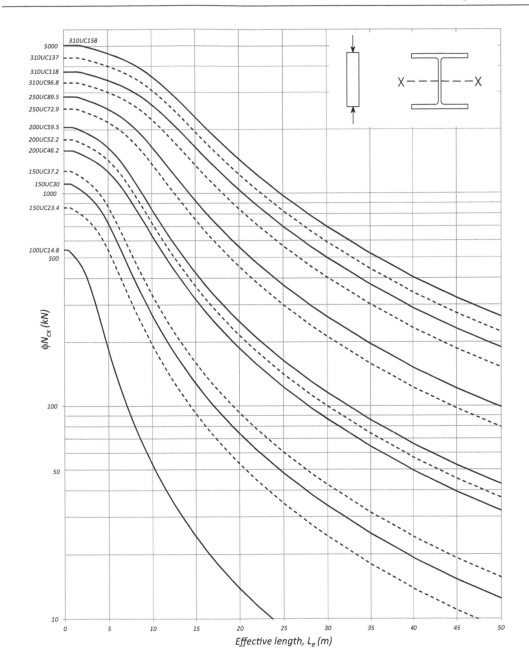

Figure 7.12 Universal columns: Members subject to axial compression (major axis) grade 300.

Table 7.23 Parallel Flange Channels

Section Name	Weight (kg/m)	d (mm)	b_f (mm)	t_f (mm)	t_w (mm)	r_1 (mm)	d_1 (mm)	A_g (mm²)	x_L (mm)	x_0 (mm)	I_x (10⁶ mm⁴)	Z_x (10³ mm³)	S_x (10³ mm³)	r_x (mm)	I_y (10⁶ mm⁴)	Z_{yR} (10³ mm³)	Z_{yL} (10³ mm³)	S_y (10³ mm³)	r_y (mm)	J (10³ mm⁴)	I_w (10⁹ mm⁶)	f_{yf} (MPa)	f_{yw} (MPa)	Form Factor k_f	Z_{ex} (10³ mm³)	$Z_{ey(A)}$ (10³ mm³)	$Z_{ey(B)}$ (10³ mm³)	s_{gf} (mm)	s_{gv} (mm)	$\phi R_{bb}/b_b$ (kN/mm)	$\phi R_{by}/b_{bf}$ (kN/mm)
380 PFC	55.2	380	100	17.5	10	14	345	7030	27.5	56.7	152	798	946	147	6.48	89.4	236	161	30.4	472	151	280	320	1	946	115	134	55	140 / 90 / 70	1.44	3.6
300 PFC	40.1	300	90	16	8	14	268	5110	27.2	56.1	72.4	483	564	119	4.04	64.4	148	117	28.1	290	58.2	300	320	1	564	82.3	96.6	55 M20		1.19	2.88
250 PFC	35.5	250	90	15	8	12	220	4520	28.6	58.5	45.1	361	421	99.9	3.64	59.3	127	107	28.4	238	35.9	300	320	1	421	88.7	89	55 M20		1.45	2.88
230 PFC	25.1	230	75	12	6.5	12	206	3200	22.6	46.7	26.8	233	271	91.4	1.76	33.6	77.8	61	23.5	108	15	300	320	1	271	45.1	50.4	45 M20	90 / 70*	1.03	2.34
200 PFC	22.9	200	75	12	6	12	176	2920	24.4	50.5	19.1	191	221	80.9	1.65	32.7	67.8	58.9	23.8	101	10.6	300	320	1	221	46.7	49.1	45 M20	90 / 70	1.02	2.16
180 PFC	20.9	180	75	11	6	12	158	2660	24.5	50.3	14.1	157	182	72.9	1.51	29.9	61.5	53.8	23.8	81.4	7.82	300	320	1	182	44.9	44.8	45 M20	70*	1.12	2.16
150 PFC	17.7	150	75	9.5	6	10	131	2250	24.9	51	8.34	111	129	60.8	1.29	25.7	51.6	46	23.9	54.9	4.59	320	320	1	129	38.5	38.5	45 M20	65 M20*	1.27	2.16
125 PFC	11.9	125	65	7.5	4.7	8	110	1520	21.8	45	3.97	63.5	73	51.1	0.658	15.2	30.2	27.2	20.8	23.1	1.64	320	320	1	72.8	22.8	22.8	35 M20	50 M16	0.955	1.69
100 PFC	8.33	100	50	6.7	4.2	8	86.6	1060	16.7	33.9	1.74	34.7	40.3	40.4	0.267	8.01	16	14.4	15.9	13.2	0.424	320	320	1	40.3	12	12	30 M16	N/A	0.916	1.51
75 PFC	5.92	75	40	6.1	3.8	8	62.8	754	13.7	27.1	0.683	18.2	21.4	30.1	0.12	4.56	8.71	8.2	12.6	8.13	0.106	320	320	1	21.4	6.84	6.84	N/A	N/A	0.905	1.37

(Continued)

Table 7.23 (Continued) Parallel Flange Channels

Section Name	φMcx (kNm)	φMsy (kNm)	φVv (kN)	φNs (kN)	φNt (kN)	Member Bending Capacity, φMbx [αm = 1] (kNm) — Effective length (m)											Member Axial Capacity (Minor Axis), φNcy (kN) — Effective length (m)											Member Axial Capacity (Major Axis), φNcx (kN) — Effective length (m)										
						1	2	4	6	8	10	12	14	16	18	20	0.5	1	2	3	4	5	6	7	8	9	10	1	2	4	6	8	10	12	14	16	18	20
380 PFC	238	29	657	1772	1772	230	192	127	89.9	68.8	55.5	46.5	39.9	35	31.2	28.1	1737	1581	1209	816	544	379	276	210	164	132	108	1772	1764	1635	1500	1349	1181	1010	852	717	606	516
300 PFC	152	22.2	415	1380	1380	144	117	76.4	54.6	42	34	28.6	24.6	21.6	19.2	17.3	1338	1200	864	547	354	243	176	133	104	82.9	67.9	1380	1346	1216	1071	904	736	591	476	388	320	268
250 PFC	114	23.9	346	1220	1220	108	89.3	61	44.9	35.1	28.7	24.2	20.9	18.4	16.4	14.8	1185	1065	771	491	318	219	159	120	93.4	74.9	61.3	1220	1169	1029	864	685	528	408	321	257	210	175
230 PFC	73.2	12.2	258	864	864	66.9	51.6	31.9	22.4	17.1	13.8	11.6	9.94	8.71	7.76	6.99	822	715	455	263	164	111	79.4	59.6	46.3	37	30.2	864	819	708	576	441	331	252	196	157	127	105
200 PFC	59.7	12.6	207	788	788	54.8	43.3	28.1	20.3	15.7	12.7	10.7	9.24	8.11	7.23	6.52	751	655	421	245	153	103	74.2	55.7	43.2	34.6	28.2	788	735	617	476	346	253	189	146	115	93.2	76.9
180 PFC	49.1	12.1	187	718	718	45.2	35.9	23.7	17.2	13.4	10.9	9.18	7.91	6.95	6.2	5.59	684	597	383	223	140	94.2	67.6	50.7	39.4	31.5	23.7	713	658	535	392	274	196	145	111	87.1	70.2	57.7
150 PFC	37.2	11.1	156	648	644	34	27	17.9	13	10.1	8.26	6.96	6	5.27	4.7	4.24	615	534	335	193	120	80.8	57.9	43.4	33.7	26.9	22	631	570	421	274	179	124	89.8	67.9	53	42.5	34.8
125 PFC	21	6.57	102	438	435	18.5	14	8.76	6.22	4.79	3.88	3.25	2.8	2.46	2.19	1.97	408	342	191	104	63.5	42.4	30.2	22.5	17.5	13.9	11.4	419	368	243	144	90.9	61.6	44.2	33.2	25.8	20.7	16.9
100 PFC	11.6	3.46	72.6	305	303	9.55	6.97	4.28	3.02	2.31	1.87	1.57	1.35	1.18	1.05	0.95	271	205	90.6	45.8	27.2	17.9	12.6	9.39	7.25	5.76	4.69	283	236	128	69.2	42	28	19.9	14.9	11.5	9.17	7.48
75 PFC	6.16	1.97	49.2	217	216	4.87	3.59	2.23	1.58	1.21	0.98	0.83	0.71	0.62	0.56	0.5	182	119	44	21.4	12.5	8.16	5.74	4.25	3.28	2.6	2.11	191	140	59.1	29.5	17.4	11.5	8.08	6	4.63	3.68	3

Note: Geometry and capacity table, Grade 300, f_u = 440 MPa

Gauge lines listed are for M24 and M20 bolts in order of preference (refer Figure 3.29). Maximum bolt size is shown in superscript where limited for each gauge. 230 PFC can also use $s_{gw} = 140^{M16}$, 180 PFC can also use $s_{gw} = 90^{M20}$, 150 PFC can also use $s_{gw} = 70^{M16}$.

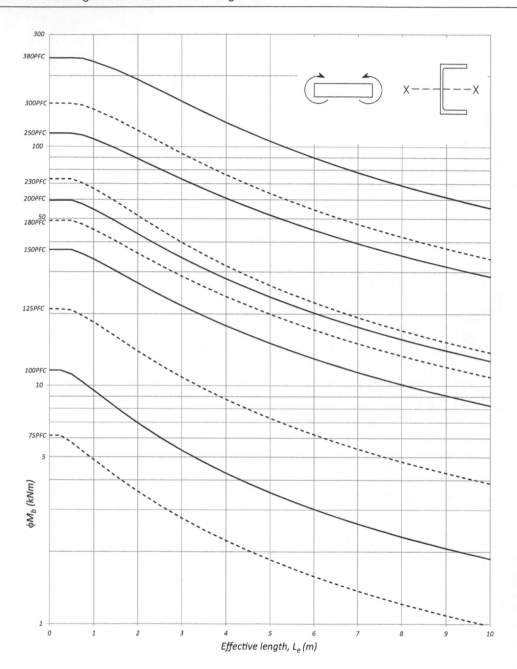

Figure 7.13 Parallel flange channels: Members subject to bending (major axis) grade 300 ($\alpha_m = 1$).

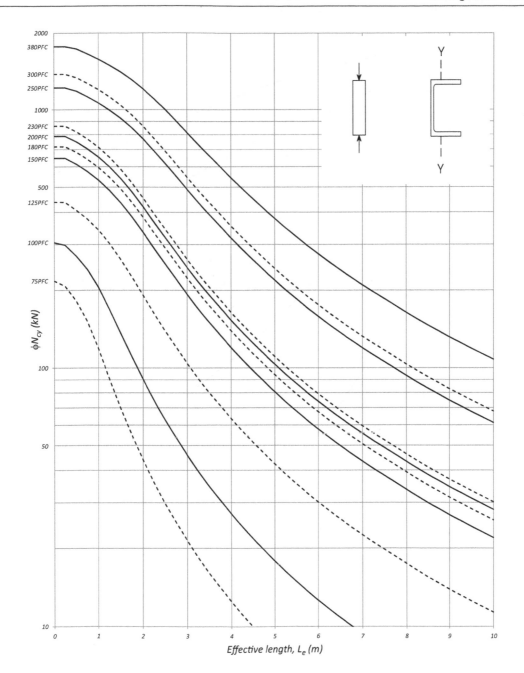

Figure 7.14 Parallel flange channels: Members subject to axial compression (minor axis) grade 300.

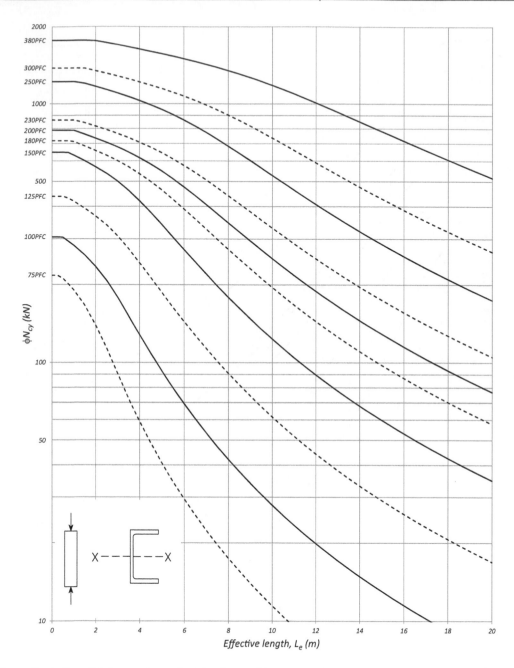

Figure 7.15 Parallel flange channels: Members subject to axial compression (major axis) grade 300.

Table 7.24 Equal Angles: 150 and 200 EA

Section Name	Weight	t	r_1	r_2	A_g	pB	I_x	$y_1 = y_4$	$Z_{x1} = Z_{x4}$	S_x	r_x	I_y	x_3	Z_{y3}	x_5	Z_{y5}	S_y	r_y	J	f_y	Form Factor k_f	Z_{ex}	Z_{ey} Load B	Z_{ey} Load D	s_{g1}	s_{g2}	s_{g3}
b x nominal thickness (mm)	(kg/m)	(mm)	(mm)	(mm)	(mm^2)	(mm)	(10^6 mm^4)	(mm)	(10^3 mm^3)	(10^3 mm^3)	(mm)	(10^6 mm^4)	(mm)	(10^3 mm^3)	(mm)	(10^3 mm^3)	(10^3 mm^3)	(mm)	(10^3 mm^4)	(MPa)		(10^3 mm^3)	(10^3 mm^3)	(10^3 mm^3)	(mm)	(mm)	(mm)
200x26 EA	76.8	26	18	5	9780	59.3	56.8	141	402	643	76.2	14.9	73.9	202	83.8	178	329	39	2250	280	1	602	267	267	75	75	120
200x20 EA	60.1	20	18	5	7660	57	45.7	141	323	511	77.2	11.8	72.9	162	80.6	147	260	39.3	1060	280	1	479	218	220	75	75	120
200x18 EA	54.4	18	18	5	6930	56.2	41.7	141	295	464	77.6	10.8	72.6	149	79.5	136	236	39.4	778	280	1	427	196	204	75	75	120
200x16 EA	48.7	16	18	5	6200	55.4	37.6	141	266	417	77.9	9.72	72.3	135	78.4	124	212	39.6	554	300	1	369	172	186	75	75	120
200x13 EA	40	13	18	5	5090	54.2	31.2	141	221	344	78.3	8.08	71.9	112	76.6	105	176	39.8	304	300	1	285	136	158	75	75	120
150x19 EA	42.1	19	13	5	5360	44.2	17.6	106	166	265	57.2	4.6	54.9	83.8	62.6	73.5	135	29.3	657	280	1	248	110	110	55 ^{M20}	55 ^{M20}	90 ^{M20}
150x16 EA	35.4	15.8	13	5	4520	43	15.1	106	142	225	57.8	3.91	54.3	71.9	60.8	64.2	115	29.4	386	300	1	212	95.7	96.3	55 ^{M20}	55 ^{M20}	90 ^{M20}
150x12 EA	27.3	12	13	5	3480	41.5	11.9	106	112	175	58.4	3.06	53.7	56.9	58.7	52.1	89.3	29.6	174	300	1	155	72.3	78.1	55 ^{M20}	55 ^{M20}	90 ^{M20}
150x10 EA	21.9	9.5	13	5	2790	40.5	9.61	106	90.6	141	58.7	2.48	53.4	46.4	57.3	43.3	72	29.8	88.9	320	0.958	114	54.5	64.9	55 ^{M20}	55 ^{M20}	90 ^{M20}

(*Continued*)

Table 7.24 (Continued) Equal Angles: 150 and 200 EA

Section Name	ϕM_{cx}	ϕM_{sy}	ϕN_s	ϕN_t	Member Axial Capacity (Minor Axis), ϕN_{cy} (kN) — Effective length (m)																Member Axial Capacity (Major Axis), ϕN_{cx} (kN) — Effective length (m)															
b x nominal thickness (mm)	(kNm)	(kNm)	(kN)	(kN)	0.5	1	1.5	2	2.5	3	3.5	4	4.5	5	5.5	6	7	8	9	10	0.5	1	2	3	4	5	6	7	8	9	10	12	14	16	18	20
200x26 EA	152	67.3	2465	2465	2464	2296	2121	1928	1712	1486	1270	1080	920	787	678	589	454	359	290	239	2465	2460	2288	2108	1908	1685	1454	1237	1048	889	759	567	436	344	278	229
200x20 EA	121	54.9	1930	1930	1930	1800	1665	1515	1347	1172	1004	854	728	624	538	467	360	285	231	190	1930	1928	1796	1657	1503	1332	1153	983	835	710	607	454	349	276	223	184
200x18 EA	108	49.4	1746	1746	1746	1629	1507	1372	1221	1063	911	776	661	566	489	424	327	259	210	173	1746	1745	1626	1501	1363	1209	1048	895	760	647	553	414	319	252	204	168
200x16 EA	100	46.4	1674	1674	1671	1555	1434	1298	1148	992	844	715	608	519	447	388	298	236	190	157	1674	1669	1551	1427	1289	1135	976	828	700	593	506	377	290	229	185	152
200x13 EA	77	36.7	1374	1374	1372	1278	1179	1068	946	818	697	591	502	429	370	321	247	195	158	130	1374	1371	1274	1173	1061	935	805	684	578	491	419	312	240	190	153	126
150x19 EA	62.5	27.7	1351	1351	1320	1196	1057	898	736	593	479	391	324	271	230	197	149	117	93.8	76.9	1351	1317	1190	1046	882	717	575	463	377	311	260	189	143	112	89.6	73.4
150x16 EA	57.2	25.8	1220	1220	1189	1073	942	793	643	515	414	337	278	232	197	169	128	99.6	79.9	65.5	1220	1187	1069	935	782	631	503	403	328	270	226	164	124	96.5	77.4	63.4
150x12 EA	41.9	19.5	940	940	916	827	728	613	499	400	322	262	216	181	153	131	99.4	77.7	62.3	51.1	940	915	825	723	607	491	393	315	256	211	177	128	97	75.7	60.7	49.7
150x10 EA	32.8	15.7	770	798	743	645	543	447	363	295	241	199	166	140	120	103	78.6	61.8	49.8	40.9	770	742	642	539	441	357	289	236	194	162	137	100	76.5	60.1	48.4	39.8

Note: Geometry and capacity table, Grade 300, $f_u = 440$ MPa

Gauge lines listed are for M24 bolts or smaller (refer Figure 3.29). Maximum bolt size is shown in superscript where limited for each gauge.

Table 7.25 Equal Angles: 100 and 125 EA

Section Name b x nominal thickness (mm)	Weight (kg/m)	t (mm)	r1 (mm)	r2 (mm)	Ag (mm²)	pB (mm)	Ix (10⁶mm⁴)	y1 = yt (mm)	Zxt = Zxt (10³mm³)	Sx (10³mm³)	rx (mm)	Iy (10⁶mm⁴)	x3 (mm)	Zy3 (10³mm³)	x5 (mm)	Zy5 (10³mm³)	Sy (10³mm³)	ry (mm)	J (10³mm⁴)	fy (MPa)	Form Factor kf	Zxx (10³mm³)	Zcy Load B (10³mm³)	Zcy Load D (10³mm³)	sp1 (mm)	sp2 (mm)	sp3 (mm)
125x16 EA	29.1	15.8	10	5	3710	36.8	8.43	88.4	95.4	153	47.7	2.2	45.4	48.5	52.1	42.3	77.8	24.4	313	300	1	143	63.4	63.4	45 M20	50 M20	75 M20
125x12 EA	22.5	12	10	5	2870	35.4	6.69	88.4	75.7	120	48.3	1.73	44.7	38.6	50.1	34.5	60.8	24.5	141	300	1	110	50.3	51.7	45 M20	50 M20	75 M20
125x10 EA	18	9.5	10	5	2300	34.4	5.44	88.4	61.6	96.5	48.7	1.4	44.4	31.5	48.7	28.8	49	24.7	71.9	320	1	83.2	38.9	43.1	45 M20	50 M20	75 M20
125x8 EA	14.9	7.8	10	5	1900	33.7	4.55	88.4	51.5	80.2	48.9	1.17	44.2	26.5	47.7	24.5	40.8	24.8	40.6	320	0.943	64.3	30.7	36.8	45 M20	50 M20	75 M20
100x12 EA	17.7	12	8	5	2260	29.2	3.29	70.7	46.6	74.5	38.2	0.857	35.8	23.9	41.3	20.8	37.9	19.5	110	300	1	69.9	31.1	31.1	N/A	N/A	55 M20
100x10 EA	14.2	9.5	8	5	1810	28.2	2.7	70.7	38.2	60.4	38.6	0.695	35.4	19.6	39.9	17.4	30.7	19.6	56.2	320	1	55.1	25.2	26.1	N/A	N/A	55 M20
100x8 EA	11.8	7.8	8	5	1500	27.5	2.27	70.7	32	50.3	38.8	0.582	35.2	16.5	38.9	14.9	25.6	19.7	31.7	320	1	43.7	20.4	22.4	N/A	N/A	55 M20
100x6 EA	9.16	6	8	5	1170	26.8	1.78	70.7	25.2	39.3	39.1	0.458	35	13.1	37.9	12.1	20	19.8	14.8	320	0.906	30.9	14.8	18.1	N/A	N/A	55 M20

Section Name b x nominal thickness (mm)	ϕM_{sx} (kNm)	ϕM_{sy} (kNm)	ϕN_s (kN)	ϕN_t (kN)
125x16 EA	38.6	17.1	1002	1002
125x12 EA	29.7	13.6	775	775
125x10 EA	24	11.2	662	658
125x8 EA	18.5	8.84	516	544
100x12 EA	18.9	8.4	610	610
100x10 EA	15.9	7.26	521	518
100x8 EA	12.6	5.88	432	429
100x6 EA	8.9	4.26	305	335

Member Axial Capacity (Minor Axis), ϕN_{cy} (kN)

Section Name	Effective length (m)															
b x nominal thickness (mm)	0.5	1	1.5	2	2.5	3	3.5	4	4.5	5	5.5	6	7	8	9	10
125x16 EA	957	839	699	549	420	324	254	203	166	137	116	98.7	74.1	57.6	46	37.6
125x12 EA	740	650	542	426	327	252	198	158	129	107	90.2	76.9	57.8	44.9	35.9	29.3
125x10 EA	631	551	456	355	270	207	162	130	106	87.7	73.8	62.9	47.2	36.7	29.3	23.9
125x8 EA	486	406	326	256	199	157	125	101	83.5	69.9	59.2	50.7	38.3	30	24	19.7
100x12 EA	566	470	356	256	185	138	106	83.6	67.5	55.6	46.6	39.6	29.5	22.9	18.2	14.9
100x10 EA	481	397	296	210	151	112	86.2	68	54.9	45.2	37.8	32.1	24	18.5	14.8	12
100x8 EA	399	329	247	176	126	94	72.1	56.9	45.9	37.8	31.7	26.9	20	15.5	12.4	10.1
100x6 EA	277	218	164	120	89.7	68.3	53.4	42.6	34.7	28.8	24.2	20.7	15.5	12	9.63	7.86

Member Axial Capacity (Major Axis), ϕN_{cx} (kN)

Section Name	Effective length (m)															
b x nominal thickness (mm)	0.5	1	2	3	4	5	6	7	8	9	10	12	14	16	18	20
125x16 EA	1002	954	833	689	536	407	312	244	195	159	132	94.6	71	55.2	44.1	36
125x12 EA	775	739	647	537	420	320	246	193	154	126	104	74.9	56.2	43.7	34.9	28.5
125x10 EA	662	630	549	452	350	265	203	159	127	103	85.5	61.2	45.9	35.7	28.5	23.3
125x8 EA	516	485	404	323	252	196	153	122	99.1	81.6	68.2	49.4	37.4	29.2	23.4	19.2
100x12 EA	608	564	466	349	248	179	133	102	80.5	65	53.5	38.1	28.4	22	17.5	14.3
100x10 EA	518	480	394	292	206	148	110	83.9	66.1	53.4	43.9	31.2	23.3	18	14.3	11.7
100x8 EA	430	398	327	243	172	123	91.6	70.2	55.3	44.6	36.7	26.1	19.5	15.1	12	9.78
100x6 EA	305	276	217	162	119	88.1	67	52.3	41.7	34	28.1	20.2	15.1	11.8	9.39	7.67

Note: Geometry and capacity table, Grade 300, f_u = 440 MPa

Gauge lines listed are for M24 bolts or smaller (refer Figure 3.29). Maximum bolt size is shown in superscript where limited for each gauge.

Table 7.26 Equal Angles: 75 and 90 EA

Section Name b x nominal thickness (mm)	Weight (kg/m)	t (mm)	r_1 (mm)	r_2 (mm)	A_g (mm²)	pB (mm)	I_x (10⁶mm⁴)	$y_1 = y_A$ (mm)	$Z_{x1} = Z_{x4}$ (10³mm³)	S_x (10³mm³)	r_x (mm)	I_y (10⁶mm⁴)	x_3 (mm)	Z_{y3} (10³mm³)	x_5 (mm)	Z_{y5} (10³mm³)	S_y (10³mm³)	r_y (mm)	J (10³mm⁴)	f_y (MPa)	Form Factor k_f	Z_{xx} (10³mm³)	Z_{xy} Load B (10³mm³)	Z_{xy} Load D (10³mm³)	s_{g1} (mm)	s_{g2} (mm)	s_{g3} (mm)
90x10 EA	12.7	9.5	8	5	1620	25.7	1.93	63.6	30.4	48.3	34.5	0.5	31.9	15.7	36.4	13.8	24.6	17.6	50.5	320	1	45	20.4	20.6	N/A	N/A	5.5 M20
90x8 EA	10.6	7.8	8	5	1350	25	1.63	63.6	25.6	40.4	34.8	0.419	31.7	13.2	35.4	11.8	20.5	17.6	28.6	320	1	36	16.7	17.8			
90x6 EA	8.22	6	8	5	1050	24.3	1.28	63.6	20.1	31.6	35	0.33	31.5	10.5	34.3	9.62	16.1	17.8	13.4	320	1	25.9	12.4	14.4			
75x10 EA	10.5	9.5	8	5	1340	22	1.08	53	20.4	32.8	28.4	0.282	26.6	10.6	31.1	9.09	16.8	14.5	41.9	320	1	30.5	13.6	13.6			
75x8 EA	8.73	7.8	8	5	1110	21.3	0.913	53	17.2	27.5	28.7	0.237	26.4	8.99	30.1	7.87	14	14.6	23.8	320	1	25.4	11.6	11.8			
75x6 EA	6.81	6	8	5	867	20.5	0.722	53	13.6	21.6	28.9	0.187	26.2	7.15	29	6.44	11	14.7	11.2	320	1	18.7	8.85	9.66			
75x5 EA	5.27	4.6	8	5	672	19.9	0.563	53	10.6	16.7	29	0.147	26.1	5.62	28.1	5.22	8.61	14.8	5.28	320	0.927	13.2	6.47	7.82	N/A	N/A	4.5 M20

Section Name b x nominal thickness (mm)	ϕM_{sx} (kNm)	ϕM_{sy} (kNm)	ϕN_s (kN)	ϕN_t (kN)	Member Axial Capacity (Minor Axis) ϕN_{cy} (kN) Effective length (m)																	Member Axial Capacity (Major Axis) ϕN_{cx} (kN) Effective length (m)															
					0.5	1	1.5	2	2.5	3	3.5	4	4.5	5	5.5	6	7	8	9	10	0.5	1	2	3	4	5	6	7	8	9	10	12	14	16	18	20
90x10 EA	13	5.88	467	463	423	335	236	162	114	83.7	63.8	50.1	40.3	33.1	27.6	23.4	17.4	13.5	10.7	8.74	460	421	331	231	157	110	80.8	61.5	48.2	38.8	31.9	22.6	16.8	13	10.3	8.41
90x8 EA	10.4	4.81	389	386	352	279	197	135	94.9	69.7	53.1	41.7	33.6	27.6	23	19.5	14.5	11.2	8.94	7.28	383	352	277	194	132	93.2	68.4	52.1	40.8	32.9	27	19.1	14.2	11	8.75	7.12
90x6 EA	7.46	3.57	302	300	275	219	155	106	75.2	55.3	42.2	33.1	26.7	21.9	18.3	15.5	11.6	8.93	7.11	5.79	298	274	216	152	104	73.1	53.7	40.9	32.1	25.8	21.2	15	11.2	8.65	6.88	5.6
75x10 EA	8.78	3.92	386	383	336	241	153	98.9	67.9	49.1	37.1	28.9	23.2	19	15.8	13.4	9.93	7.66	6.08	4.95	373	334	237	148	95.6	65.4	47.3	35.7	27.8	22.3	18.2	12.8	9.53	7.35	5.84	4.75
75x8 EA	7.32	3.34	320	318	279	201	128	82.8	56.9	41.2	31.1	24.3	19.4	15.9	13.3	11.2	8.33	6.43	5.11	4.16	310	277	198	125	80.5	55.2	39.9	30.1	23.5	18.8	15.4	10.9	8.06	6.22	4.94	4.02
75x6 EA	5.39	2.55	250	248	218	158	101	65.4	45	32.6	24.6	19.2	15.4	12.6	10.5	8.88	6.6	5.09	4.04	3.29	242	217	156	98.4	63.6	43.6	31.6	23.8	18.6	14.9	12.2	8.59	6.38	4.92	3.91	3.18
75x5 EA	3.8	1.86	179	192	151	105	69.5	47	33.2	24.4	18.6	14.7	11.8	9.71	8.12	6.88	5.13	3.96	3.16	2.57	173	150	103	67.8	45.6	32.1	23.6	18	14.1	11.4	9.34	6.62	4.93	3.81	3.03	2.47

Note: Geometry and capacity table, Grade 300, f_u = 440 MPa

Gauge lines listed are for M24 bolts or smaller (refer Figure 3.29). Maximum bolt size is shown in superscript where limited for each gauge.

Table 7.27 Equal Angles: 50, 55 and 65 EA

(Continued)

Section Name b x nominal thickness (mm)	Weight (kg/m)	t (mm)	r_1 (mm)	r_2 (mm)	A_g (mm²)	p_B (mm)	I_x (10⁶ mm⁴)	$y_1 = y_4$ (mm)	$Z_{x1} = Z_{x4}$ (10³ mm³)	S_x (10³ mm³)	r_x (mm)	I_y (10⁶ mm⁴)	x_3 (mm)	Z_{y3} (10³ mm³)	x_5 (mm)	Z_{y5} (10³ mm³)	S_y (10³ mm³)	r_y (mm)	J (10³ mm⁴)	f_y (MPa)	Form Factor k_f	Z_{ex} (10³ mm³)	Z_{ey} Load B (10³ mm³)	Z_{ey} Load D (10³ mm³)	s_{p1} (mm)	s_{p2} (mm)	s_{p3} (mm)
65x10 EA	9.02	9.5	6	3	1150	19.6	0.691	46	15	24.3	24.5	0.183	23.7	7.71	27.7	6.6	12.5	12.6	35.1	320	1	22.5	9.9	9.9	N/A	N/A	35 M16
65x8 EA	7.51	7.8	6	3	957	19	0.589	46	12.8	20.5	24.8	0.154	23.4	6.56	26.8	5.73	10.5	12.7	20	320	1	19.2	8.59	8.59	N/A	N/A	35 M16
65x6 EA	5.87	6	6	3	748	18.3	0.471	46	10.2	16.2	25.1	0.122	23.1	5.26	25.8	4.71	8.25	12.8	9.37	320	1	14.7	6.76	7.07	N/A	N/A	35 M16
65x5 EA	4.56	4.6	6	3	581	17.7	0.371	46	8.08	12.7	25.3	0.0959	23	4.18	25	3.83	6.46	12.9	4.36	320	1	10.6	5.05	5.75	N/A	N/A	35 M16
55x6 EA	4.93	6	6	3	628	15.8	0.278	38.9	7.14	11.4	21	0.0723	19.6	3.69	22.3	3.24	5.82	10.7	7.93	320	1	10.7	4.84	4.86	N/A	N/A	35 M16
55x5 EA	3.84	4.6	6	3	489	15.2	0.22	38.9	5.66	8.93	21.2	0.0571	19.4	2.94	21.5	2.66	4.57	10.8	3.71	320	1	7.88	3.7	3.98	N/A	N/A	35 M16
50x8 EA	5.68	7.8	6	3	723	15.2	0.253	35.4	7.16	11.7	18.7	0.0675	18.1	3.73	21.5	3.14	6	9.66	15.2	320	1	10.7	4.71	4.71	N/A	N/A	30 M16
50x6 EA	4.46	6	6	3	568	14.5	0.205	35.4	5.79	9.3	19	0.0536	17.8	3.01	20.5	2.61	4.76	9.71	7.21	320	1	8.69	3.92	3.92	N/A	N/A	30 M16
50x5 EA	3.48	4.6	6	3	443	13.9	0.163	35.4	4.61	7.32	19.2	0.0424	17.6	2.4	19.7	2.15	3.75	9.78	3.38	320	1	6.6	3.08	3.22	N/A	N/A	30 M16
50x3 EA	2.31	3	6	3	295	13.2	0.11	35.4	3.11	4.9	19.3	0.0289	17.6	1.65	18.7	1.55	2.53	9.9	1.01	320	0.907	3.82	1.9	2.32	N/A	N/A	30 M16

Table 7.27 (Continued) Equal Angles: 50, 55 and 65 EA

Member Axial Capacity (Minor Axis, φN_{cy}) (kN) — Effective length (m)

Section Name (b x nominal thickness (mm))	φM_{sx} (kNm)	φM_{sy} (kNm)	φN_s (kN)	φN_t (kN)	0.5	1	1.5	2	2.5	3	3.5	4	4.5	5	5.5	6	7	8	9	10
65x10 EA	6.48	2.85	331	329	277	181	107	67.2	45.4	32.6	24.5	19	15.2	12.4	10.4	8.75	6.48	5	3.97	3.22
65x8 EA	5.53	2.47	276	274	231	152	90.1	56.6	38.4	27.5	20.7	16.1	12.9	10.5	8.75	7.39	5.48	4.22	3.35	2.73
65x6 EA	4.23	1.95	215	214	181	120	71.2	44.9	30.4	21.8	16.4	12.8	10.2	8.34	6.94	5.87	4.35	3.35	2.66	2.16
65x5 EA	3.05	1.45	167	166	141	93.9	56	35.3	23.9	17.2	12.9	10.1	8.04	6.58	5.47	4.63	3.43	2.64	2.1	1.71
55x6 EA	3.08	1.39	181	180	143	82	45.1	27.6	18.4	13.1	9.83	7.62	6.07	4.96	4.12	3.48	2.57	1.98	1.57	1.28
55x5 EA	2.27	1.07	141	140	112	64.6	35.7	21.8	14.6	10.4	7.79	6.04	4.82	3.93	3.27	2.76	2.04	1.57	1.25	1.01
50x8 EA	3.08	1.36	208	207	157	82.4	43.8	26.5	17.6	12.5	9.31	7.21	5.74	4.68	3.89	3.28	2.42	1.86	1.48	1.2
50x6 EA	2.5	1.13	164	163	124	65.2	34.7	21	13.9	9.9	7.39	5.72	4.55	3.71	3.08	2.6	1.92	1.48	1.17	0.953
50x5 EA	1.9	0.887	128	127	97	51.4	27.4	16.6	11	7.83	5.84	4.52	3.6	2.94	2.44	2.06	1.52	1.17	0.928	0.754
50x3 EA	1.1	0.547	77.1	84.4	55.1	30.4	17.2	10.7	7.26	5.21	3.91	3.04	2.43	1.98	1.65	1.39	1.03	0.795	0.631	0.513

Member Axial Capacity (Major Axis, φN_{cx}) (kN) — Effective length (m)

Section Name	0.5	1	2	3	4	5	6	7	8	9	10	12	14	16	18	20
65x10 EA	315	275	176	102	64	43.2	31	23.2	18.1	14.4	11.8	8.28	6.14	4.73	3.75	3.05
65x8 EA	263	230	148	86.9	54.4	36.7	26.4	19.8	15.4	12.3	10	7.06	5.23	4.03	3.2	2.6
65x6 EA	206	180	118	69.2	43.4	29.3	21.1	15.8	12.3	9.83	8.03	5.65	4.18	3.22	2.56	2.08
65x5 EA	160	140	92.1	54.4	34.2	23.1	16.6	12.5	9.7	7.75	6.33	4.45	3.3	2.54	2.02	1.64
55x6 EA	169	142	80	43.8	26.7	17.8	12.7	9.48	7.35	5.86	4.78	3.35	2.48	1.91	1.51	1.23
55x5 EA	132	111	63.1	34.6	21.1	14.1	10.1	7.52	5.83	4.64	3.79	2.66	1.97	1.51	1.2	0.976
50x8 EA	191	155	78.7	41.5	24.9	16.5	11.7	8.75	6.77	5.39	4.39	3.08	2.27	1.75	1.39	1.13
50x6 EA	150	123	63.2	33.5	20.2	13.4	9.5	7.08	5.48	4.37	3.56	2.49	1.84	1.42	1.12	0.913
50x5 EA	117	96.1	50.1	26.6	16	10.6	7.56	5.64	4.36	3.47	2.83	1.98	1.47	1.13	0.895	0.727
50x3 EA	69.6	54.4	29.4	16.6	10.3	6.93	4.97	3.73	2.89	2.31	1.89	1.33	0.982	0.756	0.6	0.488

Note: Geometry and capacity table, Grade 300, f_u = 440 MPa

Gauge lines listed are for M24 bolts or smaller (refer Figure 3.29). Maximum bolt size is shown in superscript where limited for each gauge.

Table 7.28 Equal Angles: 40 and 45 EA

Section Name (b x nominal thickness) (mm)	Weight (kg/m)	t (mm)	r1 (mm)	r2 (mm)	Ag (mm²)	pB (mm)	Ix (10³ mm⁴)	y1=y4 (mm)	Zx1=Zx4 (10³ mm³)	Sx (10³ mm³)	rx (mm)	If (10⁶ mm⁴)	x3 (mm)	Zp3 (10³ mm³)	x5 (mm)	Zp5 (10³ mm³)	Sy (10³ mm³)	ry (mm)	J (10³ mm⁴)	fy (MPa)	Form Factor kf	Zcz (10³ mm³)	Znp Load B (10³ mm³)	Znp Load D (10³ mm³)	sp1 (mm)	sp2 (mm)	sp3 (mm)
45x6 EA	3.97	6	5	3	506	13.3	0.146	31.8	4.59	7.41	17	0.0383	16	2.39	18.8	2.04	3.79	8.71	6.32	320	1	6.88	3.06	3.06	N/A	N/A	N/A
45x5 EA	3.1	4.6	5	3	394	12.7	0.117	31.8	3.66	5.84	17.2	0.0303	15.8	1.91	18	1.68	2.99	8.76	2.96	320	1	5.39	2.47	2.52			
45x3 EA	2.06	3	5	3	263	12	0.079	31.8	2.48	3.92	17.3	0.0206	15.7	1.31	17	1.21	2.02	8.85	0.875	320	1	3.19	1.55	1.81			
40x6 EA	3.5	6	5	3	446	12	0.0997	28.3	3.53	5.75	15	0.0265	14.3	1.86	17	1.55	2.95	7.71	5.6	320	1	5.29	2.33	2.33			
40x5 EA	2.73	4.6	5	3	348	11.5	0.0801	28.3	2.83	4.55	15.2	0.0209	14	1.49	16.2	1.29	2.33	7.75	2.63	320	1	4.25	1.93	1.93			
40x3 EA	1.83	3	5	3	233	10.8	0.0545	28.3	1.93	3.06	15.3	0.0142	13.9	1.02	15.3	0.93	1.58	7.82	0.785	320	1	2.59	1.25	1.4			

Section Name (b x nominal thickness) (mm)	φMsx (kNm)	φMsy (kNm)	φNs (kN)	φNt (kN)	Member Axial Capacity (Minor Axis), φNcy (kN) Effective length (m)															
					0.5	1	1.5	2	2.5	3	3.5	4	4.5	5	5.5	6	7	8	9	10
45x6 EA	1.98	0.881	145.7	145	104	49.7	25.7	15.7	10.1	7.18	5.34	4.13	3.29	2.68	2.22	1.87	1.38	1.06	0.843	0.685
45x5 EA	1.55	0.711	113.5	113	81.3	39	20.2	12.1	7.98	5.65	4.21	3.25	2.59	2.11	1.75	1.48	1.09	0.838	0.664	0.539
45x3 EA	0.92	0.446	75.7	75.2	54.6	26.4	13.7	8.21	5.43	3.85	2.86	2.21	1.76	1.43	1.19	1	0.742	0.571	0.452	0.367
40x6 EA	1.52	0.671	128.4	127.6	84.4	36.3	18.3	10.8	7.1	5.01	3.72	2.87	2.28	1.86	1.54	1.3	0.96	0.737	0.584	0.474
40x5 EA	1.22	0.556	100.2	99.6	66.1	28.6	14.4	8.5	5.59	3.95	2.93	2.26	1.8	1.47	1.22	1.02	0.756	0.581	0.461	0.374
40x3 EA	0.746	0.36	67.1	66.7	44.6	19.4	9.78	5.79	3.81	2.69	2	1.54	1.23	0.998	0.828	0.698	0.516	0.396	0.314	0.255

Section Name (b x nominal thickness) (mm)	Member Axial Capacity (Major Axis), φNcx (kN) Effective length (m)															
	0.5	1	2	3	4	5	6	7	8	9	10	12	14	16	18	20
45x6 EA	131	102	47.9	24.6	14.7	9.68	6.85	5.1	3.94	3.13	2.55	1.79	1.32	1.01	0.804	0.653
45x5 EA	102	80.4	38	19.6	11.7	7.71	5.45	4.06	3.14	2.5	2.03	1.42	1.05	0.808	0.64	0.52
45x3 EA	68.4	53.8	25.6	13.2	7.87	5.2	3.68	2.74	2.12	1.69	1.37	0.961	0.71	0.546	0.432	0.351
40x6 EA	113	82.6	34.8	17.4	10.3	6.73	4.75	3.53	2.72	2.16	1.76	1.23	0.909	0.698	0.553	0.449
40x5 EA	88.1	65.1	27.7	13.9	8.2	5.39	3.8	2.82	2.18	1.73	1.41	0.986	0.728	0.559	0.443	0.36
40x3 EA	59.1	43.8	18.7	9.4	5.56	3.65	2.58	1.92	1.48	1.18	0.956	0.669	0.494	0.379	0.301	0.244

Note: Geometry and capacity table, Grade 300, f_u = 440 MPa

Angles are too small for structural bolts.

Table 7.29 Equal Angles: 25 and 30 EA

Section Name $b \times$ nominal thickness (mm)	Weight (kg/m)	t (mm)	r_1 (mm)	r_2 (mm)	A_g (mm²)	p_B (mm)	I_n (10⁶mm⁴)	$y_1 = y_4$ (mm)	$Z_{c1} = Z_{c4}$ (10³mm³)	S_x (10³mm³)	r_x (mm)	I_x (10⁶mm⁴)	x_3 (mm)	Z_{y3} (10³mm³)	x_5 (mm)	Z_{y5} (10³mm³)	r_y (mm)	J (10³mm⁴)	f_y (MPa)	Form Factor k_f	Z_{nn} (10³mm³)	Z_{xy} Load B (10³mm³)	Z_{xy} Load D (10³mm³)	s_{p1} (mm)	s_{p2} (mm)	s_{p3} (mm)
30x6 EA	2.56	6	5	3	326	9.53	0.0387	21.2	1.83	3.06	10.9	0.0107	10.7	0.993	13.5	0.79	5.72	4.16	320	1	2.74	1.19	1.19	N/A	N/A	N/A
30x5 EA	2.01	4.6	5	3	256	8.99	0.0316	21.2	1.49	2.45	11.1	0.0084	10.5	0.799	12.7	0.66	5.72	1.98	320	1	2.23	0.99	0.99			
30x3 EA	1.35	3	5	3	173	8.3	0.0218	21.2	1.03	1.67	11.2	0.0057	10.3	0.554	11.7	0.488	5.76	0.605	320	1	1.5	0.714	0.732			
25x6 EA	2.08	6	5	3	266	8.28	0.021	17.7	1.19	2.03	8.89	0.006	8.97	0.669	11.7	0.513	4.75	3.44	320	1	1.78	0.769	0.769			
25x5 EA	1.65	4.6	5	3	210	7.75	0.0173	17.7	0.98	1.65	9.07	0.0047	8.73	0.537	11	0.428	4.72	1.66	320	1	1.47	0.642	0.642			
25x3 EA	1.12	3	5	3	143	7.07	0.0121	17.7	0.685	1.13	9.22	0.0032	8.56	0.373	9.99	0.319	4.73	0.515	320	1	1.03	0.479	0.479			

Section Name $b \times$ nominal thickness (mm)	ϕM_{sx} (kNm)	ϕM_{sy} (kNm)	ϕN_s (kN)	ϕN_t (kN)	Member Axial Capacity (Minor Axis), ϕN_{cy} (kN) Effective length (m)																Member Axial Capacity (Major Axis), ϕN_{cx} (kN) Effective length (m)															
					0.5	1	1.5	2	2.5	3	3.5	4	4.5	5	5.5	6	7	8	9	10	0.5	1	2	3	4	5	6	7	8	9	10	12	14	16	18	20
30x6 EA	0.789	0.343	93.9	93.3	46.2	16.1	7.73	4.49	2.93	2.06	1.52	1.17	0.93	0.756	0.626	0.528	0.389	0.299	0.236	0.192	74.8	43.6	14.8	7.07	4.1	2.67	1.87	1.38	1.07	0.846	0.687	0.479	0.353	0.271	0.215	0.174
30x5 EA	0.642	0.285	73.7	73.2	36.3	12.7	6.07	3.53	2.3	1.61	1.2	0.92	0.73	0.594	0.492	0.414	0.305	0.234	0.186	0.151	59.1	35	12	5.74	3.33	2.17	1.52	1.13	0.867	0.688	0.559	0.39	0.288	0.221	0.175	0.142
30x3 EA	0.432	0.206	49.8	49.5	24.7	8.66	4.16	2.42	1.57	1.11	0.819	0.631	0.5	0.407	0.337	0.284	0.209	0.161	0.127	0.103	40.1	23.9	8.24	3.94	2.29	1.49	1.05	0.775	0.597	0.473	0.385	0.268	0.198	0.152	0.12	0.098
25x6 EA	0.513	0.221	76.6	76.1	29.6	9.45	4.45	2.57	1.67	1.17	0.863	0.664	0.526	0.427	0.354	0.298	0.22	0.168	0.133	0.108	55.4	26.9	8.37	3.92	2.26	1.46	1.03	0.758	0.582	0.462	0.375	0.261	0.193	0.148	0.117	0.095
25x5 EA	0.423	0.185	60.5	60.1	23.2	7.37	3.47	2	1.3	0.91	0.673	0.518	0.41	0.333	0.276	0.232	0.171	0.131	0.104	0.084	44.2	21.9	6.86	3.22	1.85	1.2	0.842	0.622	0.478	0.379	0.308	0.215	0.158	0.121	0.096	0.078
25x3 EA	0.297	0.138	41.2	40.9	15.8	5.04	2.37	1.37	0.89	0.622	0.46	0.354	0.281	0.228	0.189	0.159	0.117	0.09	0.071	0.058	30.4	15.3	4.81	2.26	1.3	0.845	0.592	0.437	0.336	0.267	0.217	0.151	0.111	0.085	0.068	0.055

Note: Geometry and capacity table, Grade 300, f_u = 440 MPa

Angles are too small for structural bolts.

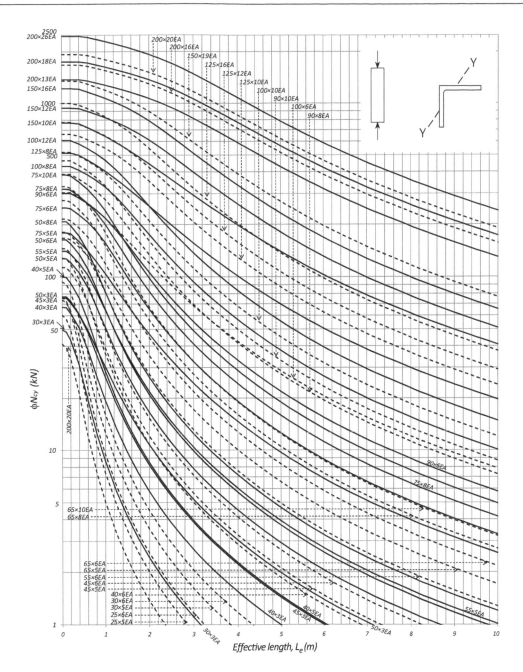

Figure 7.16 Equal angles: Members subject to axial compression (minor axis) grade 300 (solid lines labelled on the *y* axis).

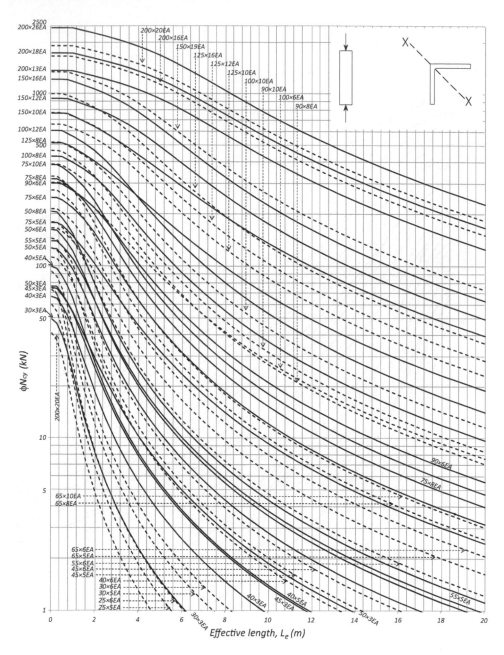

Figure 7.17 Equal angles: Members subject to axial compression (major axis) grade 300 (solid lines labelled on the *y* axis).

Table 7.30 Unequal Angles: 125 and 150 UA

Section Name b₁ x b₂ x nominal thickness (mm)	Weight (kg/m)	t (mm)	r_1 (mm)	r_2 (mm)	A_g (mm²)	p_B (mm)	n_L (mm)	I_x (10⁶ mm⁴)	Y_1 (mm)	Z_{x1} (10³ mm³)	Y_4 (mm)	Z_{x4} (10³ mm³)	Y_5 (mm)	Z_{x5} (10³ mm³)	S_x (10³ mm³)	r_x (mm)	I_y (10⁶ mm⁴)	x_2 (mm)	Z_{y2} (10³ mm³)	x_3 (mm)	Z_{y3} (10³ mm³)	x_5 (mm)	Z_{y5} (10³ mm³)	S_y (10³ mm³)	r_y (mm)	J (10³ mm⁴)	$\tan(\alpha)$	f_y (MPa)	Form Factor k_f	Z_{ex} Load A (10³ mm³)	Z_{ex} Load C (10³ mm³)	Z_{ey} Load B (10³ mm³)	Z_{ey} Load D (10³ mm³)
150x100x12 UA	22.5	12	10	5	2870	49.1	24.3	7.51	102	73.5	75.3	99.7	35.2	213	127	51.2	1.35	27.6	48.8	52.9	25.5	42	32.1	51.7	21.7	141	0.438	300	1	102	110	35.3	38.2
150x100x10 UA	18	9.5	10	5	2300	48.1	23.3	6.11	103	59.5	74.3	81.5	34.6	177	102	51.6	1.09	26.9	40.7	53	20.6	40.7	26.9	41.8	21.8	71.9	0.441	320	0.975	74.8	81.7	26	30.9
150x90x16 UA	27.9	15.8	10	5	3550	52.5	22.7	8.8	99.5	88.4	71.9	122	41.9	210	154	49.8	1.32	24.6	53.8	49.9	26.5	38.9	34	55.9	19.3	300	0.353	300	1	132	133	39.5	39.8
150x90x12 UA	21.6	12	10	5	2750	51	21.2	6.97	100	69.4	71.3	97.8	40.8	171	120	50.4	1.04	23.4	44.5	50.1	20.8	37.2	28	43.8	19.5	136	0.36	300	1	96.3	104	28.8	31.1
150x90x10 UA	17.3	9.5	10	5	2200	50	20.2	5.66	101	56.1	70.5	80.1	40.1	141	96.6	50.7	0.847	22.6	37.4	50.4	16.8	36.1	23.5	35.4	19.6	69	0.363	320	0.973	70.6	81.8	21.2	25.2
150x90x8 UA	14.3	7.8	10	5	1820	49.2	19.6	4.73	101	46.7	70.5	67.3	39.5	120	80.1	50.6	0.71	22.1	32.2	50.6	14	35.2	20.2	29.5	19.7	39	0.364	320	0.863	53.1	60.3	15.9	21
125x75x12 UA	17.7	12	8	5	2260	43.3	18.4	3.91	83.2	47	59.7	65.5	34.6	113	81.4	41.6	0.585	19.9	29.3	41.4	14.1	31.9	18.4	29.7	16.1	110	0.356	300	1	68.6	70.5	20.6	21.2
125x75x10 UA	14.2	9.5	8	5	1810	42.3	17.5	3.2	83.8	38.2	59.3	53.9	33.9	94.4	65.8	42	0.476	19.2	24.9	41.8	11.4	30.7	15.5	24.1	16.2	56.2	0.36	320	1	51.6	57.2	15.5	17.2
125x75x8 UA	11.8	7.8	8	5	1500	41.5	16.8	2.68	84.2	31.8	58.9	45.5	33.3	80.4	54.6	42.2	0.399	18.6	21.5	41.8	9.55	29.9	13.3	20.1	16.3	31.7	0.363	320	0.964	39.8	46	11.9	14.3
125x75x6 UA	9.16	6	8	5	1170	40.7	16	2.1	84.7	24.8	58.5	36	32.8	64.1	42.4	42.5	0.315	18	17.5	42.1	7.47	29	10.8	15.7	16.4	14.8	0.364	320	0.824	26.8	30.1	8.07	11.2

(Continued)

Table 7.30 (Continued) Unequal Angles: 125 and 150 UA

Section Name $b_1 \times b_2 \times$ nominal thickness (mm)	ϕM_{sx} (kNm)	ϕM_{sy} (kNm)	ϕN_s (kN)	ϕN_t (kN)	Member Axial Capacity (Minor Axis), ϕN_{cy} (kN) — Effective length (m) 0.5	1	1.5	2	2.5	3	3.5	4	4.5	5	5.5	6	7	8	9	10	Member Axial Capacity (Major Axis), ϕN_{cx} (kN) — Effective length (m) 0.5	1	2	3	4	5	6	7	8	9	10	12	14	16	18	20
150x100x12 UA	27.5	9.5	775	725	729	624	497	372	276	208	162	128	104	86	72.2	61.4	46	35.7	28.4	23.2	775	744	658	557	446	346	269	212	170	139	116	83.4	62.7	48.8	39	31.9
150x100x10 UA	21.5	7.49	646	581	593	476	365	273	206	158	124	99.8	81.5	67.8	57.1	48.8	36.7	28.5	22.8	18.7	646	610	513	415	327	256	202	162	132	109	90.9	66.1	50	39.1	31.4	25.7
150x90x16 UA	35.6	10.67	959	896	887	735	554	396	286	213	163	129	104	85.7	71.8	61	45.5	35.2	28.1	22.9	958	918	808	678	536	413	319	251	201	164	136	98	73.6	57.3	45.8	37.4
150x90x12 UA	26	7.78	743	694	688	572	434	311	225	168	129	102	82.2	67.7	56.7	48.1	35.9	27.8	22.2	18.1	743	712	628	529	420	325	252	198	159	130	108	77.6	58.3	45.4	36.3	29.7
150x90x10 UA	20.3	6.11	616	555	554	430	316	230	170	128	99.9	79.5	64.7	53.5	45	38.3	28.7	22.3	17.8	14.5	616	581	487	391	307	240	189	151	122	101	84.4	61.3	46.3	36.2	29	23.8
150x90x8 UA	15.3	4.58	452	459	412	327	247	183	137	104	81.6	65.3	53.2	44.2	37.2	31.7	23.8	18.5	14.8	12.1	452	431	366	300	240	190	152	122	99.6	82.5	69.3	50.5	38.3	30	24.1	19.8
125x75x12 UA	18.5	5.56	610	571	547	423	289	194	136	99.1	75.3	59	47.4	38.9	32.5	27.5	20.5	15.8	12.6	10.2	610	571	483	378	278	204	153	118	93.7	73.9	62.7	44.7	33.4	25.9	20.6	16.8
125x75x10 UA	14.9	4.46	521	457	465	355	239	159	111	80.9	61.3	48	38.6	31.6	26.4	22.3	16.6	12.8	10.2	8.31	521	486	409	316	231	168	126	97.2	76.9	62.3	51.3	36.6	27.3	21.2	16.9	13.8
125x75x8 UA	11.5	3.43	416	379	358	259	177	122	87.1	64.6	49.6	39.1	31.6	26	21.8	18.5	13.8	10.7	8.5	6.93	416	380	303	230	171	128	98	76.8	61.5	50.2	41.6	29.9	22.5	17.5	14	11.4
125x75x6 UA	7.72	2.32	278	295	244	183	129	91.4	66.2	49.6	38.3	30.3	24.6	20.3	17	14.5	10.8	8.37	6.68	5.45	278	258	211	164	126	95.9	74.3	58.7	47.3	38.8	32.3	23.3	17.6	13.7	11	8.97

Note: Geometry and capacity table, Grade 300, $f_u = 440$ MPa

Gauge lines for each leg should be selected from the Equal Angle table for the same leg length (refer Figure 3.29).

Table 7.31 Unequal Angles: 65, 75 and 100 UA

Section Name $b_1 \times b_2 \times$ nominal thickness (mm)	Weight (kg/m)	t (mm)	r_1 (mm)	r_2 (mm)	A_g (mm²)	pB (mm)	n_L (mm)	I_x (10⁶ mm⁴)	y_1 (mm)	Z_{x1} (10³ mm³)	y_4 (mm)	Z_{x4} (10³ mm³)	y_5 (mm)	Z_{x5} (10³ mm³)	S_x (10³ mm³)	r_x (mm)	I_y (10⁶ mm⁴)	x_2 (mm)	Z_{y2} (10³ mm³)	x_3 (mm)	Z_{y3} (10³ mm³)	x_5 (mm)	Z_{y5} (10³ mm³)	S_y (10³ mm³)	r_y (mm)	J (10³ mm⁴)	$\tan(\alpha)$	f_y (MPa)	Form Factor k_f	Z_{ex} Load A (10³ mm³)	Z_{ex} Load C (10³ mm³)	Z_{ey} Load B (10³ mm³)	Z_{ey} Load D (10³ mm³)
100x75x10 UA	12.4	9.5	8	5	1580	31.8	19.4	1.89	69.2	27.3	54.2	34.6	18.6	101	46.5	34.6	0.401	22.3	18	36.4	11	32.2	12.5	21.2	16	49.1	0.546	320	1	39.4	40.9	15.9	16.6
100x75x8 UA	10.3	7.8	8	5	1310	31.1	18.7	1.59	69.4	22.9	54.2	29.2	18.2	87	38.7	34.8	0.337	21.8	15.4	36.4	9.26	31.3	10.7	17.8	16	27.8	0.549	320	1	31.2	33.1	12.6	13.9
100x75x6 UA	7.98	6	8	5	1020	30.3	17.9	1.25	69.7	17.9	54	23.1	17.9	70	30.1	35.1	0.265	21.4	12.4	36.5	7.27	30.3	8.75	13.9	16.2	13	0.551	320	0.946	22	21.8	8.93	10.9
75x50x8 UA	7.23	7.8	7	3	921	25.2	12.8	0.586	50.8	11.5	37.8	15.5	18	32.5	20	25.2	0.106	14.2	7.46	26.4	4.01	21.7	4.88	8.19	10.7	19.5	0.43	320	1	17	17.3	5.93	6.02
75x50x6 UA	5.66	6	7	3	721	24.4	12.1	0.468	51.2	9.15	37.5	12.5	17.6	26.7	15.8	25.5	0.084	13.6	6.17	26.5	3.18	20.8	4.04	6.48	10.8	9.21	0.435	320	1	12.6	13.7	4.37	4.77
75x50x5 UA	4.4	4.6	7	3	560	23.8	11.5	0.37	51.5	7.17	37.2	9.93	17.2	21.5	12.3	25.7	0.067	13.2	5.03	26.6	2.5	20.1	3.32	5.09	10.9	4.32	0.437	320	0.956	8.89	9.65	3.1	3.75
65x50x8 UA	6.59	7.8	6	3	840	21.1	13.6	0.421	44.9	9.37	36.3	11.6	11.6	36.4	16.1	22.4	0.094	15.6	6	23.9	3.91	22.3	4.2	7.49	10.6	17.6	0.57	320	1	14.1	14.1	5.86	5.86
65x50x6 UA	5.16	6	6	3	658	20.4	12.9	0.338	45.2	7.48	36.1	9.35	11.2	30.2	12.7	22.7	0.074	15.1	4.91	23.9	3.11	21.4	3.48	5.93	10.6	8.29	0.575	320	1	10.7	11.2	4.46	4.67
65x50x5 UA	4.02	4.6	6	3	512	19.8	12.4	0.267	45.4	5.89	35.9	7.43	10.9	24.5	9.92	22.8	0.059	14.8	3.97	23.9	2.46	20.6	2.85	4.66	10.7	3.87	0.577	320	1	7.76	7.92	3.23	3.68

(Continued)

Table 7.31 (Continued) Unequal Angles: 65, 75 and 100 UA

The table below lists, for each section: ϕM_{sx}, ϕM_{sy}, ϕN_s, ϕN_t; the **Member Axial Capacity (Minor Axis), ϕN_y (kN)** at effective lengths (m) 0.5–10 (columns prefixed ϕN_y); and the **Member Axial Capacity (Major Axis), ϕN_x (kN)** at effective lengths (m) 0.5–20 (columns prefixed ϕN_x). $b_1 \times b_2 \times$ nominal thickness (mm).

Section Name	ϕM_{sx} (kNm)	ϕM_{sy} (kNm)	ϕN_s (kN)	ϕN_t (kN)	ϕN_y 0.5	ϕN_y 1	ϕN_y 1.5	ϕN_y 2	ϕN_y 2.5	ϕN_y 3	ϕN_y 3.5	ϕN_y 4	ϕN_y 4.5	ϕN_y 5	ϕN_y 5.5	ϕN_y 6	ϕN_y 7	ϕN_y 8	ϕN_y 9	ϕN_y 10	ϕN_x 0.5	ϕN_x 1	ϕN_x 2	ϕN_x 3	ϕN_x 4	ϕN_x 5	ϕN_x 6	ϕN_x 7	ϕN_x 8	ϕN_x 9	ϕN_x 10	ϕN_x 12	ϕN_x 14	ϕN_x 16	ϕN_x 18	ϕN_x 20
100x75x10 UA	11.3	4.58	455	399	405	307	205	136	94.8	69.1	52.3	41	32.9	27	22.5	19.1	14.2	10.9	8.69	7.08	449	411	323	226	154	108	79.2	60.3	47.3	38	31.2	22.1	16.5	12.7		
100x75x8 UA	8.99	3.63	377	331	336	255	170	113	78.6	57.3	43.4	34	27.3	22.4	18.7	15.8	11.7	9.07	7.21	5.87	372	341	269	189	128	90.4	66.3	50.5	39.6	31.9	26.2	18.5	13.8	10.7		
100x75x6 UA	6.28	2.57	278	257	239	173	118	81.7	58.4	43.4	33.3	26.3	21.2	17.5	14.6	12.4	9.26	7.17	5.71	4.66	274	244	183	129	91.3	66.2	49.5	38.2	30.3	24.5	20.2	14.4	10.8	8.36		
75x50x8 UA	4.9	1.71	265	233	210	120	66.2	40.5	27	19.3	14.4	11.2	8.91	7.27	6.04	5.1	3.77	2.9	2.3	1.87	253	222	145	85.7	53.8	36.4	26.1	19.6	15.3	12.2	9.96	7.01	5.19	4		
75x50x6 UA	3.63	1.26	208	182	165	95.2	52.6	32.2	21.5	15.4	11.5	8.9	7.1	5.79	4.81	4.06	3.01	2.32	1.84	1.49	199	175	115	68.3	43	29.1	20.9	15.7	12.2	9.76	7.98	5.61	4.16	3.21		
75x50x5 UA	2.56	0.89	154	141	114	66	38.4	24.2	16.5	11.9	8.91	6.94	5.55	4.54	3.78	3.19	2.37	1.82	1.45	1.18	146	123	78.6	48.7	31.7	21.9	16	12.1	9.44	7.58	6.21	4.39	3.26	2.51		
65x50x8 UA	4.06	1.69	242	212	191	108	59.4	36.3	24.2	17.3	12.9	10	7.98	6.51	5.41	4.57	3.38	2.6	2.06	1.68	228	195	116	65	40	26.8	19.2	14.3	11.1	8.87	7.24	5.08	3.76	2.9		
65x50x6 UA	3.08	1.28	190	166	149	84.9	46.6	28.4	19	13.5	10.1	7.84	6.25	5.1	4.24	3.58	2.65	2.04	1.62	1.31	179	153	92.4	52	32.1	21.5	15.4	11.5	8.93	7.13	5.82	4.09	3.03	2.33		
65x50x5 UA	2.23	0.93	147	129	117	66.8	36.8	22.5	15	10.7	8.01	6.21	4.95	4.04	3.36	2.83	2.1	1.61	1.28	1.04	139	119	72.2	40.8	25.2	16.9	12.1	9.03	7.01	5.59	4.57	3.21	2.37	1.83		

Note: Geometry and capacity table, Grade 300, $f_u = 440$ MPa

Gauge lines for each leg should be selected from the Equal Angle table for the same leg length (refer Figure 3.29).

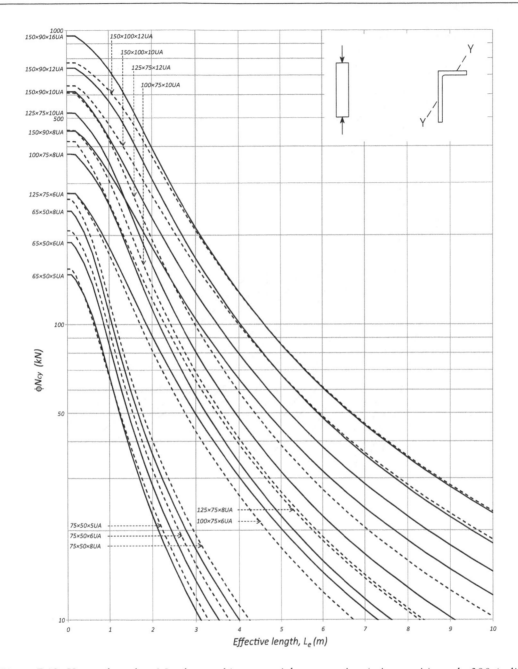

Figure 7.18 Unequal angles: Members subject to axial compression (minor axis) grade 300 (solid lines labelled on the *y* axis).

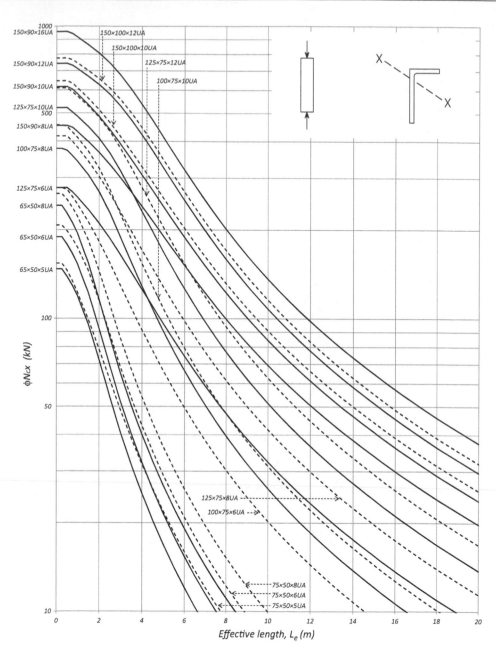

Figure 7.19 Unequal angles: Members subject to axial compression (major axis) grade 300 (solid lines labelled on the *y* axis).

Table 7.32 Circular Hollow Sections: 101.6, 114.3, 139.7 and 165.1 CHS

Section Name $d_0 \times t$ (mm)	weight (kg/m)	A_g (mm²)	I (10⁶mm⁴)	Z (10³mm³)	S (10³mm³)	r (mm)	J (10⁶mm⁴)	C (10³mm³)	Form Factor k_f	Compactness	Z_e (10³mm³)
165.1x5.4CHS	21.3	2710	8.65	105	138	56.5	17.3	209	1	C	138
165.1x5CHS	19.7	2510	8.07	97.7	128	56.6	16.1	195	1	C	128
139.7x5.4CHS	17.9	2280	5.14	73.7	97.4	47.5	10.3	147	1	C	97.4
139.7x5CHS	16.6	2120	4.81	68.8	90.8	47.7	9.61	138	1	C	90.8
114.3x5.4CHS	14.5	1850	2.75	48	64.1	38.5	5.49	96.1	1	C	64.1
114.3x4.5CHS	12.2	1550	2.34	41	54.3	38.9	4.69	82	1	C	54.3
101.6x5CHS	11.9	1520	1.77	34.9	46.7	34.2	3.55	69.9	1	C	46.7
101.6x4CHS	9.63	1230	1.46	28.8	38.1	34.5	2.93	57.6	1	C	38.1

Section Name $d_0 \times t$ (mm)	ϕM_s (kNm)	ϕV_v (kN)	ϕN_s (kN)	ϕN_t (kN)	ϕN_c (kN)	ϕT_o (kNm)	Member Axial Capacity (Any Axis), ϕN_c (kN) Effective length (m)																										
							0.1	0.2	0.4	0.6	0.8	1	1.25	1.5	1.75	2	2.25	2.5	2.75	3	3.25	3.5	3.75	4	4.5	5	5.5	6	6.5	7	7.5	8	8.5	9	9.5	10
165.1x5.4 CHS	31.1	220	610	610	610	28.3	610	610	610	610	609	605	601	595	589	582	574	565	555	544	533	520	506	491	456	418	378	339	302	268	239	214	192	173	156	142
165.1x5 CHS	28.8	203	565	565	565	26.4	565	565	565	565	564	561	556	551	545	539	531	523	514	504	494	482	469	455	423	388	351	314	280	249	222	199	178	160	145	132
139.7x5.4 CHS	21.9	185	513	513	513	19.9	513	513	513	513	510	506	501	495	487	479	470	459	447	434	420	404	386	368	328	288	251	219	191	167	148	131	117	105	94.7	85.8
139.7x5 CHS	20.4	172	477	477	477	18.6	477	477	477	477	474	471	466	460	453	446	437	427	416	404	391	376	360	343	306	270	235	205	179	157	138	123	110	98.4	88.7	80.4
114.3x5.4 CHS	14.4	150	416	416	416	13	416	416	416	415	411	407	400	393	384	374	362	349	333	316	298	278	258	238	201	169	144	123	106	92.1	80.8	71.4	63.6	56.9	51.3	46.4
114.3x4.5 CHS	12.2	126	349	349	349	11.1	349	349	349	348	345	341	336	329	322	314	304	293	281	267	252	235	219	202	171	144	123	105	90.4	78.7	69	61	54.3	48.7	43.8	39.7
101.6x5 CHS	10.5	123	342	342	342	9.43	342	342	342	340	336	332	325	317	308	297	285	270	254	237	218	199	182	165	136	113	95.2	80.9	69.6	60.4	52.9	46.7	41.5	37.2	33.5	30.3
101.6x4 CHS	8.57	100	277	277	277	7.77	277	277	277	275	272	269	263	257	250	241	231	220	207	193	178	163	149	135	112	93	78.3	66.6	57.2	49.7	43.5	38.4	34.2	30.6	27.6	24.9

Note: Geometry and capacity table, Grade C250L0 – $f_y = 250$ MPa – $f_u = 320$ MPa

Table 7.33 Circular Hollow Sections: 60.3, 76.1 and 88.9 CHS

Section Name $d_0 \times t$ (mm)	weight (kg/m)	A_z (mm²)	I (10⁶mm⁴)	Z (10³mm³)	S (10³mm³)	r (mm)	J (10⁶mm⁴)	C (10³mm³)	Form Factor k_f	Compactness	Z_e (10³mm³)
88.9x5.9CHS	12.1	1540	1.33	30	40.7	29.4	2.66	59.9	1	C	40.7
88.9x5CHS	10.3	1320	1.16	26.2	35.2	29.7	2.33	52.4	1	C	35.2
88.9x4CHS	8.38	1070	0.963	21.7	28.9	30	1.93	43.3	1	C	28.9
76.1x5.9CHS	10.2	1300	0.807	21.2	29.1	24.9	1.61	42.4	1	C	29.1
76.1x4.5CHS	7.95	1010	0.651	17.1	23.1	25.4	1.3	342	1	C	23.1
76.1x3.6CHS	6.44	820	0.54	14.2	18.9	25.7	1.08	28.4	1	C	18.9
60.3x5.4CHS	7.31	931	0.354	11.8	16.3	19.5	0.709	23.5	1	C	16.3
60.3x4.5CHS	6.19	789	0.309	10.2	14	19.8	0.618	20.5	1	C	14
60.3x3.6CHS	5.03	641	0.259	8.58	11.6	20.1	0.517	17.2	1	C	11.6

Member Axial Capacity (Any Axis), ϕN_c (kN)

Section Name $d_0 \times t$ (mm)	ϕM_s (kNm)	ϕV_v (kN)	ϕN_s (kN)	ϕN_t (kN)	ϕT_u (kNm)	Effective length (m) 0.1	0.2	0.4	0.6	0.8	1	1.25	1.5	1.75	2	2.25	2.5	2.75	3	3.25	3.5	3.75	4	4.5	5	5.5	6	6.5	7	7.5	8	8.5	9	9.5	10
88.9x5.9 CHS	9.16	125	347	347	8.09	347	347	346	342	338	332	323	312	299	284	266	246	225	203	182	163	145	130	106	86.9	72.6	61.5	52.8	45.7	40	35.3	31.4	28.1	25.3	22.9
88.9x5 CHS	7.92	107	297	297	7.07	297	297	297	294	290	285	277	268	258	245	230	213	195	176	158	142	127	114	92.2	75.9	63.5	53.8	46.1	40	35	30.9	27.4	24.5	22.1	20
88.9x4 CHS	6.5	86.7	241	241	5.85	241	241	241	238	235	231	225	218	209	199	188	174	159	145	130	117	104	93.7	76.1	62.7	52.4	44.4	38.1	33	28.9	25.5	22.7	20.3	18.3	16.5
76.1x5.9 CHS	6.55	105	293	293	5.73	293	293	291	287	282	275	265	252	236	218	197	175	154	135	118	104	91.8	81.5	65.4	53.6	44.6	37.7	32.3	28	24.5	21.6	19.2	17.1	15.4	13.9
76.1x4.5 CHS	5.2	81.8	227	227	4.62	227	227	226	223	219	214	206	197	185	171	156	139	123	108	94.8	83.5	73.9	65.7	52.8	43.2	36	30.5	26.1	22.6	19.8	17.4	15.5	13.8	12.5	11.3
76.1x3.6 CHS	4.25	66.4	185	185	3.83	185	185	184	181	178	174	168	160	151	140	128	114	101	89.2	78.5	69.2	61.3	54.5	43.8	35.9	29.9	25.3	21.7	18.8	16.4	14.5	12.9	11.5	10.3	9.36
60.3x5.4 CHS	3.67	75.4	209	209	3.17	209	209	207	202	196	189	176	161	142	122	103	87.1	73.9	63.2	54.5	47.5	41.7	36.8	29.4	24	19.9	16.8	14.4	12.4	10.9	9.58	8.51	7.6	6.84	6.18
60.3x4.5 CHS	3.15	63.9	178	178	2.77	178	178	175	172	167	160	150	137	122	105	89.4	75.7	64.4	55.1	47.6	41.4	36.3	32.1	25.6	20.9	17.4	14.7	12.6	10.9	9.49	8.36	7.43	6.64	5.97	5.4
60.3x3.6 CHS	2.61	51.9	144	144	2.32	144	144	143	140	136	131	123	113	100	87.1	74.3	63.1	53.7	46	39.7	34.6	30.4	26.9	21.4	17.5	14.5	12.3	10.5	9.09	7.94	7	6.21	5.56	5	4.52

Note: Geometry and capacity table, Grade C250L0 – f_y = 250 MPa – f_u = 320 MPa

Table 7.34 Circular Hollow Sections: 26.9, 33.7, 42.4 and 48.3 CHS

Section Name d_0 x t (mm)	weight (kg/m)	A_s (mm²)	I (10⁶ mm⁴)	Z (10³ mm³)	S (10³ mm³)	r (mm)	J (10⁶ mm⁴)	C (10³ mm³)	Form Factor k_f	Compactness	Z_e (10³ mm³)
48.3x4 CHS	4.37	557	0.138	5.7	7.87	15.7	0.275	11.4	1	C	7.87
48.3x3.2 CHS	3.56	453	0.116	4.8	6.52	16	0.232	9.59	1	C	6.52
42.4x4 CHS	3.79	483	0.0899	4.24	5.92	13.6	0.18	8.48	1	C	5.92
42.4x3.2 CHS	3.09	394	0.0762	3.59	4.93	13.9	0.152	7.19	1	C	4.93
33.7x4 CHS	2.93	373	0.0419	2.49	3.55	10.6	0.0838	4.97	1	C	3.55
33.7x3.2 CHS	2.41	307	0.036	2.14	2.99	10.8	0.0721	4.28	1	C	2.99
26.9x4 CHS	2.26	288	0.0194	1.45	2.12	8.22	0.0389	2.89	1	C	2.12
26.9x3.2 CHS	1.87	238	0.017	1.27	1.81	8.46	0.0341	2.53	1	C	1.81
26.9x2.6 CHS	1.56	198	0.0148	1.1	1.54	8.64	0.0296	2.2	1	C	1.54

Member Axial Capacity (Any Axis) ϕN_c (kN)

Section Name d_0 x t (mm)	ϕM_s (kNm)	ϕV_v (kN)	ϕN_s (kN)	ϕN_t (kN)	ϕT_u (kNm)	0.1	0.2	0.4	0.6	0.8	1	1.25	1.5	1.75	2	2.25	2.5	2.75	3	3.25	3.5	3.75	4	4.5	5	5.5	6	6.5	7	7.5	8	8.5	9	9.5	10
48.3x4 CHS	1.77	45.1	125	125	1.54	125	125	123	119	113	106	93.8	79.4	65	52.7	43	35.5	29.7	25.2	21.7	18.8	16.4	14.5	11.5	9.4	7.81	6.58	5.63	4.87	4.25	3.75	3.33	2.97	2.67	2.41
48.3x3.2 CHS	1.47	36.7	102	102	1.3	102	102	99.8	96.6	92.3	86.6	77.3	65.9	54.3	44.2	36.2	29.9	25.1	21.3	18.3	15.8	13.9	12.2	9.74	7.93	6.59	5.56	4.75	4.11	3.59	3.16	2.81	2.51	2.26	2.04
42.4x4 CHS	1.33	39.1	109	109	1.15	109	108	105	101	94.3	85.6	71.8	57.2	44.9	35.6	28.7	23.5	19.6	16.6	14.2	12.3	10.8	9.52	7.57	6.16	5.11	4.31	3.68	3.18	2.78	2.45	2.17	1.94	1.75	1.58
42.4x3.2 CHS	1.11	31.9	88.7	88.7	0.97	88.7	88.5	86.1	82.4	77.4	70.7	59.9	48.1	38	30.2	24.4	20	16.7	14.1	12.1	10.5	9.18	8.1	6.44	5.24	4.35	3.67	3.14	2.71	2.37	2.09	1.85	1.65	1.49	1.34
33.7x4 CHS	0.799	30.2	83.9	83.9	0.671	83.9	83.2	79.5	73.7	65.1	55.9	40	29.4	22.2	17.3	13.8	11.3	9.38	7.92	6.78	5.87	5.13	4.52	3.59	2.92	2.42	2.04	1.74	1.51	1.31	1.16	1.03	0.917	0.824	0.745
33.7x3.2 CHS	0.673	24.9	69.1	69.1	0.578	69.1	68.5	65.6	61	54.2	45.3	33.9	25	19	14.8	11.8	9.63	8.01	6.76	5.79	5.01	4.38	3.86	3.06	2.49	2.07	1.74	1.49	1.29	1.12	0.988	0.877	0.783	0.704	0.636
26.9x4 CHS	0.477	23.3	64.8	64.8	0.39	64.8	63.5	59	51.3	40.2	29.4	20	14.2	10.6	8.2	6.53	5.32	4.42	3.73	3.19	2.75	2.41	2.12	1.68	1.37	1.13	0.955	0.815	0.704	0.614	0.541	0.48	0.428	0.385	0.348
26.9x3.2 CHS	0.407	19.3	53.6	53.6	0.342	53.6	52.6	49	43	34.3	25.4	17.4	12.4	9.27	7.17	5.7	4.65	3.86	3.26	2.78	2.41	2.1	1.85	1.47	1.2	0.991	0.835	0.713	0.616	0.537	0.473	0.42	0.375	0.337	0.304
26.9x2.6 CHS	0.347	16	44.6	44.6	0.297	44.6	43.8	40.9	36.2	29.2	21.8	15	10.7	8.02	6.21	4.94	4.03	3.35	2.82	2.41	2.09	1.82	1.61	1.28	1.04	0.859	0.724	0.618	0.534	0.466	0.41	0.364	0.325	0.292	0.264

Note: Geometry and capacity table, Grade C250L0 – f_y = 250 MPa – f_u = 320 MPa

Table 7.35 Circular Hollow Sections: 406.4, 457 and 508 CHS

Section Name d_0 x t (mm)	weight (kg/m)	A_g (mm²)	I (10⁶ mm⁴)	Z (10³ mm³)	S (10³ mm³)	r (mm)	J (10⁶ mm⁴)	C (10³ mm³)	Form Factor k_f	Compactness	Z_e (10³ mm³)
508x12.7 CHS	155	19762	606	2390	3120	175	1210	4770	1	N	3050
508x9.5 CHS	117	14878	462	1820	2360	176	925	3640	1	N	2170
508x6.4 CHS	79.2	10085	317	1250	1610	177	634	2500	0.857	N	1290
457x12.7 CHS	139	17727	438	1920	2510	157	876	3830	1	N	2500
457x9.5 CHS	105	13356	334	1460	1900	158	669	2930	1	N	1790
457x6.4 CHS	71.1	9060	230	1010	1300	159	460	2010	0.904	N	1090
406.4x12.7 CHS	123	15708	305	1500	1970	139	609	3000	1	C	1970
406.4x9.5 CHS	93	11846	233	1150	1500	140	467	2300	1	N	1450
406.4x6.4 CHS	63.1	8042	161	792	1020	141	322	1580	0.96	N	895

Member Axial Capacity (Any Axis), ϕN_c (kN)

Section Name d_0 x t (mm)	ϕM_s (kNm)	ϕV_v (kN)	ϕN_s (kN)	ϕN_t (kN)	ϕT_u (kNm)	Effective length (m) 0.5	1	1.5	2	2.5	3	3.5	4	4.5	5	5.5	6	6.5	7	7.5	8	8.5	9	9.5	10	11	12	13	14	16	18	20	22	24	26
508x12.7 CHS	961	2241	6225	6225	902	6225	6225	6225	6225	6190	6153	6112	6067	6018	5964	5906	5842	5773	5698	5618	5532	5440	5341	5235	5121	4873	4596	4297	3985	3369	2820	2364	1997	1703	1467
508x9.5 CHS	684	1687	4687	4687	688	4687	4687	4687	4687	4661	4633	4603	4570	4533	4493	4449	4402	4350	4295	4235	4171	4102	4029	3950	3866	3681	3475	3252	3019	2556	2142	1798	1519	1296	1116
508x6.4 CHS	406	1144	3177	2723	472	2723	2723	2723	2723	2714	2699	2684	2667	2649	2629	2607	2583	2558	2531	2501	2470	2437	2401	2363	2322	2233	2133	2023	1904	1654	1413	1200	1022	876	757
457x12.7 CHS	788	2010	5584	5584	724	5584	5584	5584	5570	5534	5494	5451	5402	5349	5289	5225	5154	5077	4994	4904	4806	4701	4587	4466	4336	4054	3750	3436	3128	2567	2109	1749	1468	1247	1072
457x9.5 CHS	564	1515	4207	4207	553	4207	4207	4207	4197	4170	4141	4108	4072	4032	3988	3940	3887	3830	3768	3701	3629	3550	3466	3376	3280	3070	2843	2608	2377	1954	1607	1333	1119	951	817
457x6.4 CHS	343	1027	2854	2580	380	2580	2580	2580	2577	2562	2545	2527	2507	2485	2461	2434	2405	2374	2340	2304	2265	2222	2177	2128	2076	1962	1837	1703	1567	1307	1085	905	762	649	558
406.4x12.7 CHS	621	1781	4948	4948	567	4948	4948	4948	4919	4881	4839	4791	4738	4678	4612	4539	4459	4371	4275	4171	4057	3935	3803	3662	3515	3205	2892	2592	2317	1856	1505	1238	1035	877	752
406.4x9.5 CHS	457	1343	3731	3731	434	3731	3731	3731	3710	3682	3651	3615	3575	3531	3481	3427	3368	3303	3232	3154	3070	2979	2881	2776	2667	2436	2201	1975	1767	1417	1149	946	791	670	575
406.4x6.4 CHS	282	912	2533	2432	299	2432	2432	2432	2420	2403	2383	2361	2336	2309	2278	2245	2208	2168	2124	2077	2025	1969	1909	1845	1777	1633	1484	1338	1202	968	787	649	543	460	395

Note: Geometry and capacity table, Grade C350L0 – f_y = 350 MPa – f_u = 430 MPa

Table 7.36 Circular Hollow Sections: 273.1, 323.9 and 355.6 CHS

Section Name d_0 x t (mm)	weight (kg/m)	A_g (mm²)	I (10^6 mm⁴)	Z (10^3 mm³)	S (10^3 mm³)	r (mm)	J (10^6 mm⁴)	C (10^3 mm³)	Form Factor k_f	Compactness	Z_e (10^3 mm³)
355.6x12.7 CHS	107	13681	201	1130	1490	121	403	2260	1	C	1490
355.6x9.5 CHS	81.1	10329	155	871	1140	122	310	1740	1	N	1130
355.6x6.4 CHS	55.1	7021	107	602	781	123	214	1200	1	N	710
323.9x12.7 CHS	97.5	12416	151	930	1230	110	301	1860	1	C	1230
323.9x9.5 CHS	73.7	9383	116	717	939	111	232	1430	1	C	939
323.9x6.4 CHS	50.1	6384	80.5	497	645	112	161	994	1	N	601
273.1x12.7 CHS	81.6	10389	88.3	646	862	92.2	177	1290	1	C	862
273.1x9.3 CHS	60.5	7707	67.1	492	647	93.3	134	983	1	C	647
273.1x6.4 CHS	42.1	5362	47.7	349	455	94.3	95.4	699	1	N	441
273.1x4.8 CHS	31.8	4046	36.4	267	346	94.9	72.8	533	1	N	312

(Continued)

Table 7.36 (Continued) Circular Hollow Sections: 273.1, 323.9 and 355.6 CHS

Section Name	ϕM_s	ϕV_v	ϕN_s	ϕN_t	ϕT_u	Member Axial Capacity (Any Axis), ϕN_c (kN) Effective length (m)																										
d_0 x t (mm)	(kNm)	(kN)	(kN)	(kN)	(kNm)	0.5	1	1.5	2	2.5	3	3.5	4	4.5	5	5.5	6	6.5	7	7.5	8	8.5	9	9.5	10	11	12	13	14	16	18	20	22	24	26
355.6x12.7 CHS	469	1551	4310	4310	428	4310	4310	4301	4265	4225	4178	4124	4064	3996	3920	3836	3742	3639	3525	3401	3266	3122	2971	2815	2658	2351	2068	1818	1602	1261	1013	830	692	585	501
355.6x9.5 CHS	356	1171	3254	3254	329	3254	3254	3248	3221	3191	3156	3116	3071	3021	2965	2902	2833	2756	2672	2580	2480	2373	2260	2144	2026	1795	1581	1391	1227	967	777	637	530	449	384
355.6x6.4 CHS	224	796	2212	2212	228	2212	2212	2208	2190	2170	2147	2120	2090	2056	2018	1976	1930	1879	1823	1761	1695	1623	1548	1470	1390	1234	1088	959	846	667	536	439	366	310	265
323.9x12.7 CHS	387	1408	3911	3911	351	3911	3911	3894	3857	3814	3763	3705	3639	3565	3481	3387	3283	3167	3039	2900	2752	2598	2441	2285	2133	1852	1608	1401	1227	959	768	628	522	441	378
323.9x9.5 CHS	296	1064	2956	2956	271	2956	2956	2944	2916	2884	2846	2803	2754	2699	2637	2567	2490	2404	2309	2206	2096	1981	1864	1747	1633	1420	1234	1076	943	737	590	483	402	339	291
323.9x6.4 CHS	189	724	2011	2011	188	2011	2011	2003	1984	1963	1938	1909	1876	1839	1798	1751	1700	1642	1579	1511	1437	1360	1281	1202	1124	979	852	743	652	510	409	334	278	235	201
273.1x12.7 CHS	272	1178	3273	3273	244	3273	3273	3241	3200	3151	3092	3024	2946	2856	2753	2637	2507	2364	2212	2056	1901	1750	1609	1478	1359	1153	986	851	740	574	458	374	310	262	224
273.1x9.3 CHS	204	874	2428	2428	186	2428	2428	2405	2375	2340	2297	2248	2191	2126	2052	1968	1874	1771	1660	1546	1431	1320	1215	1118	1028	873	747	645	562	436	348	284	236	199	170
273.1x6.4 CHS	139	608	1689	1689	132	1689	1689	1674	1654	1629	1600	1567	1528	1484	1433	1376	1312	1242	1167	1088	1009	932	859	791	728	619	530	458	399	309	247	201	167	141	121
273.1x4.8 CHS	98	459	1274	1274	101	1274	1274	1263	1248	1230	1208	1183	1154	1122	1084	1042	994	942	885	827	768	710	654	603	555	472	405	349	304	236	189	154	128	108	92

Note: Geometry and capacity table, Grade C350L0 – f_y = 350 MPa – f_u = 430 MPa

Table 7.37 Circular Hollow Sections: 168.3 and 219.1 CHS

Section Name $d_0 \times t$ (mm)	weight (kg/m)	A_g (mm²)	I (10³ mm⁴)	Z (10³ mm³)	S (10³ mm³)	r (mm)	J (10⁶ mm⁴)	C (10³ mm³)	Form Factor k_f	Compactness	Z_e (10³ mm³)
219.1x8.2 CHS	42.6	5433	30.3	276	365	74.6	60.5	552	1	C	365
219.1x6.4 CHS	33.6	4277	24.2	221	290	75.2	48.4	442	1	C	290
219.1x4.8 CHS	25.4	3232	18.6	169	220	75.8	37.1	339	1	N	210
168.3x7.1 CHS	28.2	3596	11.7	139	185	57	23.4	278	1	C	185
168.3x6.4 CHS	25.6	3255	10.7	127	168	57.3	21.4	254	1	C	168
168.3x4.8 CHS	19.4	2466	8.25	98	128	57.8	16.5	196	1	C	128

Member Axial Capacity (Any Axis), ϕN_c (kN)

Section Name $d_0 \times t$ (mm)	ϕM_s (kNm)	ϕV_v (kN)	ϕN_s (kN)	ϕN_t (kN)	ϕT_u (kNm)	Effective length (m)																											
						0.5	1	1.5	2	2.5	3	3.5	4	4.5	5	5.5	6	6.5	7	7.5	8	8.5	9	9.5	10	11	12	13	14	16	18	20	22	24	26
219.1x8.2 CHS	115	616	1711	1711	104	1711	1705	1680	1649	1611	1565	1510	1446	1371	1285	1190	1089	988	892	804	724	653	591	536	488	409	347	298	258	200	159	129	107	90.6	78
219.1x6.4 CHS	91.4	485	1347	1347	83.5	1347	1342	1323	1299	1269	1234	1192	1142	1084	1017	943	865	786	711	641	578	521	472	428	390	327	277	238	207	160	127	103	85.9	72.4	62
219.1x4.8 CHS	66.2	366	1018	1018	64	1018	1014	1000	982	960	934	902	865	822	773	718	659	600	543	490	442	399	362	328	299	251	213	183	158	122	97.5	79.4	65.9	55.6	48
168.3x7.1 CHS	58.3	408	1133	1133	52.6	1133	1118	1093	1060	1017	964	899	821	736	648	567	494	432	379	334	297	265	238	215	195	162	137	117	102	78.3	62.2	50.6	42	35.4	30
168.3x6.4 CHS	52.9	369	1025	1025	48	1025	1013	990	960	922	874	816	747	670	591	517	451	394	346	306	271	242	218	196	178	148	125	107	92.9	71.6	56.9	46.3	38.4	32.4	28
168.3x4.8 CHS	40.3	280	777	777	37	777	767	750	728	700	664	621	569	512	453	397	347	303	266	235	209	187	168	151	137	114	96.5	82.6	71.5	55.2	43.8	35.7	29.6	24.9	21

Note: Geometry and capacity table, Grade C350L0 – f_y = 350 MPa – f_u = 430 MPa

Table 7.38 Circular Hollow Sections: 101.6, 114.3, 139.7 and 165.1 CHS

Section Name d_0 x t (mm)	weight (kg/m)	A_g (mm²)	I (10⁶mm⁴)	Z (10³mm³)	S (10³mm³)	r (mm)	J (10⁶mm⁴)	C (10³mm³)	Form Factor k_f	Compactness	Z_e (10³mm³)
165.1x3.5CHS	13.9	1780	5.8	70.3	91.4	57.1	11.6	141	1	N	86.6
165.1x3CHS	12	1530	5.02	60.8	78.8	57.3	10	122	1	N	71.9
139.7x3.5CHS	11.8	1500	3.47	49.7	64.9	48.2	6.95	99.5	1	N	63.7
139.7x3CHS	10.1	1290	3.01	43.1	56.1	48.3	6.02	86.2	1	N	53.3
114.3x3.6CHS	9.83	1250	1.92	33.6	44.1	39.2	3.84	67.2	1	C	44.1
114.3x3.2CHS	8.77	1120	1.72	30.2	39.5	39.3	3.45	60.4	1	N	39.5
101.6x3.2CHS	7.77	989	1.2	23.6	31	34.8	2.4	47.2	1	C	31
101.6x2.6CHS	6.35	809	0.991	19.5	25.5	35	1.98	39	1	N	25.1

Member Axial Capacity (Any Axis), ϕN_c (kN)

Effective length (m)

Section Name d_0 x t (mm)	ϕM_s (kNm)	ϕV_v (kN)	ϕN_t (kN)	ϕN_s (kN)	ϕJ_o (kNm)	0.1	0.2	0.4	0.6	0.8	1	1.25	1.5	1.8	2	2.25	2.5	2.75	3	3.25	3.5	3.75	4	4.5	5	5.5	6	6.5	7	7.5	8	8.5	9	9.5	10
165.1x3.5CHS	27.3	201.9	560.7	560.7	26.6	561	561	561	561	558	554	548	541	533	525	515	504	491	477	462	445	427	407	365	322	281	245	214	188	166	147	132	118	107	96.7
165.1x3CHS	22.6	173.5	482	482	23	482	482	482	482	480	476	471	465	459	451	443	433	423	411	398	383	368	351	315	278	243	212	185	163	144	128	114	102	92.3	83.7
139.7x3.5CHS	20.1	170.1	472.5	472.5	18.8	472	473	473	472	468	463	457	449	440	429	418	404	389	372	353	333	312	290	248	211	180	154	133	116	102	90.3	80.4	72	64.9	58.8
139.7x3CHS	16.8	146.3	406.4	406.4	16.3	406	406	406	406	402	398	393	386	378	369	359	348	335	320	304	287	269	250	214	182	155	133	115	100	88.1	78	69.4	62.2	56	50.8
114.3x3.6CHS	13.9	141.8	393.8	393.8	12.7	394	394	394	391	386	381	373	363	352	339	323	306	286	264	242	220	200	181	148	123	103	87.7	75.4	65.4	57.3	50.6	45	40.2	36.2	32.8
114.3x3.2CHS	12.4	127	352.8	352.8	11.4	353	353	353	350	346	341	334	326	316	304	290	274	256	237	217	198	179	163	134	111	93	79	67.9	58.9	51.6	45.5	40.5	36.2	32.6	29.5
101.6x3.2CHS	9.77	112.2	311.5	311.5	8.92	312	312	311	308	304	298	290	281	269	256	240	222	202	182	164	146	131	117	95	78.2	65.3	55.4	47.5	41.1	36	31.8	28.2	25.3	22.7	20.6
101.6x2.6CHS	7.91	91.7	254.8	254.8	7.38	255	255	255	252	248	244	238	230	221	210	197	182	166	150	135	121	108	97	79	64.6	54	45.8	39.3	34	29.8	26.3	23.4	20.9	18.8	17

Note: Geometry and capacity table, Grade C350L0 – f_y = 350 MPa – f_u = 430 MPa

Table 7.39 Circular Hollow Sections: 48.3, 60.3, 76.1 and 88.9 CHS

Section Name d_0 x t (mm)	weight (kg/m)	A_g (mm²)	I (10⁶ mm⁴)	Z (10³ mm³)	S (10³ mm³)	r (mm)	J (10⁶ mm⁴)	C (10³ mm³)	Form Factor k_f	Compactness	Z_e (10³ mm³)
88.9x3.2 CHS	6.76	862	0.792	17.8	23.5	30.3	1.58	35.6	1	C	23.5
88.9x2.6 CHS	5.53	705	0.657	14.8	19.4	30.5	1.31	29.6	1	C	19.4
76.1x3.2 CHS	5.75	733	0.488	12.8	17	25.8	0.976	25.6	1	C	17
76.1x2.3 CHS	4.19	533	0.363	9.55	12.5	26.1	0.727	19.1	1	C	12.5
60.3x2.9 CHS	4.11	523	0.216	7.16	9.56	20.3	0.432	14.3	1	C	9.56
60.3x2.3 CHS	3.29	419	0.177	5.85	7.74	20.5	0.353	11.7	1	C	7.74
48.3x2.9 CHS	3.25	414	0.107	4.43	5.99	16.1	0.214	8.86	1	C	5.99
48.3x2.3 CHS	2.61	332	0.0881	3.65	4.87	16.3	0.176	7.3	1	C	4.87

Section Name d_0 x t (mm)	ϕM_s (kNm)	ϕV_v (kN)	ϕN_s (kN)	ϕN_t (kN)	ϕT_u (kNm)	Member Axial Capacity (Any Axis), ϕN_c (kN) — Effective length (m)																												
						0.1	0.2	0.4	0.6	0.8	1	1.25	1.5	1.8	2	2.25	2.5	2.75	3	3.25	3.5	3.75	4	4.5	5	5.5	6	6.5	7	7.5	8	8.5	9	9.5	10
88.9x3.2 CHS	7.4	97.8	271.5	271.5	6.74	272	272	271	267	262	256	247	236	222	206	187	168	148	131	115	101	90	80	64	52.5	43.7	37	31.7	27.4	24	21.1	18.8	16.8	15.1	13.7
88.9x2.6 CHS	6.11	79.9	222.1	222.1	5.59	222	222	221	218	214	210	202	193	182	169	154	138	123	108	95	84	74	66	53	43.4	36.2	30.6	26.2	22.7	19.9	17.5	15.6	13.9	12.5	11.3
76.1x3.2 CHS	5.36	83.1	230.9	230.9	4.85	231	231	229	225	219	213	202	188	172	153	133	115	99	85	74	65	57	50	40	32.8	27.3	23	19.7	17.1	14.9	13.1	11.7	10.4	9.38	8.49
76.1x2.3 CHS	3.94	60.4	167.9	167.9	3.61	168	168	167	164	160	155	147	138	126	112	98	85	73	63	55	48	42	37	30	24.4	20.3	17.1	14.7	12.7	11.1	9.77	8.68	7.76	6.98	6.31
60.3x2.9 CHS	3.01	59.3	164.7	164.7	2.71	165	165	162	157	151	143	130	114	97	80	66	55	46	39	34	29	26	23	18	14.7	12.2	10.3	8.8	7.61	6.65	5.86	5.2	4.65	4.18	3.78
60.3x2.3 CHS	2.44	47.5	132	132	2.21	132	132	130	126	121	115	105	92	78	65	54	45	38	32	28	24	21	19	15	12	9.96	8.41	7.19	6.22	5.43	4.79	4.25	3.8	3.42	3.09
48.3x2.9 CHS	1.89	46.9	130.4	130.4	1.67	130	130	126	121	113	103	86	69	54	43	35	28	24	20	17	15	13	11	9.1	7.4	6.14	5.18	4.42	3.83	3.34	2.94	2.61	2.33	2.1	1.9
48.3x2.3 CHS	1.53	37.6	104.6	104.6	1.38	105	104	101	97	91	83	70	56	44	35	28	23	19	16	14	12	11	9.4	7.5	6.08	5.04	4.25	3.64	3.14	2.74	2.42	2.15	1.92	1.72	1.56

Note: Geometry and capacity table, Grade C350L0 – f_y = 350 MPa – f_u = 430 MPa

Table 7.40 Circular Hollow Sections: 26.9, 33.7 and 42.4 CHS

Section Name $d_0 \times t$ (mm)	weight (kg/m)	A_g (mm²)	I ($10^6 mm^4$)	Z ($10^3 mm^3$)	S ($10^3 mm^3$)	r (mm)	J ($10^6 mm^4$)	C ($10^3 mm^3$)	Form Factor k_f	Compactness	Z_e ($10^3 mm^3$)
42.4x2.6 CHS	2.55	325	0.0646	3.05	4.12	14.1	0.129	6.1	1	C	4.12
42.4x2 CHS	1.99	254	0.0519	2.45	3.27	14.3	0.104	4.9	1	C	3.27
33.7x2.6 CHS	1.99	254	0.0309	1.84	2.52	11	0.0619	3.67	1	C	2.52
33.7x2 CHS	1.56	199	0.0251	1.49	2.01	11.2	0.0502	2.98	1	C	2.01
26.9x2.3 CHS	1.4	178	0.0136	1.01	1.4	8.74	0.0271	2.02	1	C	1.4
26.9x2 CHS	1.23	156	0.0122	0.907	1.24	8.83	0.0244	1.81	1	C	1.24

Member Axial Capacity (Any Axis), ϕN_c (kN)

Section Name $d_0 \times t$ (mm)	ϕM_s (kNm)	ϕV_v (kN)	ϕN_s (kN)	ϕN_t (kN)	ϕT_u (kNm)	0.1	0.2	0.4	0.6	0.8	1	1.25	1.5	1.8	2	2.25	2.5	2.75	3	3.25	3.5	3.75	4	4.5	5	5.5	6	6.5	7	7.5	8	8.5	9	9.5	10
42.4x2.6 CHS	1.3	36.9	102.4	102.4	1.15	102	102	98	93	85	74	58	44	34	26	21	17	14	12	10	9	7.9	6.9	5.5	4.48	3.72	3.13	2.68	2.31	2.02	1.78	1.58	1.41	1.27	1.15
42.4x2 CHS	1.03	28.8	80	80	0.926	80	80	77	73	66	58	46	35	27	21	17	14	12	9.7	8.3	7.2	6.3	5.6	4.4	3.6	2.99	2.52	2.15	1.86	1.62	1.43	1.27	1.13	1.02	0.92
33.7x2.6 CHS	0.794	28.8	80	80	0.694	80	79	74	67	56	44	31	22	17	13	10	8.3	6.9	5.9	5	4.3	3.8	3.3	2.7	2.15	1.78	1.5	1.28	1.11	0.97	0.85	0.76	0.68	0.61	0.55
33.7x2 CHS	0.633	22.6	62.7	62.7	0.563	63	62	59	53	45	35	25	18	13	10	8.3	6.8	5.6	4.8	4.1	3.5	3.1	2.7	2.2	1.75	1.45	1.22	1.04	0.9	0.79	0.69	0.61	0.55	0.49	0.45
26.9x2.3 CHS	0.441	20.2	56.1	56.1	0.381	56	55	50	41	30	21	14	10	7.5	5.8	4.6	3.7	3.1	2.6	2.2	1.9	1.7	1.5	1.2	0.96	0.79	0.67	0.57	0.49	0.43	0.38	0.34	0.3	0.27	0.24
26.9x2 CHS	0.391	17.7	49.1	49.1	0.343	49	48	44	37	27	19	13	9	6.7	5.2	4.1	3.3	2.8	2.3	2	1.7	1.5	1.3	1.1	0.86	0.71	0.6	0.51	0.44	0.39	0.34	0.3	0.27	0.24	0.22

Note: Geometry and capacity table, Grade C350L0 – f_y = 350 MPa – f_u = 430 MPa

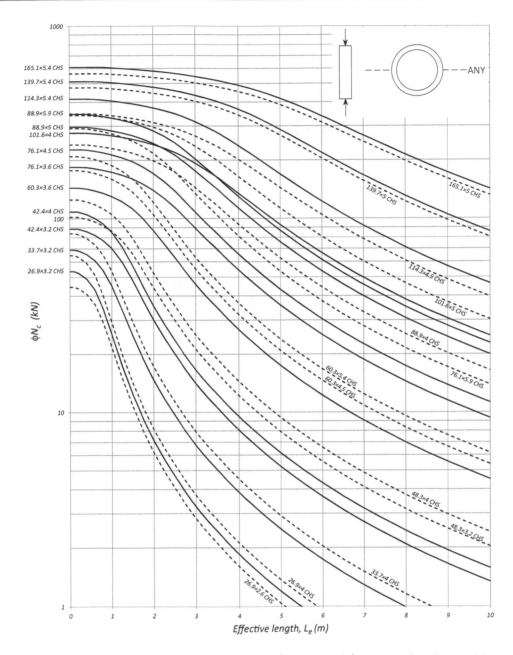

Figure 7.20 Circular hollow sections: Members subject to axial compression (any axis) grade
C250L0 (solid lines labelled on the *y* axis).

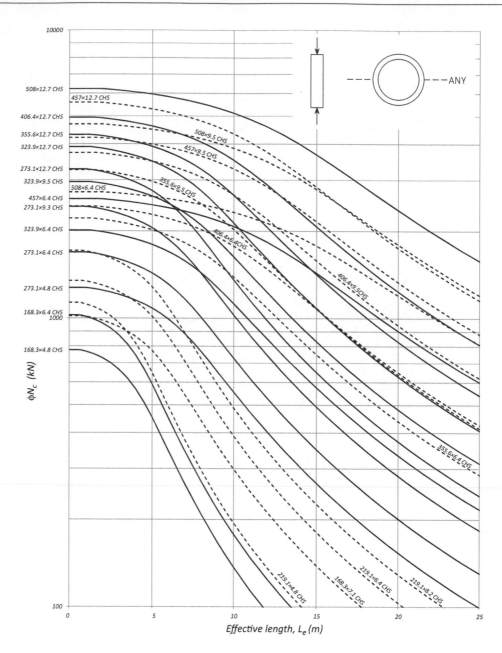

Figure 7.21 Circular hollow sections: Members subject to axial compression (any axis) grade C350L0 (solid lines labelled on the *y* axis). 1 of 2.

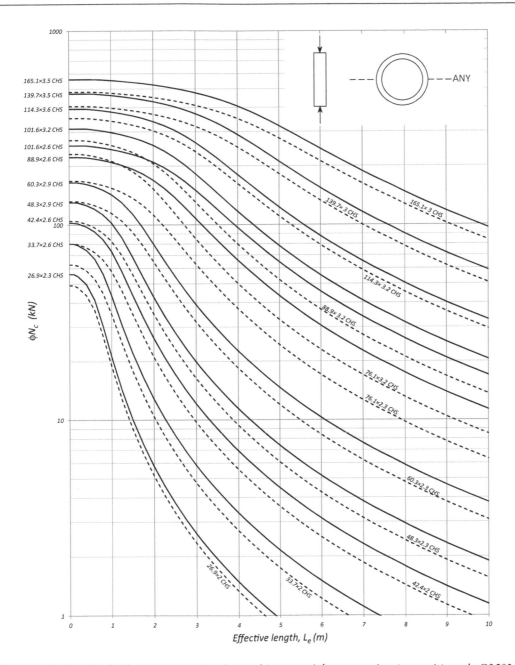

Figure 7.22 Circular hollow sections: Members subject to axial compression (any axis) grade C350L0 (solid lines labelled on the *y* axis). 2 of 2.

Table 7.41 Rectangular Hollow Sections: 65 and 75 RHS

Section Name $D \times b \times t$ (mm)	Weight (kg/m)	D (mm)	b (mm)	t (mm)	r_m (mm)	A_g (mm²)	I_x (10^6 mm⁴)	Z_x (10^3 mm³)	S_x (10^3 mm³)	r_x (mm)	I_y (10^6 mm⁴)	Z_y (10^3 mm³)	S_y (10^3 mm³)	r_y (mm)	J (10^6 mm⁴)	C (10^3 mm³)	Form Factor k_f	X Compactness	Z_{ex} (10^3 mm³)	Y Compactness	Z_{ey} (10^3 mm³)
75x25x2.5 RHS	3.6	75	25	2.5	3.75	459	0.285	7.6	10.1	24.9	0.0487	3.89	4.53	10.3	0.144	7.14	1	C	10	N	4.33
75x25x2 RHS	2.93	75	25	2	3	374	0.238	6.36	8.31	25.3	0.0414	3.31	3.77	10.5	0.12	6.04	0.96	C	8.3	S	3.18
75x25x1.6 RHS	2.38	75	25	1.6	2.4	303	0.197	5.26	6.81	25.5	0.0347	2.78	3.11	10.7	0.0993	5.05	0.81	C	6.8	S	2.22
65x35x4 RHS	5.35	65	35	4	8	681	0.328	10.1	13.3	22	0.123	7.03	8.58	13.4	0.32	12.5	1	C	13	C	8.58
65x35x3 RHS	4.25	65	35	3	4.5	541	0.281	8.65	11	22.8	0.106	6.04	7.11	14	0.259	10.4	1	C	11	C	7.11
65x35x2.5 RHS	3.6	65	35	2.5	3.75	459	0.244	7.52	9.45	23.1	0.0926	5.29	6.13	14.2	0.223	9.1	1	C	9.5	C	6.13
65x35x2 RHS	2.93	65	35	2	3	374	0.204	6.28	7.8	23.4	0.0778	4.44	5.07	14.4	0.184	7.62	1	C	7.8	N	4.69

Member Axial Capacity (Major Axis), ϕN_{cx} (kN)

Section Name	Effective length (m)										
	0.5	1	2	3	4	5	6	7	8	9	10
75x25x2.5 RHS	142	132	92.1	50	29.4	19.2	13.5	9.96	7.67	6.09	4.96
75x25x2 RHS	112	104	75	41.7	24.6	16.1	11.3	8.36	6.44	5.12	4.16
75x25x1.6 RHS	76.6	72.5	55.9	33.4	20.1	13.1	9.25	6.86	5.29	4.2	3.42
65x35x4 RHS	209	191	117	59.3	34.4	22.4	15.7	11.6	8.94	7.09	5.77
65x35x3 RHS	166	153	97.5	50.3	29.3	19	13.4	9.88	7.61	6.04	4.91
65x35x2.5 RHS	141	130	84.2	43.7	25.5	16.6	11.6	8.6	6.63	5.26	4.28
65x35x2 RHS	115	106	69.7	36.5	21.3	13.8	9.71	7.18	5.53	4.39	3.57

Section Name $D \times b \times t$ (mm)	ϕM_{sx} (kNm)	ϕM_{sy} (kNm)	ϕV_v (kN)	ϕN_s (kN)	ϕN_t (kN)	ϕT_u (kNm)	Member Axial Capacity (Minor Axis), ϕN_{cy} (kN) Effective length (m)										
							0.5	1	2	3	4	5	6	7	8	9	10
75x25x2.5 RHS	3.18	1.36	61.5	145	145	1.35	126	71.7	20.4	9.3	5.29	3.41	2.38	1.76	1.35	1.07	0.869
75x25x2 RHS	2.62	1	49.9	118	113	1.14	100	59.5	17.2	7.85	4.47	2.89	2.02	1.49	1.14	0.905	0.735
75x25x1.6 RHS	2.15	0.7	40.4	95	78	0.954	70.3	46.8	14.4	6.58	3.75	2.42	1.69	1.25	0.96	0.761	0.618
65x35x4 RHS	4.19	2.7	82.4	215	215	2.36	199	147	50.1	23	13.1	8.5	5.94	4.39	3.37	2.67	2.17
65x35x3 RHS	3.47	2.24	64	170	170	1.97	159	122	43.2	19.9	11.4	7.35	5.14	3.8	2.92	2.31	1.88
65x35x2.5 RHS	2.98	1.93	54.2	145	145	1.72	135	105	37.6	17.4	9.92	6.41	4.49	3.31	2.55	2.02	1.64
65x35x2 RHS	2.46	1.48	44.1	118	118	1.44	110	86.1	31.5	14.5	8.3	5.37	3.75	2.77	2.13	1.69	1.37

Note: Geometry and capacity table, Grade C350L0 – f_y = 350 MPa – f_u = 430 MPa

Table 7.42 Rectangular Hollow Sections: 50 RHS

Section Name D x b x t (mm)	Weight (kg/m)	D (mm)	b (mm)	t (mm)	rm (mm)	Ag (mm²)	Ix (10⁶ mm⁴)	rx (mm)	Sx (10³ mm³)	Zx (10³ mm³)	Iy (10⁶ mm⁴)	Zy (10³ mm³)	Sy (10³ mm³)	ry (mm)	J (10⁶ mm⁴)	C (10³ mm³)	Form Factor kf	X Compactness	Zex (10³ mm³)	Y Compactness	Zey (10³ mm³)
50x25x3 RHS	3.07	50	25	3	4.5	391	0.112	16.9	5.86	4.47	0.0367	2.93	3.56	9.69	0.0964	5.18	1	C	5.9	C	3.56
50x25x2.5 RHS	2.62	50	25	2.5	3.75	334	0.0989	17.2	5.11	3.95	0.0328	2.62	3.12	9.91	0.0843	4.6	1	C	5.1	C	3.12
50x25x2 RHS	2.15	50	25	2	3	274	0.0838	17.5	4.26	3.35	0.0281	2.25	2.62	10.1	0.0706	3.92	1	C	4.3	C	2.62
50x25x1.6 RHS	1.75	50	25	1.6	2.4	223	0.0702	17.7	3.53	2.81	0.0237	1.9	2.17	10.3	0.0585	3.29	1	C	3.5	N	2.05
50x20x3 RHS	2.83	50	20	3	4.5	361	0.0951	16.2	5.16	3.81	0.0212	2.12	2.63	7.67	0.062	3.88	1	C	5.2	C	2.63
50x20x2.5 RHS	2.42	50	20	2.5	3.75	309	0.0848	16.6	4.51	3.39	0.0192	1.92	2.32	7.89	0.055	3.49	1	C	4.5	C	2.32
50x20x2 RHS	1.99	50	20	2	3	254	0.0723	16.9	3.78	2.89	0.0167	1.67	1.96	8.11	0.0466	3	1	C	3.8	C	1.96
50x20x1.6 RHS	1.63	50	20	1.6	2.4	207	0.0608	17.1	3.14	2.43	0.0142	1.42	1.63	8.29	0.0389	2.55	1	C	3.1	N	1.54

Section Name D x b x t (mm)	φMsx (kNm)	φMsy (kNm)	φVv (kN)	φNs (kN)	φNt (kN)	φTu (kNm)	Member Axial Capacity (Minor Axis), φNcy (kN) Effective length (m)											Member Axial Capacity (Major Axis), φNcx (kN) Effective length (m)										
							0.5	1	2	3	4	5	6	7	8	9	10	0.5	1	2	3	4	5	6	7	8	9	10
50x25x3 RHS	1.85	1.12	47.5	123	123	0.979	106	55.5	15.5	7.03	4	2.58	1.8	1.33	1.02	0.808	0.656	118	99.6	44	20.7	11.9	7.68	5.37	3.97	3.06	2.42	1.97
50x25x2.5 RHS	1.61	0.983	40.5	105	105	0.869	90.8	49.1	13.8	6.27	3.57	2.3	1.61	1.19	0.911	0.722	0.586	101	85.8	38.8	18.3	10.5	6.79	4.75	3.51	2.7	2.14	1.74
50x25x2 RHS	1.34	0.825	33.1	86.2	86.2	0.741	74.9	41.5	11.7	5.34	3.04	1.96	1.37	1.01	0.775	0.614	0.498	82.6	70.9	32.7	15.5	8.89	5.76	4.03	2.98	2.29	1.82	1.48
50x25x1.6 RHS	1.11	0.646	27	70.3	70.3	0.622	61.4	34.8	9.94	4.52	2.57	1.66	1.16	0.855	0.657	0.521	0.423	67.4	58.1	27.2	12.9	7.41	4.8	3.36	2.48	1.91	1.52	1.23
50x20x3 RHS	1.63	0.828	46.9	113.7	113.7	0.733	87	34.1	9.07	4.1	2.33	1.5	1.05	0.772	0.593	0.469	0.381	108	89.9	37.7	17.6	10.1	6.53	4.57	3.37	2.6	2.06	1.67
50x20x2.5 RHS	1.42	0.731	40	97.3	97.3	0.66	75.8	30.8	8.2	3.71	2.11	1.36	0.948	0.699	0.537	0.425	0.345	92.7	78	33.7	15.8	9.05	5.86	4.1	3.03	2.33	1.85	1.5
50x20x2 RHS	1.19	0.617	32.7	79.9	79.9	0.567	63.3	26.5	7.11	3.22	1.83	1.18	0.822	0.606	0.465	0.369	0.299	76.3	64.7	28.6	13.4	7.7	4.98	3.49	2.58	1.98	1.57	1.28
50x20x1.6 RHS	0.989	0.485	26.6	65.3	65.3	0.482	52.3	22.5	6.06	2.74	1.56	1.01	0.701	0.517	0.397	0.314	0.255	62.4	53.1	23.8	11.2	6.43	4.16	2.92	2.15	1.66	1.31	1.07

Note: Geometry and capacity table, Grade C350L0 – f_y = 350 MPa – f_u = 430 MPa

Table 7.43 Rectangular Hollow Sections: 400 RHS

Section Name D x b x t (mm)	Weight (kg/m)	D (mm)	b (mm)	t (mm)	r_m (mm)	A_g (mm²)	I_x (10³ mm⁴)	Z_x (10³ mm³)	S_x (10³ mm³)	r_x (mm)	I_y (10⁶ mm⁴)	Z_y (10³ mm³)	S_y (10³ mm³)	r_y (mm)	J (10⁶ mm⁴)	C (10³ mm³)	Form Factor k_f	X Compactness	Z_{ex} (10³ mm³)	Y Compactness	Z_{ey} (10³ mm³)
400x300x16 RHS	161	400	300	16	32	20497	453	2260	2750	149	290	1940	2260	119	586	3170	1	C	2750	N	2230
400x300x12.5 RHS	128	400	300	12.5	25	16338	370	1850	2230	151	238	1590	1830	121	471	2590	0.996	C	2230	S	1580
400x300x10 RHS	104	400	300	10	20	13257	306	1530	1820	152	197	1320	1500	122	384	2130	0.877	N	1600	S	1120
400x300x8 RHS	84.2	400	300	8	16	10724	251	1260	1490	153	162	1080	1220	123	312	1750	0.715	S	1140	S	800
400x200x16 RHS	136	400	200	16	32	17297	335	1670	2140	139	113	1130	1320	80.8	290	2000	1	C	2140	N	1300
400x200x12.5 RHS	109	400	200	12.5	25	13838	277	1380	1740	141	94	940	1080	82.4	236	1650	0.996	C	1740	S	936
400x200x10 RHS	88.4	400	200	10	20	11257	230	1150	1430	143	78.6	786	888	83.6	194	1370	0.855	C	1430	S	658
400x200x8 RHS	71.6	400	200	8	16	9124	190	949	1170	144	65.2	652	728	84.5	158	1130	0.745	N	1150	S	464

Section Name D x b x t (mm)	ϕM_{sx} (kNm)	ϕM_{sy} (kNm)	ϕV_v (kN)	ϕN_s (kN)	ϕN_t (kN)	ϕT_u (kNm)	Member Axial Capacity (Minor Axis), ϕN_{sy} (kN) — Effective length (m)											Member Axial Capacity (Major Axis), ϕN_{cx} (kN) — Effective length (m)										
							1	2	4	6	8	10	12	14	16	18	20	1	2	4	6	8	10	12	14	16	18	20
400x300x16 RHS	1114	903	2787	8301	7840	770	8301	8169	7666	6828	5590	4232	3150	2391	1864	1490	1217	8301	8239	7903	7380	6617	5599	4500	3551	2821	2277	1870
400x300x12.5 RHS	903	640	2218	6591	6249	629	6591	6491	6105	5466	4519	3453	2583	1964	1533	1226	1002	6591	6544	6285	5882	5297	4512	3649	2891	2301	1859	1528
400x300x10 RHS	648	454	1798	4709	5071	518	4709	4651	4409	4018	3435	2723	2081	1598	1252	1004	821	4709	4685	4521	4270	3912	3427	2855	2308	1858	1510	1245
400x300x8 RHS	462	324	1454	3105	4102	425	3105	3080	2949	2743	2442	2044	1627	1276	1011	815	668	3105	3099	3008	2874	2687	2435	2119	1778	1465	1206	1002
400x200x16 RHS	867	527	2725	7005	6616	486	6970	6723	5809	4225	2743	1837	1302	968	748	595	484	7005	6937	6617	6108	5356	4387	3429	2664	2100	1688	1383
400x200x12.5 RHS	705	379	2170	5582	5293	401	5557	5368	4672	3455	2268	1524	1081	804	621	494	403	5582	5531	5283	4891	4313	3561	2801	2183	1724	1387	1137
400x200x10 RHS	579	266	1759	3898	4306	333	3889	3776	3372	2658	1837	1256	897	669	517	412	335	3898	3873	3725	3497	3167	2723	2223	1771	1415	1145	942
400x200x8 RHS	466	188	1422	2733	3490	275	2752	2681	2438	2012	1462	1022	735	550	426	339	277	2753	2741	2650	2511	2314	2046	1724	1406	1138	928	766

Note: Geometry and capacity table, Grade C450L0 – f_y = 450 MPa – f_u = 500 MPa

Table 7.44 Rectangular Hollow Sections: 300 and 350 RHS

Section Name $D \times b \times t$ (mm)	Weight (kg/m)	D (mm)	b (mm)	t (mm)	r_m (mm)	A_g (mm²)	I_x (10⁶ mm⁴)	Z_x (10³ mm³)	S_x (10³ mm³)	r_x (mm)	I_y (10⁶ mm⁴)	Z_y (10³ mm³)	S_y (10³ mm³)	r_y (mm)	J (10⁶ mm⁴)	C (10³ mm³)	Form Factor k_f	X Compactness	Z_{ex} (10³ mm³)	Y Compactness	Z_{ey} (10³ mm³)
350x250x16 RHS	136	350	250	16	32	17297	283	1620	1990	128	168	1340	1580	98.5	355	2230	1	C	1990	C	1580
350x250x12.5 RHS	109	350	250	12.5	25	13838	233	1330	1620	130	139	1110	1290	100	287	1840	1	C	1620	N	1200
350x250x10 RHS	88.4	350	250	10	20	11257	194	1110	1330	131	116	927	1060	101	235	1520	0.943	N	1320	S	865
350x250x8 RHS	71.6	350	250	8	16	9124	160	914	1090	132	95.7	766	869	102	191	1250	0.833	N	928	S	614
300x200x16 RHS	111	300	200	16	32	14097	161	1080	1350	107	85.7	857	1020	78	193	1450	1	C	1350	C	1020
300x200x12.5 RHS	89	300	200	12.5	25	11338	135	899	1110	109	72	720	842	79.7	158	1210	1	C	1110	C	842
300x200x10 RHS	72.7	300	200	10	20	9257	113	754	921	111	60.6	606	698	80.9	130	1010	1	C	921	N	628
300x200x8 RHS	59.1	300	200	8	16	7524	93.9	626	757	112	50.4	504	574	81.9	106	838	0.903	N	746	S	447
300x200x6 RHS	45	300	200	6	12	5732	73	487	583	113	39.3	393	443	82.8	81.4	651	0.753	S	474	S	288

Section Name $D \times b \times t$ (mm)	ϕM_{sx} (kNm)	ϕM_{sy} (kNm)	ϕV_v (kN)	ϕN_s (kN)	ϕN_t (kN)	ϕT_u (kNm)	Member Axial Capacity (Minor Axis) ϕN_{cy} (kN) — Effective length (m)											Member Axial Capacity (Major Axis) ϕN_{cx} (kN) — Effective length (m)										
							1	2	4	6	8	10	12	14	16	18	20	1	2	4	6	8	10	12	14	16	18	20
350x250x16 RHS	806	640	2402	7005	6616	542	7004	6825	6214	5144	3752	2627	1891	1416	1097	874	712	7005	6915	6544	5939	5039	3959	3007	2303	1803	1445	1181
350x250x12.5 RHS	656	486	1918	5605	5293	447	5605	5465	4991	4162	3066	2158	1556	1166	903	720	587	5605	5536	5247	4779	4082	3233	2468	1895	1485	1190	974
350x250x10 RHS	535	350	1558	4299	4306	369	4299	4202	3865	3282	2480	1770	1283	963	747	596	486	4299	4252	4045	3713	3220	2598	2009	1552	1220	979	802
350x250x8 RHS	376	249	1261	3078	3490	304	3078	3020	2810	2454	1939	1425	1046	789	614	490	400	3078	3052	2921	2715	2413	2014	1599	1253	992	799	655
300x200x16 RHS	547	413	2017	5709	5392	352	5674	5461	4659	3293	2104	1402	992	737	569	453	368	5709	5589	5165	4438	3414	2464	1794	1349	1047	835	682
300x200x12.5 RHS	450	341	1617	4592	4337	294	4567	4402	3785	2723	1757	1174	831	618	477	380	309	4592	4500	4171	3609	2809	2042	1492	1124	873	696	568
300x200x10 RHS	373	254	1317	3749	3541	245	3730	3598	3111	2264	1471	985	698	519	401	319	260	3749	3677	3418	2977	2343	1717	1259	949	738	589	480
300x200x8 RHS	302	181	1069	2752	2878	204	2742	2656	2342	1787	1201	813	578	431	333	265	216	2752	2707	2537	2253	1835	1383	1027	779	607	485	396
300x200x6 RHS	192	117	813	1748	2193	158	1747	1700	1537	1251	894	620	445	333	257	205	167	1748	1727	1637	1492	1275	1011	772	593	465	372	305

Note: Geometry and capacity table, Grade C450L0 – f_y = 450 MPa – f_u = 500 MPa

Table 7.45 Rectangular Hollow Sections: 250 RHS

Section Name $D \times b \times t$ (mm)	Weight (kg/m)	D (mm)	b (mm)	t (mm)	r_w (mm)	A_g (mm²)	I_x (10^6 mm⁴)	Z_x (10^3 mm³)	S_x (10^3 mm³)	r_x (mm)	I_y (10^6 mm⁴)	Z_y (10^3 mm³)	S_y (10^3 mm³)	r_y (mm)	J (10^6 mm⁴)	C (10^3 mm³)	Form Factor k_f	X Compactness	Z_{cx} (10^3 mm³)	Y Compactness	Z_{ey} (10^3 mm³)
250x150x16 RHS	85.5	250	150	16	32	10897	80.2	641	834	85.8	35.8	478	583	57.3	88.2	836	1	C	834	C	583
250x150x12.5 RHS	69.4	250	150	12.5	25	8838	68.5	548	695	88	30.8	411	488	59	73.4	710	1	C	695	C	488
250x150x10 RHS	57	250	150	10	20	7257	58.3	466	582	89.6	26.3	351	409	60.2	61.2	602	1	C	582	N	404
250x150x9 RHS	51.8	250	150	9	18	6598	53.7	430	533	90.2	24.3	324	375	60.7	56	554	1	C	533	N	352
250x150x8 RHS	46.5	250	150	8	16	5924	48.9	391	482	90.8	22.2	296	340	61.2	50.5	504	1	C	482	N	299
250x150x6 RHS	35.6	250	150	6	12	4532	38.4	307	374	92	17.5	233	264	62.2	39	395	0.843	N	368	S	191
250x150x5 RHS	29.9	250	150	5	10	3814	32.7	262	317	92.6	15	199	224	62.6	33	337	0.762	N	275	S	144

Section Name $D \times b \times t$ (mm)	ϕM_{sx} (kNm)	ϕM_{sy} (kNm)	ϕV_y (kN)	ϕN_s (kN)	ϕN_t (kN)	ϕT_u (kNm)	Member Axial Capacity (Minor Axis), ϕN_{cy} (kN) — Effective length (m)											Member Axial Capacity (Major Axis), ϕN_{cx} (kN) — Effective length (m)										
							1	2	4	6	8	10	12	14	16	18	20	1	2	4	6	8	10	12	14	16	18	20
250x150x16 RHS	338	236	1631	4413	4168	203	4336	4048	2858	1568	923	602	423	313	241	191	156	4398	4258	3749	2853	1912	1293	920	685	529	421	343
250x150x12.5 RHS	281	198	1315	3580	3381	173	3521	3301	2390	1339	791	516	363	269	207	164	134	3569	3460	3069	2376	1616	1098	783	583	451	359	292
250x150x10 RHS	236	164	1076	2939	2776	146	2894	2719	2002	1138	675	441	310	229	177	140	114	2932	2845	2535	1987	1365	932	665	496	383	305	249
250x150x9 RHS	216	143	976	2672	2524	135	2632	2476	1834	1049	623	407	286	212	163	130	106	2666	2588	2310	1818	1254	857	612	456	353	281	229
250x150x8 RHS	195	121	875	2399	2266	122	2364	2226	1660	956	568	371	261	193	149	118	96.3	2394	2325	2079	1643	1138	779	556	415	321	256	208
250x150x6 RHS	149	77.4	668	1734	1547	96	1531	1457	1163	728	443	291	205	152	117	93.1	75.8	1547	1509	1379	1151	849	598	431	323	250	199	163
250x150x5 RHS	111	58.3	561	1459	1177	81.9	1167	1116	918	604	373	247	174	129	99.6	79.2	64.5	1177	1152	1063	910	696	501	364	274	212	169	138

Note: Geometry and capacity table, Grade C450L0 – f_y = 450 MPa – f_u = 500 MPa

Table 7.46 Rectangular Hollow Sections: 152 and 200 RHS

Section Name $D \times b \times t$ (mm)	Weight (kg/m)	D (mm)	b (mm)	t (mm)	r_m (mm)	A_g (mm²)	I_x (10⁶ mm⁴)	Z_x (10³ mm³)	S_x (10³ mm³)	r_x (mm)	I_y (10⁶ mm⁴)	Z_y (10³ mm³)	S_y (10³ mm³)	r_y (mm)	J (10⁶ mm⁴)	C (10³ mm³)	Form Factor k_f	X Compactness	Z_{ex} (10³ mm³)	Y Compactness	Z_{ey} (10³ mm³)
200x100x10 RHS	41.3	200	100	10	20	5257	24.4	244	318	68.2	8.18	164	195	39.4	21.5	292	1	C	318	C	195
200x100x9 RHS	37.7	200	100	9	18	4798	22.8	228	293	68.9	7.64	153	180	39.9	19.9	272	1	C	293	C	180
200x100x8 RHS	33.9	200	100	8	16	4324	20.9	209	267	69.5	7.05	141	165	40.4	18.1	250	1	C	267	N	163
200x100x6 RHS	26.2	200	100	6	12	3332	16.7	167	210	70.8	5.69	114	130	41.3	14.2	200	0.967	C	210	S	110
200x100x5 RHS	22.1	200	100	5	10	2814	14.4	144	179	71.5	4.92	98.3	111	41.8	12.1	172	0.855	C	179	S	82.2
200x100x4 RHS	17.9	200	100	4	8	2281	11.9	119	147	72.1	4.07	81.5	91	42.3	9.89	142	0.745	N	144	S	58
152x76x6 RHS	19.4	152	76	6	12	2468	6.91	90.9	116	52.9	2.33	61.4	71.5	30.7	5.98	108	1	C	116	N	70.2
152x76x5 RHS	16.4	152	76	5	10	2094	6.01	79	99.8	53.6	2.04	53.7	61.6	31.2	5.13	94.3	1	C	99.8	N	55.2

Section Name $D \times b \times t$ (mm)	ϕM_{sx} (kNm)	ϕM_{sy} (kNm)	ϕV_v (kN)	ϕN_s (kN)	ϕN_t (kN)	ϕT_u (kNm)	Member Axial Capacity (Minor Axis), ϕN_{cy} (kN) Effective length (m)											Member Axial Capacity (Major Axis), ϕN_{cx} (kN) Effective length (m)										
							1	2	4	6	8	10	12	14	16	18	20	1	2	4	6	8	10	12	14	16	18	20
200x100x10 RHS	129	79	833	2129	2011	71	2038	1746	798	377	216	140	98.1	72.5	55.8	44.2	36	2107	2006	1607	1012	616	405	286	212	163	130	106
200x100x9 RHS	119	72.9	758	1943	1835	66.1	1863	1603	745	353	202	131	91.8	67.9	52.2	41.4	33.6	1923	1833	1477	939	573	377	266	197	152	121	98.4
200x100x8 RHS	108	66	681	1751	1654	60.8	1681	1452	686	326	187	121	84.8	62.7	48.2	38.2	31.1	1734	1654	1339	857	525	346	244	181	139	111	90.2
200x100x6 RHS	85.1	44.6	522	1305	1275	48.6	1257	1100	546	261	150	97.3	68.2	50.4	38.8	30.8	25	1294	1238	1020	674	417	276	194	144	111	88.5	72
200x100x5 RHS	72.5	33.3	440	974	1076	41.8	944	843	459	224	129	83.9	58.8	43.5	33.5	26.5	21.6	968	931	792	556	354	235	167	124	95.5	75.9	61.8
200x100x4 RHS	58.3	23.5	355	688	873	34.5	670	610	366	184	107	69.3	48.6	36	27.7	22	17.9	685	662	579	432	285	192	136	101	78.4	62.3	50.8
152x76x6 RHS	47	28.4	389	1000	944	26.2	928	694	238	109	62.5	40.4	28.2	20.9	16	12.7	10.3	978	902	589	308	179	117	81.9	60.6	46.7	37.1	30.1
152x76x5 RHS	40.4	22.4	329	848	801	22.9	789	597	208	95.8	54.7	35.4	24.7	18.3	14	11.1	9.04	830	768	508	267	156	102	71.3	52.8	40.7	32.3	26.2

Note: Geometry and capacity table, Grade C450L0 – f_y = 450 MPa – f_u = 500 MPa

Table 7.47 Rectangular Hollow Sections: 150 RHS

Section Name $D \times b \times t$ (mm)	Weight (kg/m)	D (mm)	b (mm)	t (mm)	r_m (mm)	A_g (mm²)	I_x (10^6 mm⁴)	Z_x (10^3 mm³)	S_x (10^3 mm³)	r_x (mm)	I_y (10^6 mm⁴)	Z_y (10^3 mm³)	S_y (10^3 mm³)	r_y (mm)	J (10^6 mm⁴)	C (10^3 mm³)	Form Factor k_f	X Compactness	Z_{ex} (10^3 mm³)	Y Compactness	Z_{ey} (10^3 mm³)
150x100x10 RHS	33.4	150	100	10	20	4257	11.6	155	199	52.2	6.14	123	150	38	14.3	211	1	C	199	C	150
150x100x9 RHS	30.6	150	100	9	18	3898	10.9	145	185	52.9	5.77	115	140	38.5	13.2	197	1	C	185	C	140
150x100x8 RHS	27.7	150	100	8	16	3524	10.1	134	169	53.5	5.36	107	128	39	12.1	182	1	C	169	C	128
150x100x6 RHS	21.4	150	100	6	12	2732	8.17	109	134	54.7	4.36	87.3	102	40	9.51	147	1	C	134	N	101
150x100x5 RHS	18.2	150	100	5	10	2314	7.07	94.3	115	55.3	3.79	75.7	87.3	40.4	8.12	127	1	C	115	N	78.5
150x100x4 RHS	14.8	150	100	4	8	1881	5.87	78.2	94.6	55.9	3.15	63	71.8	40.9	6.64	105	0.903	N	93.2	S	55.9
150x50x6 RHS	16.7	150	50	6	12	2132	5.06	67.5	91.2	48.7	0.86	34.4	40.9	20.1	2.63	64.3	1	C	91.2	N	40.4
150x50x5 RHS	14.2	150	50	5	10	1814	4.44	59.2	78.9	49.5	0.765	30.6	35.7	20.5	2.3	56.8	1	C	78.9	N	31.8

Section Name $D \times b \times t$ (mm)	ϕM_{sx} (kNm)	ϕM_{sy} (kNm)	ϕV_v (kN)	ϕN_s (kN)	ϕN_t (kN)	ϕT_u (kNm)	Member Axial Capacity (Minor Axis), ϕN_{cy} (kN) — Effective length (m)											Member Axial Capacity (Major Axis), ϕN_{cx} (kN) — Effective length (m)										
							1	2	4	6	8	10	12	14	16	18	20	1	2	4	6	8	10	12	14	16	18	20
150x100x10 RHS	80.6	60.8	611	1724	1628	51.3	1645	1389	607	285	163	106	74	54.7	42.1	33.4	27.1	1685	1551	999	518	302	196	138	102	78.5	62.3	50.6
150x100x9 RHS	74.9	56.7	559	1579	1491	47.9	1508	1280	569	268	153	99.3	69.5	51.4	39.5	31.4	25.5	1544	1425	931	486	283	184	129	95.8	73.7	58.5	47.6
150x100x8 RHS	68.4	51.8	504	1427	1348	44.2	1365	1165	526	248	142	92.1	64.5	47.7	36.7	29.1	23.6	1397	1291	854	449	262	170	120	88.5	68.2	54.1	44
150x100x6 RHS	54.3	40.9	389	1107	1045	35.7	1061	914	426	202	116	75	52.5	38.8	29.8	23.7	19.3	1085	1006	680	362	212	138	96.8	71.7	55.2	43.8	35.6
150x100x5 RHS	46.6	31.8	329	937	885	30.9	900	777	367	174	100	64.8	45.4	33.5	25.8	20.5	16.6	919	854	583	313	183	119	83.7	62	47.8	37.9	30.8
150x100x4 RHS	37.7	22.6	267	688	720	25.5	664	585	300	144	83	53.8	37.7	27.9	21.4	17	13.8	677	634	458	256	151	98.7	69.3	51.3	39.6	31.4	25.6
150x50x6 RHS	36.9	16.4	374	864	816	15.6	715	335	91.3	41.4	23.5	15.2	10.6	7.81	6	4.75	3.85	841	764	455	228	132	85.9	60.2	44.6	34.3	27.2	22.1
150x50x5 RHS	32	12.9	316	735	694	13.8	613	295	80.7	36.6	20.8	13.4	9.37	6.91	5.3	4.2	3.41	716	653	396	200	116	75.4	52.9	39.1	30.1	23.9	19.4

Note: Geometry and capacity table, Grade C450L0 – f_y = 450 MPa – f_u = 500 MPa

Table 7.48 Rectangular Hollow Sections: 127 and 150 RHS

Section geometry and properties

Section Name D x b x t (mm)	Weight (kg/m)	D (mm)	b (mm)	t (mm)	r_m (mm)	A_g (mm²)	I_x (10⁶mm⁴)	Z_x (10³mm³)	S_x (10³mm³)	r_x (mm)	I_y (10⁶mm⁴)	Z_y (10³mm³)	S_y (10³mm³)	r_y (mm)	J (10⁶mm⁴)	C (10³mm³)	Form Factor k_f	X Compactness	Z_{ex} (10³mm³)	Y Compactness	Z_{ey} (10³mm³)
150x50x4 RHS	11.6	150	50	4	8	1481	3.74	49.8	65.4	50.2	0.653	26.1	29.8	21	1.93	48.2	0.877	C	65.4	S	22.7
150x50x3 RHS	8.96	150	50	3	4.5	1141	2.99	39.8	51.4	51.2	0.526	21.1	23.5	21.5	1.5	38.3	0.713	C	51.4	S	14.5
150x50x2.5 RHS	7.53	150	50	2.5	3.75	959	2.54	33.9	43.5	51.5	0.452	18.1	19.9	21.7	1.28	32.8	0.633	C	43.5	S	10.9
150x50x2 RHS	6.07	150	50	2	3	774	2.08	27.7	35.3	51.8	0.372	14.9	16.3	21.9	1.04	26.9	0.553	N	31.6	S	7.64
127x51x6 RHS	14.7	127	51	6	12	1868	3.28	51.6	68.9	41.9	0.761	29.8	35.8	20.2	2.2	54.9	1	C	68.9	C	35.8
127x51x5 RHS	12.5	127	51	5	10	1594	2.89	45.6	59.9	42.6	0.679	26.6	31.3	20.6	1.93	48.6	1	C	59.9	N	30.6
127x51x3.5 RHS	9.07	127	51	3.5	7	1155	2.2	34.7	44.6	43.7	0.526	20.6	23.4	21.3	1.44	37.2	0.905	C	44.6	S	18.5

Capacity table

Section Name D x b x t (mm)	ϕM_{sx} (kNm)	ϕM_{sy} (kNm)	ϕV_v (kN)	ϕN_s (kN)	ϕN_t (kN)	ϕT_u (kNm)	Minor 0.5	1	2	3	4	5	6	7	8	9	10	Major 1	2	4	6	8	10	12	14	16	18	20
150x50x4 RHS	26.5	9.19	257	526	567	11.7	454	245	68.7	31.2	17.8	11.5	8.04	5.91	4.54	3.59	2.92	515	476	315	166	96.9	63	44.2	32.7	25.2	20	16.3
150x50x3 RHS	20.8	5.87	195	329	436	9.31	295	185	55	25.1	14.3	9.22	6.44	4.75	3.65	2.89	2.35	324	305	226	129	76.6	50.1	35.2	26.1	20.1	16	13
150x50x2.5 RHS	17.6	4.41	164	246	367	7.97	223	151	46.8	21.4	12.2	7.87	5.5	4.06	3.12	2.47	2.01	243	230	178	108	64.6	42.4	29.8	22.1	17	13.5	11
150x50x2 RHS	12.8	3.09	132	173	296	6.54	160	115	38.2	17.5	9.99	6.45	4.51	3.33	2.56	2.03	1.65	172	164	133	85.1	52.2	34.4	24.2	18	13.9	11	8.96
127x51x6 RHS	27.9	14.5	315	757	715	13.3	628	296	80.8	36.6	20.8	13.4	9.37	6.91	5.31	4.2	3.41	729	637	315	151	86.7	56.2	39.3	29.1	22.4	17.8	14.4
127x51x5 RHS	24.3	12.4	267	646	610	11.8	540	261	71.6	32.5	18.5	11.9	8.3	6.13	4.71	3.73	3.03	622	547	276	133	76.4	49.5	34.7	25.6	19.7	15.6	12.7
127x51x3.5 RHS	18.1	7.49	192	423	442	9.04	365	197	55.2	25.1	14.3	9.2	6.42	4.74	3.64	2.88	2.34	411	368	205	100	57.9	37.6	26.3	19.5	15	11.9	9.67

Member Axial Capacity (Minor Axis), ϕN_{cy} (kN) — Effective length (m); Member Axial Capacity (Major Axis), ϕN_{cx} (kN) — Effective length (m)

Note: Geometry and capacity table, Grade C450L0 – f_y = 450 MPa – f_u = 500 MPa

Table 7.49 Rectangular Hollow Sections: 102 and 125 RHS

Section Name $D \times b \times t$ (mm)	Weight (kg/m)	D (mm)	b (mm)	t (mm)	r_m (mm)	A_g (mm²)	I_x (10⁶ mm⁴)	Z_x (10³ mm³)	S_x (10³ mm³)	r_x (mm)	I_y (10⁶ mm⁴)	Z_y (10³ mm³)	S_y (10³ mm³)	r_y (mm)	J (10⁶ mm⁴)	C (10³ mm³)	Form Factor k_f	X Compactness	Z_{ex} (10³ mm³)	Y Compactness	Z_{ey} (10³ mm³)
125x75x6 RHS	16.7	125	75	6	12	2132	4.16	66.6	84.2	44.2	1.87	50	59.1	29.6	4.44	86.2	1	C	84.2	C	59.1
125x75x5 RHS	14.2	125	75	5	10	1814	3.64	58.3	72.7	44.8	1.65	43.9	51.1	30.1	3.83	75.3	1	C	72.7	N	50.5
125x75x4 RHS	11.6	125	75	4	8	1481	3.05	48.9	60.3	45.4	1.39	37	42.4	30.6	3.16	63	1	C	60.3	N	37.4
125x75x3 RHS	8.96	125	75	3	4.5	1141	2.43	38.9	47.3	46.1	1.11	29.5	33.3	31.1	2.43	49.5	0.845	N	46.5	S	24.2
125x75x2.5 RHS	7.53	125	75	2.5	3.75	959	2.07	33	40	46.4	0.942	25.1	28.2	31.4	2.05	42.1	0.763	N	34.7	S	18.2
125x75x2 RHS	6.07	125	75	2	3	774	1.69	27	32.5	46.7	0.771	20.6	22.9	31.6	1.67	34.4	0.624	S	24.8	S	13
102x76x6 RHS	14.7	102	76	6	12	1868	2.52	49.4	61.9	36.7	1.59	42	50.5	29.2	3.38	69.8	1	C	61.9	S	50.5
102x76x5 RHS	12.5	102	76	5	10	1594	2.22	43.5	53.7	37.3	1.41	37	43.9	29.7	2.91	61.2	1	C	53.7	C	43.9
102x76x3.5 RHS	9.07	102	76	3.5	7	1155	1.68	33	39.9	38.2	1.07	28.2	32.6	30.5	2.14	46.1	1	C	39.9	N	29.8

Section Name $D \times b \times t$ (mm)	ϕM_{sx} (kNm)	ϕM_{sy} (kNm)	ϕV_v (kN)	ϕN_t (kN)	ϕN_s (kN)	ϕT_u (kNm)	Member Axial Capacity (Minor Axis), ϕN_{cy} (kN) Effective length (m)											Member Axial Capacity (Major Axis), ϕN_{cx} (kN) Effective length (m)										
							0.5	1	2	3	4	5	6	7	8	9	10	0.5	1	2	3	4	5	6	7	8	9	10
125x75x6 RHS	34.1	23.9	317	864	816	20.9	850	797	579	325	192	125	88.1	65.3	50.3	39.9	32.5	861	835	742	576	393	267	190	142	110	87.3	71.1
125x75x5 RHS	29.4	20.5	269	735	694	18.3	723	680	500	284	169	110	77.4	57.4	44.2	35.1	28.6	733	711	634	497	341	233	166	124	95.8	76.3	62.1
125x75x4 RHS	24.4	15.1	219	600	567	15.3	591	556	415	239	142	92.8	65.3	48.4	37.3	29.6	24.1	599	581	520	411	285	195	139	104	80.3	63.9	52.1
125x75x3 RHS	18.8	9.8	167	390	436	12	386	368	293	183	111	73.2	51.6	38.3	29.5	23.4	19.1	390	381	348	291	214	151	109	81.7	63.3	50.4	41.1
125x75x2.5 RHS	14.1	7.37	140	296	367	10.2	294	281	231	153	94.5	62.4	44	32.7	25.2	20	16.3	296	290	268	229	176	126	91.9	69.1	53.6	42.8	34.9
125x75x2 RHS	10	5.27	113	196	296	8.36	194	187	161	116	75.3	50.3	35.7	26.5	20.5	16.3	13.3	196	192	180	160	131	98.7	73.4	55.6	43.4	34.7	28.3
102x76x6 RHS	25.1	20.5	255	757	715	17	744	696	500	278	164	107	75.2	55.7	42.9	34.1	27.7	751	719	598	402	250	166	117	86.9	67	53.3	43.4
102x76x5 RHS	21.7	17.8	218	646	610	14.9	635	596	434	244	144	94.3	66.3	49.1	37.8	30.1	24.4	641	615	515	351	220	146	103	76.5	59	46.9	38.2
102x76x3.5 RHS	16.2	12.1	157	468	442	11.2	461	434	323	185	110	71.9	50.6	37.5	28.9	22.9	18.7	465	447	378	262	166	110	78.1	58	44.8	35.6	29

Note: Geometry and capacity table, Grade C450L0 – f_y = 450 MPa – f_u = 500 MPa

Table 7.50 Rectangular Hollow Sections: 100 RHS

Section Name $D \times b \times t$ (mm)	Weight (kg/m)	D (mm)	b (mm)	t (mm)	t_m (mm)	A_g (mm²)	I_x (10³ mm⁴)	r_x (mm)	S_x (10³ mm³)	Z_x (10³ mm³)	I_y (10⁶ mm⁴)	r_y (mm)	J (10⁶ mm⁴)	C (10³ mm³)	Form Factor k_f	X Compactness	Z_{ex} (10³ mm³)	Y Compactness	Z_{ey} (10³ mm³)
100x50x6 RHS	12	100	50	6	12	1532	1.71	33.4	45.3	34.2	0.567	19.2	1.53	40.9	1	C	45.3	C	27.7
100x50x5 RHS	10.3	100	50	5	10	1314	1.53	34.1	39.8	30.6	0.511	19.7	1.35	36.5	1	C	39.8	C	24.4
100x50x4 RHS	8.49	100	50	4	8	1081	1.31	34.8	33.4	26.1	0.441	20.2	1.13	31.2	1	C	33.4	N	20.3
100x50x3.5 RHS	7.53	100	50	3.5	7	959	1.18	35.1	29.9	23.6	0.4	20.4	1.01	28.2	1	C	29.9	N	17.1
100x50x3 RHS	6.6	100	50	3	4.5	841	1.06	35.6	26.7	21.3	0.361	20.7	0.886	25	0.967	C	26.7	S	13.9
100x50x2.5 RHS	5.56	100	50	2.5	3.75	709	0.912	35.9	22.7	18.2	0.311	20.9	0.754	21.5	0.856	C	22.7	S	10.4
100x50x2 RHS	4.5	100	50	2	3	574	0.75	36.2	18.5	15	0.257	21.2	0.616	17.7	0.746	N	18.2	S	7.33
100x50x1.6 RHS	3.64	100	50	1.6	2.4	463	0.613	36.4	15	12.3	0.211	21.3	0.501	14.5	0.661	N	12.5	S	5.19

Section Name $D \times b \times t$ (mm)	ϕM_{sx} (kNm)	ϕM_{sy} (kNm)	ϕV_v (kN)	ϕN_s (kN)	ϕN_t (kN)	ϕT_u (kNm)	Member Axial Capacity (Minor Axis), ϕN_{cy} (kN) Effective length (m)											Member Axial Capacity (Major Axis), ϕN_{cx} (kN) Effective length (m)										
							0.5	1	2	3	4	5	6	7	8	9	10	0.5	1	2	3	4	5	6	7	8	9	10
100x50x6 RHS	18.3	11.2	244	621	586	9.94	593	503	223	105	60	38.9	27.2	20.1	15.5	12.3	9.96	614	583	462	286	173	114	80	59.3	45.7	36.3	29.6
100x50x5 RHS	16.1	9.88	208	532	503	8.87	510	436	200	94.3	54.1	35	24.5	18.1	13.9	11.1	8.99	527	502	402	253	154	101	71.4	53	40.8	32.5	26.4
100x50x4 RHS	13.5	8.22	170	438	414	7.58	420	363	171	81.4	46.7	30.3	21.2	15.7	12.1	9.56	7.77	434	414	335	215	132	86.7	61.1	45.3	35	27.8	22.6
100x50x3.5 RHS	12.1	6.93	151	388	367	6.85	373	323	155	73.6	42.3	27.4	19.2	14.2	10.9	8.65	7.03	385	367	299	193	119	78.2	55.1	40.9	31.5	25.1	20.4
100x50x3 RHS	10.8	5.63	131	329	322	6.08	317	278	138	66.2	38.1	24.7	17.3	12.8	9.83	7.8	6.34	327	313	258	171	106	70.3	49.6	36.8	28.4	22.6	18.4
100x50x2.5 RHS	9.19	4.21	110	246	271	5.22	238	213	116	56.5	32.6	21.1	14.8	10.9	8.43	6.69	5.43	244	235	200	141	89.8	59.8	42.3	31.4	24.3	19.3	15.7
100x50x2 RHS	7.37	2.97	88.9	173	219	4.3	169	154	92.4	46.5	27	17.5	12.3	9.09	6.99	5.55	4.51	173	167	146	109	72.2	48.7	34.6	25.7	19.9	15.8	12.9
100x50x1.6 RHS	5.06	2.1	71.7	124	177	3.52	121	112	72.2	37.5	21.9	14.2	9.97	7.38	5.69	4.51	3.67	124	120	107	83.8	57.5	39.3	28	20.9	16.1	12.9	10.5

Note: Geometry and capacity table, Grade C450L0 – f_y = 450 MPa – f_u = 500 MPa

Table 7.51 Rectangular Hollow Sections: 75 and 76 RHS

Section Name $D \times b \times t$ (mm)	Weight (kg/m)	D (mm)	b (mm)	t (mm)	r_m (mm)	A_g (mm²)	I_x (10⁶mm⁴)	Z_x (10³mm³)	S_x (10³mm³)	r_x (mm)	I_y (10⁶mm⁴)	Z_y (10³mm³)	S_y (10³mm³)	r_y (mm)	J (10⁶mm⁴)	C (10³mm³)	Form Factor k_f	X Compactness	Z_{ex} (10³mm³)	Y Compactness	Z_{ey} (10³mm³)
76x38x4 RHS	6.23	76	38	4	8	793	0.527	13.9	18.1	25.8	0.176	9.26	11.1	14.9	0.466	16.6	1	C	18.1	C	11.1
76x38x3 RHS	4.9	76	38	3	4.5	625	0.443	11.7	14.8	26.6	0.149	7.82	9.09	15.4	0.373	13.6	1	C	14.8	N	8.92
76x38x2.5 RHS	4.15	76	38	2.5	3.75	529	0.383	10.1	12.7	26.9	0.129	6.81	7.81	15.6	0.32	11.8	1	C	12.7	N	7
75x50x6 RHS	9.67	75	50	6	12	1232	0.8	21.3	28.1	25.5	0.421	16.9	21.1	18.5	1.01	29.3	1	C	28.1	C	21.1
75x50x5 RHS	8.35	75	50	5	10	1064	0.726	19.4	24.9	26.1	0.384	15.4	18.8	19	0.891	26.4	1	C	24.9	C	18.8
75x50x4 RHS	6.92	75	50	4	8	881	0.63	16.8	21.1	26.7	0.335	13.4	16	19.5	0.754	22.7	1	C	21.1	C	16
75x50x3 RHS	5.42	75	50	3	4.5	691	0.522	13.9	17.1	27.5	0.278	11.1	12.9	20	0.593	18.4	1	C	17.1	N	12.8
75x50x2.5 RHS	4.58	75	50	2.5	3.75	584	0.45	12	14.6	27.7	0.24	9.6	11	20.3	0.505	15.9	1	C	14.6	N	9.95
75x50x2 RHS	3.72	75	50	2	3	474	0.372	9.91	12	28	0.199	7.96	9.06	20.5	0.414	13.1	0.904	N	11.8	S	7.07
75x50x1.6 RHS	3.01	75	50	1.6	2.4	383	0.305	8.14	9.75	28.2	0.164	6.56	7.4	20.7	0.337	10.8	0.799	N	8.26	S	5.01

(Continued)

Table 7.51 (Continued) Rectangular Hollow Sections: 75 and 76 RHS

Section Name $D \times b \times t$ (mm)	Weight (kg/m)	D (mm)	b (mm)	t (mm)	r_m (mm)	A_g (mm²)	I_x (10^6 mm⁴)	Z_x (10^3 mm³)	S_x (10^3 mm³)	r_x (mm)	I_y (10^6 mm⁴)	Z_y (10^3 mm³)	S_y (10^3 mm³)	r_y (mm)	J (10^6 mm⁴)	C (10^3 mm³)	Form Factor k_f	X Compactness	Z_{ex} (10^3 mm³)	Y Compactness	Z_{ey} (10^3 mm³)
75x38x4 RHS	6.23	76	38	4	8	793	0.527	13.9	18.1	25.8	0.176	9.26	11.1	14.9	0.466	16.6	1	C	18.1	C	11.1
75x38x3 RHS	4.9	76	38	3	4.5	625	0.443	11.7	14.8	26.6	0.149	7.82	9.09	15.4	0.373	13.6	1	C	14.8	N	8.92
76x38x2.5 RHS	4.15	76	38	2.5	3.75	529	0.383	10.1	12.7	26.9	0.129	6.81	7.81	15.6	0.32	11.8	1	C	12.7	N	7
75x50x6 RHS	9.67	75	50	6	12	1232	0.8	21.3	28.1	25.5	0.421	16.9	21.1	18.5	1.01	29.3	1	C	28.1	C	21.1
75x50x5 RHS	8.35	75	50	5	10	1064	0.726	19.4	24.9	26.1	0.384	15.4	18.8	19	0.891	26.4	1	C	24.9	C	18.8
75x50x4 RHS	6.92	75	50	4	8	881	0.63	16.8	21.1	26.7	0.335	13.4	16	19.5	0.754	22.7	1	C	21.1	C	16
75x50x3 RHS	5.42	75	50	3	4.5	691	0.522	13.9	17.1	27.5	0.278	11.1	12.9	20	0.593	18.4	1	C	17.1	N	12.8
75x50x2.5 RHS	4.58	75	50	2.5	3.75	584	0.45	12	14.6	27.7	0.24	9.6	11	20.3	0.505	15.9	1	C	14.6	N	9.95
75x50x2 RHS	3.72	75	50	2	3	474	0.372	9.91	12	28	0.199	7.96	9.06	20.5	0.414	13.1	0.904	N	11.8	S	7.07
75x50x1.6 RHS	3.01	75	50	1.6	2.4	383	0.305	8.14	9.75	28.2	0.164	6.56	7.4	20.7	0.337	10.8	0.799	N	8.26	S	5.01

Note: Geometry and capacity table, Grade C450L0 – $f_y = 450$ MPa – $f_u = 500$ MPa

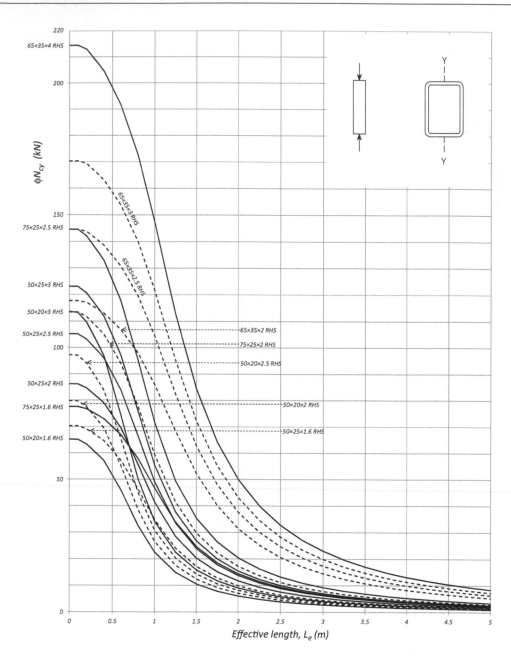

Figure 7.23 Rectangular hollow sections: Members subject to axial compression (minor axis) grade C350L0 (solid lines labelled on the *y* axis). 1 of 2.

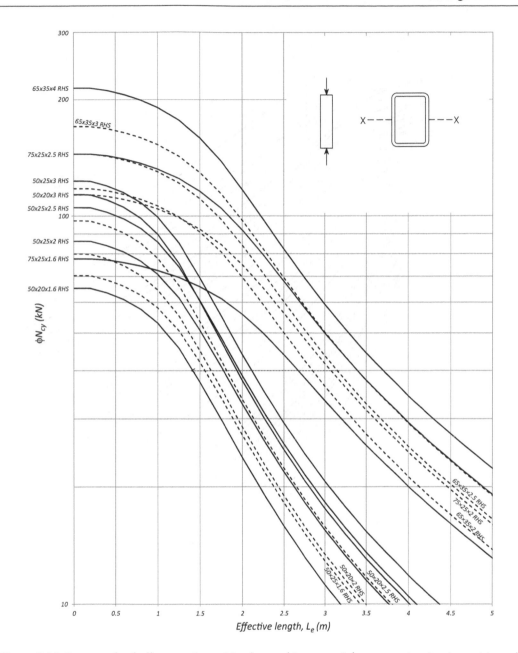

Figure 7.24 Rectangular hollow sections: Members subject to axial compression (major axis) grade C350L0 (solid lines labelled on the *y* axis). 2 of 2.

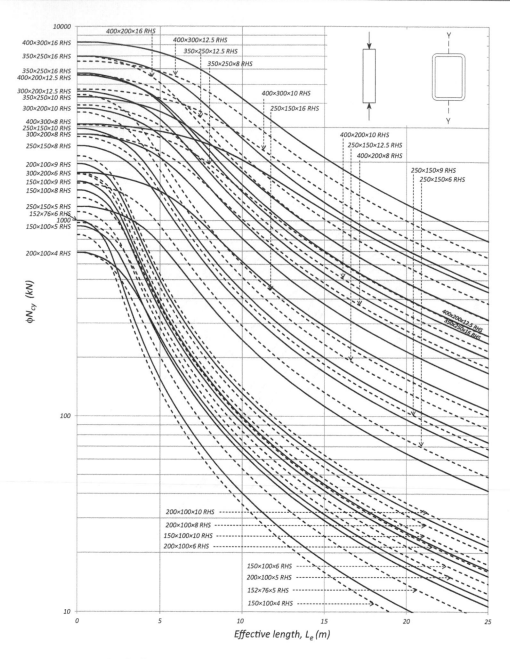

Figure 7.25 Rectangular hollow sections: Members subject to axial compression (minor axis) grade C450L0 (solid lines labelled on the *y* axis). 1 of 2.

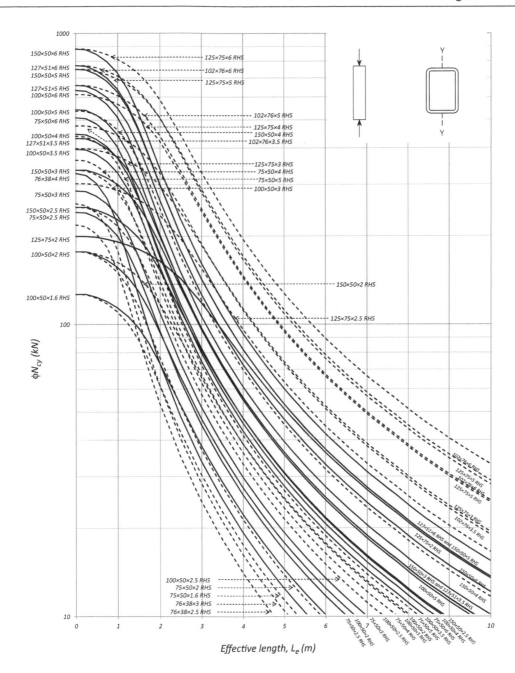

Figure 7.26 Rectangular hollow sections: Members subject to axial compression (minor axis) grade C450L0 (solid lines labelled on the *y* axis). 2 of 2.

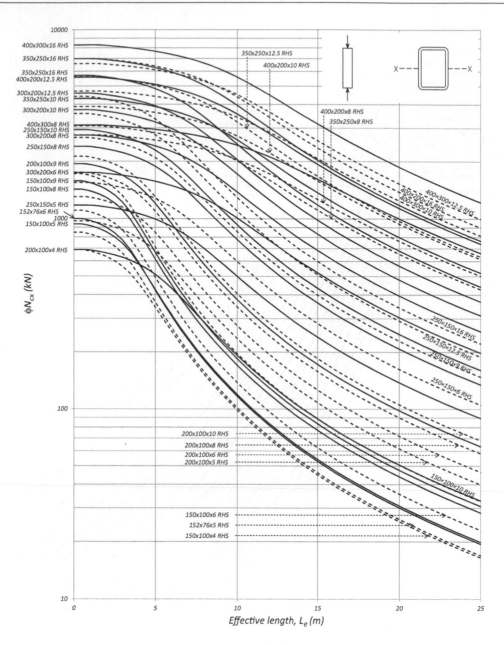

Figure 7.27 Rectangular hollow sections: Members subject to axial compression (major axis) grade C450L0 (solid lines labelled on the *y* axis). 1 of 2.

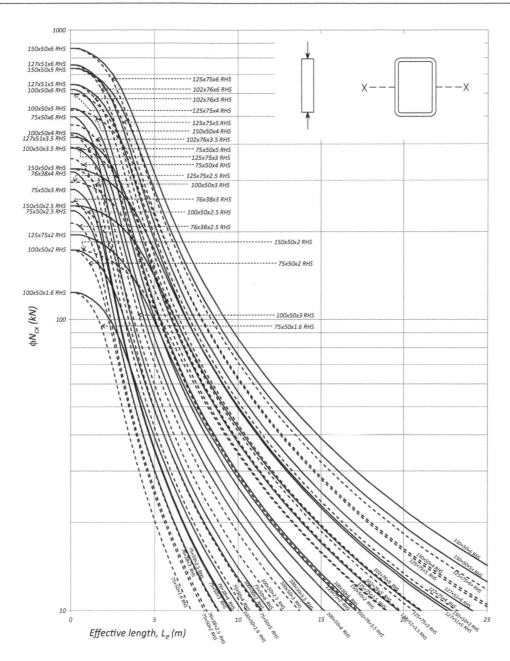

Figure 7.28 Rectangular hollow sections: Members subject to axial compression (major axis) grade C450L0 (solid lines labelled on the *y* axis). 2 of 2.

Table 7.52 Square Hollow Sections: 50 SHS

Section Name D x t (mm)	Weight (kg/m)	D (mm)	t (mm)	r_m (mm)	A_g (mm²)	I_x (10⁶ mm⁴)	Z_x (10³ mm³)	Z_u (10³ mm³)	S_x (10³ mm³)	r_x (mm)	J (10⁶ mm⁴)	C (10³ mm³)	Form Factor k_f	X Compactness	Z_{ex} (10³ mm³)
50x6 SHS	7.32	50	6	12	932	0.275	11	9.45	14.5	17.2	0.518	17.7	1	C	14.5
50x5 SHS	6.39	50	5	10	814	0.257	10.3	8.51	13.2	17.8	0.469	16.3	1	C	13.2
50x4 SHS	5.35	50	4	8	681	0.229	9.15	7.33	11.4	18.3	0.403	14.3	1	C	11.4
50x3 SHS	4.25	50	3	4.5	541	0.195	7.79	5.92	9.39	19	0.321	11.8	1	C	9.39
50x2.5 SHS	3.6	50	2.5	3.75	459	0.169	6.78	5.09	8.07	19.2	0.275	10.2	1	C	8.07
50x2 SHS	2.93	50	2	3	374	0.141	5.66	4.2	6.66	19.5	0.226	8.51	1	C	6.66
50x1.6 SHS	2.38	50	1.6	2.4	303	0.117	4.68	3.44	5.46	19.6	0.185	7.03	1	N	5.1

Section Name D x t (mm)	ϕM_s (kNm)	ϕV_v (kN)	ϕN_s (kN)	ϕN_t (kN)	ϕT_u (kNm)	Member Axial Capacity (Any Axis), ϕN_c (kN) Effective length (m)																					
						0.25	0.5	0.75	1	1.5	2	2.5	3	3.5	4	4.5	5	5.5	6	6.5	7	7.5	8	8.5	9	9.5	10
50x6 SHS	4.57	85.1	294	294	3.35	292	281	264	240	169	108	72.1	51	37.9	29.3	23.3	19	15.7	13.3	11.3	9.81	8.57	7.55	6.7	5.98	5.38	4.86
50x5 SHS	4.16	74.7	256	256	3.08	255	246	232	213	155	100	67.1	47.6	35.4	27.3	21.7	17.7	14.7	12.4	10.6	9.16	8	7.05	6.26	5.59	5.03	4.54
50x4 SHS	3.59	62.7	215	215	2.7	214	206	195	180	134	87.9	59.1	42	31.2	24.1	19.2	15.6	13	10.9	9.36	8.09	7.07	6.23	5.53	4.94	4.44	4.01
50x3 SHS	2.96	49.3	170	170	2.23	170	164	156	145	111	74.3	50.3	35.8	26.7	20.6	16.4	13.4	11.1	9.35	8	6.92	6.04	5.32	4.72	4.22	3.8	3.43
50x2.5 SHS	2.54	42	145	145	1.93	144	140	133	123	94.9	64.2	43.5	31	23.1	17.8	14.2	11.6	9.6	8.1	6.93	5.99	5.23	4.61	4.09	3.66	3.29	2.97
50x2 SHS	2.1	34.3	118	118	1.61	117	114	109	101	78.5	53.6	36.5	26	19.4	15	11.9	9.71	8.06	6.8	5.81	5.03	4.39	3.87	3.44	3.07	2.76	2.5
50x1.6 SHS	1.61	28	95.5	95.5	1.33	95.3	92.4	88.1	82.2	64.1	43.8	29.8	21.3	15.9	12.3	9.76	7.95	6.6	5.57	4.76	4.12	3.6	3.17	2.82	2.52	2.26	2.05

Note: Geometry and capacity table, Grade C350L0 – f_y = 350 MPa – f_u = 430 MPa

Table 7.53 Square Hollow Sections: 35 and 40 SHS

Section Name $D \times t$ (mm)	Weight (kg/m)	D (mm)	t (mm)	r_m (mm)	A_g (mm²)	I_x (10⁶ mm⁴)	Z_x (10³ mm³)	Z_u (10³ mm³)	S_x (10³ mm²)	r_x (mm)	J (10⁶ mm⁴)	C (10³ mm³)	Form Factor k_f	X Compactness	Z_{ex} (10³ mm³)
40x4 SHS	4.09	40	4	8	521	0.105	5.26	4.36	6.74	14.2	0.192	8.33	1	C	6.74
40x3 SHS	3.3	40	3	4.5	421	0.0932	4.66	3.61	5.72	14.9	0.158	7.07	1	C	5.72
40x2.5 SHS	2.82	40	2.5	3.75	359	0.0822	4.11	3.13	4.97	15.1	0.136	6.21	1	C	4.97
40x2 SHS	2.31	40	2	3	294	0.0694	3.47	2.61	4.13	15.4	0.113	5.23	1	C	4.13
40x1.6 SHS	1.88	40	1.6	2.4	239	0.0579	2.9	2.15	3.41	15.6	0.0927	4.36	1	C	3.41
35x3 SHS	2.83	35	3	4.5	361	0.0595	3.4	2.67	4.23	12.8	0.102	5.18	1	C	4.23
35x2.5 SHS	2.42	35	2.5	3.75	309	0.0529	3.02	2.33	3.69	13.1	0.0889	4.58	1	C	3.69
35x2 SHS	1.99	35	2	3	254	0.0451	2.58	1.95	3.09	13.3	0.0741	3.89	1	C	3.09
35x1.6 SHS	1.63	35	1.6	2.4	207	0.0379	2.16	1.62	2.57	13.5	0.0611	3.26	1	C	2.57

Section Name $D \times t$ (mm)	ϕM_s (kNm)	ϕV_v (kN)	ϕN_s (kN)	ϕN_t (kN)	ϕT_u (kNm)	Member Axial Capacity (Any Axis) ϕN_c (kN) Effective length (m)																					
						0.25	0.5	0.75	1	1.5	2	2.5	3	3.5	4	4.5	5	5.5	6	6.5	7	7.5	8	8.5	9	9.5	10
40x4 SHS	2.12	47.8	164	164	1.57	162	153	139	119	71.2	42.7	28	19.7	14.6	11.3	8.95	7.28	6.04	5.09	4.35	3.76	3.28	2.89	2.57	2.29	2.06	1.86
40x3 SHS	1.8	38.1	133	133	1.34	131	125	115	99.4	62.1	37.7	24.8	17.5	13	9.99	7.94	6.46	5.36	4.52	3.86	3.34	2.92	2.57	2.28	2.04	1.83	1.65
40x2.5 SHS	1.57	32.7	113	113	1.17	112	107	98.1	85.6	54.1	33	21.7	15.3	11.3	8.75	6.95	5.66	4.69	3.96	3.38	2.92	2.55	2.25	1.99	1.78	1.6	1.45
40x2 SHS	1.3	26.9	92.5	92.5	0.988	91.6	87.4	80.8	71	45.6	28	18.4	13	9.65	7.44	5.91	4.81	3.99	3.37	2.88	2.49	2.17	1.91	1.7	1.52	1.36	1.23
40x1.6 SHS	1.07	22	75.3	75.3	0.824	74.6	71.3	66	58.3	37.9	23.3	15.4	10.9	8.05	6.21	4.94	4.02	3.33	2.81	2.4	2.08	1.81	1.6	1.42	1.27	1.14	1.03
35x3 SHS	1.33	32.5	114	114	0.979	112	105	92.4	74.6	41.3	24.4	15.9	11.2	8.27	6.37	5.06	4.11	3.41	2.88	2.46	2.12	1.85	1.63	1.45	1.29	1.16	1.05
35x2.5 SHS	1.16	28	97.3	97.3	0.866	95.8	89.8	80	65.4	36.8	21.8	14.2	10	7.41	5.71	4.53	3.69	3.06	2.58	2.2	1.9	1.66	1.46	1.3	1.16	1.04	0.941
35x2 SHS	0.973	23.1	79.9	79.9	0.735	78.7	74	66.1	54.5	31.1	18.4	12	8.45	6.26	4.83	3.83	3.12	2.59	2.18	1.86	1.61	1.41	1.24	1.1	0.981	0.881	0.797
35x1.6 SHS	0.81	19	65.3	65.3	0.616	64.3	60.5	54.4	45.2	26	15.5	10.1	7.11	5.27	4.06	3.22	2.62	2.18	1.83	1.57	1.35	1.18	1.04	0.923	0.825	0.741	0.67

Note: Geometry and capacity table, Grade C350L0 – f_y = 350 MPa – f_u = 430 MPa

Table 7.54 Square Hollow Sections: 20, 25 and 30 SHS

Section Name D x t (mm)	Weight (kg/m)	D (mm)	t (mm)	r_m (mm)	A_g (mm²)	I_x (10⁶ mm⁴)	Z_x (10³ mm³)	Z_m (10³ mm³)	S_x (10³ mm³)	r_x (mm)	J (10⁶ mm⁴)	C (10³ mm³)	Form Factor k_f	X Compactness	Z_{ex} (10³ mm³)
30x3 SHS	2.36	30	3	4.5	301	0.035	2.34	1.87	2.96	10.8	0.0615	3.58	1	C	2.96
30x2.5 SHS	2.03	30	2.5	3.75	259	0.0316	2.1	1.65	2.61	11	0.054	3.2	1	C	2.61
30x2 SHS	1.68	30	2	3	214	0.0272	1.81	1.39	2.21	11.3	0.0454	2.75	1	C	2.21
30x1.6 SHS	1.38	30	1.6	2.4	175	0.0231	1.54	1.16	1.84	11.5	0.0377	2.32	1	C	1.84
25x3 SHS	1.89	25	3	4.5	241	0.0184	1.47	1.21	1.91	8.74	0.0333	2.27	1	C	1.91
25x2.5 SHS	1.64	25	2.5	3.75	209	0.0169	1.35	1.08	1.71	8.99	0.0297	2.07	1	C	1.71
25x2 SHS	1.36	25	2	3	174	0.0148	1.19	0.926	1.47	9.24	0.0253	1.8	1	C	1.47
25x1.6 SHS	1.12	25	1.6	2.4	143	0.0128	1.02	0.78	1.24	9.44	0.0212	1.54	1	C	1.24
20x2 SHS	1.05	20	2	3	134	0.00692	0.692	0.554	0.877	7.2	0.0121	1.06	1	C	0.877
20x1.6 SHS	0.873	20	1.6	2.4	111	0.00608	0.608	0.474	0.751	7.39	0.0103	0.924	1	C	0.751

(*Continued*)

Table 7.54 (Continued) Square Hollow Sections: 20, 25 and 30 SHS

| Section Name | ϕM_s | ϕV_v | ϕN_s | ϕN_t | ϕT_u | Member Axial Capacity (Any Axis) ϕN_c (kN) | | | | | | | | | | | | | | | | | | | | | | |
| D x t (mm) | (kNm) | (kN) | (kN) | (kN) | (kNm) | Effective length (m) | | | | | | | | | | | | | | | | | | | | | | |
						0.25	0.5	0.75	1	1.5	2	2.5	3	3.5	4	4.5	5	5.5	6	6.5	7	7.5	8	8.5	9	9.5	10
30x3 SHS	0.932	26.9	94.8	94.8	0.677	92.3	83.9	69.3	50.4	25.3	14.7	9.53	6.68	4.95	3.81	3.02	2.46	2.04	1.72	1.47	1.27	1.1	0.973	0.863	0.771	0.692	0.626
30x2.5 SHS	0.822	23.3	81.6	81.6	0.605	79.5	72.6	60.5	44.5	22.6	13.1	8.5	5.96	4.41	3.4	2.7	2.19	1.82	1.53	1.31	1.13	0.986	0.868	0.77	0.688	0.618	0.558
30x2 SHS	0.696	19.4	67.3	67.3	0.52	65.7	60.3	50.9	38.1	19.6	11.4	7.39	5.19	3.84	2.96	2.35	1.91	1.58	1.33	1.14	0.984	0.858	0.756	0.67	0.599	0.538	0.486
30x1.6 SHS	0.58	16	55.2	55.2	0.438	53.9	49.6	42.2	32	16.6	9.64	6.27	4.4	3.26	2.51	1.99	1.62	1.34	1.13	0.966	0.835	0.728	0.641	0.569	0.508	0.457	0.413
25x3 SHS	0.602	21.3	75.9	75.9	0.429	72.7	62.4	44.7	28.8	13.6	7.8	5.05	3.54	2.62	2.01	1.6	1.3	1.07	0.905	0.773	0.667	0.582	0.512	0.454	0.406	0.365	0.329
25x2.5 SHS	0.539	18.7	65.8	65.8	0.391	63.2	54.8	40.1	26.2	12.4	7.15	4.63	3.24	2.4	1.84	1.46	1.19	0.985	0.83	0.709	0.612	0.534	0.47	0.417	0.372	0.334	0.302
25x2 SHS	0.463	15.7	54.7	54.7	0.34	52.7	46.1	34.5	22.8	10.9	6.27	4.06	2.85	2.1	1.62	1.28	1.04	0.865	0.728	0.622	0.537	0.469	0.413	0.366	0.327	0.294	0.265
25x1.6 SHS	0.391	13	45.1	45.1	0.291	43.5	38.3	29.1	19.5	9.36	5.39	3.49	2.45	1.81	1.39	1.1	0.897	0.744	0.626	0.535	0.462	0.403	0.355	0.315	0.281	0.252	0.228
20x2 SHS	0.276	11.9	42.1	42.1	0.2	39.5	30.8	18.7	11.3	5.2	2.97	1.92	1.34	0.99	0.76	0.605	0.491	0.407	0.343	0.292	0.252	0.22	0.194	0.172	0.153	0.138	0.124
20x1.6 SHS	0.237	10	35	35	0.175	32.9	26.1	16.2	9.82	4.55	2.6	1.68	1.18	0.87	0.67	0.529	0.43	0.356	0.3	0.256	0.221	0.193	0.17	0.15	0.134	0.121	0.109

Note: Geometry and capacity table, Grade C350L0 – f_y = 350 MPa – f_u = 430 MPa

Table 7.55 Square Hollow Sections: 350 and 400 SHS

Section Name D x t (mm)	Weight (kg/m)	D (mm)	t (mm)	r_m (mm)	A_g (mm²)	I_x (10^6 mm⁴)	Z_x (10^3 mm³)	Z_a (10^3 mm³)	S_x (10^3 mm³)	r_x (mm)	J (10^6 mm⁴)	C (10^3 mm³)	Form Factor k_f	X Compactness	Z_{ss} (10^3 mm³)
400x16 SHS	186	400	16	32	23697	571	2850	2140	3370	155	930	4350	1	N	3320
400x12.5 SHS	148	400	12.5	25	18838	464	2320	1720	2710	157	744	3520	0.994	S	2310
400x10 SHS	120	400	10	20	15257	382	1910	1400	2210	158	604	2890	0.785	S	1650
350x16 SHS	161	350	16	32	20497	372	2130	1610	2530	135	614	3250	1	C	2530
350x12.5 SHS	128	350	12.5	25	16338	305	1740	1300	2040	137	493	2650	1	N	1900
350x10 SHS	104	350	10	20	13257	252	1440	1060	1670	138	401	2180	0.904	S	1350
350x8 SHS	84.2	350	8	16	10724	207	1180	865	1370	139	326	1790	0.715	S	971

Section Name D x t (mm)	ϕM_s (kNm)	ϕV_v (kN)	ϕN_s (kN)	ϕN_t (kN)	ϕT_u (kNm)	Member Axial Capacity (Any Axis), ϕN_c (kN) — Effective length (m)																					
						0.5	1	1.5	2	2.5	3	4	4.5	5	6	7	8	9	10	12	14	16	18	20	22	24	26
400x16 SHS	1345	2826	9064	9597	1057	9597	9597	9597	9537	9462	9377	9175	9055	8922	8615	8246	7807	7295	6718	5489	4377	3497	2830	2328	1945	1647	1412
400x12.5 SHS	936	2250	7206	7584	855	7584	7584	7584	7539	7481	7416	7261	7169	7067	6833	6551	6216	5825	5383	4428	3547	2840	2302	1895	1584	1342	1150
400x10 SHS	668	1824	5836	4850	702	4850	4850	4850	4838	4807	4773	4693	4647	4596	4478	4340	4177	3988	3770	3263	2723	2236	1838	1524	1279	1087	934
350x16 SHS	1025	2442	7840	8301	790	8301	8301	8282	8211	8131	8038	7812	7676	7525	7171	6738	6221	5636	5022	3884	3000	2358	1893	1552	1291	1092	935
350x12.5 SHS	770	1950	6249	6617	644	6617	6617	6604	6549	6486	6414	6239	6134	6017	5743	5409	5010	4556	4074	3168	2454	1931	1552	1271	1059	895	767
350x10 SHS	547	1584	5071	4854	530	4854	4854	4851	4813	4771	4723	4608	4539	4463	4285	4069	3812	3514	3187	2534	1987	1573	1268	1040	868	734	629
350x8 SHS	393	1283	4102	3105	435	3105	3105	3105	3091	3069	3044	2985	2950	2912	2824	2719	2594	2449	2282	1912	1552	1252	1019	841	703	596	512

Note: Geometry and capacity table, Grade C450L0 – f_y = 450 MPa – f_u = 500 MPa

Table 7.56 Square Hollow Sections: 250 and 300 SHS

Section Name D x t (mm)	Weight (kg/m)	D (mm)	t (mm)	r_m (mm)	A_g (mm²)	I_x (10⁶ mm⁴)	Z_x (10³ mm³)	Z_a (10³ mm³)	S_x (10³ mm³)	r_x (mm)	J (10⁶ mm⁴)	C (10³ mm³)	Form Factor k_f	X Compactness	Z_{ex} (10³ mm³)
300x16 SHS	136	300	16	32	17297	226	1510	1160	1810	114	378	2310	1	C	1810
300x12.5 SHS	109	300	12.5	25	13838	187	1240	937	1470	116	305	1900	1	C	1470
300x10 SHS	88.4	300	10	20	11257	155	1030	769	1210	117	250	1570	1	N	1080
300x8 SHS	71.6	303	8	16	9124	128	853	628	991	118	203	1290	0.84	S	768
250x16 SHS	111	250	16	32	14097	124	992	774	1210	93.8	212	1530	1	C	1210
250x12.5 SHS	89	250	12.5	25	11338	104	830	634	992	95.7	173	1270	1	C	992
250x10 SHS	72.7	250	10	20	9257	87.1	697	523	822	97	142	1060	1	N	811
250x9 SHS	65.9	250	9	18	8398	79.8	639	477	750	97.5	129	972	1	N	699
250x8 SHS	59.1	250	8	16	7524	72.3	578	429	676	98	116	878	1	N	586
250x6 SHS	45	250	6	12	5732	56.2	450	330	521	99	88.7	681	0.753	S	380

(*Continued*)

Table 7.56 (Continued) Square Hollow Sections: 250 and 300 SHS

Section Name	ϕM_s	ϕV_v	ϕN_s	ϕN_t	ϕT_u	Member Axial Capacity (Any Axis) ϕN_c (kN)																					
						Effective length (m)																					
$D \times t$ (mm)	(kNm)	(kN)	(kN)	(kN)	(kNm)	0.25	0.5	0.75	1	1.5	2	2.5	3	3.5	4	4.5	5	5.5	6	6.5	7	7.5	8	8.5	9	9.5	10
300x16 SHS	733	2058	7005	6616	561	7005	7005	6957	6880	6790	6684	6419	6259	6077	5643	5112	4511	3903	3346	2466	1864	1450	1158	945	786	664	568
300x12.5 SHS	595	1650	5605	5293	462	5605	5605	5569	5509	5439	5357	5152	5028	4889	4555	4145	3677	3196	2750	2034	1540	1199	958	782	650	549	470
300x10 SHS	437	1344	4559	4306	382	4559	4559	4531	4483	4427	4361	4197	4099	3987	3720	3393	3018	2630	2266	1680	1273	991	792	647	538	454	389
300x8 SHS	311	1091	3104	3490	313	3104	3104	3093	3065	3032	2995	2902	2846	2784	2637	2457	2244	2007	1768	1345	1031	808	647	529	441	373	319
250x16 SHS	490	1674	5709	5392	372	5709	5702	5631	5545	5440	5315	4996	4797	4569	4027	3422	2847	2358	1965	1407	1051	813	648	528	438	370	316
250x12.5 SHS	402	1350	4592	4337	309	4592	4588	4532	4466	4384	4288	4042	3889	3713	3295	2820	2359	1961	1638	1175	878	680	542	441	367	309	265
250x10 SHS	328	1104	3749	3541	258	3749	3747	3702	3649	3584	3507	3312	3191	3052	2720	2339	1964	1637	1369	984	736	570	454	370	307	259	222
250x9 SHS	283	1002	3401	3212	236	3401	3400	3359	3311	3253	3184	3009	2900	2775	2477	2135	1795	1498	1253	901	674	522	416	339	282	238	203
250x8 SHS	237	899	3047	2878	213	3047	3046	3011	2968	2916	2855	2700	2603	2493	2229	1924	1621	1353	1133	815	610	472	376	307	255	215	184
250x6 SHS	154	685	1748	2193	165	1748	1748	1736	1717	1694	1668	1602	1562	1517	1409	1276	1127	975	836	616	466	362	289	236	196	166	142

Note: Geometry and capacity table, Grade C450L0 – $f_y = 450$ MPa – $f_u = 500$ MPa

Table 7.57 Square Hollow Sections: 200 SHS

Section Name D x t (mm)	Weight (kg/m)	D (mm)	t (mm)	r_m (mm)	A_g (mm²)	I_x (10^6 mm⁴)	Z_x (10^3 mm³)	Z_n (10^3 mm³)	S_x (10^3 mm³)	r_x (mm)	J (10^6 mm⁴)	C (10^3 mm³)	Form Factor k_f	X Compactness	Z_{ex} (10^3 mm³)
200x16 SHS	85.5	200	16	32	10897	58.6	586	469	728	73.3	103	914	1	C	728
200x12.5 SHS	69.4	200	12.5	25	8838	50	500	389	607	75.2	85.2	772	1	C	607
200x10 SHS	57	200	10	20	7257	42.5	425	324	508	76.5	70.7	651	1	C	508
200x9 SHS	51.8	200	9	18	6598	39.2	392	297	465	77.1	64.5	599	1	C	465
200x8 SHS	46.5	200	8	16	5924	35.7	357	268	421	77.6	58.2	544	1	N	415
200x6 SHS	35.6	200	6	12	4532	28	280	207	327	78.6	44.8	425	0.952	S	272
200x5 SHS	29.9	200	5	10	3814	23.9	239	175	277	79.1	37.8	362	0.785	S	207

Section Name D x t (mm)	ϕM_s (kNm)	ϕV_v (kN)	ϕN_s (kN)	ϕN_t (kN)	ϕT_u (kNm)	Member Axial Capacity (Any Axis), ϕN_c (kN) — Effective length (m)																				
						0.25	0.5	0.75	1	1.5	2	2.5	3	3.5	4	4.5	5	5.5	6	6.5	7	7.5	8	8.5	9	9.5	10
200x16 SHS	295	1290	4413	4168	222	4413	4378	4298	4194	4065	3907	3486	3221	2928	2338	1837	1457	1174	964	681	505	390	310	252	209	177	151
200x12.5 SHS	246	1050	3580	3381	188	3580	3554	3491	3411	3312	3190	2868	2664	2436	1966	1555	1237	999	821	580	431	332	264	215	178	151	129
200x10 SHS	206	864	2939	2776	158	2939	2919	2869	2806	2727	2630	2375	2214	2033	1652	1313	1047	846	696	492	365	282	224	183	152	128	109
200x9 SHS	188	786	2672	2524	146	2672	2655	2610	2553	2482	2396	2168	2024	1862	1518	1209	965	781	642	454	337	260	207	169	140	118	101
200x8 SHS	168	707	2399	2266	132	2399	2384	2344	2294	2231	2154	1953	1825	1681	1375	1097	876	709	583	413	307	237	188	153	127	107	91.7
200x6 SHS	110	541	1748	1734	103	1748	1738	1711	1677	1634	1582	1447	1362	1264	1050	847	681	553	456	323	240	185	147	120	99.7	84.1	71.9
200x5 SHS	83.8	456	1213	1459	88	1213	1210	1193	1173	1149	1120	1045	998	943	817	682	560	460	382	272	203	157	125	102	84.6	71.3	61

Note: Geometry and capacity table, Grade C450L0 – f_y = 450 MPa – f_u = 500 MPa

Table 7.58 Square Hollow Sections: 125 and 150 SHS

Section Name D x t (mm)	Weight (kg/m)	D (mm)	t (mm)	r_m (mm)	A_g (mm²)	I_x (10⁶ mm⁴)	Z_x (10³ mm³)	Z_n (10³ mm³)	S_x (10³ mm³)	r_x (mm)	J (10⁶ mm⁴)	C (10³ mm³)	Form Factor k_f	X Compactness	Z_{xx} (10³ mm³)
150x10 SHS	41.3	150	10	20	5257	16.5	220	173	269	56.1	28.4	341	1	C	269
150x9 SHS	37.7	150	9	18	4798	15.4	205	159	248	56.6	26.1	316	1	C	248
150x8 SHS	33.9	150	8	16	4324	14.1	188	144	226	57.1	23.6	289	1	C	226
150x6 SHS	26.2	150	6	12	3332	11.3	150	113	178	58.2	18.4	229	1	N	175
150x5 SHS	22.1	150	5	10	2814	9.7	129	96.2	151	58.7	15.6	197	1	N	135
125x10 SHS	33.4	125	10	20	4257	8.93	143	114	178	45.8	15.7	223	1	C	178
125x9 SHS	30.6	125	9	18	3898	8.38	134	106	165	46.4	14.5	208	1	C	165
125x8 SHS	27.7	125	8	16	3524	7.75	124	96.8	151	46.9	13.3	192	1	C	151
125x6 SHS	21.4	125	6	12	2732	6.29	101	76.5	120	48	10.4	154	1	C	120
125x5 SHS	18.2	125	5	10	2314	5.44	87.1	65.4	103	48.5	8.87	133	1	N	101

(Continued)

Table 7.58 (Continued) Square Hollow Sections: 125 and 150 SHS

Section Name D x t (mm)	φMs (kNm)	φVv (kN)	φNs (kN)	φNt (kN)	φTu (kNm)	Member Axial Capacity (Any Axis), φNc (kN) — Effective length (m)																					
						0.25	0.5	0.75	1	1.5	2	2.5	3	3.5	4	4.5	5	5.5	6	6.5	7	7.5	8	8.5	9	9.5	10
150x10 SHS	109	624	2129	2011	82.9	2129	2089	2028	1945	1837	1701	1347	1160	992	729	550	428	341	279	196	145	112	88.6	72.1	59.8	50.4	43
150x9 SHS	100	570	1943	1835	76.8	1943	1908	1853	1778	1682	1560	1242	1072	918	676	510	397	317	259	182	135	104	82.3	66.9	55.5	46.8	40
150x8 SHS	91.5	515	1751	1654	70.2	1751	1720	1671	1605	1520	1412	1130	978	839	618	467	364	290	237	167	123	95	75.4	61.4	50.9	42.9	36.6
150x6 SHS	70.9	397	1350	1275	55.6	1350	1327	1290	1241	1178	1099	889	773	666	493	373	291	232	190	133	98.7	76	60.4	49.1	40.7	34.3	29.3
150x5 SHS	54.7	336	1140	1076	47.9	1140	1121	1091	1050	998	932	757	660	569	422	320	249	199	163	114	84.7	65.3	51.8	42.2	35	29.5	25.2
125x10 SHS	72.1	504	1724	1628	54.2	1721	1671	1599	1498	1362	1190	829	684	569	406	303	235	187	152	107	78.9	60.7	48.2	39.2	32.5	27.3	23.4
125x9 SHS	66.8	462	1579	1491	50.5	1576	1532	1467	1377	1256	1103	774	640	533	381	285	220	175	143	100	74.1	57	45.3	36.8	30.5	25.7	21.9
125x8 SHS	61.2	419	1427	1348	46.7	1425	1386	1329	1249	1142	1007	712	590	491	352	263	203	162	132	92.5	68.4	52.7	41.8	34	28.2	23.7	20.3
125x6 SHS	48.6	325	1107	1045	37.4	1106	1076	1034	975	896	796	571	475	397	285	213	165	131	107	75	55.5	42.7	33.9	27.6	22.9	19.3	16.4
125x5 SHS	40.9	276	937	885	32.3	937	912	877	828	763	680	491	409	342	246	184	142	113	92.5	64.8	48	36.9	29.3	23.8	19.8	16.6	14.2

Note: Geometry and capacity table, Grade C450L0 – $f_y = 450$ MPa – $f_u = 500$ MPa

Table 7.59 Square Hollow Sections: 100 and 125 SHS

Section Name $D \times t$ (mm)	Weight (kg/m)	D (mm)	t (mm)	r_m (mm)	A_g (mm²)	I_x (10^6 mm⁴)	Z_x (10^3 mm³)	Z_a (10^3 mm³)	S_x (10^3 mm³)	t_x (mm)	J (10^6 mm⁴)	C (10^3 mm³)	Form Factor k_f	X Compactness	Z_{ex} (10^3 mm³)
125x4 SHS	14.8	125	4	8	1881	4.52	72.3	53.6	84.5	49	7.25	110	1	N	73.2
100x10 SHS	25.6	100	10	20	3257	4.11	82.2	68.1	105	35.5	7.5	130	1	C	105
100x9 SHS	23.5	100	9	18	2998	3.91	78.1	63.6	98.6	36.1	7	123	1	C	98.6
100x8 SHS	21.4	100	8	16	2724	3.66	73.2	58.6	91.1	36.7	6.45	114	1	C	91.1
100x6 SHS	16.7	100	6	12	2132	3.04	60.7	47.1	73.5	37.7	5.15	93.6	1	C	73.5
100x5 SHS	14.2	100	5	10	1814	2.66	53.1	40.5	63.5	38.3	4.42	81.4	1	C	63.5
100x4 SHS	11.6	100	4	8	1481	2.23	44.6	33.5	52.6	38.8	3.63	68	1	N	51.9
100x3 SHS	8.96	100	3	4.5	1141	1.77	35.4	26	41.2	39.4	2.79	53.2	0.952	S	34.4
100x2.5 SHS	7.53	100	2.5	3.75	959	1.51	30.1	21.9	34.9	39.6	2.35	45.2	0.787	S	26.1
100x2 SHS	6.07	100	2	3	774	1.23	24.6	17.8	28.3	39.9	1.91	36.9	0.624	S	18.8

(Continued)

Table 7.59 (Continued) Square Hollow Sections: 100 and 125 SHS

Section Name D x t (mm)	φMₛ (kNm)	φVᵥ (kN)	φNₛ (kN)	φNₜ (kN)	φTᵤ (kNm)	Member Axial Capacity (Any Axis), φNc (kN) Effective length (m)																					
						0.25	0.5	0.75	1	1.5	2	2.5	3	3.5	4	4.5	5	5.5	6	6.5	7	7.5	8	8.5	9	9.5	10
125x4 SHS	29.6	225	762	720	26.7	762	762	753	742	714	675	623	557	481	405	338	283	239	204	175	152	134	118	105	94.1	84.7	76.7
100x10 SHS	42.5	384	1319	1246	31.6	1319	1307	1282	1249	1157	1022	846	667	520	411	331	271	226	191	164	142	124	109	97.3	87	78.3	70.8
100x9 SHS	39.9	354	1214	1147	29.9	1214	1204	1181	1152	1070	951	793	629	493	390	314	258	215	182	156	135	118	104	92.5	82.8	74.5	67.4
100x8 SHS	36.9	323	1103	1042	27.7	1103	1095	1074	1049	977	872	733	586	460	365	294	242	202	171	146	127	111	97.7	86.8	77.7	69.9	63.2
100x6 SHS	29.8	253	864	816	22.7	864	857	842	823	770	693	589	476	377	300	242	199	166	141	121	104	91.4	80.6	71.6	64.1	57.7	52.2
100x5 SHS	25.7	216	735	694	19.8	735	730	717	702	658	594	509	414	329	262	212	174	146	123	106	91.6	80.1	70.7	62.8	56.2	50.6	45.8
100x4 SHS	21	177	600	567	16.5	600	596	586	573	539	488	420	344	274	219	177	146	122	103	88.5	76.7	67.1	59.2	52.6	47.1	42.4	38.3
100x3 SHS	13.9	135	440	436	12.9	440	438	431	422	398	365	319	265	214	172	140	115	96.4	81.7	70.1	60.7	53.2	46.9	41.7	37.3	33.6	30.4
100x2.5 SHS	10.6	114	306	367	11	306	305	301	296	282	263	238	206	172	141	116	96.3	80.8	68.7	59	51.2	44.8	39.6	35.2	31.5	28.4	25.7
100x2 SHS	7.61	92.2	196	296	8.97	196	196	193	191	184	175	162	146	128	108	91.1	76.6	64.8	55.3	47.7	41.5	36.4	32.1	28.6	25.6	23.1	20.9

Note: Geometry and capacity table, Grade C450L0 – f_y = 450 MPa – f_u = 500 MPa

Table 7.60 Square Hollow Sections: 75, 89 and 90 SHS

Section Name D x t (mm)	Weight (kg/m)	D (mm)	t (mm)	r_m (mm)	A_g (mm²)	I_x ($10^6 mm^4$)	Z_x ($10^3 mm^3$)	Z_a ($10^3 mm^3$)	S_x ($10^3 mm^3$)	r_x (mm)	J ($10^6 mm^4$)	C ($10^3 mm^3$)	Form Factor k_f	X Compactness	Z_{xx} ($10^3 mm^3$)
90x2.5 SHS	6.74	90	2.5	3.75	859	1.09	24.1	17.6	28	35.6	1.7	36.2	0.878	S	22.3
90x2 SHS	5.45	90	2	3	694	0.889	19.7	14.3	22.8	35.8	1.38	29.6	0.696	S	16
89x6 SHS	14.7	89	6	12	1868	2.06	46.4	36.4	56.7	33.2	3.55	71.8	1	C	56.7
89x5 SHS	12.5	89	5	10	1594	1.82	40.8	31.5	49.2	33.8	3.06	62.8	1	C	49.2
89x3.5 SHS	9.07	89	3.5	7	1155	1.38	31	23.3	36.5	34.6	2.25	47.2	1	N	35.8
89x2 SHS	5.38	89	2	3	686	0.858	19.3	14	22.3	35.4	1.33	29	0.704	S	15.7
75x6 SHS	12	75	6	12	1532	1.16	30.9	24.7	38.4	27.5	2.04	48.2	1	C	38.4
75x5 SHS	10.3	75	5	10	1314	1.03	27.5	21.6	33.6	28	1.77	42.6	1	C	33.6
75x4 SHS	8.49	75	4	8	1081	0.882	23.5	18.1	28.2	28.6	1.48	36.1	1	C	28.2
75x3.5 SHS	7.53	75	3.5	7	959	0.797	21.3	16.1	25.3	28.8	1.32	32.5	1	C	25.3

(Continued)

Table 7.60 (Continued) Square Hollow Sections: 75, 89 and 90 SHS

Section Name D x t (mm)	φM_s (kNm)	φV_v (kN)	φN_s (kN)	φN_t (kN)	φT_u (kNm)	Member Axial Capacity (Any Axis, φN_c) (kN) Effective length (m)																					
						0.25	0.5	0.75	1	1.5	2	2.5	3	3.5	4	4.5	5	5.5	6	6.5	7	7.5	8	8.5	9	9.5	10
75x3 SHS	8.99	99.4	341	322	6.97	305	303	298	291	273	246	210	170	135	107	86.9	71.4	59.6	50.5	43.3	37.5	32.8	28.9	25.7	23	20.7	18.7
75x2.5 SHS	6.89	84	287	271	5.98	196	195	192	189	179	166	148	126	104	84.7	69.3	57.3	48	40.8	35	30.4	26.6	23.5	20.9	18.7	16.8	15.2
75x2 SHS	4.9	68.2	195	219	4.91	757	748	732	711	650	560	449	345	266	209	167	137	114	96.4	82.6	71.5	62.5	55.1	49	43.8	39.4	35.6
65x6 SHS	11.1	153	523	494	8.31	646	639	625	608	558	484	391	303	234	184	148	121	101	85.2	72.9	63.2	55.2	48.7	43.3	38.7	34.8	31.5
65x5 SHS	9.84	132	451	426	7.44	468	463	454	442	407	357	291	227	177	139	112	91.6	76.3	64.6	55.3	47.9	41.9	36.9	32.8	29.3	26.4	23.9
65x4 SHS	8.34	109	373	352	6.37	196	195	192	188	179	165	147	124	102	82.4	67.2	55.6	46.5	39.5	33.9	29.4	25.7	22.7	20.2	18.1	16.3	14.7
65x3 SHS	6.72	85	292	276	5.1	621	608	590	565	490	384	280	205	154	120	95.8	78.2	65	54.9	46.9	40.6	35.5	31.3	27.8	24.8	22.3	20.2
65x2.5 SHS	5.55	72	247	233	4.4	532	522	507	486	425	336	247	182	137	107	85	69.4	57.7	48.7	41.7	36.1	31.5	27.8	24.7	22.1	19.9	17.9
65x2 SHS	3.97	58.6	196	189	3.62	438	430	418	401	353	283	210	155	117	91.2	72.9	59.5	49.5	41.8	35.7	30.9	27	23.8	21.2	18.9	17	15.4
65x1.6 SHS	2.84	47.5	125	153	2.96	388	382	371	357	315	253	189	139	105	82	65.5	53.5	44.5	37.6	32.1	27.8	24.3	21.4	19	17	15.3	13.8

Note: Geometry and capacity table, Grade C450L0 – f_y = 450 MPa – f_u = 500 MPa

Table 7.61 Square Hollow Sections: 65 and 75 SHS

Section Name D x t (mm)	Weight (kg/m)	D (mm)	t (mm)	r_m (mm)	A_g (mm²)	I_x (10⁶ mm⁴)	Z_x (10³ mm³)	Z_n (10³ mm³)	S_x (10³ mm³)	r_x (mm)	J (10⁶ mm⁴)	C (10³ mm³)	Form Factor k_f	X Compactness	Z_{xx} (10³ mm³)
75x3 SHS	6.6	75	3	4.5	841	0.716	19.1	14.2	22.5	29.2	1.15	28.7	1	N	22.2
75x2.5 SHS	5.56	75	2.5	3.75	709	0.614	16.4	12	19.1	29.4	0.971	24.6	1	N	17
75x2 SHS	4.5	75	2	3	574	0.505	13.5	9.83	15.6	29.7	0.79	20.2	0.841	S	12.1
65x6 SHS	10.1	65	6	12	1292	0.706	21.7	17.8	27.5	23.4	1.27	34.2	1	C	27.5
65x5 SHS	8.75	65	5	10	1114	0.638	19.6	15.6	24.3	23.9	1.12	30.6	1	C	24.3
65x4 SHS	7.23	65	4	8	921	0.552	17	13.2	20.6	24.5	0.939	26.2	1	C	20.6
65x3 SHS	5.66	65	3	4.5	721	0.454	14	10.4	16.6	25.1	0.733	21	1	C	16.6
65x2.5 SHS	4.78	65	2.5	3.75	609	0.391	12	8.91	14.1	25.3	0.624	18.1	1	N	13.7
65x2 SHS	3.88	65	2	3	494	0.323	9.94	7.29	11.6	25.6	0.509	14.9	0.978	S	9.8
65x1.6 SHS	3.13	65	1.6	2.4	399	0.265	8.16	5.94	9.44	25.8	0.414	12.2	0.774	S	7.01

(Continued)

Table 7.61 (Continued) Square Hollow Sections: 65 and 75 SHS

Section Name	φM_s	φV_v	φN_s	φN_t	φT_u	Member Axial Capacity (Any Axis), φN_c (kN)																					
						Effective length (m)																					
$D \times t$ (mm)	(kNm)	(kN)	(kN)	(kN)	(kNm)	0.25	0.5	0.75	1	1.5	2	2.5	3	3.5	4	4.5	5	5.5	6	6.5	7	7.5	8	8.5	9	9.5	10
75x3 SHS	8.99	99.4	341	322	6.97	341	335	326	313	278	225	169	125	94.8	73.8	59	48.2	40.1	33.8	29	25.1	21.9	19.3	17.1	15.3	13.8	12.5
75x2.5 SHS	6.89	84	287	271	5.98	287	282	275	265	235	191	144	107	80.9	63	50.4	41.1	34.2	28.9	24.7	21.4	18.7	16.5	14.7	13.1	11.8	10.7
75x2 SHS	4.9	68.2	219	219	4.91	195	193	189	183	166	142	112	85.6	65.6	51.4	41.2	33.7	28.1	23.7	20.3	17.6	15.4	13.6	12	10.8	9.69	8.77
65x6 SHS	11.1	153	523	494	8.31	523	508	487	458	368	260	179	129	96	74.3	59.1	48.2	40	33.8	28.9	25	21.8	19.2	17.1	15.3	13.7	12.4
65x5 SHS	9.84	132	451	426	7.44	451	439	421	397	323	231	161	115	86.1	66.7	53.1	43.3	36	30.3	26	22.5	19.6	17.3	15.3	13.7	12.3	11.1
65x4 SHS	8.34	109	373	352	6.37	373	363	349	330	273	198	139	99.8	74.7	57.8	46.1	37.6	31.2	26.3	22.5	19.5	17	15	13.3	11.9	10.7	9.68
65x3 SHS	6.72	85	292	276	5.1	292	285	274	260	217	161	113	81.7	61.2	47.4	37.8	30.8	25.6	21.6	18.5	16	14	12.3	10.9	9.77	8.79	7.94
65x2.5 SHS	5.55	72	247	233	4.4	247	241	232	220	185	137	96.9	70	52.5	40.7	32.4	26.4	22	18.5	15.9	13.7	12	10.6	9.38	8.38	7.54	6.82
65x2 SHS	3.97	58.6	196	189	3.62	196	191	185	176	149	112	79.9	57.9	43.4	33.7	26.9	21.9	18.2	15.4	13.1	11.4	9.94	8.76	7.78	6.95	6.25	5.65
65x1.6 SHS	2.84	47.5	125	153	2.96	125	123	120	115	102	83	62.4	46.3	35.1	27.3	21.8	17.8	14.8	12.5	10.7	9.29	8.12	7.15	6.35	5.68	5.11	4.62

Note: Geometry and capacity table, Grade C450L0 – f_y = 450 MPa – f_u = 500 MPa

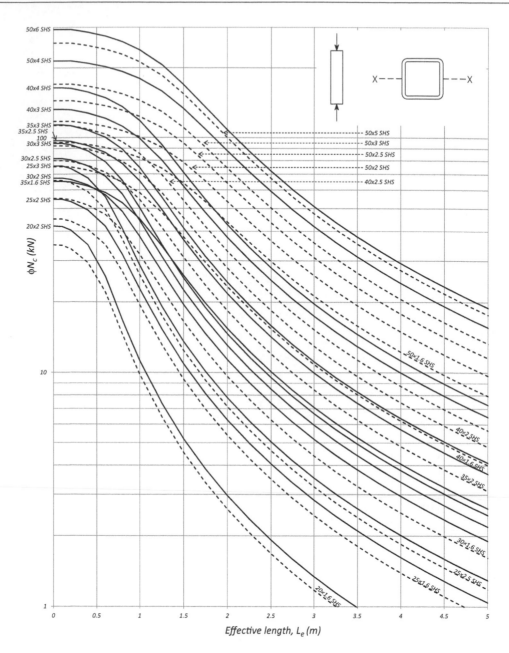

Figure 7.29 Square hollow sections: Members subject to axial compression (major and minor axes) grade C350L0 (solid lines labelled on the *y* axis).

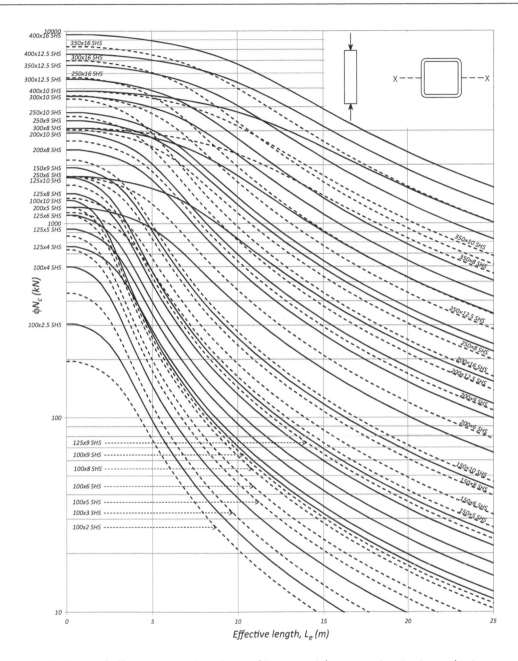

Figure 7.30 Square hollow sections: Members subject to axial compression (major and minor axes) grade C450L0 (solid lines labelled on the *y* axis). 1 of 2.

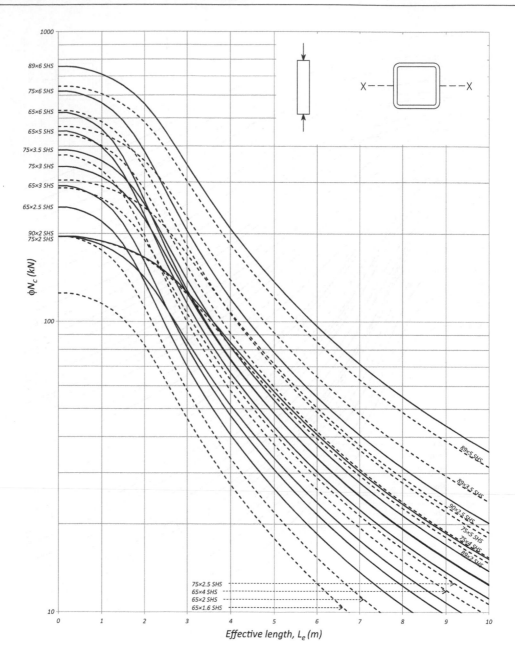

Figure 7.31 Square hollow sections: Members subject to axial compression (major and minor axes) grade C450L0 (solid lines labelled on the *y* axis). 2 of 2.

7.3.2 Concrete catalogues and capacity tables

7.3.2.1 Concrete reinforcement

Reinforcing bars are available in lengths varying from 6.0 to 15.0 m depending on the bar diameter. Standard reinforcing bars are grade D500N (deformed with a normal ductility class). Round bars of class R250N (round with a normal ductility class) are often used for shear ligatures. Mesh is commonly available in a standard sheet size of 6 × 2.4 m and is a low ductility class. Reinforcing mesh is most commonly used to prevent shrinkage cracking for slabs on grade. Commonly available reinforcement types and properties are summarised in Table 7.62.

7.3.2.2 Slabs

Slab capacities are shown in Table 7.63 for commonly used slab thicknesses, compressive strength, cover and reinforcement arrangements. Slab capacity tables are based on the theory shown in Section 4.3. Punching shear capacities are provided in Table 7.64 for both a standard wheel dimension and a 100 mm × 100 mm load dimension.

7.3.2.3 Beams

Beam capacity tables are provided in Table 7.65 for typical geometry, compressive strength, cover and reinforcement arrangements. Beam capacity tables are based on the theory shown in Section 4.2.3 and 4.2.4.

7.3.2.4 Columns

Column interaction diagrams a provided in Figures 7.32 through 7.39 for typical column and pile cross-sections. The capacities are provided for typical reinforcement arrangements with two options for compressive strength. Interaction diagrams are based on the theory shown in Section 4.4. Confinement bars are not shown.

Table 7.62 Available reinforcement properties

Deformed reinforcing bar diameters (mm) Grade D500N (AS/NZS 4671)	10, 12, 16, 20, 24, 28, 32, 36, 40
Round reinforcing bar diameters (mm) Grade R250N (AS/NZS 4671)	6, 10, 12, 16, 20, 24, 28, 30, 32, 36
Square mesh Grade D500L (AS/NZS 4671) (diameter × spacing)	SL52 (4.77 mm × 200 mm), SL62 (6.0 mm × 200 mm), SL72 (6.75 mm × 200 mm), SL81 (7.6 mm × 100 mm), SL82 (7.6 mm × 200 mm), SL92 (8.6 mm × 200 mm), SL102 (9.5 mm × 200 mm)
Rectangular mesh Grade D500L (AS/NZS 4671) (diameter × spacing, transverse diameter × transverse spacing)	RL718 (6.75 mm × 100 mm, 7.6 mm × 200 mm) RL818 (7.6 mm × 100 mm, 7.6 mm × 200 mm) RL918 (8.6 mm × 100 mm, 7.6 mm × 200 mm) RL1018 (9.5 mm × 100 mm, 7.6 mm × 200 mm) RL1118 (10.7 mm × 100 mm, 7.6 mm × 200 mm) RL1218 (11.9 mm × 100 mm, 7.6 mm × 200 mm)
Round dowels diameters (mm) Grade 300 (AS/NZS 3679.1) Standard lengths = 400 mm, 500 mm and 600 mm	16, 20, 24, 27, 33, 36
Square dowels width (mm) Grade 300 (AS/NZS 3679.1) Standard lengths = 400 mm, 500 mm and 600 mm	16, 20, 25, 32, 40, 50

Source: OneSteel.

Table 7.63 Slab bending and shear capacity table [per metre width]

Capacity	f'_c (MPa)	Thickness (mm)	Cover (mm)	Reinforcement					
				N12-200	N12-150	N16-200	N16-150	N20-200	N20-150
ϕM_{uo} (kNm/m)	32	100	30	13.3	17.2	21.2	22.3	20.8	22.0
			50	8.78	10.9	10.1	10.6	10.0	10.5
		150	30	24.6	32.3	41.3	53.4	60.0	69.5
			50	20.1	26.3	33.3	42.7	45.7	46.6
		200	30	35.9	47.4	61.4	80.3	91.5	117.9
			50	31.4	41.3	53.4	69.5	78.9	101.2
		250	30	47.2	62.5	81.5	107.1	122.9	159.8
			50	42.7	56.4	73.5	96.3	110.3	143.0
		300	30	–	77.5	101.6	133.9	154.3	201.7
			50	–	71.5	93.6	123.1	141.7	184.9
	40	100	30	13.5	17.6	22.0	26.0	24.7	25.5
			50	9.01	11.6	12.1	12.2	11.6	12.2
		150	30	24.8	32.7	42.1	54.8	61.9	78.9
			50	20.3	26.7	34.0	44.0	49.3	55.4
		200	30	36.2	47.8	62.2	81.6	93.3	121.1
			50	31.6	41.8	54.1	70.9	80.7	104.4
		250	30	–	62.9	82.3	108.4	124.7	163.0
			50	–	56.8	74.2	97.7	112.1	146.3
		300	30	–	–	102.4	135.2	156.1	204.9
			50	–	71.9	94.3	124.5	143.5	188.2
ϕV_u (kN/m)	32	100	30	43.6	48.0	48.9	53.8	51.8	57.1
			50	31.9	35.2	33.8	37.2	33.1	36.4
		150	30	66.1	72.8	77.2	85.0	86.2	94.8
			50	57.9	63.7	67.0	73.7	73.8	81.3
		200	30	83.4	91.8	98.7	108.7	111.9	123.1
			50	76.9	84.7	90.7	99.9	102.3	112.6
		250	30	97.3	107.1	116.0	127.7	132.4	145.8
			50	92.1	101.3	109.5	120.5	124.7	137.3
		300	30	–	119.7	130.2	143.3	149.3	164.3
			50	–	114.9	124.8	137.4	142.9	157.3
	40	100	30	47.0	51.7	52.6	57.9	55.9	61.5
			50	34.4	37.9	36.4	40.1	35.6	39.2
		150	30	71.2	78.4	83.2	91.6	92.8	102.2
			50	62.4	68.7	72.1	79.4	79.6	87.6
		200	30	89.9	98.8	106.4	117.1	120.5	132.6
			50	82.9	91.2	97.7	107.6	110.2	121.3
		250	30	–	115.4	125.0	137.6	142.7	157.0
			50	–	109.1	118.0	129.9	134.3	147.9
		300	30	–	–	140.2	154.3	160.8	177.0
			50	–	123.8	134.5	148.0	153.9	169.4

Notes:

1. Capacity of various slab widths can be calculated by multiplying by actual widths. For example, 2 m wide slab capacity is the tabulated value multiplied by 2 m.

2. Tabulated values are for $\beta_2 = 1.0$ (no axial load), $\beta_3 = 1.0$ (>$2d_o$ from support face). Listed shear capacities can be doubled for shear at the support location ($\beta_2 = 2.0$).

3. Values are not provided for arrangements which fail strain code requirements.

Table 7.64 Slab punching shear capacity table

Capacity	f_c' (MPa)	Thickness (mm)	Cover (mm)	100 × 100 Square Load			250 × 400 Rectangle (Wheel) Load		
				N12	N16	N20	N12	N16	N20
ϕV_{uo} (kN)	32	100	30	49.4	44.8	40.4	116.8	110.2	101
			50	28.2	24.5	21	74.3	65.7	57.4
		150	30	121	114.3	107.7	248.8	240.3	228.9
			50	89.1	83.2	77.5	195.7	185	174.5
		200	30	219.5	210.7	201.9	407.8	397.3	383.7
			50	176.9	168.9	161	344.1	331.2	318.5
		250	30	345	334	323.1	593.7	581.2	565.5
			50	291.6	281.4	271.4	519.4	504.4	489.5
		300	30	497.4	484.2	471.2	806.5	792	774.1
			50	433.2	420.9	408.7	721.6	704.4	687.4
	40	100	30	55.2	50.1	45.2	130.6	123.2	112.9
			50	31.6	27.4	23.5	83.1	73.5	64.1
		150	30	135.3	127.7	120.4	278.2	268.6	255.9
			50	99.6	93.1	86.7	218.8	206.9	195.1
		200	30	245.4	235.5	225.8	455.9	444.1	429
			50	197.8	188.8	180	384.7	370.3	356.1
		250	30	385.7	373.4	361.3	663.7	649.8	632.2
			50	326	314.6	303.5	580.7	563.9	547.3
		300	30	556.1	541.4	526.8	901.7	885.5	865.5
			50	484.4	470.6	457	806.8	787.6	768.6

Notes:

1. Cover dimension and reinforcement are for the tension bars.

2. No moment is assumed to be transferred into the slab from the loaded area.

3. Average depth to reinforcement calculated assuming wheel load applied in conservative orientation.

Table 7.65 Beam bending and shear capacity table

Capacity	f_c' (MPa)	Geometry (mm)	Cover (mm)	Tension reinforcement (Bottom side of beam)					
				2N12	2N16	3N16	4N16	4N20	6N20
ϕM_{uo} (kNm)	32	200 (B) × 300 (D)	30	22.9	39.2	56.5	–	–	–
			50	21.1	36.0	51.7	–	–	–
		200 (B) × 400 (D)	30	32.0	55.3	80.7	–	–	–
			50	30.2	52.0	75.8	–	–	–
		300 (B) × 600 (D)	30	–	–	131.1	172.9	262.1	–
			50	–	–	126.3	166.4	252.1	–
		400 (B) × 600 (D)	30	–	–	132.3	174.8	267.0	389.6
			50	–	–	127.4	168.4	256.9	374.5
		500 (B) × 750 (D)	30	–	–	-	–	345.3	509.2
			50	–	–	-	–	335.2	494.1
	40	200 (B) × 300 (D)	30	23.1	39.8	57.9	–	–	–
			50	21.3	36.5	53.0	–	–	–
		200 (B) × 400 (D)	30	32.2	55.8	82.0	–	–	–
			50	30.4	52.6	77.2	–	–	–
		300 (B) ×600 (D)	30	–	–	132.0	174.5	174.5	–
			50	–	–	127.2	168.0	168.0	–
		400 (B) × 600 (D)	30	–	–	–	176.0	269.9	396.1
			50	–	–	–	169.6	259.8	381.0
		500 (B) × 750 (D)	30	–	–	–	–	347.6	514.4
			50	–	–	–	–	337.5	499.3
ϕV_u (kN)	32	200 (B) × 300 (D)	30	160.4	162.6	167.4	–	–	–
			50	148.8	150.7	155.3	–	–	–
		200 (B) × 400 (D)	30	171.0	175.1	180.6	–	–	–
			50	162.4	166.3	171.7	–	–	–
		300 (B) × 600 (D)	30	–	–	281.3	287.9	296.9	–
			50	–	–	273.0	279.5	288.4	–
		400 (B) × 600 (D)	30	–	–	295.1	303.0	314.4	329.0
			50	–	–	286.7	294.6	305.9	320.3
		500 (B) × 750 (D)	30	–	–	–	–	391.2	408.3
			50	–	–	–	–	383.6	400.7
	40	200 (B) × 300 (D)	30	162.5	165.2	170.3	–	–	–
			50	150.8	153.2	158.1	–	–	–
		200 (B) × 400 (D)	30	173.5	178.0	184.0	–	–	–
			50	164.8	169.2	175.0	–	–	–
		300 (B) × 600 (D)	30	–	–	286.3	293.4	303.3	–
			50	–	–	278.0	285.0	294.8	–
		400 (B) × 600 (D)	30	–	–	–	309.7	322.2	337.9
			50	–	–	–	301.3	313.6	329.2
		500 (B) × 750 (D)	30	–	–	–	–	400.3	418.7
			50	–	–	–	–	392.7	411.1

Notes:

1. Tabulated values are for $\beta_2 = 1.0$ (no axial load), $\beta_3 = 1.0$ ($>2d_o$ from support face). Capacity includes concrete and steel strength, therefore values cannot be doubled to calculate the capacity for $\beta_3 = 2.0$.

2. Shear reinforcement is closed loops with N12-150 for the 300 mm deep beams and N12-200 for all other beams.

3. Values are not provided for arrangements which fail strain code requirements.

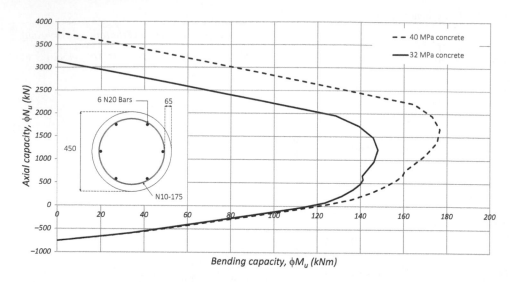

Figure 7.32 450 diameter column interaction diagram.

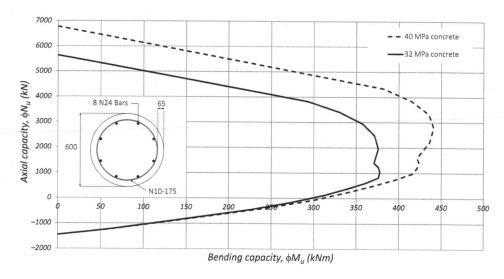

Figure 7.33 600 diameter column interaction diagram.

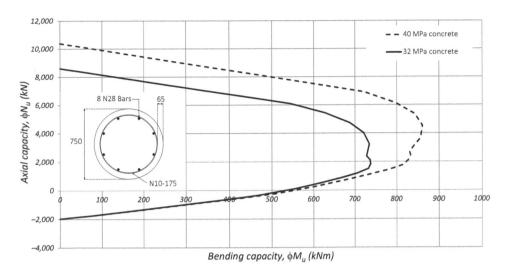

Figure 7.34 750 diameter column interaction diagram.

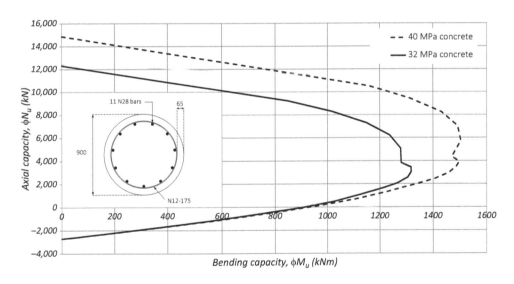

Figure 7.35 900 diameter column interaction diagram.

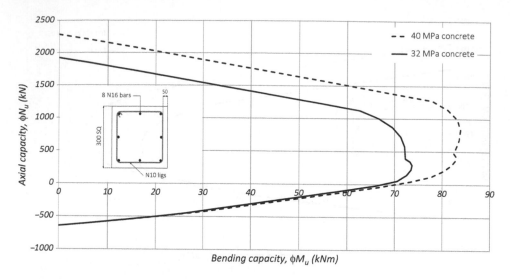

Figure 7.36 300 x 300 square column interaction diagram.

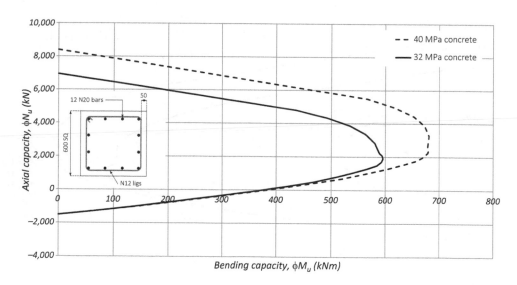

Figure 7.37 600 x 600 square column interaction diagram.

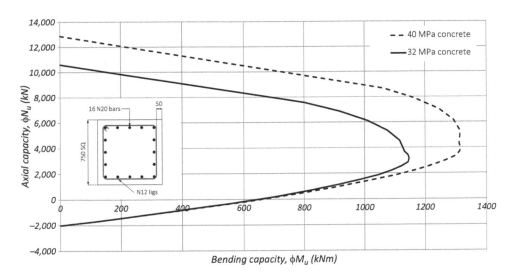

Figure 7.38 750 x 750 square column interaction diagram.

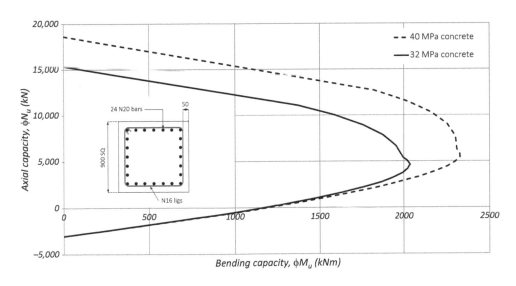

Figure 7.39 900 x 900 Square column interaction diagram.

Chapter 8

Vendor catalogues

The websites listed in this chapter provide commonly used vendor catalogues for the mining and oil and gas industries.

Vendors	Product	Reference
OneSteel	Steel sections	www.onesteel.com
OneSteel	Hollow sections	www.onesteel.com
OneSteel	Reinforcement and mesh	www.onesteel.com
OneSteel	Piping	www.onesteel.com
Nobles	Shackles	www.nobles.com.au
Hilti	Post installed anchor bolts	www.hilti.com.au
Parchem	Grout	www.parchem.com.au
Parchem	Waterstops	www.parchem.com.au
Parchem	Filler board	www.parchem.com.au
Unistrut	Cable ladder	www.unistrut.com.au
Webforge	Grating, treads and handrail	www.webforge.com.au
Blue Scope Steel	Checker plate	www.bluescopesteel.com.au
Parchem	Polyethylene, damp-proof membrane	www.parchem.com.au
Danley	Dowels and joints	www.danley.com.au
Parchem	Void formers, Eccovoid	www.parchem.com.au
Geofabrics	Geotextiles	www.geofabrics.com.au
Geofabrics	Subsoil drain	www.geofabrics.com.au
Regupol	Transportation mats	www.regupol.com.au

Chapter 9

Notations and abbreviations

Abbreviation	Description
ACI	American Concrete Institute
AISC	American Institute of Steel Construction
ARR	Average risk rating
AS	Australian Standard
AS/NZS	Joint Australian and New Zealand Standard
ASI	Australian Steel Institute
BCA	Building Code of Australia
CBR	California bearing ratio
CC	Construction category
CF	Cold formed
CHS	Circular hollow section
CP	Code of Practice (British Standard)
CP	Complete penetration
DIN	Deutsches Institut für Normung (German Standard)
EA	Equal angle
EDC	Earthquake design category
EF	Each face
EIS	Environmental impact statement
EL	Elevation
ES	Each side
EXP	Expansion
FEA	Finite element analysis
FEM	Finite element method
GP	General purpose
HDPE	High-density polyethylene
HR	Hot rolled
IRR	Individual risk rating

ISO	International Organization for Standardization
LODMAT	Lowest one-day mean ambient temperature
NA	Neutral axis
NCC	National Construction Code
NZS	New Zealand Standard
OPE	Operational
PAU	Pre-assembled unit
PFC	Parallel flange channel
PIP	Process Industry Practices
PNA	Plastic neutral axis
PTFE	Polytetrafluoroethylene (Teflon)
RC	Rated capacity
RHS	Rectangular hollow section
SHS	Square hollow section
SLS	Serviceability limit state
SP	Structural purpose
SPT	Standard penetration test
SUS	Sustained
TB	Tensioned bearing
TB	Top and bottom
TF	Tensioned friction
TOC	Top of concrete
TOS	Top of steel
TQ	Technical query
UA	Unequal angle
UB	Universal beam
UC	Universal column
ULS	Ultimate limit state
UNO	Unless noted otherwise
WB	Welded beam
WC	Welded column
WLL	Working load limit

References

TEXTBOOK AND GUIDE REFERENCES

1. Australian Steel Institute. 2014. *Structural Steelwork Fabrication and Erection Code of Practice*. 1st ed. Sydney: Australian Steel Institute.
2. Bowles, JE. 1997. *Foundation Analysis and Design*. International ed. Singapore: McGraw-Hill.
3. Cement Concrete & Aggregates Australia. 2009. *Guide to Industrial Floors and Pavements – Design, Construction and Specification*. 3rd ed. Australia: Cement Concrete & Aggregates Australia (CCAA).
4. Fisher, JM and Kloiber, PE. 2006. *Base Plate and Anchor Rod Design*. 2nd ed. Chicago: American Institute of Steel Construction.
5. Gorenc, BE, Tinyou, R and Syam, AA. 2005. *Steel Designers' Handbook*. 7th ed. Sydney: University of New South Wales Press.
6. Hogan, TJ and Munter, SA. 2007. *Structural Steel Connections Series – Simple Connections Suite*. 1st ed. Sydney: Australian Steel Institute.
7. Hogan, TJ and van der Kreek, N. 2009. *Structural Steel Connections Series – Rigid Connections Suite*. 1st ed. Sydney: Australian Steel Institute.
8. Jameson, G. 2012. *Guide to Pavement Technology Part 2: Pavement Structural Design*. 2nd ed. Sydney: Austroads.
9. Marks, LS. 1996. *Marks' Standard Handbook for Mechanical Engineers*. 10th ed. New York: McGraw-Hill.
10. National Transport Commission. 2004. *Load Restraint Guide*. 2nd ed. Sydney: Roads & Traffic Authority NSW.
11. NAASRA (National Association of Australian State Road Authorities). 1987. *A Guide to the Structural Design of Road Pavements*. Sydney: NAASRA.
12. Queensland Government Transport and Main Roads. 2013. *Guideline for Excess Dimension in Queensland*. 8th ed. Brisbane, Queensland: Queensland Government.
13. Rawlinsons Quantity Surveyors and Construction Cost Consultants. 2017. *Rawlinsons Australian Construction Handbook*. 35th ed. Rivervale, Western Australia: Rawlinsons Publishing.
14. Syam, A. 1999. *Design Capacity Tables for Structural Steel*. 3rd ed. Sydney: Australian Institute of Steel Construction.
15. Tomlinson, MJ. 2001. *Foundation Design and Construction*. 7th ed. London: Pearson Education.
16. Tomlinson, MJ. 2007. *Pile Design and Construction Practice*. 5th ed. London: Taylor & Francis.

17. Woolcock, ST, Kitipornchai, S, Bradford, MA and Haddad, GA. 2011. *Design of Portal Frame Buildings*. 4th ed. Sydney: Australian Steel Institute.
18. Young, WC, Budynas, RG and Sadegh, AM. 2012. *Roark's Formulas for Stress and Strain*. 8th ed. New York: McGraw-Hill.
19. Zaragoza, JR. 1997. *Economic Structural Steelwork*. 4th ed. Sydney: Australian Steel Institute.

JOURNAL REFERENCES AND MAPS

20. Birrcher, D, Tuchscherer, R, Huizinga, M, Bayrak, O, Wood, SL and Jirsa, JO. 2008. Strength and serviceability design of reinforced concrete deep beams. FHWA/TX-09/0-5253-1. Austin: Centre for Transportation Research at the University of Texas.
21. Fox, E. 2000. A climate-based design depth of moisture change map of Queensland and the use of such maps to classify sites under AS 2870–1996. *Australian Geomechanics Journal*, 35 (4), 53–60.
22. Hansen, BJ. 1961. The ultimate resistance of rigid piles against transversal forces. *Danish Geotechnical Institute Bulletin* 12, 5–9.
23. Hansen, BJ. 1970. A revised and extended formula for bearing capacity. *Danish Geotechnical Institute Bulletin* 28, 3–11.
24. Ingold, TS. 1979. The effects of compaction on retaining walls. *Gèotechnique*, 29, 265–283.
25. McCue, K, Gibson, G, Michael-Leiba, M, Love, D, Cuthbertson, R and Horoschun, G. 1991. Earthquake hazard map of Australia. Symonston: Australian Seismological Centre, Geoscience Australia.
26. Mitchell, PW. 2008. Footing design for residential type structures in arid climates. *Australian Aeromechanics*, 43 (4), 51–68.
27. Murray, TM. 1983. Design of lightly loaded steel column base plates. *Engineering Journal*, 4, 143–152.
28. Mutton, BR and Trahair, NS. 1973. Stiffness requirements for lateral bracing. *Journal of the Structural Division*, 99 (ST10), 2167–2182.
29. Mutton, BR and Trahair, NS. 1975. Design requirements for column braces. *Civil Engineering Transactions*, CE17 (1), 30–36.
30. Ranzi, G and Kneen, P. 2002. Design of pinned column base plates. *Steel Construction*, 36 (2), 3–11.
31. Terzaghi, K. 1955. Evaluation of coefficients of subgrade reaction. *Gèotechnique*, 5/4, 297–326.
32. Thornton, WA. 1990. Design of base plates for wide flange columns – A concatenation method. *Engineering Journal*, 20 (4), 143–152.
33. Walsh, PF and Walsh, SF. 1986. Structure/reactive-clay model for a microcomputer. CSIRO, Division of Building Research Report R 86/9. Canberra: Commonwealth Scientific and Industrial Research Organisation.

INTERNATIONAL STANDARDS AND CODES OF PRACTICE

34. ACI 302.1R. 1996. Guide for concrete floor and slab construction. Farmington Hills, MI: American Concrete Institute.
35. ACI 351.3R. 2004. Foundations for dynamic equipment. Farmington Hills, MI: American Concrete Institute.
36. ACI 543R. 2000. Design, manufacture, and installation of concrete piles. Farmington Hills, MI: American Concrete Institute.
37. API 620. 2013. Design and construction of large, welded, low-pressure storage tanks. Washington, DC: American Petroleum Institute.
38. API 650. 2013. Welded tanks for oil storage. Washington, DC: American Petroleum Institute.
39. ASME B31.3. 2014. Process piping. New York: American Society of Mechanical Engineers.

40. ASME B31.8. 2014. Gas transmission and distribution piping systems. New York: American Society of Mechanical Engineers.
41. BS 7385-2. 1993. Evaluation and measurement for vibration in buildings. Guide to damage levels from groundborne vibration. London: BSI Group.
42. Concrete Society TR 34. 2003. Concrete industrial ground floors – A guide to design and construction. 3rd ed. Berkshire, UK: Concrete Society.
43. CP 2012-1. 1974. Code of practice for foundations for machinery. Foundations for reciprocating machines. London: BSI Group.
44. EN 1997. 1997. Eurocode 7: Geotechnical design. Brussels: European Committee for Standardization.
45. ISO 1940-1. 2003. Mechanical vibration – Balance quality requirements for rotors in a constant (rigid) state – Part 1: Specification and verification of balance tolerances. Geneva: International Organization for Standardization.
46. PCA Circular Concrete Tanks without Prestressing. 1993. Skokie, IL: Portland Cement Association.
47. PIP STC01015. Process Industry Practices – Structural design criteria. Austin, TX: Process Industry Practices.
48. UFC 3-260-03. Unified Facilities Criteria – Airfield pavement evaluation. Washington, DC: National Institute of Building Sciences.

AUSTRALIAN AND NEW ZEALAND STANDARDS AND CODES OF PRACTICE

AS 1101.3. 2005. Graphical symbols for general engineering – Part 3 – Welding and non-destructive examination. Sydney: SAI Global.
AS/NZS 1170.0. 2002. Structural design actions – General principles. Sydney: SAI Global.
AS/NZS 1170.1. 2002. Structural design actions – Permanent, imposed and other actions. Sydney: SAI Global.
AS/NZS 1170.2. 2011. Structural design actions – Wind actions. Sydney: SAI Global.
AS/NZS 1170.3. 2003. Structural design actions – Snow and ice actions. Sydney: SAI Global.
AS 1170.4. 2007. Structural design actions – Earthquake actions in Australia. Sydney: SAI Global.
NZS 1170.5. 2004. Structural design actions – Earthquake actions – New Zealand. Sydney: SAI Global.
AS 1210. 2010. Pressure vessels. Sydney: SAI Global.
AS 1379. 2007. The specification and manufacture of concrete. Sydney: SAI Global.
AS/NZS 1418 Set (1–18). Cranes, hoists and winches. Sydney: SAI Global.
AS/NZS 1554 Set (1–7). Structural steel welding. Sydney: SAI Global.
AS 1657. 2013. Fixed platforms, walkways, stairways and ladders – Design, construction and installation. Sydney: SAI Global.
AS 1940. 2004. The storage and handling of flammable and combustible liquids. Sydney: SAI Global.
AS 2067. 2008. Substations and high voltage installations exceeding 1 kV a.c. Sydney: SAI Global.
AS 2159. 2009. Piling – Design and installation. Sydney: SAI Global.
AS 2327.1. 2003. Composite structures – Simply supported beams. Sydney: SAI Global.
AS 2741. 2002. Shackles. Sydney: SAI Global.
AS 2870. 2011. Residential slabs and footings. Sydney: SAI Global.
NZS 3106. 2009. Design of concrete structures for the storage of liquids. Sydney: SAI Global.
AS 3600. 2009. Concrete structures. Sydney: SAI Global.
AS 3610. 1995. Formwork for concrete. Sydney: SAI Global.
AS/NZS 3678. 2011. Structural steel – Hot rolled plates, floorplates and slabs. Sydney: SAI Global.
AS/NZS 3679.1. 2010. Structural steel – Hot rolled bars and sections. Sydney: SAI Global.
AS 3700. 2011. Masonry structures. Sydney: SAI Global.
AS 3735. 2001. Concrete structures for retaining liquids. Sydney: SAI Global.
AS 3780. 2008. The storage and handling of corrosive substances. Sydney: SAI Global.

AS 3850. 2003. Tilt-up concrete construction. Sydney: SAI Global.

AS 3990. 1993. Mechanical equipment – Steelwork. Sydney: SAI Global.

AS 3995. 1994. Design of steel lattice towers and masts. Sydney: SAI Global.

AS 3996. 2006. Access covers and grates. Sydney: SAI Global.

AS 4100. 1998. Steel structures. Sydney: SAI Global.

AS/NZS 4452. 1997. The storage and handling of toxic substances. Sydney: SAI Global.

AS/NZS 4600. 2005. Cold-formed steel structures. Sydney: SAI Global.

AS/NZS 4671. 2001. Steel reinforcing materials. Sydney: SAI Global.

AS 4678. 2002. Earth-retaining structures. Sydney: SAI Global.

AS 4991. 2004. Lifting devices. Sydney: SAI Global.

AS 5100 Set (1–7). Bridge design. Sydney: SAI Global.

Australian Dangerous Goods Code – Edition 7.4. 2016. Melbourne: National Transport Commission Australia.

NOHSC 1015. 2001. National standard – Storage and handling of workplace dangerous goods. Canberra: Safe Work Australia.

Index